MONOGRAPHS AND RESEARCH NOTES IN MATHEMATICS

Delay Differential Evolutions Subjected to Nonlocal Initial Conditions

Monica-Dana Burlică
"Gheorghe Asachi" Technical University of Iași
Romania

Mihai Necula
"Alexandru Ioan Cuza" University of Iași
Romania

Daniela Roșu
"Gheorghe Asachi" Technical University of Iași
Romania

Ioan I. Vrabie
"Alexandru Ioan Cuza" University of Iași and
"Octav Mayer" Mathematics Institute of the
Romanian Academy
Iași, Romania

CRC Press is an imprint of the
Taylor & Francis Group, an **informa** business
A CHAPMAN & HALL BOOK

MONOGRAPHS AND RESEARCH NOTES IN MATHEMATICS

Series Editors

John A. Burns
Thomas J. Tucker
Miklos Bona
Michael Ruzhansky

Published Titles

Application of Fuzzy Logic to Social Choice Theory, John N. Mordeson, Davender S. Malik and Terry D. Clark

Blow-up Patterns for Higher-Order: Nonlinear Parabolic, Hyperbolic Dispersion and Schrödinger Equations, Victor A. Galaktionov, Enzo L. Mitidieri, and Stanislav Pohozaev

Complex Analysis: Conformal Inequalities and the Bieberbach Conjecture, Prem K. Kythe

Computational Aspects of Polynomial Identities: Volume I, Kemer's Theorems, 2nd Edition Alexei Kanel-Belov, Yakov Karasik, and Louis Halle Rowen

A Concise Introduction to Geometric Numerical Integration, Fernando Casas and Sergio Blanes

Cremona Groups and Icosahedron, Ivan Cheltsov and Constantin Shramov

Delay Differential Evolutions Subjected to Nonlocal Initial Conditions Monica-Dana Burlică, Mihai Necula, Daniela Roșu, and Ioan I. Vrabie

Diagram Genus, Generators, and Applications, Alexander Stoimenow

Difference Equations: Theory, Applications and Advanced Topics, Third Edition Ronald E. Mickens

Dictionary of Inequalities, Second Edition, Peter Bullen

Introduction to Abelian Model Structures and Gorenstein Homological Dimensions Marco A. Pérez

Iterative Optimization in Inverse Problems, Charles L. Byrne

Line Integral Methods for Conservative Problems, Luigi Brugnano and Felice Iavernaro

Lineability: The Search for Linearity in Mathematics, Richard M. Aron, Luis Bernal González, Daniel M. Pellegrino, and Juan B. Seoane Sepúlveda

Modeling and Inverse Problems in the Presence of Uncertainty, H. T. Banks, Shuhua Hu, and W. Clayton Thompson

Monomial Algebras, Second Edition, Rafael H. Villarreal

Nonlinear Functional Analysis in Banach Spaces and Banach Algebras: Fixed Point Theory Under Weak Topology for Nonlinear Operators and Block Operator Matrices with Applications, Aref Jeribi and Bilel Krichen

Partial Differential Equations with Variable Exponents: Variational Methods and Qualitative Analysis, Vicenţiu D. Rădulescu and Dušan D. Repovš

A Practical Guide to Geometric Regulation for Distributed Parameter Systems Eugenio Aulisa and David Gilliam

Published Titles Continued

Reconstruction from Integral Data, Victor Palamodov

Signal Processing: A Mathematical Approach, Second Edition, Charles L. Byrne

Sinusoids: Theory and Technological Applications, Prem K. Kythe

Special Integrals of Gradshteyn and Ryzhik: the Proofs – Volume I, Victor H. Moll

Special Integrals of Gradshteyn and Ryzhik: the Proofs – Volume II, Victor H. Moll

Stochastic Cauchy Problems in Infinite Dimensions: Generalized and Regularized Solutions, Irina V. Melnikova

Submanifolds and Holonomy, Second Edition, Jürgen Berndt, Sergio Console, and Carlos Enrique Olmos

The Truth Value Algebra of Type-2 Fuzzy Sets: Order Convolutions of Functions on the Unit Interval, John Harding, Carol Walker, and Elbert Walker

Forthcoming Titles

Actions and Invariants of Algebraic Groups, Second Edition, Walter Ferrer Santos and Alvaro Rittatore

Analytical Methods for Kolmogorov Equations, Second Edition, Luca Lorenzi

Geometric Modeling and Mesh Generation from Scanned Images, Yongjie Zhang

Groups, Designs, and Linear Algebra, Donald L. Kreher

Handbook of the Tutte Polynomial, Joanna Anthony Ellis-Monaghan and Iain Moffat

Microlocal Analysis on R^n and on NonCompact Manifolds, Sandro Coriasco

Practical Guide to Geometric Regulation for Distributed Parameter Systems, Eugenio Aulisa and David S. Gilliam

Symmetry and Quantum Mechanics, Scott Corry

CRC Press
Taylor & Francis Group
6000 Broken Sound Parkway NW, Suite 300
Boca Raton, FL 33487-2742

© 2016 by Taylor & Francis Group, LLC
CRC Press is an imprint of Taylor & Francis Group, an Informa business

No claim to original U.S. Government works

Printed on acid-free paper
Version Date: 20160504

International Standard Book Number-13: 978-1-4987-4644-1 (Hardback)

This book contains information obtained from authentic and highly regarded sources. Reasonable efforts have been made to publish reliable data and information, but the author and publisher cannot assume responsibility for the validity of all materials or the consequences of their use. The authors and publishers have attempted to trace the copyright holders of all material reproduced in this publication and apologize to copyright holders if permission to publish in this form has not been obtained. If any copyright material has not been acknowledged please write and let us know so we may rectify in any future reprint.

Except as permitted under U.S. Copyright Law, no part of this book may be reprinted, reproduced, transmitted, or utilized in any form by any electronic, mechanical, or other means, now known or hereafter invented, including photocopying, microfilming, and recording, or in any information storage or retrieval system, without written permission from the publishers.

For permission to photocopy or use material electronically from this work, please access www.copyright.com (http://www.copyright.com/) or contact the Copyright Clearance Center, Inc. (CCC), 222 Rosewood Drive, Danvers, MA 01923, 978-750-8400. CCC is a not-for-profit organization that provides licenses and registration for a variety of users. For organizations that have been granted a photocopy license by the CCC, a separate system of payment has been arranged.

Trademark Notice: Product or corporate names may be trademarks or registered trademarks, and are used only for identification and explanation without intent to infringe.

Visit the Taylor & Francis Web site at
http://www.taylorandfrancis.com

and the CRC Press Web site at
http://www.crcpress.com

Dedicated

to our Distinguished Mentor,
Professor Viorel Barbu,
on the occasion of his 75th birthday.

Contents

Preface	**xiii**
Motivation	**xvii**
Symbol Description	**xxv**

1 Preliminaries — **1**

1.1	Topologies on Banach spaces	1
1.2	A Lebesgue-type integral for vector-valued functions	3
	1.2.1 The Bochner integral	3
	1.2.2 The L^p spaces	5
1.3	The superposition operator	8
1.4	Compactness theorems	9
	1.4.1 Generalities	9
	1.4.2 Topological fixed-point theorems	10
	1.4.3 Compactness in function spaces	11
1.5	Multifunctions	14
	1.5.1 Generalities	14
	1.5.2 Continuity properties	15
	1.5.3 Superposition multifunctions	17
	1.5.4 Fixed-point theorems for multifunctions	22
1.6	C_0-semigroups	23
	1.6.1 Generalities	23
	1.6.2 Generation theorems	24
1.7	Mild solutions	25
	1.7.1 Types of solutions	25
	1.7.2 Compactness of the solution operator	27
1.8	Evolutions governed by m-dissipative operators	28
	1.8.1 Semi-inner products	28
	1.8.2 Dissipative operators and evolution equations	30
	1.8.3 Compactness of the solution operator	34
1.9	Examples of m-dissipative operators	36
	1.9.1 Sobolev spaces	36
	1.9.2 The Laplace operator	37
	1.9.3 C_0-groups generators	38

vii

viii   *Contents*

    1.9.4   The nonlinear diffusion operator   39
    1.9.5   The p-Laplace operator   42
1.10 Strong solutions   43
1.11 Nonautonomous evolution equations   44
1.12 Delay evolution equations   45
    1.12.1 The autonomous case   46
    1.12.2 The quasi-autonomous case   49
1.13 Integral inequalities   51
1.14 Brezis–Browder ordering principle   53
1.15 Bibliographical notes and comments   54

2   Local Initial Conditions   59

2.1   An existence result for ODEs with delay   59
    2.1.1   A Weierstrass-type boundedness result   60
    2.1.2   The main local existence theorem   61
2.2   An application to abstract hyperbolic problems   62
    2.2.1   The main abstract result   63
    2.2.2   A semilinear wave equation with delay   64
2.3   Local existence: The case f Lipschitz   66
2.4   Local existence: The case f continuous   69
2.5   Local existence: The case f compact   71
2.6   Global existence   72
2.7   Examples   76
    2.7.1   Spring mass system with delay   77
    2.7.2   A delayed glucose level–dependent dosage   78
2.8   Global existence of bounded C^0-solutions   80
2.9   Three more examples   82
    2.9.1   The nonlinear diffusion equation with delay   83
    2.9.2   A singular transport equation   84
    2.9.3   A semilinear damped wave equation with delay   86
2.10 Bibliographical notes and comments   88

3   Nonlocal Initial Conditions: The Autonomous Case   91

3.1   The problem to be studied   91
3.2   The case f and g Lipschitz   94
    3.2.1   The main result   97
    3.2.2   The nondelayed case   98
3.3   Proofs of the main theorems   99
    3.3.1   Proof of Theorem 3.2.1   99
    3.3.2   Proof of Theorem 3.2.2   101
3.4   The transport equation in $\mathbb{R}^d$ 103
3.5   The damped wave equation with nonlocal initial conditions .   107
3.6   The case f Lipschitz and g continuous   111

Contents

3.7	Parabolic problems governed by the p-Laplacian		115
3.8	Bibliographical notes and comments		116
	3.8.1	The nondelayed case	116
	3.8.2	The delayed case	118

4 Nonlocal Initial Conditions: The Quasi-Autonomous Case 121

4.1	The quasi-autonomous case with f and g Lipschitz		121
	4.1.1	Periodic solutions	123
	4.1.2	Anti-periodic solutions	124
	4.1.3	The nondelayed case	127
4.2	Proofs of Theorems 4.1.1, 4.1.2		129
	4.2.1	An auxiliary existence result	129
	4.2.2	A boundedness lemma	133
	4.2.3	Proof of Theorem 4.1.1	134
	4.2.4	Proof of Theorem 4.1.2	135
4.3	Nonlinear diffusion with nonlocal initial conditions		136
	4.3.1	The main result	137
	4.3.2	The periodic and the anti-periodic problem	138
4.4	Continuity with respect to the data		139
	4.4.1	Statement of the main result	140
	4.4.2	Proof of the main result	141
	4.4.3	The nondelayed case	144
	4.4.4	A glance at periodic problems	145
4.5	The case f continuous and g Lipschitz		149
	4.5.1	Statement of the main result	149
	4.5.2	Excursion to the nondelayed case	150
	4.5.3	Proof of the main result	152
4.6	An example involving the p-Laplacian		154
4.7	The case f Lipschitz and g continuous		156
	4.7.1	The general assumptions and the main theorem	156
	4.7.2	An auxiliary lemma	157
	4.7.3	The fixed-point argument	158
	4.7.4	Proof of the main theorem	162
4.8	The case A linear, f compact, and g nonexpansive		163
	4.8.1	Statement of the main result	163
	4.8.2	The nondelayed case	164
	4.8.3	Proof of the main result	166
4.9	The case f Lipschitz and compact, g continuous		166
	4.9.1	The main assumptions and some preliminaries	167
	4.9.2	The main theorem	168
4.10	The damped wave equation revisited		173
	4.10.1	The first existence result	174
	4.10.2	The second existence result	175
4.11	Further investigations in the case $\ell = \omega$		176

x   *Contents*

    4.11.1 The assumptions 177
    4.11.2 The main result 177
    4.11.3 Periodic solutions 178
    4.11.4 Anti-periodic solutions 178
    4.11.5 The nondelayed case 179
    4.11.6 Proof of the main result 181
  4.12 The nonlinear diffusion equation revisited 184
    4.12.1 Statement of the main result 184
    4.12.2 Proof of the main result 186
  4.13 Bibliographical notes and comments 189

5 Almost Periodic Solutions   **193**

  5.1 Almost periodic functions 193
  5.2 The main results 195
  5.3 Auxiliary lemmas 196
  5.4 Proof of Theorem 5.2.1 200
  5.5 The ω-limit set 203
  5.6 The transport equation in one dimension 207
  5.7 An application to the damped wave equation 210
  5.8 Bibliographical notes and comments 213

6 Evolution Systems with Nonlocal Initial Conditions   **215**

  6.1 Single-valued perturbed systems 215
  6.2 The main result 217
  6.3 The idea of the proof 220
  6.4 An auxiliary lemma 221
  6.5 Proof of Theorem 6.2.1 224
  6.6 Application to a reaction–diffusion system in $L^2(\Omega)$ 229
  6.7 Nonlocal initial conditions with linear growth 231
  6.8 The idea of the proof 237
  6.9 Auxiliary results 238
  6.10 Proof of Theorem 6.7.1 244
  6.11 A nonlinear reaction–diffusion system in $L^1(\Omega)$ 245
  6.12 Bibliographical notes and comments 248

7 Delay Evolution Inclusions   **251**

  7.1 The problem to be studied 251
  7.2 The main results and the idea of the proof 253
    7.2.1 The idea of the proof of Theorem 7.2.3 255
  7.3 Proof of Theorem 7.2.1 256
  7.4 A nonlinear parabolic differential inclusion 267
  7.5 The nonlinear diffusion in $L^1(\Omega)$ 272

Contents

xi

7.6	The case when F has affine growth	276
7.7	Proof of Theorem 7.6.1	280
7.8	A differential inclusion governed by the p-Laplacian	286
7.9	A nonlinear diffusion inclusion in $L^1(\Omega)$	288
7.10	Bibliographical notes and comments	290

8 Multivalued Reaction–Diffusion Systems

293

8.1	The problem to be studied	293
8.2	The main result	294
8.3	Idea of the proof of Theorem 8.2.1	299
8.4	A first auxiliary lemma	301
8.5	The operator Γ_ε	302
8.6	Proof of Theorem 8.2.1	306
8.7	A reaction–diffusion system in $L^1(\Omega)$	308
8.8	A reaction–diffusion system in $L^2(\Omega)$	311
8.9	Bibliographical notes and comments	313

9 Viability for Nonlocal Evolution Inclusions

315

9.1	The problem to be studied	315
9.2	Necessary conditions for viability	316
9.3	Sufficient conditions for viability	318
9.4	A sufficient condition for null controllability	322
9.5	The case of nonlocal initial conditions	326
9.6	An approximate equation	328
9.7	Proof of Theorem 9.5.1	331
9.8	A comparison result for the nonlinear diffusion	331
9.9	Bibliographical notes and comments	336

Bibliography

339

Index

359

Preface

This book contains several very recent results referring to the existence, boundedness, regularity, and asymptotic behavior of global solutions for differential equations and inclusions, with or without delay, subjected to nonlocal implicit initial conditions. An explanation of the choice of this topic – which, as usual, cannot completely lack bias and thus subjectivity – is included in the section immediately following the Preface.

The book contains ten chapters, preceded by a detailed List of Symbols, Motivation, and followed by a bibliography and index. As far as the references list is concerned, it should be noted that, although it is not exhaustive, a great effort was made to include the original source of each notion or result used and, whenever possible, to add one or more complementary references to some easy-to-find monographs where additional information on the subject considered can be found. Although the material is self-contained, a good knowledge of the basics of linear – or even nonlinear – functional analysis would be helpful.

Each chapter ends with a section entitled "Bibliographical notes and comments," offering an as accurate as possible historical perspective to the notions and results discussed therein. Moreover, each chapter belonging to the main body of the book, i.e., Chapters 2–9, contains several illustrative examples of both semilinear or fully nonlinear ODEs and PDEs subjected to various local or nonlocal initial conditions, each one being important on its own, and thus offering a good insight into the possibilities of the abstract developed theory.

Chapter 1, Preliminaries collects the basic concepts and results in functional analysis, topology, integration theory, superposition operators, multifunctions, C_0-semigroups, and linear evolution equations, nonlinear evolution equations in abstract Banach spaces, m-dissipative partial differential operators, delay evolution equations subjected to initial local conditions, integral inequalities with or without delay, and the Brezis–Browder Ordering Principle – all needed for a good understanding of the entire book.

Chapter 2, Local Initial Conditions is mainly concerned with some simple but fundamental results on abstract delay evolution equations subjected to local initial conditions already introduced in Chapter 1. However, it should be noted that we analyze the case of locally Lipschitz perturbations as well as the case of continuous perturbations by a direct approach, borrowed from classical ODE theory and avoid the use of the abstract ma-

xiii

xiv   *Preface*

chinery presented in Section 1.12 of Chapter 1. Some illustrative examples
from mathematical modeling are included. We have to stress here that this
book was not intended as a purely applied mathematics book. Therefore, for
almost all ODEs and PDEs presented and that are justified as mathematical models, we did not provide all the specific details referring to the aspects
involved.

Chapter 3, Nonlocal Initial Conditions: The Autonomous Case includes several abstract results on the existence, uniqueness, and global asymptotic stability result referring to autonomous nonlinear delay evolution equations, as well as three illustrative examples. It should be noted that we decided
to split this problem into the autonomous and the quasi-autonomous case for
two reasons. First, because we believe that it is more convenient for readers
interested only in the autonomous case to have at their disposal a simpler,
even indirect approach based on the abstract theory already developed in Section 1.12. Second, we consider that this split could be a very good lesson of
what we have to do when we are facing the dilemma of putting into balance
what we can win and what we can lose by using either a direct approach,
as done in Chapter 4, or by simply imbedding the problem into an abstract
one by taking advantage of the already invented mathematical tools in order
to solve it. More precisely, in many circumstances, approaching a problem
by means of a very general abstract theory, we could miss some important
aspects which can be revealed only by a direct analysis. This certainly does
not happen in the autonomous case discussed here where if, no matter which
of the two approaches we choose, we get the very same result. In contrast, in
the quasi-autonomous one, once we have chosen to follow the abstract way
offered by the results in Section 1.12, we have to impose a stronger continuity
assumption of f – see (iii) in Theorem 1.12.5, which proves to be unnecessary
in a direct approach.

**Chapter 4, Nonlocal Initial Conditions: The Quasi-Autonomous
Case** refers to quasi-autonomous nonlinear delay evolution equations, extending the results already proved in Chapter 3 in the autonomous case. As we
have mentioned, here we are using rather elementary methods which, in spite
of some inherent technicalities, lead to more general results than those obtained by reducing the problem to an abstract one studied in Section 1.12.
Although this book, as its title stresses, concentrates mainly on the analysis of evolution equations and inclusions with delay, subjected to nonlocal
implicit initial conditions, whenever some consequences referring to the non-
delay case are of interest we did not hesitate to include them in the most
appropriate places. Moreover, the hypotheses used are general enough to handle not only classical initial-value problems, but also various other problems
subjected to periodic, anti-periodic, and mean-value conditions, with or without delay. Many of the results presented are new and extend related theorems
established in the last few years by the authors of this research. Finally, a

Preface   xv

global existence result referring to an infinite delay evolution equation subjected to a nonlocal condition is derived.

Chapter 5, Almost Periodic Solutions is devoted to some sufficient conditions for an evolution equation with delay, subjected to nonlocal initial conditions, to have almost periodic solutions. The existence of ω-limit sets and of compact attractors is proved under the natural condition that the m-dissipative operator on the right-hand side has a compact resolvent. The effectiveness of the abstract results obtained is proved by the two examples included: one referring to a first-order semilinear hyperbolic problem and one referring to a nonlinear damped wave equation, both with delay and subjected to nonlocal initial conditions.

Chapter 6, Evolution Systems with Nonlocal Initial Conditions is dedicated to the presentation of some abstract reaction–diffusion systems with delay subjected to nonlocal initial conditions. For the two cases considered, i.e., when the initial history constraint function has either affine or linear growth, some sufficient conditions for the global existence, uniqueness, and uniform asymptotic stability of C^0-solutions are established. Two applications to specific nonlinear reaction–diffusion systems are discussed in detail.

Chapter 7, Delay Evolution Inclusions is concerned with the study of a broad class of nonlinear functional differential evolution inclusions with delay subjected to nonlocal initial conditions. We start with a class of problems in which the key assumptions are a flow-invariance condition on the reaction term and the compactness of the generated semigroup. This class incorporates several important multivalued variants of the functional differential evolution equations with delay that we analyzed in Chapter 3. Two examples referring to some parabolic partial differential inclusions driven either by the Laplace operator in $L^1(\Omega)$ or by the p-Laplace operator in $L^2(\Omega)$, both subjected to various nonlocal initial conditions on the history, are also analyzed. In the next sections, we complement the preceding results by some variants obtained under the usual affine growth condition on the perturbed term rather than the flow-invariance hypothesis already used before. Here also, the m-dissipative operator involved is assumed to generate a compact semigroup. In order to show that the hypotheses of this latter abstract result are sometimes easier to verify than the flow-invariance condition, some applications to nonlinear PDE inclusions with delay subjected to nonlocal initial conditions are discussed.

Chapter 8, Multivalued Reaction–Diffusion Systems reconsiders, within the general frame of multivalued right-hand sides, the main problems studied in Chapter 6. Of course, the analysis here rests heavily on an interplay between multivalued function theory and topological methods specific to nonlinear semigroup theory. The chapter ends with two examples of specific reaction–diffusion systems with delay subjected to initial nonlocal implicit conditions. Although some fixed-point methods are also used in this case,

xvi   *Preface*

there are some important differences between the proof of the main result in this chapter and the proof of the corresponding main result in Chapter 6.

Chapter 9, Viability for Nonlocal Evolution Inclusions is dedicated to the existence of C^0-solutions whose graphs are included in the graph of an a priori given multifunction. The main difficulty here is that, due to the nonlocal initial condition, it is impossible to approach the problem directly as happens in almost all viability results for initial-value problems. In order to illustrate the effectiveness of the abstract results, three significant particular cases of practical interest are discussed.

Finally, we hope that this research monograph will offer an incentive for further study in this interesting topic.

Acknowledgments. The support of the Romanian National Authority for Scientific Research, CNCS–UEFISCDI, project number PN-II-ID-PCE-2011-3-0052, was essential in carrying out the writing of this monograph.

We also would like to express our warmest thanks to Professor Dr. Irene Benedetti from Università di Perugia, Italy; Professor Dr. Aleksander Ćwiszewski and Professor Dr. Wojciech Kryszewski, both from Nicolaus Copernicus University in Toruń, Poland; Professor Dr. Mark McKibben from West Chester University of Pennsylvania, U.S.A.; and Professor Dr. Octavia–Maria Nica (Bolojan) from "Babeş-Bolyai" University in Cluj-Napoca, Romania. Their comments, remarks, and suggestions, made after an extremely careful reading of the manuscript, led to a real improvement of the presentation.

We express our deep gratitude to the Taylor & Francis team and especially to the editors of the monograph series of CRC Press, Taylor & Francis Group, Sarfraz Khan, Alexander Edwards, Stephanie Morkert and Karen Simon, whose guidance during the whole process of finishing the manuscript was extremely helpful. Finally, our thanks also go to the copywriter Katy Smith who wrote the promotional material for our book.

Monica-Dana Burlică, Mihai Necula, Daniela Roşu and Ioan I. Vrabie

Motivation

Why study delay equations?

Differential equations and systems are mainly motivated by applications. In particular, many mathematical models take the abstract form of a differential equation or system whose study can reveal important features concerning the evolution of the modeled phenomenon. More than this, a good and accurate model of the uncontrolled evolution of some real system in the absence of any exterior intervention could suggest what kind of feedback can be used in order to make the system behave according to some performance criteria we wish to achieve. For instance, the controlled evolution of levels both of glucose in the bloodstream and in the gastrointestinal system can be described, at a first approximation, by the following simple system of ordinary differential equations

$$\begin{cases} u'(t) = -au(t) + f(t, u(t), v(t)), & t \in I, \\ v'(t) = au(t) - bv(t) + g(t, u(t), v(t)), & t \in I, \end{cases} \tag{M.1.1}$$

where $I = [0, T]$, $u(t)$ is the glucose level in the gastrointestinal system, a is the instantaneous rate of absorption of the glucose, $v(t)$ is the glucose level in the bloodstream which increases like $u(t)$, and b is the instantaneous rate of elimination of the glucose from the bloodstream. Both $f(t, u(t), v(t))$ and $g(t, u(t), v(t))$ are feedback operators representing the instantaneous dosage of the glucose in the gastrointestinal system as well as in the bloodstream.[1] Moreover, one presumes we are able to make some precise instantaneous measurements on both $u(t)$ and $v(t)$ in order to intervene instantaneously with the appropriate dosages $f(t, u(t), v(t))$ and $g(t, u(t), v(t))$. Of course, due to the obvious fact that the "instantaneous measurements" coupled with "instantaneous reactions" are unrealistic requirements, this is practically impossible. For this reason, we have to replace the above model with a more appropriate one which takes into account the whole history of both u and v on a given interval $[t - \tau, t]$ rather than the values obtained at the exact time t of intervention of both f and g into the system. So, we are led either to a model

---

[1] Clearly $f(t, u(t), v(t))$ contributes to the rate of change variation of $u(t)$ caused by the total mass of carbohydrates in the food, while $g(t, u(t), v(t))$ modifies the rate change of $v(t)$ due to the influence of some drugs: insulin, etc.

xviii *Motivation*

expressed by a system of differential-difference equations

$$\begin{cases} u'(t) = -au(t) + f(t, u(t-\tau), v(t-\tau)), & t \in I, \\ v'(t) = au(t) - bv(t) + g(t, u(t-\tau), v(t-\tau)), & t \in I, \end{cases} \quad \text{(M.1.2)}$$

or to another one consisting of delay differential equations

$$\begin{cases} u'(t) = -au(t) + f\left(t, \int_{-\tau}^{0} u(t+s)\,ds, \int_{-\tau}^{0} v(t+s)\,ds\right), & t \in I, \\ v'(t) = au(t) - bv(t) + g\left(t, \int_{-\tau}^{0} u(t+s)\,ds, \int_{-\tau}^{0} v(t+s)\,ds\right), & t \in I. \end{cases} \quad \text{(M.1.3)}$$

In case (M.1.2), in order to know what to do at a certain time t, we need to measure both u and v at $t - \tau$, which can be done by measuring the quantity of glucose ingested at $t - \tau$ and by doing a blood test measuring the glucose instantaneous blood level, again at $t - \tau$, and then to react, i.e., to intervene appropriately. Of course, due to technical constraints, customarily the necessary intervention is performed with some "small a priori accepted" delay τ, i.e., at the time t.

In the last case, (M.1.3), τ equals 3 months[2] and the cumulative history

$$\int_{-\tau}^{0} v(t+s)\,ds$$

can be measured by the hemoglobin A1c (or glycosylated hemoglobin) test.

We can write all three cases considered above in the unified form

$$\begin{cases} u'(t) = -au(t) + F(t, u_t, v_t), & t \in I, \\ v'(t) = au(t) - bv(t) + G(t, u_t, v_t), & t \in I, \end{cases} \quad \text{(M.1.4)}$$

where u_t is the continuous function defined on $[-\tau, 0]$ by $u_t(s) = u(t+s)$ for each $s \in [-\tau, 0]$ and v_t is similarly defined.

Indeed, in case (M.1.1), the dosages are defined by

$$\begin{cases} F(t, u_t, v_t) = f(t, u_t(0), v_t(0)), \\ G(t, u_t, v_t) = g(t, u_t(0), v_t(0)), \end{cases}$$

in case (M.1.2), by

$$\begin{cases} F(t, u_t, v_t) = f(t, u_t(-\tau), v_t(-\tau)), \\ G(t, u_t, v_t) = g(t, u_t(-\tau), v_t(-\tau)), \end{cases}$$

while in case (M.1.3), by

$$\begin{cases} F(t, u_t, v_t) = f\left(t, \int_{-\tau}^{0} u_t(s)\,ds, \int_{-\tau}^{0} v_t(s)\,ds\right), \\ G(t, u_t, v_t) = g\left(t, \int_{-\tau}^{0} u_t(s)\,ds, \int_{-\tau}^{0} v_t(s)\,ds\right). \end{cases}$$

---

[2] The average lifetime of red blood cells that bind hemoglobin is 3 months.

Of course, we can imagine various combinations of the three cases described above, as for instance

$$F(t, u_t, v_t) = f\left(t, u_t(-\tau), \int_{-\tau}^{0} v_t(s)\, ds\right)$$

and so on.

Why study delay differential inclusions?

Sometimes, it would be more realistic to assume that the dosage $G(t, u_t, v_t)$ of the glucose into the bloodstream is not a single-valued function. It may happen that, in certain circumstances, and at some moments t and for some critical values of the histories u_t and/or v_t, from physiological reasons, this dosage is not unique but rather somewhere between two limits: a lower limit $G_1(t, u_t, v_t)$ and an upper limit $G_2(t, u_t, v_t)$. More precisely, if at some t, u_t and v_t are "optimal," then the reaction of the organism, i.e., its feedback, could be to choose at random a dosage $G(t, u_t, v_t)$ of the glucose in the bloodstream somewhere between two given values $G_1(t, u_t, v_t) < G_2(t, u_t, v_t)$. Every choice in the interval $[G_1(t, u_t, v_t), G_2(t, u_t, v_t)]$ being considered is acceptable by the organism itself.

So, in order to handle this new situation as well, we are led to consider a more general model:

$$\begin{cases} u'(t) = -au(t) + F\left(t, u_t, v_t\right), & t \in I, \\ v'(t) = au(t) - bv(t) + g(t), & t \in I, \\ G_1\left(t, u_t, v_t\right) \leq g(t) \leq G_2\left(t, u_t, v_t\right), & t \in I, \end{cases} \qquad \text{(M.2.1)}$$

in which the second differential equation appearing in (M.1.4) was replaced by a differential inclusion.

Why use nonlocal initial conditions?

In the process of obtaining descriptions of the evolution of some chemical, physical, biological, economical (and the list could continue) phenomena that are as accurate as possible, there are at least two main types of problems encountered, which differ from each other.

The first one is an initial-value or Cauchy problem and consists of finding

xx   *Motivation*

a function $u = u(\cdot, a, \xi)$, defined on an interval J, that satisfies a differential
equation + an initial condition

$$
\begin{cases}
u'(t) = f(t, u(t)), & t \in J, \\
u(a) = \xi,
\end{cases}
\tag{M.3.1}
$$

where $a \in J$ is the so-called initial time and ξ, belonging to the state space,
which could be $\mathbb{R}$, $\mathbb{R}^d$ or an infinite-dimensional Banach space X, is the so-
called initial state. This is a very simple deterministic model which amounts
to saying that, once we know the initial state ξ of the system at the initial
time a and the law which describes the dependence of the instantaneous rate
of change $u'(t)$ of the state u on the instantaneous state $u(t)$, we are able to
predict the complete evolution of the system on the entire interval on which
it exists.

The second one is a so-called periodic problem which involves finding a C^1
function u satisfying:

$$
\begin{cases}
u'(t) = f(t, u(t)), & t \in \mathbb{R}_+, \\
u(t) = u(t + T) & t \in \mathbb{R}_+.
\end{cases}
\tag{M.3.2}
$$

Of course, in this case, in order to have a solution, it is almost necessary to
assume at least that f is T-periodic with respect to its first argument, i.e.,
$f(t, u) = f(t+T, u)$ for each $t \in \mathbb{R}_+$ and each u in the state space. In the latter
case, as far as the solution of the problem is concerned, although describing a
deterministic model, we cannot speak about the initial time or the initial state,
notions which, in this case, are not a priori given. Nevertheless, due to some
mathematical reasons, in many situations, we are led to fix an arbitrary initial
time and to treat the problem (M.3.2) – at least in the first stage – as an initial-
value problem.[3] We emphasize, however, that in spite of the fact that the two
types of problems are completely different from the mathematical viewpoint,
they can be compared to each other from the perspective of how accurate
the description of the phenomenon is that each offers. We can say that an
initial-value problem better describes the evolution of a certain phenomenon
than a periodic problem or conversely. Sometimes, starting from an initial-
value problem, it may happen that, for some initial data, the corresponding
solutions are periodic, or at least in the initial history, i.e., for $t \in [-\tau, 0]$, u
satisfies the T-periodicity-like condition $u(t) = u(t + T)$, even though f fails
to be periodic with respect to its first argument.

There are, however, cases for which the prediction offered by an initial-
value problem is more accurate if instead of a single initial datum given at a

---

[3]This idea goes back to Poincaré who introduced the operator – owing his name and
therefore denoted by P– which assigns to each initial-value ξ of a Cauchy problem the value
of the corresponding solution $u(\cdot, 0, \xi)$ at a final time T, i.e., $P(\xi) = u(T, 0, \xi)$. Clearly, each
fixed point ξ of P is the initial data for a solution satisfying $u(T, 0, \xi) = \xi = u(0, 0, \xi)$ and
conversely. If, in addition, $f(t + T, u) = f(t, u)$ for each $t \in \mathbb{R}_+$ and u in the state space,
then $u(\cdot, 0, \xi)$ can be continued up to a T-periodic solution defined on $\mathbb{R}_+$.

Motivation

single initial time $t = a$, more data, at certain different times, strictly greater than a, are collected and their weighted average is used. For the sake of simplicity, let us assume that $J = \mathbb{R}_+$, $a = 0$ and we have the possibility to measure the values of U – the *exact solution*[4] – at some points $0 < t_1 < t_2 < \cdots < t_n$. It should be emphasized that the exact solution U does not coincide with the solution u of the mathematical model (M.3.1), simply because in the construction of (M.3.1) it is impossible to take into consideration all the data involved in the evolution of the phenomenon considered.

Then, one may approximate the exact solution U of the system by the solution v of the differential model below whose initial data, $v(0)$, is assumed to be the weighted average of the measured data $U(t_1), U(t_2), \ldots U(t_n)$, gathered at the specific moments $0 < t_1 < t_2 < \cdots < t_n$. We denote this weighted average by $g(U)$, i.e.,

$$g(U) = \sum_{k=1}^{n} \alpha_k U(t_k),$$

where $\alpha_k \in (0, 1)$, $k = 1, 2, \ldots n$, are such that

$$\sum_{k=1}^{n} \alpha_k = 1.$$

Thus, instead of the initial-value problem (M.3.1) with $J = \mathbb{R}_+$, $a = 0$ and $u(0) = \xi = U(0)$, it is more convenient to consider a variant, i.e.,

$$\begin{cases} v'(t) = f(t, v(t)), & t \in \mathbb{R}_+, \\ v(0) = g(U(\cdot)). \end{cases} \tag{M.3.3}$$

Of course, (M.3.3) is simply (M.3.1) with $\xi = g(U(\cdot))$. In some practical circumstances,[5] the problem is that it is very hard, or even impossible, to make accurate measurements. Therefore, we have to choose a different approach. A possible strategy would be to replace $U(t_k)$ by $v(t_k)$ for $k = 1, 2, \ldots, n$. Empirical studies have shown that the "model" thus obtained, i.e.,

$$\begin{cases} v'(t) = f(t, v(t)), & t \in \mathbb{R}_+, \\ v(0) = \sum_{k=1}^{n} \alpha_k v(t_k) = g(v(\cdot)), \end{cases} \tag{M.3.4}$$

even less exact than the preceding one, is still reliable enough to be taken into consideration as an acceptable alternative. So, we are led to consider the nonlocal initial-value problem above as a substitute for (M.3.3).

---

[4]In fact, we are able to measure the *observed solution*, which usually differs from the exact one but furnishes the main features of the exact solution.

[5]As for instance when we are trying to measure some real data in the deep ocean, or to gather meteorological data in order to formulate a forecast.

xxii   *Motivation*

Clearly, the problem (M.3.4), involving a nonlocal initial condition, is completely different from its classical initial-value counterpart (M.3.1) and is sufficiently complicated. Indeed, the difficulties come from the fact that we must find v satisfying not only the differential equation, but an implicit constraint as well, i.e., $v(0) = g(v(\cdot))$, which in its simplest case reduces to the T-periodicity condition, $v(0) = v(T)$.

Therefore, the mathematical machinery which is appropriate in the study of (M.3.1) is no longer useful in the case of (M.3.4). In addition, the problem of finding suitable methods for analyzing (M.3.4) is not at all simple and could be, in many cases, rather challenging.

Putting it all together

Recalling that, at least for closed-loop systems, delay equations reflect more accurately the reality than the nondelayed ones, we can conclude that almost all mathematical models describing the controlled evolution of a phenomenon, as those mentioned before, can be reformulated in a more general abstract form as evolution equations with delay subjected to nonlocal initial conditions of the form:

$$\begin{cases} u'(t) \in Au(t) + f(t, u_t), & t \in \mathbb{R}_+, \\ u(t) = g(u(\cdot))(t), & t \in [-\tau, 0]. \end{cases} \tag{M.4.1}$$

Here $A : D(A) \subseteq X \rightsquigarrow X$ is an m-dissipative operator in the (infinite dimensional) Banach space X. Usually, $X = L^p(\Omega)$, $1 \le p \le \infty$, and A is a nonlinear partial differential operator (e.g., the nonlinear diffusion operator, or the transport operator, or the wave operator) representing a both heuristically and experimentally determined natural law upon which the uncontrolled evolution of the phenomenon takes place. Namely, it describes the instantaneous rate of change of u depending on its spatial configuration within a domain $\Omega \subseteq \mathbb{R}^d$ at any time t. Furthermore, the delay $\tau \ge 0$ is arbitrary but fixed, $f : \mathbb{R}_+ \times C([-\tau, 0]; X) \to X$ is continuous and represents a distributed feedback, i.e., an exterior reaction term taking into account the history u_t of the function u rather than its instantaneous state $u(t)$ for $t \in \mathbb{R}_+$. In all the cases considered subsequently, the nonlocal constraint on the initial delay of the state, i.e., the possible nonlinear function $g : C_b(\mathbb{R}_+; \overline{D(A)}) \to C([-\tau, 0]; X)$, is nonexpansive or even continuous and has affine growth, i.e., there exists $m_0 \ge 0$ such that

$$\|g(u(\cdot))\|_{C([-\tau, 0]; X)} \le \|u\|_{C_b(\mathbb{R}_+; X)} + m_0 \tag{M.4.2}$$

for each $u \in C_b(\mathbb{R}_+; \overline{D(A)})$ – the set of all continuous and bounded functions from $\mathbb{R}_+$ to $\overline{D(A)}$.[6] The simplest example of such a function satisfying (M.4.2) is given by g constant, i.e., $g(u) \equiv \psi$, where $\psi \in C([-\tau, 0]; X)$ and $\psi(0) \in \overline{D(A)}$. In this case, (M.4.1) is a delay Cauchy problem, i.e., a delay evolution equation subjected to a preassigned initial history condition $u_{|[-\tau,0]} = \psi$. We emphasize that there are many other possible choices of g leading to periodic problems, anti-periodic problems, mean-value initial problems, to mention the most important ones.

Final comments

As far as the general frame here considered is concerned, the reason why we have chosen to study problems of this kind rests heavily on the fact that they combine into a unitary frame a very large variety of evolution problems subjected either to initial local conditions with or without delay, or even to purely nonlocal initial conditions – see Definition 3.1.2 – with or without delay. For instance, problems without delay subjected to either nonlocal or local initial conditions of the general form

$$\begin{cases} u'(t) \in Au(t) + f(t, u(t)), & t \in \mathbb{R}_+, \\ u(0) = g(u(\cdot)), \end{cases} \qquad \text{(M.5.1)}$$

are particular cases of the general problem (M.4.1). Indeed, (M.5.1) contains, as specific instances, both classical initial-value problems (just take g constant, i.e., $g \equiv \xi$ for each $u \in C_b(\mathbb{R}_+; \overline{D(A)})$ or T-periodic problems (just take $g(u) = u(T)$ for each $u \in C_b(\mathbb{R}_+; \overline{D(A)})$), choices allowed, as we will see, by our general assumptions.

As we have mentioned in the Preface, our purpose is to present in book form some very recent results many of which are due to the authors of this monograph. These results concern evolution equations of the form (M.4.1) subjected to nonlocal initial conditions with delay or even differential inclusions subjected to similar initial conditions, i.e.,

$$\begin{cases} u'(t) \in Au(t) + f(t), & t \in \mathbb{R}_+, \\ f(t) \in F(t, u_t), & t \in \mathbb{R}_+, \\ u(t) = g(u(\cdot))(t), & t \in [-\tau, 0], \end{cases} \qquad \text{(M.5.2)}$$

which are more general than (M.2.1).

Besides, in order to illustrate the applicative power of the abstract results included, referring to (M.5.1) and (M.5.2), we have inserted various examples of ODEs and PDEs subjected to nonlocal initial conditions.

---

[6]We notice that, starting with Chapter 2, for the sake of simplicity, we will write $g(u)$ and $g(u)(t)$ instead of $g(u(\cdot))$ and $g(u(\cdot))(t)$, respectively.

xxiv   *Motivation*

Now, a very final comment for all those who are skeptical in accepting the existence of real phenomena for which the future determines the past. We hope that the example of the writing of a book could be convincing enough. Clearly, in this process, the history, which identifies with all chapters but the last, is determined by the future, which consists of the last chapter. It is easy to realize that, once going on with the writing, one needs to come back and to change the history according to the needs of the present and of the future. During this process, one can also see a certain delay in the decision. More than this, the history is completely determined only if the writing of the last chapter of the book is finished. Think on the celebrated Halmos sequence: $1, 1, 2, 1, 2, 3, 1, 2, 3, 4, \ldots$ describing the writing in spirals: first Chapter 1, then Chapter 1 and Chapter 2, and so on. See Halmos [137, §6. p. 131]. This shows why, for nonlocal initial-value problems, the global existence is, in some sense, a part of the problem.

Symbol Description

$AC([a,b])$ — the space of absolutely continuous functions from $[a,b]$ to $\mathbb{R}$

$\mathrm{conv}\,(M)$ — the convex hull of the set M, i.e., the set of all convex combinations of elements in M

$C([a,b];X)$ — the space of all continuous functions from $[a,b]$ to X endowed with the sup-norm $\|\cdot\|_{C([a,b];X)}$

$C(I;\overline{D(A)})$ — the closed subset of $C(I;X)$ containing all $u\in C(I;X)$ satisfying $u(t)\in\overline{D(A)}$ for each $t\in I$

$C_b(I;X)$ — the space of all bounded and continuous functions from I to X, equipped with the sup-norm

$C_b(I;\overline{D(A)})$ — the closed subset in $C_b(I;X)$ consisting of all elements $u\in C_b(I;X)$ satisfying $u(t)\in\overline{D(A)}$ for each $t\in I$

$\widetilde{C}_b(I;X)$ — $=C_b(I;X)$, with $I=[a,+\infty)$, endowed with the family of seminorms $\{\|\cdot\|_k;\ k\in\mathbb{N},\ k\geq a\}$

$\mathcal{D}$ — $=\{u\in\mathcal{X}\ u(0)\in\overline{D(A)}\}$

$D(\xi,r)$ — the closed ball with center ξ and radius r

$F:K\rightsquigarrow X$ — denotes a multifunction $F:K\to 2^X$

$\mathrm{Fin}\,(X^*)$ — the class of all finite subsets in X^*

$J:X\rightsquigarrow X^*$ — the duality mapping

$J(x)$ — $=\{x^*\in X^*;\ (x,x^*)=\|x\|^2=\|x^*\|^2\}$

$|\kappa|$ — $=\kappa_1+\kappa_2+\cdots+\kappa_n$, for $\kappa_1,\kappa_2,\ldots,\kappa_n\in\mathbb{N}$

$\mathcal{L}(X)$ — the space of all linear bounded $S:X\to X$

$L^1(0,T;X)$ — $=L^1([0,T];X)$

$L^p(\Omega)$ — $=L^p(\Omega;\mathbb{R})$

$L^p(\Omega;X)$ — $=L^p(\Omega,\mu;X)$ if μ is the Lebesgue measure

$L^p(\Omega;\mu,X)$ — $=\mathcal{L}^p(\Omega,\mu;X)/\sim$ with $\sim$ the μ-a.e. equality on Ω

$\mathcal{L}^p(\Omega,\mu;X)$ — the set of all strongly measurable functions, f, from Ω to X with $\|f\|^p$ μ-integrable

$H^1(\Omega)$ — $=W^{1,2}(\Omega)$

$H^2(\Omega)$ — $=W^{2,2}(\Omega)$

$H^{-1}(\Omega)$ — $=[H_0^1(\Omega)]^*$

$H_0^1(\Omega)$ — $=W_0^{1,2}(\Omega)$

$\mathcal{H}_0^1(\Omega)$ — $=C([-\tau,0];H_0^1(\Omega)$

$\mathbb{N}$ — the set of all positive integers

$\mathbb{N}^*$ — the set of all positive integers without 0

Q_+ — $=\mathbb{R}_+\times\Omega$

Q_+^ψ — $=[0,T_\psi)\times(0,1)$

Q_τ — $=[-\tau,0]\times\Omega$

Ω — a nonempty and open subset in $\mathbb{R}^n$

$|\Omega|$ — the Lebesgue measure of Ω

(Ω,Σ,μ) — a measure space

$\mathbb{R}$ — the set of real numbers

$R(\lambda;A)$ — $=(\lambda I-A)^{-1}$

$\mathbb{R}_+$ — the set of nonnegative real numbers

$S(\xi,r)$ — the open ball with center ξ and radius r

Σ_+ — $=\mathbb{R}_+\times\Sigma$

u_t — $u_t(s)=u(t+s)$ for each $s\in[-\tau,0]$

$W^{m,p}(\Omega)$ — the space of all functions $u:\Omega\to\mathbb{R}$ which, together with their partial derivatives up to the order m, in the sense of distributions over Ω, belong to $L^p(\Omega)$

$W_0^{m,p}(\Omega)$	the closure of $C_0^\infty(\Omega)$ in $W^{m,p}(\Omega)$	$(x,y)_+$	$= \lim\limits_{h\downarrow 0} \dfrac{\|x+hy\|^2 - \|x\|^2}{2h}$
$\mathcal{X}$	$= C([-\tau,0];X)$		
$\mathcal{X}_1$	$= C([-\tau,0];L^1(\Omega))$	$(x,y)_-$	$= \lim\limits_{h\uparrow 0} \dfrac{\|x+hy\|^2 - \|x\|^2}{2h}$
$\mathcal{X}_2$	$= C([-\tau,0];L^2(\Omega))$		
$\mathcal{X}_p$	$= C([-\tau,0];L^p(\Omega))$	$[x,y]_+$	$= \lim\limits_{h\downarrow 0} \dfrac{\|x+hy\| - \|x\|}{h}$
(x,x^*)	$= x^*(x)$		
X^*	the topological dual of the Banach space X	$[x,y]_-$	$= \lim\limits_{h\uparrow 0} \dfrac{\|x+hy\| - \|x\|}{h}.$

Chapter 1

Preliminaries

Overview

In this chapter, we collect the auxiliary notions and results that are needed for a good understanding of the whole book. So, we included here basic facts about strong and weak topologies on Banach spaces, the Bochner integral and usual function spaces, compactness theorems as for instance the infinite dimensional version of the Arzelà–Ascoli Theorem and two sufficient conditions for weak compactness in $L^1(\Omega, \mu; X)$, one for Ω of finite measure and one for Ω having σ-finite measure. We give a brief introduction to the theory of C_0-semigroups, m-dissipative operators, the nonlinear evolutions governed by them, evolution systems, m-dissipative linear as well as nonlinear partial differential operators. Some basic facts on delay evolution equations subjected to initial conditions, as well as on differential and integral inequalities found their place in this chapter.

1.1 Topologies on Banach spaces

We denote by X a real Banach space[1] with norm $\|\cdot\|$ and by X^* its *topological dual*, i.e., the vector space of all linear continuous functionals from X to $\mathbb{R}$, which, endowed with the *dual norm* $\|x^*\| = \sup_{\|x\| \le 1} |(x, x^*)|$, for $x^* \in X^*$, is a real Banach space too. As usual, if $x \in X$ and $x^* \in X^*$, (x, x^*) denotes $x^*(x)$. Let $\mathrm{Fin}\,(X^*)$ be the class of all finite subsets in X^* and let $F \in \mathrm{Fin}\,(X^*)$. The function $\|\cdot\|_F : X \to \mathbb{R}$, defined by

$$\|x\|_F = \max\{|(x, x^*)|;\ x^* \in F\}$$

for each $x \in X$, is a seminorm on X.

The family of seminorms $\{\|\cdot\|_F;\ F \in \mathrm{Fin}\,(X^*)\}$ defines the so-called

---

[1]Sometimes, we will assume that X is a real vector space endowed with a topology, which is not defined by a norm, but in all those cases we will clearly specify that.

2   Delay Differential Evolutions Subjected to Nonlocal Initial Conditions

weak topology. Equipped with this topology, X is a separated locally convex topological vector space, denoted by X_w.

In order to avoid confusion, whenever we refer to weak topology concepts, we shall use the name of the concept in question preceded or followed by the word weak (weakly). For instance, a subset B in X is called weakly closed if it is closed in the weak topology. If B is norm or strongly closed, we simply say that B is closed.

Throughout, if $\xi \in X$ and $r > 0$, we denote by $D(\xi, r)$ the closed ball with center ξ and radius r.

THEOREM 1.1.1 (Mazur) *The weak closure of a convex subset in a Banach space coincides with its strong closure.*

See Mazur [181] or Hille and Phillips [143, Theorem 2.9.3, p. 36].

If $M \subseteq X$, conv (M) denotes the *convex hull* of M, i.e., the set of all convex combinations of elements in M.

COROLLARY 1.1.1 *If* $\lim_n x_n = x$ *weakly in* X, *then there exists* $(y_n)_n$, *with* $y_n \in \operatorname{conv} \{x_k; \ k \geq n\}$, *such that* $\lim_n y_n = x$.

See Hille and Phillips [143, Corollary to Theorem 2.9.3, p. 36].

DEFINITION 1.1.1 A Banach space X is called *uniformly convex* if for each $\varepsilon \in (0, 2]$ there exists $\delta(\varepsilon) > 0$ such that, for each $x, y \in X$ with $\|x\| \leq 1$, $\|y\| \leq 1$ and $\|x - y\| \geq \varepsilon$, we have $\|x + y\| \leq 2(1 - \delta(\varepsilon))$.

DEFINITION 1.1.2 A Banach space X is called *reflexive* if the natural mapping $x \mapsto x^{**}$, defined by $(x^{**}, x^*) = (x^*, x)$ for each $x^* \in X^*$, is an isomorphism between X and X^{**} – the topological dual of X^*.

THEOREM 1.1.2 (Pettis) *A Banach space is reflexive if and only if its topological dual is reflexive.*

See Pettis [215] or Hille and Phillips [143, Corollary 2, p. 38].

THEOREM 1.1.3 (Milman) *Every uniformly convex space is reflexive.*

See Milman [183], Pettis [216] or Yosida [268, Theorem 2, p. 127]. An immediate consequence of Theorems 1.1.2 and 1.1.3 is

COROLLARY 1.1.2 *A Banach space whose topological dual is uniformly convex is reflexive.*

We recall that the *duality mapping*[2] $J : X \rightsquigarrow X^*$ is defined by

$$J(x) = \left\{ x^* \in X^*; \ (x, x^*) = \|x\|^2 = \|x^*\|^2 \right\}$$

for each $x \in X$. In view of the Hahn–Banach Theorem, it follows that, for each $x \in X$, $J(x)$ is nonempty.

---

[2]Whenever F is a multi-valued mapping from a set D to a set E, i.e., $F : D \to 2^E$, we denote this by $F : D \rightsquigarrow E$.

Preliminaries   3

THEOREM 1.1.4 (Kato) *If the dual of X is uniformly convex, then the duality mapping $J : X \rightsquigarrow X^*$ is single-valued and uniformly continuous on bounded subsets in X.*

See Kato [152] or Barbu [21, Proposition 1.5, p. 14].

DEFINITION 1.1.3 Let (X, d) be a metric space and $C \subseteq X$ a nonempty set. A mapping $Q : C \rightarrow X$ is said to be *nonexpansive* or *a contraction* if the inequality $d(Qx, Qy) \le d(x, y)$ holds for all $x, y \in C$. The mapping Q is called *a strict contraction* if there exists $\ell \in (0, 1)$ such that $d(Qx, Qy) \le \ell d(x, y)$ holds for all $x, y \in C$.

1.2 A Lebesgue-type integral for vector-valued functions

In this section we present the main concepts and results concerning a Lebesgue-type integral for vector-valued functions.

1.2.1 The Bochner integral

A measure space (Ω, Σ, μ) is called *σ-finite* if there exists $\{\Omega_n; n \in \mathbb{N}\} \subseteq \Sigma$ such that $\mu(\Omega_n) < +\infty$ for each $n \in \mathbb{N}$ and $\Omega = \cup_{n \in \mathbb{N}} \Omega_n$. It is called *finite* if $\mu(\Omega) < \infty$. The measure space (Ω, Σ, μ) is called *complete* if the measure μ is *complete*, i.e., if each subset of a null μ-measure set is measurable, i.e., belongs to the σ-field Σ.

Let X be a Banach space with norm $\| \cdot \|$ and (Ω, Σ, μ) a measure space with a σ-finite and complete measure.

DEFINITION 1.2.1 A function $x : \Omega \rightarrow X$ is called

(i) *countably-valued* if there exist: $\{\Omega_n; n \in \mathbb{N}\} \subseteq \Sigma$ and $\{x_n; n \in \mathbb{N}\} \subseteq X$, with $\Omega_k \cap \Omega_p = \emptyset$ for each $k \ne p$, $\Omega = \cup_{n \ge 0} \Omega_n$ and such that $x(\theta) = x_n$ for all $\theta \in \Omega_n$

(ii) *strongly measurable* if there exists a sequence of countably-valued functions convergent to x μ-a.e. on Ω. If X is finite-dimensional, instead of strongly measurable, we simply say *measurable*.

THEOREM 1.2.1 *A function $x : \Omega \rightarrow X$ is strongly measurable if and only if there exists a sequence of countably-valued functions from Ω to X, which is uniformly μ-a.e. convergent on Ω to x.*

See Vrabie [253, Theorem 1.1.3, p. 3 and Remark 1.1.2, p. 4].

4   *Delay Differential Evolutions Subjected to Nonlocal Initial Conditions*

Since the two families $\{\Omega_n; n \in \mathbb{N}\}$ and $\{x_n; n \in \mathbb{N}\}$ in the definition of a countably-valued function are not unique, in the sequel, a pair of sets,

$$(\{\Omega_n; n \in \mathbb{N}\}, \{x_n; n \in \mathbb{N}\}),$$

with the above properties, is called *a representation of the countably-valued function* x. Since Ω has σ-finite measure, for each countably-valued function $x : \Omega \to X$ there exists at least one representation such that, for each $n \in \mathbb{N}$, $\mu(\Omega_n) < +\infty$. A representation of this sort is called a *σ-finite representation*.

DEFINITION 1.2.2 Let $x : \Omega \to X$ be a countably-valued function, and let $\mathcal{R} = (\{\Omega_n; n \in \mathbb{N}\}, \{x_n; n \in \mathbb{N}\})$ be one of its σ-finite representations. We say that $\mathcal{R}$ is *Bochner integrable* (B-integrable) on Ω with respect to μ, if

$$\sum_{n=0}^{\infty} \mu(\Omega_n)\|x_n\| < +\infty.$$

REMARK 1.2.1 If $\mathcal{R}$ and $\mathcal{R}'$ are two σ-finite representations of the very same countably-valued function $x : \Omega \to X$, the series $\sum_{n=0}^{\infty} \mu(\Omega_n)x_n$ and $\sum_{n=0}^{\infty} \mu(\Omega'_n)x'_n$ are either both convergent, or both divergent and, in the former case, they have the same sum. So, $\mathcal{R}$ is B-integrable on Ω with respect to μ if and only if $\mathcal{R}'$ is so.

This remark enables us to introduce:

DEFINITION 1.2.3 The countably-valued function $x : \Omega \to X$ is called *Bochner integrable* on Ω with respect to μ if it has a σ-finite representation

$$\mathcal{R} = (\{\Omega_n; n \in \mathbb{N}\}, \{x_n; n \in \mathbb{N}\})$$

which is B-integrable on Ω with respect to μ in the sense of Definition 1.2.2. In this case, the vector

$$\sum_{n=0}^{\infty} \mu(\Omega_n)x_n = \int_{\Omega} x(\theta)\, d\mu(\theta) = \int_{\Omega} x\, d\mu,$$

which does not depend on the choice of $\mathcal{R}$ (see Remark 1.2.1), is called *the Bochner integral* on Ω of the function x with respect to μ.

DEFINITION 1.2.4 A function $x : \Omega \to X$ is called *Bochner integrable* on Ω with respect to μ if it is strongly measurable and there exists a sequence $(x_k)_k$ of countably-valued, and Bochner integrable functions on Ω with respect to μ, such that

$$\lim_{k} \int_{\Omega} \|x(\theta) - x_k(\theta)\|\, d\mu(\theta) = 0.$$

Preliminaries 5

PROPOSITION 1.2.1 *If $x : \Omega \to X$ is Bochner integrable on Ω with respect to μ and $(x_k)_k$ is a sequence with the properties in Definition 1.2.4, then there exists*

$$\lim_k \int_\Omega x_k \, d\mu$$

in the norm topology of X. In addition, if $(y_k)_k$ is another sequence of countably-valued functions with the property that

$$\lim_k \int_\Omega \|x(\theta) - y_k(\theta)\| \, d\mu(\theta) = 0,$$

then

$$\lim_k \int_\Omega x_k(\theta) \, d\mu(\theta) = \lim_k \int_\Omega y_k(\theta) \, d\mu(\theta).$$

See Vrabie [253, Proposition 1.2.1, p. 5].

DEFINITION 1.2.5 Let $x : \Omega \to X$ be a Bochner integrable function on Ω. The vector

$$\lim_k \int_\Omega x_k \, d\mu = \int_\Omega x(\theta) \, d\mu(\theta) = \int_\Omega x \, d\mu$$

which, according to Proposition 1.2.1, exists and does not depend on the choice of the sequence $(x_k)_k$ in Definition 1.2.4, is called *the Bochner integral* of the function x on Ω with respect to μ.

THEOREM 1.2.2 (Bochner) *A function $x : \Omega \to X$ is Bochner integrable on Ω with respect to μ if and only if x is strongly measurable and the real function $\|x\|$ is integrable on Ω with respect to μ.*

See Bochner [37] or Vrabie [253, Theorem 1.2.1, p. 6].

1.2.2 The L^p spaces

We denote by $\mathcal{L}^p(\Omega, \mu; X)$ the set of all functions $f : \Omega \to X$, which are strongly measurable on Ω and $\|f\|^p$ is integrable on Ω with respect to μ. Let us define $\| \cdot \|_{\mathcal{L}^p(\Omega,\mu;X)} : \mathcal{L}^p(\Omega, \mu; X) \to \mathbb{R}_+$ by

$$\|f\|_{\mathcal{L}^p(\Omega,\mu;X)} = \left(\int_\Omega \|f\|^p \, d\mu \right)^{1/p}$$

for each $f \in \mathcal{L}^p(\Omega, \mu; X)$. This is a seminorm on $\mathcal{L}^p(\Omega, \mu; X)$. The relation "$\sim$" defined by $f \sim g$ if $f(\theta) = g(\theta)$ μ-a.e. for $\theta \in \Omega$ is an equivalence on $\mathcal{L}^p(\Omega, \mu; X)$. Let $L^p(\Omega, \mu; X)$ be the quotient space $\mathcal{L}^p(\Omega, \mu; X)/ \sim$. One may easily see that if $f \sim g$, then $\|f\|_{\mathcal{L}^p(\Omega,\mu;X)} = \|g\|_{\mathcal{L}^p(\Omega,\mu;X)}$. As a consequence, the mapping $\| \cdot \|_{L^p(\Omega,\mu;X)} : L^p(\Omega, \mu; X) \to \mathbb{R}_+$, given by

$$\|\hat{f}\|_{L^p(\Omega,\mu;X)} = \|f\|_{\mathcal{L}^p(\Omega,\mu;X)}$$

for each $\hat{f} \in L^p(\Omega, \mu; X)$, is well-defined (i.e., it does not depend on the choice of $f \in \hat{f}$) and it is a norm on $L^p(\Omega, \mu; X)$. Endowed with this norm, $L^p(\Omega, \mu; X)$ is a Banach space.

Next, let $\mathcal{L}^\infty(\Omega, \mu; X)$ be the space of all strongly measurable functions $f : \Omega \to X$ satisfying

$$\|f\|_{\mathcal{L}^\infty(\Omega,\mu;X)} = \inf\{\alpha \in \overline{\mathbb{R}}; \ \|f(\theta)\| \le \alpha \ \text{a.e. for} \ \theta \in \Omega\} < +\infty.$$

The mapping $\| \cdot \|_{\mathcal{L}^\infty(\Omega,\mu;X)} : \mathcal{L}^\infty(\Omega, \mu; X) \to \mathbb{R}_+$, defined as above, is a seminorm. Let $L^\infty(\Omega, \mu; X) = \mathcal{L}^\infty(\Omega, \mu; X)/\sim$, where "$\sim$" is the μ-a.e. equality on Ω and let $\| \cdot \|_{L^\infty(\Omega,\mu;X)} : L^\infty(\Omega, \mu; X) \to \mathbb{R}_+$, given by

$$\|\hat{f}\|_{L^\infty(\Omega,\mu;X)} = \|f\|_{\mathcal{L}^\infty(\Omega,\mu;X)}$$

for each $\hat{f} \in L^\infty(\Omega, \mu; X)$. Obviously $\| \cdot \|_{L^\infty(\Omega,\mu;X)}$ is well-defined and, in addition, is a norm on $L^\infty(\Omega, \mu; X)$, with respect to which this is a Banach space. For simplicity, we denote by f either a fixed element in $\mathcal{L}^p(\Omega, \mu; X)$ or its corresponding equivalence class in $L^p(\Omega, \mu; X)$. If $X = \mathbb{R}$, we denote by $L^p(\Omega, \mu) = L^p(\Omega, \mu; \mathbb{R})$.

THEOREM 1.2.3 (Lebesgue) *Let $(f_n)_n$ be a sequence in $L^1(\Omega, \mu; X)$ with*

$$\lim_n f_n(\theta) = f(\theta)$$

μ-a.e. for $\theta \in \Omega$. If there exists $\ell \in L^1(\Omega, \mu)$ such that

$$\|f_n(\theta)\| \le \ell(\theta)$$

for $n = 1, 2, \dots$ and μ-a.e. for $\theta \in \Omega$, then $f \in L^1(\Omega, \mu; X)$ and $\lim_n f_n = f$ in the norm of $L^1(\Omega, \mu; X)$.

For the proof of Theorem 1.2.3, known as the Lebesgue Dominated Convergence Theorem, see Lebesgue [167] or Dinculeanu [102].

We also need the following specific form of the Fatou Lemma.

LEMMA 1.2.1 (Fatou) *Let (Ω, Σ, μ) be a σ-finite and complete measure space and let $(f_n)_n$ be a sequence of measurable functions from Ω to $\mathbb{R}_+$, but not necessarily μ-integrable. Then*

$$\int_\Omega \liminf_k f_k(\theta) \, d\mu(\theta) \le \liminf_k \int_\Omega f_k(\theta) \, d\mu(\theta).$$

See Fatou [114] or Dunford and Schwartz [105, Theorem 19, p. 152].

The next result gives a simple but precise description of the topological dual of $L^p(\Omega, \mu; X)$ for certain classes of Banach spaces.

THEOREM 1.2.4 *If either X is reflexive or X^* is separable, then, for each $p \in [1, +\infty)$, $(L^p(\Omega, \mu; X))^*$ can be identified with $L^q(\Omega, \mu; X^*)$, where $\frac{1}{p} + \frac{1}{q} = 1$ if $p > 1$ and $q = \infty$ if $p = 1$.*

Preliminaries 7

See Dinculeanu [102, Corollary 1, p. 252]. Some extensions and variants of Theorem 1.2.4 can be found in Edwards [110, Theorem 8.18.2, p. 588, Remarks, p. 589 and Theorem 8.20.5, p. 607].

A remarkable consequence of Theorem 1.2.4 is stated below.

COROLLARY 1.2.1 *If X is reflexive and $p \in (1, +\infty)$, then $L^p(\Omega, \mu; X)$ is reflexive. If X is separable, then, for each $p \in [1, +\infty)$, $L^p(\Omega, \mu; X)$ is separable.*

Now, let $(\Omega_i, \Sigma_i, \mu_i)$, $i = 1, 2$, be two measure spaces and let us define *the product measure space* (Ω, Σ, μ) as the measure space for which $\Omega = \Omega_1 \times \Omega_2$, Σ is the smallest σ-field containing all the sets $E_1 \times E_2$ with $E_i \in \Sigma_i$, $i = 1, 2$ and such that $\mu(E_1 \times E_2) = \mu_1(E_1)\mu_2(E_2)$ for each $E_i \in \Sigma_i$, $i = 1, 2$.

THEOREM 1.2.5 (Fubini) *Let $(\Omega_i, \Sigma_i, \mu_i)$, $i = 1, 2$, be finite measure spaces and let (Ω, Σ, μ) be their product space. Let X be a Banach space and let $f \in L^1(\Omega, \mu; X)$. Then, for μ_1-a.e. $s \in \Omega_1$, $t \mapsto f(s, t)$ belongs to $L^1(\Omega_2, \mu_2; X)$, the function $s \mapsto \int_{\Omega_2} f(s, t) \, d\mu_2(t)$ belongs to $L^1(\Omega_1, \mu_1; X)$, and*

$$\int_{\Omega_1} \int_{\Omega_2} f(s, t) \, d\mu_2(t) \, d\mu_1(s) = \int_{\Omega} f(\theta) \, d\mu(\theta).$$

For the proof, see Dunford and Schwartz [105, Theorem 9, p. 190].

In the theorem below, μ is the Lebesgue measure on $\mathbb{R}^d$. In order to simplify the notation, whenever Ω is a Lebesgue strongly measurable subset in $\mathbb{R}^d$ and μ is the Lebesgue measure on Ω, we denote by $L^p(\Omega; X) = L^p(\Omega, \mu; X)$. If, in addition, $X = \mathbb{R}$, a further simplification is made, i.e., we denote by $L^p(\Omega) = L^p(\Omega; \mathbb{R})$. Finally, if $\Omega = [a, b]$, we simply write $L^p(a, b; X)$ instead of $L^p([a, b]; X)$.

THEOREM 1.2.6 (Clarkson) *If $\Omega \subseteq \mathbb{R}^d$, $d \geq 1$, is nonempty, bounded and Lebesgue measurable and $p \in (1, +\infty)$, then the space $L^p(\Omega)$, endowed with its usual norm, is uniformly convex.*

See Clarkson [83] or Ciorănescu [82, Teorema 4.1, p. 113].

If $\Omega = \mathbb{R}$ or $\Omega = [a, +\infty)$, μ is the Lebesgue measure and $p \in [1, +\infty]$, we denote by $L^p_{\mathrm{loc}}(\Omega, X)$ the space of equivalence classes of functions $f : \Omega \to X$ with respect of the a.e. equality on Ω, whose restrictions to $\Omega_k = \{t \in \Omega; |t| \leq k\}$, $k = 1, 2, \dots$, belong to $L^p(\Omega_k, \mu; X)$. The space $L^p_{\mathrm{loc}}(\Omega, X)$, endowed with the family of seminorms $\{\| \cdot \|_k; \ k = 1, 2, \dots\}$, where

$$\|f\|_k = \|f\|_{L^p(\Omega_k, \mu; X)},$$

for $k = 1, 2, \dots$ and $f \in L^p_{\mathrm{loc}}(\Omega, X)$, is a Fréchet space.

1.3 The superposition operator

We include next some sufficient conditions for the so-called superposition operator, i.e., the operator obtained by composing a fixed continuous real function by elements in a given function space, be well-defined and continuous. Let Ω be a bounded domain in $\mathbb{R}^d$, $d \geq 1$, $h : \mathbb{R}_+ \times \overline{\Omega} \times \mathbb{R} \to \mathbb{R}$, and $p \in [1, +\infty]$.

DEFINITION 1.3.1 The function $h_p : \mathbb{R}_+ \times D_p \subset \mathbb{R}_+ \times L^p(\Omega) \to L^1(\Omega)$ defined by

$$h_p(t, u)(x) = h(t, x, u(x))$$

for each $t \in \mathbb{R}_+$, $u \in D_p$, and a.e. $x \in \Omega$, where

$$D_p = \{u \in L^p(\Omega); \ h(t, \cdot, u(\cdot)) \in L^1(\Omega) \text{ for each } t \in \mathbb{R}_+\}$$

is called *the superposition operator on $L^p(\Omega)$ associated with the function h*.

We say that the superposition operator on $L^p(\Omega)$ associated with h is *well-defined on $L^p(\Omega)$*, if $D_p = L^p(\Omega)$, and h_p maps $\mathbb{R}_+ \times D_p$ in $L^p(\Omega)$. We say that h_p is *well-defined on $C(\overline{\Omega})$*, if $C(\overline{\Omega}) \subset D_p$ and h_p maps $\mathbb{R}_+ \times C(\overline{\Omega})$ in $C(\overline{\Omega})$.

LEMMA 1.3.1 *Assume that $h : \mathbb{R}_+ \times \overline{\Omega} \times \mathbb{R} \to \mathbb{R}$ satisfies*

(h_1) *for each $(t, x) \in \mathbb{R}_+ \times \overline{\Omega}$, $u \mapsto h(t, x, u)$ is continuous*

(h_2) *for each $u \in \mathbb{R}$, $(t, x) \mapsto h(t, x, u)$ is measurable*

(h_3) *for each $T > 0$ and $r > 0$, the restriction of h to $[0, T] \times \Omega \times [-r, r]$ is bounded.*

Let $p \in [1, +\infty]$, and let $h_p : \mathbb{R}_+ \times D_p \subset \mathbb{R}_+ \times L^p(\Omega) \to L^1(\Omega)$ be the superposition operator on $L^p(\Omega)$ associated with h. Then

(i) *$L^\infty(\Omega) \subset D_p$. If $p = +\infty$, $D_\infty = L^\infty(\Omega)$, and h_∞ is well-defined on $L^\infty(\Omega)$. In addition, for each $t \in \mathbb{R}_+$, the function $u \mapsto h_\infty(t, u)$ is continuous from $L^\infty(\Omega)$ into itself and, for each $u \in L^\infty(\Omega)$, the function $t \mapsto h_\infty(t, u)$ is strongly measurable. Also in this case, i.e., $p = +\infty$, if h is jointly continuous, then h_∞ is well-defined on $C(\overline{\Omega})$ and continuous from $\mathbb{R}_+ \times C(\overline{\Omega})$ to $C(\overline{\Omega})$.*

(ii) *If $p \in [1, +\infty)$, and for each $T > 0$ there exists $a_T > 0$ and $b_T \in \mathbb{R}$, such that*

$$|h(t, x, u)| \leq a_T |u|^p + b_T \qquad (1.3.1)$$

for each $(t, x, u) \in [0, T] \times \overline{\Omega} \times \mathbb{R}$, then $D_p = L^p(\Omega)$, and, for each $t \in \mathbb{R}_+$, the function $u \mapsto h_p(t, u)$ is continuous from $L^p(\Omega)$ to $L^1(\Omega)$. In addition, for each $u \in L^p(\Omega)$, the function $t \mapsto h_p(t, u)$ is strongly measurable.

Preliminaries 9

(*iii*) *If $p = 1$, and h satisfies (1.3.1) then, for each $r \in [1, +\infty]$, the restriction of h_1 to $\mathbb{R}_+ \times L^r(\Omega)$ coincides with h_r which is well-defined on $L^r(\Omega)$. In addition, for each $t \in \mathbb{R}_+$, the function $u \mapsto h_1(t, u)$ is continuous from $L^r(\Omega)$ into itself and, for each $u \in L^r(\Omega)$, the function $t \mapsto h_1(t, u)$ is strongly measurable from $\mathbb{R}_+$ to $L^r(\Omega)$. Finally, for each $T > 0$, we have*

$$\|h_1(t, u)\|_{L^r(\Omega)} \le a_T \|u\|_{L^r(\Omega)} + |b_T| \mu(\Omega)^{1/r}$$

for each $t \in [0, T]$, and $u \in L^r(\Omega)$, where a_T and b_T are given by (ii), and $\mu(\Omega)$ is the Lebesgue measure of Ω.

For the proof of Lemma 1.3.1 we refer to Vrabie [253, Lemma A.6.1, p. 313].

1.4 Compactness theorems

1.4.1 Generalities

Here we gather several compactness results which will be used later.

DEFINITION 1.4.1 A subset C of a topological space $(X, \mathcal{T})$ is called

(i) *relatively compact*, if each generalized sequence in C has at least one generalized convergent subsequence

(ii) *compact*, if it is relatively compact and closed

(iii) *sequentially relatively compact*, if each sequence in C has at least one convergent subsequence

(iv) *sequentially compact*, if it is sequentially relatively compact and closed.

If (X, d) is a metric space, $C \subseteq X$ is called *precompact* or *totally bounded* if for each $\varepsilon > 0$ there exists a finite family of closed balls of radius ε whose union includes C.

REMARK 1.4.1 As each metric space satisfies the First Axiom of Countability, i.e., each point has an at most countable fundamental system of neighborhoods, in a metric space, a subset is (relatively) compact if and only if it is sequentially (relatively) compact.

REMARK 1.4.2 Clearly, a subset $C \subseteq X$ is precompact if and only if, for each $\varepsilon > 0$ there exists a finite family of closed balls centered in points of C and having radii ε, whose union includes C.

10   *Delay Differential Evolutions Subjected to Nonlocal Initial Conditions*

THEOREM 1.4.1 (Hausdorff) *If (X, d) is a complete metric space, then a subset of it is relatively compact if and only if it is precompact.*

See Hausdorff [138, II, p. 312 and III, p. 313].

THEOREM 1.4.2 (Mazur) *The closed convex hull of a compact subset in a Banach space is compact.*

See Mazur [181] or Dunford and Schwartz [105, Theorem 6, p. 416].

THEOREM 1.4.3 (Krein–Šmulian) *The closed convex hull of a weakly relatively compact set in a Banach space is weakly compact.*

See Krein and Šmulian [158] or Dunford and Schwartz [105, Theorem 4, p. 434].

1.4.2   Topological fixed-point theorems

We begin by recalling the celebrated Schauder Fixed-Point Theorem, i.e.,

THEOREM 1.4.4 (Schauder) *Let X be a Banach space. If $K \subseteq X$ is nonempty, closed and convex, the mapping $\mathcal{M} : K \to K$ is continuous and $\mathcal{M}(K)$ is relatively compact, then $\mathcal{M}$ has at least one fixed point, i.e., there exists $\xi \in K$ such that $\mathcal{M}(\xi) = \xi$.*

See Schauder [232] or Dunford and Schwartz [105, Theorem 5, p. 456]. A more general result, i.e., the Tychonoff Fixed-Point Theorem, is stated below.

THEOREM 1.4.5 (Tychonoff) *Let X be a separated locally convex topological vector space and let K be a nonempty, convex and closed subset in X. If $\mathcal{M} : K \to K$ is continuous and $\mathcal{M}(K)$ is relatively compact, then it has at least one fixed point, i.e., there exists $\xi \in K$ such that $\mathcal{M}(\xi) = \xi$.*

See Tychonoff [245] or Edwards [110, Theorem 3.6.1, p. 161].

DEFINITION 1.4.2 Let X be a Banach space and $C \subseteq X$. By a *compact operator* $\mathcal{M} : C \to X$ we mean an operator $\mathcal{M}$ such that, for each bounded set $B \subseteq C$, $\mathcal{M}(B)$ is relatively compact.

We also need the Schaefer Fixed-Point Theorem below, which is a variant of the Leray–Schauder Principle.

THEOREM 1.4.6 (Schaefer) *Let X be a Banach space and let $\mathcal{M} : X \to X$ be a continuous, compact operator and let*

$$\mathcal{E}(\mathcal{M}) = \{x \in X; \exists \lambda \in [\, 0, 1\,], \text{ such that } x = \lambda \mathcal{M}(x)\}.$$

If $\mathcal{E}(\mathcal{M})$ is bounded, then $\mathcal{M}$ has at least one fixed point.

Preliminaries 11

See Schaefer [233] and Granas and Dugundji [132, Theorem 5.1, p. 123 and Theorem 5.4, p. 124].

THEOREM 1.4.7 *The weak closure of a weakly relatively compact set in a Banach space X coincides with its weak sequential closure.*

See Edwards [110, Theorem 8.12.1, p. 549].

THEOREM 1.4.8 (Eberlein–Šmulian) *A subset in a Banach space is weakly compact if and only if it is weakly sequentially compact.*

See Eberlein [109], Šmulian [241] or Edwards [110, Theorem 8.12.1, p. 549 and Theorem 8.12.7, p. 551].

THEOREM 1.4.9 *Let X be reflexive. A subset in X is weakly relatively sequentially compact if and only if it is norm bounded.*

See Hille and Phillips [143, Theorem 2.10.3, p. 38].

1.4.3 Compactness in function spaces

DEFINITION 1.4.3 A family $\mathcal{F}$ in $C([a,b];X)$ is called *equicontinuous* on $[a,b]$ if for each $t \in [a,b]$ and each $\varepsilon > 0$ there exists $\delta(t,\varepsilon) > 0$ such that, for each $s \in [a,b]$ satisfying $|t - s| \leq \delta(t,\varepsilon)$, we have $|f(t) - f(s)| \leq \varepsilon$ for each $f \in \mathcal{F}$.

We recall the infinite dimensional version of the Arzelà–Ascoli Theorem.

THEOREM 1.4.10 (Arzelà–Ascoli) *Let X be a Banach space. A subset $\mathcal{F}$ in $C([a,b];X)$ is relatively compact if and only if*

(i) $\mathcal{F}$ *is equicontinuous on $[a,b]$*

(ii) *there exists a dense subset D in $[a,b]$ such that, for each $t \in D$,*

$$\mathcal{F}(t) = \{f(t);\ f \in \mathcal{F}\}$$

is relatively compact in X.

See Vrabie [253, Theorem A.2.1, p. 296].

DEFINITION 1.4.4 A subset $\mathcal{F} \subseteq L^1(\Omega, \mu; X)$ is called *uniformly integrable* or *equi-absolutely-continuous*[3] if for each $\varepsilon > 0$ there exists $\delta(\varepsilon) > 0$ such that

$$\int_E \|f(t)\|\, d\mu(t) \leq \varepsilon$$

for each $f \in \mathcal{F}$ and each $E \in \Sigma$ satisfying $\mu(E) \leq \delta(\varepsilon)$.

---

[3]The term *equi-absolutely-continuous*, widely used by several authors, for instance by Roubíček [226], is in fact more accurate. Nevertheless, the most circulated name for this property is that of *uniformly integrable*. See Diestel-Uhl [101, Definition 10, p. 74].

12   *Delay Differential Evolutions Subjected to Nonlocal Initial Conditions*

REMARK 1.4.3 Let $\mathcal{F} \subseteq L^1(\Omega, \mu; X)$. It is easy to see that:

(*i*) if (Ω, Σ, μ) is of *totally bounded type*, i.e., for each $\varepsilon > 0$ there exists a finite covering $\{\Omega_k; \ k = 1, 2, \ldots, n(\varepsilon)\} \subseteq \Sigma$ of Ω with $\mu(\Omega_k) \leq \varepsilon$ for $k = 1, 2, \ldots, n(\varepsilon)$ and $\mathcal{F}$ is uniformly integrable, then it is norm bounded in $L^1(\Omega, \mu; X)$

(*ii*) if $\mathcal{F}$ is bounded in $L^p(\Omega, \mu; X)$ for some $p > 1$, then it is uniformly integrable

(*iii*) if there exists a nonnegative function $\ell \in L^1(\Omega, \mu; \mathbb{R})$ such that

$$\|f(\omega)\| \leq \ell(\omega)$$

for each $f \in \mathcal{F}$ and a.e. $\omega \in \Omega$, then $\mathcal{F}$ is uniformly integrable.

REMARK 1.4.4 If $\mathcal{F} \subseteq L^1(a, b; X)$ is uniformly integrable, then it is bounded.

The next result is a simple extension of a theorem due to Dunford [104].

THEOREM 1.4.11 (Dunford) *Let (Ω, Σ, μ) a finite measure space and let X be a reflexive Banach space. Then $\mathcal{F} \subseteq L^1(\Omega, \mu; X)$ is weakly compact if and only if it is bounded and uniformly integrable.*

See Diestel and Uhl [101, Theorem 1, p. 101].

THEOREM 1.4.12 (Diestel) *Let (Ω, Σ, μ) be a finite measure space and let X be a Banach space. Let $\mathcal{F} \subseteq L^1(\Omega, \mu; X)$ be bounded and uniformly integrable. If for each $\gamma > 0$ there exist a weakly compact subset $C_\gamma \subseteq X$ and a measurable subset $\Omega_\gamma \subseteq \Omega$ with $\mu(\Omega \setminus \Omega_\gamma) \leq \gamma$ and $f(\Omega_\gamma) \subseteq C_\gamma$ for all $f \in \mathcal{F}$, then $\mathcal{F}$ is weakly relatively compact in $L^1(\Omega, \mu; X)$.*

See Diestel [99] or Diestel and Uhl [101, p. 117].

COROLLARY 1.4.1 *If $C \subseteq X$ is weakly compact, then*

$$\{f \in L^1(a, b; X); \ f(t) \in C \text{ a.e. for } t \in [a, b]\}$$

is weakly relatively compact in $L^1(a, b; X)$.

An easy extension of Theorem 1.4.12 established in Vrabie [255] is:

THEOREM 1.4.13 *Let (Ω, Σ, μ) be a σ-finite measure space, let $\{\Omega_k; \ k \in \mathbb{N}\}$ be a subfamily of Σ such that*

$$\begin{cases} \mu(\Omega_k) < +\infty & \text{for } k = 0, 1, \ldots, \\ \Omega_k \subseteq \Omega_{k+1} & \text{for } k = 0, 1, \ldots, \\ \bigcup_{k=0}^{\infty} \Omega_k = \Omega, \end{cases}$$

Preliminaries 13

and let X be a Banach space. Let $\mathcal{F} \subseteq L^1(\Omega, \mu; X)$ be bounded and uniformly integrable in $L^1(\Omega_k, \mu; X)$, for $k = 0, 1, \dots$ and

$$\lim_k \int_{\Omega \setminus \Omega_k} \|f(\theta)\| \, d\mu(\theta) = 0 \tag{1.4.1}$$

uniformly for $f \in \mathcal{F}$. If for each $\gamma > 0$ and each $k \in \mathbb{N}$, there exist a weakly compact subset $C_{\gamma,k} \subseteq X$ and a measurable subset $\Omega_{\gamma,k} \subseteq \Omega_k$ with $\mu(\Omega_k \setminus \Omega_{\gamma,k}) \leq \gamma$ and $f(\Omega_{\gamma,k}) \subseteq C_{\gamma,k}$ for all $f \in \mathcal{F}$, then $\mathcal{F}$ is weakly relatively compact in $L^1(\Omega, \mu; X)$.

Proof. By Theorem 1.4.12, it follows that, for each $k \in \mathbb{N}$, the family $\mathcal{F}$ is weakly relatively compact in $L^1(\Omega_k, \mu; X)$. Let $(f_n)_n$ be an arbitrary sequence in $\mathcal{F}$. By the remark above, we conclude that, for each $k \in \mathbb{N}$, there exists at least one subsequence $(f_{n_k})_k$ which is weakly convergent in $L^1(\Omega_k, \mu; X)$. As usual, if $E \in \Sigma$, then χ_E denotes the characteristic function of E. Since the family $\{\Omega_k; \ k \in \mathbb{N}\}$ is monotone nondecreasing with respect to the inclusion, there exists a strongly measurable function $f : \Omega \to X$ such that the diagonal subsequence, $(f_{n_n})_n$, is weakly convergent in $L^1(\Omega_k; \mu)$ for $k = 0, 1, \dots$, to $\chi_{\Omega_k} f$. Clearly $f \in L^1(\Omega, \mu; X)$. Indeed, this follows from the fact that f is strongly measurable and $(f_{n_n})_n$ is bounded in $L^1(\Omega, \mu; X)$. Let $f^* \in \left[L^1(\Omega, \mu; X) \right]^*$. Denoting by $(\cdot, \cdot)$ the pairing between $L^1(\Omega, \mu; X)$ and its dual, we have

$$|(f_{n_n} - f, f^*)| \leq |(\chi_{\Omega_k}(f_{n_n} - f), f^*)| + |(\chi_{\Omega \setminus \Omega_k}(f_{n_n} - f), f^*)|$$

$$\leq |(\chi_{\Omega_k}(f_{n_n} - f), f^*)| + \|\chi_{\Omega \setminus \Omega_k}(f_{n_n} - f)\|_{L^1(\Omega, \mu; X)} \|f^*\|_{[L^1(\Omega, \mu; X)]^*}$$

$$\leq |(\chi_{\Omega_k}(f_{n_n} - f), f^*)|$$

$$+ \left[\|\chi_{\Omega \setminus \Omega_k} f_{n_n}\|_{L^1(\Omega, \mu; X)} + \|\chi_{\Omega \setminus \Omega_k} f\|_{L^1(\Omega, \mu; X)} \right] \|f^*\|_{[L^1(\Omega, \mu; X)]^*}$$

for each $n \in \mathbb{N}$ and $k \in \mathbb{N}$. Let $\gamma > 0$ be arbitrary. Since $(f_{n_n})_n$ is in $\mathcal{F}$ and $f \in L^1(\Omega_k, \mu; X)$, from (1.4.1), we conclude that there exists $k(\gamma) \in \mathbb{N}$ such that

$$\begin{cases} \|\chi_{\Omega \setminus \Omega_k} f_{n_n}\|_{L^1(\Omega, \mu; X)} \leq \dfrac{\gamma}{4\|f^*\|_{[L^1(\Omega, \mu; X)]^*}} \\[2mm] \|\chi_{\Omega \setminus \Omega_k} f\|_{L^1(\Omega, \mu; X)} \leq \dfrac{\gamma}{4\|f^*\|_{[L^1(\Omega, \mu; X)]^*}} \end{cases}$$

for each $k \in \mathbb{N}$ with $k \geq k(\gamma)$ and each $n \in \mathbb{N}$. Let us fix $k \geq k(\gamma)$. Since $\lim_n f_{n_n} = f$ weakly in $L^1(\Omega_k, \mu; X)$, it follows that there exists $n = n(\gamma)$ such that

$$|(\chi_{\Omega_k}(f_{n_n} - f), f^*)| \leq \frac{\gamma}{2}$$

for each $n \in \mathbb{N}$, $n \geq n(\gamma)$. From the inequalities above, we conclude that

$$|(f_{n_n} - f, f^*)| \leq \gamma$$

for each $n \in \mathbb{N}$, $n \geq n(\gamma)$. Thus $\lim_n f_{n_n} = f$ weakly in $L^1(\Omega, \mu; X)$ and accordingly $\mathcal{F}$ is weakly relatively sequentially compact in $L^1(\Omega, \mu; X)$. By Eberlein–Šmulian Theorem 1.4.8, it follows that $\mathcal{F}$ is weakly relatively compact in $L^1(\Omega, \mu; X)$ and this completes the proof. $\qquad\square$

1.5 Multifunctions

In this section we include several basic notions and results referring to multifunctions, i.e., to functions whose values are sets. Let $\mathcal{C}$ and X be topological spaces and let $F : \mathcal{C} \rightsquigarrow X$ be a given multifunction, i.e., a function $F : \mathcal{C} \to 2^X$.

1.5.1 Generalities

DEFINITION 1.5.1 By a *selection* of the multifunction $F : \mathcal{C} \rightsquigarrow X$ we mean a function $f : \mathcal{C} \to X$ satisfying $f(x) \in F(x)$ for each $x \in \mathcal{C}$.

DEFINITION 1.5.2 If X is a Banach space, the multifunction $F : \mathcal{C} \rightsquigarrow X$ is said to be (*strongly-weakly*) *upper semicontinuous* (*u.s.c.*) at $\xi \in \mathcal{C}$ if for every (weakly) open neighborhood V of $F(\xi)$ there exists an open neighborhood U of ξ such that $F(\eta) \subseteq V$ for each $\eta \in U \cap \mathcal{C}$. We say that F is (*strongly-weakly*) *u.s.c. on* $\mathcal{C}$ if it is (strongly-weakly) u.s.c. at each $\xi \in \mathcal{C}$.

The next two results will prove useful later.

LEMMA 1.5.1 *Let us assume that $F : \mathcal{C} \rightsquigarrow X$ is a nonempty and (weakly) compact-valued, (strongly-weakly) u.s.c. multifunction. Then, for each compact subset C of $\mathcal{C}$, the set $\cup_{\xi \in C} F(\xi)$ is (weakly) compact. In particular, in both cases, for each compact subset C of $\mathcal{C}$, there exists $M > 0$ such that $\|\eta\| \leq M$ for each $\xi \in C$ and each $\eta \in F(\xi)$.*

Proof. Let C be a (weakly) compact subset in $\mathcal{C}$ and let $\{D_\sigma;\ \sigma \in \Gamma\}$ be an arbitrary (weakly) open covering of $\cup_{\xi \in C} F(\xi)$. Since F is (weakly) compact-valued, for each $\xi \in C$ there exists $n(\xi) \in \mathbb{N}$ such that

$$F(\xi) \subseteq \bigcup_{1 \leq k \leq n(\xi)} D_{\sigma_k}.$$

But F is (strongly-weakly) u.s.c., and therefore there exists an open neighborhood $U(\xi)$ of ξ such that

$$F(U(\xi) \cap C) \subseteq \bigcup_{1 \leq k \leq n(\xi)} D_{\sigma_k}.$$

The family $\{U(\xi);\ \xi \in C\}$ is an open covering of C. As C is compact, there exists a finite family $\{\xi_1, \xi_2, \dots, \xi_p\}$ in C such that

$$F(C) \subseteq \bigcup_{1 \leq j \leq p} F(U(\xi_j) \cap C) \subseteq \bigcup_{1 \leq j \leq p} \bigcup_{1 \leq k \leq n(\xi_j)} D_{\sigma_k},$$

and this completes the proof. $\qquad\square$

Preliminaries 15

THEOREM 1.5.1 *Let X, Y be two Banach spaces and let $\mathcal{C}$ be a nonempty subset in Y. Let $F: \mathcal{C} \rightsquigarrow X$ be a nonempty, closed and convex-valued, strongly-weakly u.s.c. multifunction.*[4] *Let $u_m : [a, b] \to \mathcal{C}$ and $f_m \in L^1(a, b; X)$ be such that $f_m(t) \in F(u_m(t))$ for each $m \in \mathbb{N}$ and a.e. for $t \in [a, b]$.*

If $\lim_m u_m(t) = u(t)$ a.e. for $t \in [a, b]$ and $\lim_m f_m = f$ weakly in $L^1(a, b; X)$, then $f(t) \in F(u(t))$ a.e. for $t \in [a, b]$.

Proof. By Corollary 1.1.1, there exists a sequence $(g_m)_m$ of convex combinations of $\{f_k; \ k \geq m\}$, i.e., $g_m \in \mathrm{conv}\{f_m, f_{m+1}, \dots\}$ for each $m \in \mathbb{N}$, which converges strongly in $L^1(a, b; X)$ to f. By a classical result due to Lebesgue, we know that there exists a subsequence (g_{m_p}) of (g_m) which converges almost everywhere on $[a, b]$ to f. Denote by $\mathcal{T}$ the set of all $s \in [a, b]$ such that both $\left(g_{m_p}(s)\right)_p$ and $(u_m(s))_m$ are convergent to $f(s)$ and to $u(s)$, respectively, and in addition, $f_m(s) \in F(u_m(s))$ for each $m \in \mathbb{N}$. Clearly $[a, b] \setminus \mathcal{T}$ has null measure. Let $s \in \mathcal{T}$ and let E be an open half-space in X including $F(u(s))$. Since F is strongly-weakly u.s.c. at $u(s)$, $(u_m(s))_m$ converges to $u(s)$ and E is a weak neighborhood of $F(u(s))$, there exists $m(E)$ belonging to $\mathbb{N}$, such that $F(u_m(s)) \subseteq E$ for each $m \geq m(E)$. From the relation above, taking into account that $f_m(s) \in F(u_m(s))$ for each $m \in \mathbb{N}$ and a.e. for $s \in [a, b]$, we easily conclude that

$$g_{m_p}(s) \in \overline{\mathrm{conv}} \left(\bigcup_{m \geq m(E)} F(u_m(s)) \right)$$

for each $p \in \mathbb{N}$ with $m_p \geq m(E)$. Passing to the limit for $p \to +\infty$ in the relation above we deduce that $f(s) \in \overline{E}$. Since $F(u(s))$ is closed and convex, it is the intersection of all closed half-spaces which include it. So, in as much as E was arbitrary, we finally get $f(s) \in F(u(s))$ for each $s \in \mathcal{T}$ and this completes the proof. $\qquad \square$

1.5.2 Continuity properties

Let $I \subseteq \mathbb{R}$ a nonempty interval, let X, Y be two Banach spaces, and let $\mathcal{C}$ be a nonempty subset in Y.

DEFINITION 1.5.3 A multifunction $F: I \times \mathcal{C} \rightsquigarrow X$ is said to be *almost strongly-weakly u.s.c.* if for each $\gamma > 0$ there exists a Lebesgue measurable subset $E_\gamma \subseteq I$ whose Lebesgue measure $\lambda(E_\gamma) \leq \gamma$ and such that F is strongly-weakly u.s.c. from $(I \setminus E_\gamma) \times \mathcal{C}$ to X.

REMARK 1.5.1 If the sequence $(\varepsilon_n)_n$ is strictly decreasing to 0, we can always choose the sequence $(E_{\varepsilon_n})_n$, where E_{ε_n} corresponds to ε_n as specified in Definition 1.5.3, such that $E_{\varepsilon_{n+1}} \subseteq E_{\varepsilon_n}$, for $n = 0, 1, \dots$.

---

[4]If F is u.s.c., it is strongly-weakly u.s.c. too and thus the conclusion of Theorem 1.5.1 holds true also in this case.

16   Delay Differential Evolutions Subjected to Nonlocal Initial Conditions

DEFINITION 1.5.4 Let $G : I \rightsquigarrow X$ be a multifunction with nonempty values. By an *a.e. selection* of G we mean a function $f \colon I \to X$ satisfying $f(t) \in G(t)$ a.e. for $t \in I$.

LEMMA 1.5.2 *If $F : I \times \mathcal{C} \rightsquigarrow X$ is a nonempty, closed and convex-valued, almost strongly-weakly u.s.c. multifunction, then, for each continuous function $v : I \to \mathcal{C}$, there exists a strongly measurable selection $f \in L^\infty_{\text{loc}}(I, X)$ of the multifunction $G : I \rightsquigarrow X$, defined by $G(t) = F(t, v(t))$ for $t \in I$.*

Proof. We begin with the case in which F is strongly-weakly u.s.c. Since I can be represented by a finite or at most countable union of compact intervals, we may assume with no loss of generality that I is compact. So, let us consider $I = [a, b]$ and let $v : [a, b] \to \mathcal{C}$ be a continuous function, and let $(\Delta_n)_n$ be a sequence of partitions of $[a, b]$, $\Delta_n : a = t_0^n < t_1^n \cdots < t_{n-1}^n = b$, with $\lim_n \max_{i=\overline{1,n-1}} \left(t_i^n - t_{i-1}^n \right) = 0$. Let $v_n : [a, b] \to \mathcal{C}$ and $f_n : [a, b] \to X$ be defined by

$$v_n(t) = \begin{cases} v(t_i^n) & \text{for } i = 1, 2, \ldots, n - 1 \text{ and } t \in [t_{i-1}^n, t_i^n), \\ v(t_{n-1}^n) & \text{for } t = t_{n-1}^n, \end{cases}$$

and by

$$f_n(t) = \begin{cases} e_i^n & \text{for } i = 1, 2, \ldots n - 1 \text{ and } t \in [t_{i-1}^n, t_i^n), \\ e_{n-1}^n & \text{for } t = t_{n-1}^n, \end{cases}$$

respectively, where, for each $n = 1, 2, \ldots$ and each $i = 1, 2, \ldots n - 1$, e_i^n is a fixed element in $F(t_i^n, v(t_i^n))$. As v is continuous, $C = \{v(t); \ t \in [a, b]\}$ is compact and thus, by Lemma 1.5.1, $F([a, b] \times C)$ is weakly compact. Hence, the family $\{f_n; \ n = 1, 2, \ldots\}$ is bounded and has its values in a weakly compact subset in X. Therefore it satisfies the hypotheses of Diestel Theorem 1.4.12. Accordingly, it is weakly relatively compact in $L^1(a, b; X)$. So, we may assume with no loss of generality that

$$\lim_n f_n = f$$

weakly in $L^1(a, b; X)$. On the other hand, since v is continuous, we have

$$\lim_n v_n(t) = v(t)$$

uniformly on $[a, b]$. We are then in the hypotheses of Theorem 1.5.1 and this completes the proof in the case in which F is strongly-weakly u.s.c.

If F is almost strongly-weakly u.s.c., let $\varepsilon > 0$ and let $E_\varepsilon \subseteq I$ be such that $\mu(E_\varepsilon) < \varepsilon$ and F restricted to $(I \setminus E_\varepsilon) \times \mathcal{C}$ is strongly-weakly u.s.c. From now on we have to repeat the very same arguments to conclude that $f(t) \in F(t, u(t))$ a.e. for $t \in [a, b] \cap E_\varepsilon$. As $\varepsilon > 0$ is arbitrary, this completes the proof. $\square$

Preliminaries 17

DEFINITION 1.5.5 Let X and Y be two Banach spaces. A function (multifunction) $f: X \to Y$ $(F: X \rightsquigarrow Y)$ is said to have *affine growth* if there exist ℓ and m such that

$$\|f(u)\| \leq \ell \|u\| + m \qquad (\|h\| \leq \ell \|u\| + m)$$

for each $u \in X$ (and each $h \in F(u)$). We say that f has *linear growth* if it has affine growth and $m = 0$.

REMARK 1.5.2 If $f: X \to Y$ is Lipschitz of constant $\ell > 0$, then it has affine growth. Indeed, in this case, we have

$$\|f(u)\| \leq \|f(u) - f(0)\| + \|f(0)\| \leq \ell \|u\| + m,$$

where $m = \|f(0)\|$. In particular, if f is nonexpansive, it has affine growth. Moreover, if f is Lipschitz and $f(0) = 0$, it has linear growth.

PROPOSITION 1.5.1 *Let X and Y be two Banach spaces, $K \subseteq X$ a closed subset in X, and let $F : K \rightsquigarrow Y$ be a multifunction with nonempty values, whose graph is strongly×weakly sequentially closed.*

If $\bigcup_{u \in K} F(u)$ is weakly relatively compact, then F is strongly-weakly u.s.c.

Proof. Let us assume by contradiction that F is not strongly-weakly u.s.c. This means that there exists at least one weakly closed subset $C \subseteq Y$ such that

$$F^{-1}(C) = \{x \in K;\ F(x) \cap C \neq \emptyset\}$$

is not closed. This means that there exists $(x_k)_k$ in $F^{-1}(C)$ with $\lim_k x_k = x$ and such that $F(x) \cap C = \emptyset$. For each $k \in \mathbb{N}$, pick a $y_k \in F(x_k) \cap C$. Since $(y_k)_k$ is in $\bigcup_{u \in X} F(u)$, which is weakly relatively compact, it follows that there exists $y \in Y$ such that, on a subsequence at least, $\lim_k y_k = y$ weakly in Y. Since the graph of F is strongly×weakly sequentially closed, we deduce that $y \in F(x)$. As C is weakly closed and $y_k \in C$ for each $k \in \mathbb{N}$, it follows that $y \in C$. Thus $y \in F(x) \cap C$ which is in contradiction with the assumption $F(x) \cap C = \emptyset$. This contradiction can be eliminated only if F is strongly-weakly u.s.c., and this completes the proof. $\qquad \square$

1.5.3 Superposition multifunctions

The next lemma, proved in Vrabie [255], is a slight extension of a result of Paicu [203] which, in turn, extends a previous result in Cârjă, Necula and Vrabie [74, Problem 2.6.1, p. 4.6]. We introduce first:

DEFINITION 1.5.6 Let X be a topological space. A function $f : X \to \mathbb{R}$ is called *lower semicontinuous at $x \in X$* or *l.s.c at $x \in X$* if

$$\liminf_{y \to x} f(y) \geq f(x).$$

18   *Delay Differential Evolutions Subjected to Nonlocal Initial Conditions*

It is called *lower semicontinuous on* X or *l.s.c on* X if it is l.s.c. at each $x \in X$.

A function $f \colon X \to \mathbb{R}$ is called *upper semicontinuous at* $x \in X$ or *u.s.c at* $x \in X$ if

$$\limsup_{y \to x} f(y) \leq f(x).$$

It is called *upper semicontinuous on* X or *u.s.c on* X if it is u.s.c. at each $x \in X$.

LEMMA 1.5.3 *Let $f_1, f_2 : \mathbb{R}^k \to \mathbb{R}$ be two functions having affine growth, f_1 l.s.c. on $\mathbb{R}^k$, f_2 u.s.c. on $\mathbb{R}^k$, with $f_1(u) \leq f_2(u)$ for each $u \in \mathbb{R}^k$. Let $\Omega \subseteq \mathbb{R}^d$, $d \geq 1$, be a nonempty measurable subset of finite Lebesgue measure, let $p \in [\,1, +\infty)$ and let $F : [L^p(\Omega)]^k \rightsquigarrow L^p(\Omega)$ be defined by*

$$F(u) = \{f \in L^p(\Omega); \ f(x) \in [\,f_1(u(x)), f_2(u(x))\,] \ \text{a.e. for} \ x \in \Omega\}$$

for each $u \in [L^p(\Omega)]^k$.

Then, the multifunction F is nonempty, convex and weakly compact-valued and its graph is strongly×weakly sequentially closed. Moreover, the restriction of F to any weakly compact subset in $[L^p(\Omega)]^k$ is strongly-weakly u.s.c. As a consequence, for each $p > 1$, F is strongly-weakly u.s.c. on $[L^p(\Omega)]^k$.

Proof. We begin by observing that the multifunction $G : \mathbb{R}^k \rightsquigarrow \mathbb{R}$, defined by

$$G(r) = [\,f_1(r), f_2(r)\,],$$

for each $r \in \mathbb{R}^k$, has nonempty, convex and compact values and is u.s.c. Obviously G has nonempty, convex and compact values. As f_1 and f_2 have affine growth, it follows that G is locally bounded on $\mathbb{R}^k$. Moreover, since f_1 is l.s.c. and f_2 is u.s.c. on $\mathbb{R}^k$, we deduce that G has a closed graph. Thus G is u.s.c. on $\mathbb{R}^k$.

Let now $u \in [L^p(\Omega)]^k$. As f_1 is the supremum of all continuous functions which are less or equal than f_1 and f_2 is the infimum of all continuous functions which are greater or equal than f_2, it follows that both $f_1(u(\cdot))$ and $f_2(u(\cdot))$ are measurable on Ω. In addition, taking into account that f_1 and f_2 have affine growth, it follows that there exist $\ell > 0$ and $m > 0$ such that

$$\max\{|f_1(u(x))|, |f_2(u(x))|\} \leq \ell|u(x)| + m \tag{1.5.1}$$

a.e. for $x \in \Omega$. But Ω is of finite Lebesgue measure and so $x \mapsto \ell|u(x)| + m$ belongs to $L^p(\Omega)$. From the Lebesgue Theorem, we deduce that $f_1(u(\cdot)), f_2(u(\cdot)) \in L^p(\Omega)$. As $F(u(\cdot)) = [\,f_1(u(\cdot)), f_2(u(\cdot))\,]$, it follows that F has nonempty values. Obviously, F has convex values. Moreover, from (1.5.1), we deduce that each $f \in F(u)$, satisfies

$$|f(x)| \leq \ell|u(x)| + m$$

a.e. for $x \in \Omega$. If $p > 1$, as $F(u)$ is bounded in $L^p(\Omega)$ and the latter is reflexive, it follows that $F(u)$ is weakly compact. If $p = 1$, from the preceding inequality

Preliminaries 19

and Remark 1.4.3, we conclude that $F(u)$ is uniformly integrable. Since $F(u)$ is obviously bounded, from Theorem 1.4.11, we get that $F(u)$ is weakly compact in $L^1(\Omega)$. So, for each $p \in [\, 1, +\infty)$, $F(u)$ is weakly compact in $L^p(\Omega)$.

Using the fact that G has nonempty, convex and compact values and is u.s.c., from Theorem 1.5.1, it follows that F has a strongly$\times$weakly sequentially closed graph in $[L^p(\Omega)]^k \times L^p(\Omega)$.

Finally, let K be a weakly compact subset in $[L^p(\Omega)]^k$. Then, it follows that $F(K)$ is weakly compact in $L^p(\Omega)$ being bounded (if $p > 1$) and uniformly integrable, (if $p = 1$). From Proposition 1.5.1, we conclude that $F_{|K} : K \rightsquigarrow L^p(\Omega)$ is strongly-weakly u.s.c. on K. If $p > 1$, we can take $K = D(0, r)$ with $r > 0$ arbitrary and so, F is strongly-weakly u.s.c. on $L^p(\Omega)$. The proof is complete. $\qquad\square$

If Ω is a nonempty open and bounded subset in $\mathbb{R}^d$, $d \geq 1$, $p \in [\, 1, +\infty)$ and $\tau \geq 0$, we denote by

$$\mathfrak{X}_p = C([\, -\tau, 0\,]; L^p(\Omega)). \tag{1.5.2}$$

The following lemma is related to a sufficient condition for a superposition multifunction to be strongly-weakly u.s.c. in Cârjă, Necula and Vrabie [74, Problem 2.6.1, p. 46].

LEMMA 1.5.4 *Let Ω be a nonempty open and bounded subset in $\mathbb{R}^d$, $d \geq 1$, let $p \in [\, 1, +\infty)$, $\tau \geq 0$, let $\mathfrak{X}_p$ be given by (1.5.2) and let $f_i : \mathbb{R}_+ \times \Omega \times \mathfrak{X}_p \to \mathbb{R}$, $i = 1, 2$, be two given functions satisfying the following conditions:*

(F_1) *$f_1(t, x, v) \leq f_2(t, x, v)$ for each $(t, x, v) \in D(f_1, f_2)$, where*
$D(f_1, f_2) = \mathbb{R}_+ \times \Omega \times \mathfrak{X}_p$

(F_2) *there exist two nonnegative functions $\alpha, \beta \in L^1(\mathbb{R}_+) \cap L^\infty(\mathbb{R}_+)$ such that*

$$|f_i(t, x, v)| \leq \alpha(t)\|v\|_{\mathfrak{X}_p} + \beta(t)$$

for $i = 1, 2$ and each $(t, x, v) \in D(f_1, f_2)$

(F_3) *f_1 is l.s.c. and f_2 is u.s.c.*

Let $F_0 : \mathbb{R}_+ \times \mathfrak{X}_p \rightsquigarrow L^p(\Omega)$ be defined by

$$F_0(t, v) = \{f \in L^p(\Omega); \; f(x) \in [\, f_1(t, x, v), f_2(t, x, v)\,] \text{ a.e. for } x \in \Omega\} \tag{1.5.3}$$

for each $(t, v) \in \mathbb{R}_+ \times \mathfrak{X}_p$.

Then F_0 is nonempty, convex and weakly compact-valued and its graph is strongly$\times$weakly sequentially closed. Moreover, the restriction of the multifunction F_0 to any weakly compact subset in $\mathbb{R}_+ \times \mathfrak{X}_p$ is strongly-weakly u.s.c. As a consequence, if $p > 1$, then F_0 is strongly-weakly u.s.c. on $\mathbb{R}_+ \times \mathfrak{X}_p$.

Proof. Let us observe that the multifunction

$$G : \mathbb{R}_+ \times \Omega \times \mathfrak{X}_p \rightsquigarrow \mathbb{R},$$

20   *Delay Differential Evolutions Subjected to Nonlocal Initial Conditions*

defined by
$$G(t, x, v) = [\, f_1(t, x, v), f_2(t, x, v)\,],$$
for each $(t, x, v) \in \mathbb{R}_+ \times \Omega \times \mathcal{X}_p$, has nonempty, convex and compact values and is u.s.c. Obviously G has nonempty, convex and compact values. As f_i, $i = 1, 2$, have affine growth – see (F_2) – it follows that G is locally bounded. Since f_1 is l.s.c. and f_2 is u.s.c. on $\mathbb{R}_+ \times \Omega \times \mathcal{X}_p$, it follows that G has a closed graph. As G maps bounded subsets in the domain into compact subsets in the range and has a closed graph, we deduce that the multifunction G is u.s.c. on $\mathbb{R}_+ \times \Omega \times \mathcal{X}_p$.

Let now $v \in \mathcal{X}_p$ be arbitrary, but fixed. As f_1 is the supremum of all continuous functions which are less or equal than f_1 and f_2 is the infimum of all continuous functions which are greater or equal than f_2, it readily follows that $(t, x) \mapsto f_i(t, x, v)$, $i = 1, 2$, are measurable on Ω. In addition, we have

$$\max\{|f_1(t, x, v)|, |f_2(t, x, v)|\} \le \alpha(t)\|v\|_{\mathcal{X}_p} + \beta(t) \qquad (1.5.4)$$

a.e. for $(t, x) \in \mathbb{R}_+ \times \Omega$, where α and β are given by (F_2). But Ω, having finite Lebesgue measure, the function $x \mapsto \alpha(t)\|v\|_{\mathcal{X}_p} + \beta(t)$ belongs to $L^p(\Omega)$ a.e. for $t \in \mathbb{R}_+$. From the Lebesgue Theorem, we deduce that $(t, x) \mapsto f_i(t, x, v)$, $i = 1, 2$, belong to $L^p(\mathbb{R}_+; L^p(\Omega))$. So, F_0, given by (1.5.3), has nonempty and convex values. Moreover, from (1.5.4), we deduce that each $f \in F_0(t, v)$, satisfies
$$|f(x)| \le \widetilde{\alpha}\|v\|_{\mathcal{X}_p} + \widetilde{\beta}$$
a.e. for $x \in \Omega$, where $\widetilde{\alpha} = \|\alpha\|_{L^\infty(\mathbb{R}_+)}$ and $\widetilde{\beta} = \|\beta\|_{L^\infty(\mathbb{R}_+)}$. If $p > 1$, as $F_0(t, v)$ is bounded in $L^p(\Omega)$ and the latter is reflexive, it follows that $F_0(t, v)$ is weakly compact. If $p = 1$, from the last inequality, we conclude that $F_0(t, v)$ is uniformly integrable. Since, also in this case $p = 1$, $F_0(t, v)$ is bounded, from Dunford Theorem 1.4.11, we get that $F_0(t, v)$ is weakly compact în $L^1(\Omega)$. Therefore, for each $p \in [\, 1, +\infty)$ and each $(t, v) \in \mathbb{R}_+ \times \mathcal{X}_p$, $F_0(t, v)$ is weakly compact in $L^p(\Omega)$.

Since G has nonempty, convex and compact values and is u.s.c., by Theorem 1.5.1, it follows that F_0 has a strongly×weakly sequentially closed graph in the product space $[\, \mathbb{R}_+ \times \mathcal{X}_p\,] \times L^p(\Omega)$.

Now, let K be a weakly compact subset in $\mathbb{R}_+ \times \mathcal{X}_p$. Then, it follows that $F_0(K)$ is weakly compact in $L^p(\Omega)$ being bounded (if $p > 1$) and uniformly integrable, (if $p = 1$). It then follows that the restriction of F_0 to K, $F_{0|K} : K \rightsquigarrow L^p(\Omega)$, is strongly-weakly u.s.c. on K. If $p > 1$, we can take $K = D(0, r)$, with $r > 0$ arbitrary and so, F_0 is strongly-weakly u.s.c. on $L^p(\Omega)$. The proof is complete. $\qquad \square$

DEFINITION 1.5.7 Let Ω be a nonempty measurable subset in $\mathbb{R}^d$, $d \ge 1$, let $p \in [\, 1, +\infty)$, $f : \mathbb{R}_+ \times \mathcal{X}_p \times \mathcal{X}_p \to L^p(\Omega)$ and let $(t, u, v) \in \mathbb{R}_+ \times \mathcal{X}_p \times \mathcal{X}_p$. We say that f is *lower semicontinuous with respect to the usual order at* (t, u, v) or *l.s.c. with respect to the usual order at* (t, u, v) if

$$\liminf_{(\widetilde{t}, \widetilde{u}, \widetilde{v}) \to (t, u, v)} f(\widetilde{t}, \widetilde{u}, \widetilde{v})(x) \ge f(t, u, v)(x)$$

Preliminaries 21

a.e. for $x \in \Omega$.

We say that f is *upper semicontinuous with respect to the usual order at* (t, u, v) or *u.s.c. with respect to the usual order at* (t, u, v) if

$$\limsup_{(\widetilde{t}, \widetilde{u}, \widetilde{v}) \to (t, u, v)} f(\widetilde{t}, \widetilde{u}, \widetilde{v})(x) \leq f(t, u, v)(x)$$

a.e. for $x \in \Omega$.

We say that f is *l.s.c. with respect to the usual order on its domain* if it is l.s.c. with respect to the usual order at each $(t, u, v) \in \mathbb{R}_+ \times \mathfrak{X}_p \times \mathfrak{X}_p$.

We say that f is *u.s.c. with respect to the usual order on its domain* if it is u.s.c. with respect to the usual order at each $(t, u, v) \in \mathbb{R}_+ \times \mathfrak{X}_p \times \mathfrak{X}_p$.

We also need the following variant of Lemma 1.5.4.

LEMMA 1.5.5 *Let $\Omega \subseteq \mathbb{R}^d$, $d \geq 1$, be nonempty open and bounded, let $p \in [1, +\infty)$ and let $f_i : \mathbb{R}_+ \times \mathfrak{X}_p \times \mathfrak{X}_p \to L^p(\Omega)$, $i = 1, 2$, be two given functions satisfying the following conditions:*

$(\widetilde{F_1})$ *$f_1(t, u, v)(x) \leq f_2(t, u, v)(x)$ for each $(t, u, v) \in D(f_1, f_2)$ a.e. for $x \in \Omega$, where $D(f_1, f_2) = \mathbb{R}_+ \times \mathfrak{X}_p \times \mathfrak{X}_p$, $i = 1, 2$*

$(\widetilde{F_2})$ *there exist two nonnegative functions $\alpha, \beta \in L^1(\mathbb{R}_+) \cap L^\infty(\mathbb{R}_+)$ such that*

$$\|f_i(t, u, v)\|_{L^p(\Omega)} \leq \alpha(t) \max\{\|u\|_{\mathfrak{X}_p}, \|v\|_{\mathfrak{X}_p}\} + \beta(t)$$

for $i = 1, 2$ and each $(t, u, v) \in D(f_1, f_2)$

$(\widetilde{F_2^1})$ *in the particular case $p = 1$, for each $r > 0$, there exist two nonnegative functions $\alpha_r \in L^1(\mathbb{R}_+)$ and $\eta_r \in L^1(\Omega)$ such that*

$$|f_i(t, u, v)(x)| \leq \alpha_r(t) \eta_r(x)$$

for $i = 1, 2$ and each $(t, u, v) \in D(f_1, f_2)$ with

$$\max\{\|u\|_{\mathfrak{X}_1}, \|v\|_{\mathfrak{X}_1}\} \leq r$$

and a.e. for $x \in \Omega$

$(\widetilde{F_3})$ *f_1 is l.s.c. and f_2 is u.s.c. with respect to the usual order on their whole domain $D(f_1, f_2)$ in the sense of Definition 1.5.7.*

Let $F_0 : \mathbb{R}_+ \times \mathfrak{X}_p \times \mathfrak{X}_p \rightsquigarrow L^p(\Omega)$ *be defined by*

$$F_0(t, u, v) = \{f \in L^p(\Omega); \ f(x) \in [\, f_1(t, u, v)(x), f_2(t, u, v)(x)\,] \text{ a.e. in } \Omega\}$$

for each $(t, u, v) \in \mathbb{R}_+ \times \mathfrak{X}_p \times \mathfrak{X}_p$.

Then F_0 is nonempty, convex and weakly compact-valued and its graph is strongly$\times$weakly sequentially closed. Moreover, the restriction of the multifunction F_0 to any bounded subset in $\mathbb{R}_+ \times \mathfrak{X}_p \times \mathfrak{X}_p$ is strongly-weakly u.s.c. As a consequence, F_0 is strongly-weakly u.s.c. on its domain $\mathbb{R}_+ \times \mathfrak{X}_p \times \mathfrak{X}_p$.

22 *Delay Differential Evolutions Subjected to Nonlocal Initial Conditions*

Proof. From $(\widetilde{F}_1)$, $(\widetilde{F}_2)$ and $(\widetilde{F}_2^1)$ combined with Dunford Theorem 1.4.11 in the case $p = 1$, it readily follows that F_0 has nonempty, convex and weakly compact values. Moreover, using $(\widetilde{F}_3)$, Corollary 1.1.1 and Krein–Šmulian Theorem 1.4.3, we conclude that F_0 has a strongly×weakly sequentially closed graph.

As, by $(\widetilde{F}_1)$, $(\widetilde{F}_2)$ and $(\widetilde{F}_2^1)$, F_0 maps bounded subsets in the domain into weakly compact subsets in the range, by Proposition 1.5.1, we deduce that the multifunction F_0 is strongly-weakly u.s.c. on $\mathbb{R}_+ \times \mathfrak{X}_p \times \mathfrak{X}_p$. $\qquad\square$

1.5.4 Fixed-point theorems for multifunctions

We will also need the following general fixed-point theorem for multifunctions due to Ky Fan [113] and Glicksberg [127].

THEOREM 1.5.2 (Ky Fan–Glicksberg) *Let K be a nonempty, convex and compact set in a separated locally convex space and let $\Gamma : K \rightsquigarrow K$ be a nonempty, closed and convex-valued multifunction, whose graph is closed. Then Γ has at least one fixed point, i.e., there exists $f \in K$ such that $f \in \Gamma(f)$.*

A very useful variant of Theorem 1.5.2, is

THEOREM 1.5.3 *Let K be a nonempty, convex and closed set in a separated locally convex space and let $\Gamma : K \rightsquigarrow K$ be a nonempty, closed and convex-valued multifunction, whose graph is closed. If $\Gamma(K) = \cup_{x \in K} \Gamma(x)$ is relatively compact, then Γ has at least one fixed point, i.e., there exists $f \in K$ such that $f \in \Gamma(f)$.*

Proof. Since K is closed, convex and $\Gamma(K) \subseteq K$, we have

$$\overline{\operatorname{conv}\Gamma(K)} \subseteq \overline{\operatorname{conv}K} = K.$$

So,

$$\Gamma(\overline{\operatorname{conv}\Gamma(K)}) \subseteq \Gamma(K) \subseteq \overline{\operatorname{conv}\Gamma(K)},$$

which shows that the set $\mathcal{C} = \overline{\operatorname{conv}\Gamma(K)}$, which by Mazur's Theorem 1.4.2 is compact, is nonempty, closed, convex and $\Gamma(\mathcal{C}) \subseteq \mathcal{C}$. So, we are in the hypotheses of Theorem 1.5.2, with K replaced by $\mathcal{C} \subseteq K$, from which the conclusion. $\qquad\square$

By Theorem 1.4.7, in a Banach space, the weak closure of a weakly relatively compact set coincides with its weak sequential closure. So, Theorem 1.5.2 implies

THEOREM 1.5.4 *Let K be a nonempty, convex and weakly compact set in Banach space and let $\Gamma : K \rightsquigarrow K$ be a nonempty, closed and convex-valued multifunction, whose graph is sequentially closed. Then Γ has at least one fixed point, i.e., there exists $f \in K$ such that $f \in \Gamma(f)$.*

In the single-valued case, Theorem 1.5.4 is due to Arino, Gautier and Penot [12].

Preliminaries 23

1.6 C_0-semigroups

1.6.1 Generalities

Let X be a Banach space and let $\mathcal{L}(X)$ be the Banach space of all linear bounded operators from X to X endowed with the operator norm $\|\cdot\|_{\mathcal{L}(X)}$, defined by $\|U\|_{\mathcal{L}(X)} = \sup_{\|x\|\leq 1}\|Ux\|$ for each $U \in \mathcal{L}(X)$.

DEFINITION 1.6.1 A family of operators, $\{S(t) : X \to X;\ t \in \mathbb{R}_+\}$, in $\mathcal{L}(X)$ is a C_0-*semigroup on* X, if

(i) $S(0) = I$

(ii) $S(t + s) = S(t)S(s)$ for each $t, s \geq 0$

(iii) $\lim_{t\downarrow 0} S(t)x = x$, for each $x \in X$.

A family $\{S(t) : X \to X;\ t \in \mathbb{R}\}$ in $\mathcal{L}(X)$, satisfying (i), (ii) for each $t, s \in \mathbb{R}$ and (iii) with $t \to 0$ instead of $t \downarrow 0$, is called a C_0-*group*.

DEFINITION 1.6.2 By definition, *the infinitesimal generator*, or simply *generator* of the semigroup of linear operators $\{S(t) : X \to X;\ t \in \mathbb{R}_+\}$ is the operator $A : D(A) \subseteq X \to X$, defined by

$$\begin{cases} D(A) = \left\{x \in X; \text{there exists } \lim_{t\downarrow 0} \frac{1}{t}\left(S(t)x - x\right)\right\} \\ Ax = \lim_{t\downarrow 0} \frac{1}{t}\left(S(t)x - x\right) \quad \text{for each} \quad x \in D(A). \end{cases}$$

Equivalently, we say that A *generates* $\{S(t) : X \to X;\ t \in \mathbb{R}_+\}$.

THEOREM 1.6.1 *If* $\{S(t) : X \to X;\ t \in \mathbb{R}_+\}$ *is a* C_0-*semigroup, then there exist* $M \geq 1$ *and* $\omega \in \mathbb{R}$ *such that*

$$\|S(t)\|_{\mathcal{L}(X)} \leq Me^{\omega t} \tag{1.6.1}$$

for each $t \in \mathbb{R}_+$.

See Vrabie [253, Theorem 2.3.1, p. 41].

A C_0-semigroup satisfying (1.6.1) is called *of type* (M, ω). If $M = 1$ and $\omega = 0$ the C_0-semigroup is called *of contractions* or *of nonexpansive mappings*.

THEOREM 1.6.2 *Let* $\{S(t) : X \to X;\ t \in \mathbb{R}_+\}$ *be a* C_0-*semigroup of type* (M, ω). *Then there exists a norm on* X, *equivalent to the initial one, such that, with respect to this new norm, the* C_0-*semigroup is of type* $(1, \omega)$.

See Vrabie [253, Lemma 3.3.1, p. 57].

THEOREM 1.6.3 *Let* $A : D(A) \subseteq X \to X$ *be the infinitesimal generator of a* C_0-*semigroup* $\{S(t) : X \to X;\ t \in \mathbb{R}_+\}$. *Then*

24   *Delay Differential Evolutions Subjected to Nonlocal Initial Conditions*

(i) *for each $x \in X$ and each $t \in \mathbb{R}_+$, we have*

$$\lim_{h \downarrow 0} \frac{1}{h} \int_t^{t+h} S(\theta) x \, d\theta = S(t) x$$

(ii) *for each $x \in X$ and each $t > 0$, we have*

$$\int_0^t S(\theta) x \, d\theta \in D(A) \quad \text{and} \quad A \left(\int_0^t S(\theta) x \, d\theta \right) = S(t) x - x$$

(iii) *for each $x \in D(A)$ and each $t \in \mathbb{R}_+$, we have $S(t)x \in D(A)$. In addition, the mapping $t \mapsto S(t)x$ is of class C^1 on $\mathbb{R}_+$ and satisfies*

$$\frac{d}{dt}(S(t)x) = AS(t)x = S(t)Ax$$

(iv) *for each $x \in D(A)$ and each $0 \leq s \leq t < +\infty$, we have*

$$\int_s^t AS(\theta) x \, d\theta = \int_s^t S(\theta) Ax \, d\theta = S(t) x - S(s) x.$$

THEOREM 1.6.4 *Let $A : D(A) \subseteq X \to X$ be the infinitesimal generator of a C_0-semigroup $\{S(t) : X \to X; \ t \in \mathbb{R}_+\}$. Then $D(A)$ is dense in X and A is a closed operator.*

1.6.2   Generation theorems

If $A : D(A) \subseteq X \to X$ is a linear operator, *the resolvent set $\rho(A)$ is the set of all numbers λ*, called *regular values*, for which the range of $\lambda I - A$, i.e., $R(\lambda I - A) = (\lambda I - A)(D(A))$, is dense in X and $R(\lambda; A) = (\lambda I - A)^{-1}$ is continuous from $R(\lambda I - A)$ to X.

THEOREM 1.6.5 (Hille–Yosida) *A linear operator $A : D(A) \subseteq X \to X$ is the infinitesimal generator of a C_0-semigroup of contractions if and only if*

(i) *A is densely defined and closed*

(ii) *$(0, +\infty) \subseteq \rho(A)$ and, for each $\lambda > 0$, we have*

$$\|R(\lambda; A)\|_{\mathcal{L}(X)} \leq \frac{1}{\lambda}.$$

See Hille [142], Yosida [267] or Vrabie [253, Theorem 3.1.1, p. 51].

THEOREM 1.6.6 (Feller–Miyadera–Phillips) *The linear operator $A : D(A) \subseteq X \to X$ is the infinitesimal generator of a C_0-semigroup of type (M, a) if and only if*

Preliminaries 25

(i) *A is densely defined and closed*

(ii) *$(a, +\infty) \subseteq \rho(A)$ and, for each $\lambda > a$ and each $n \in \mathbb{N}$, we have*

$$\|R(\lambda; A)^n\|_{\mathcal{L}(X)} \leq \frac{M}{(\lambda - a)^n}.$$

See Feller [115], Miyadera [187], Phillips [217] or Vrabie [253, Theorem 3.3.1, p. 56].

Let $J : X \rightsquigarrow X^*$ be the duality mapping on X.

DEFINITION 1.6.3 A linear operator $A : D(A) \subseteq X \to X$ is *dissipative* if for each $x \in X$ there exists $x^* \in J(x)$ such that $(Ax, x^*) \leq 0$.

THEOREM 1.6.7 *A linear operator $A : D(A) \subseteq X \to X$ is dissipative if and only if, for each $x \in D(A)$ and $\lambda > 0$, we have*

$$\lambda\|x\| \leq \|(\lambda I - A)x\|.$$

See Vrabie [253, Theorem 3.4.1, p. 59].

THEOREM 1.6.8 (Lumer–Phillips) *Let $A : D(A) \subseteq X \to X$ be a densely defined linear operator. Then A generates a C_0-semigroup of contractions on X if and only if*

(i) *A is dissipative*

(ii) *there exists $\lambda > 0$ such that $\lambda I - A$ is surjective.*

Moreover, if A generates a C_0-semigroup of contractions, then $\lambda I - A$ is surjective for any $\lambda > 0$ and we have $(Ax, x^*) \leq 0$ for each $x \in D(A)$ and each $x^* \in J(x)$.

See Lummer and Phillips [177] or Vrabie [253, Theorem 3.4.2, p. 60].

1.7 Mild solutions

In this section we include some facts referring to the relationship between C_0-semigroups and ordinary differential equations in Banach spaces.

1.7.1 Types of solutions

First, from (iii) in Theorem 1.6.3, it follows that, if $A : D(A) \subseteq X \to X$ is the infinitesimal generator of a C_0-semigroup $\{S(t) : X \to X; \ t \in \mathbb{R}_+\}$, then, for each $a \geq 0$ and $\xi \in D(A)$, the function $u : [a, +\infty) \to X$, defined by

26 *Delay Differential Evolutions Subjected to Nonlocal Initial Conditions*

$u(t) = S(t-a)\xi$ for each $t \geq a$, is the unique C^1-solution of the homogeneous abstract Cauchy problem

$$\begin{cases} u'(t) = Au(t), \\ u(a) = \xi. \end{cases} \tag{1.7.1}$$

For this reason, it is quite natural to consider that, for each $\xi \in X$, the function u, defined as above, is a solution for (1.7.1) in a generalized sense. The aim of this section is to extend this concept of generalized solution to the nonhomogeneous problem

$$\begin{cases} u'(t) = Au(t) + f(t) \\ u(a) = \xi, \end{cases} \tag{1.7.2}$$

where A is as before, $\xi \in X$ and $f \in L^1(a,b;X)$.

DEFINITION 1.7.1 The function $u : [a,b] \to X$ is called *classical or C^1-solution* of the problem (1.7.2), if u is continuous on $[a,b]$, continuously differentiable on $(a,b]$, $u(t) \in D(A)$ for each $t \in (a,b]$ and it satisfies $u'(t) = Au(t) + f(t)$ for each $t \in (a,b]$ and $u(a) = \xi$.

DEFINITION 1.7.2 The function $u : [a,b] \to X$ is called *absolutely continuous or strong solution* of the problem (1.7.2), if u is absolutely continuous on $[a,b]$, $u' \in L^1(a,b;X)$, $u(t) \in D(A)$ a.e. for $t \in (a,b)$ and it satisfies $u'(t) = Au(t) + f(t)$ a.e. for $t \in (a,b)$ and $u(a) = \xi$.

REMARK 1.7.1 Each classical solution of (1.7.2) is a strong solution of the same problem, but not conversely.

The next result, known as the *Duhamel Principle*, is fundamental in understanding how to extend the concept of a generalized solution to nonhomogeneous problems of the form (1.7.2).

THEOREM 1.7.1 *Each strong solution of (1.7.2) is given by the so-called variation of constants formula*

$$u(t) = S(t-a)\xi + \int_a^t S(t-s)f(s)\,ds, \tag{1.7.3}$$

for each $t \in [a,b]$. In particular, each classical solution of the problem (1.7.2) is given by (1.7.3).

See Vrabie [253, Theorem 8.1.1, p. 184].

Simple examples show that, when X is infinite-dimensional and A is unbounded, the problem (1.7.2) may fail to have any strong solution, no matter how regular the datum f is. See Vrabie [253, Example 8.1.1, p. 185]. This observation justifies why, in the case of infinite-dimensional spaces X, the variation of constants formula can be promoted to the rank of definition. Namely, we introduce

DEFINITION 1.7.3 The function $u : [a,b] \to X$, defined by (1.7.3), is called a *mild solution* of the problem (1.7.2) on $[a,b]$.

Preliminaries 27

1.7.2 Compactness of the solution operator

We will next recall a necessary and sufficient condition that a given set of mild solutions be relatively compact in $C([a,b];X)$.

DEFINITION 1.7.4 The operator $\mathcal{M} : X \times L^1(a,b;X) \to C([a,b];X)$, defined by $\mathcal{M}(\xi, f) = u$, where u is the unique mild solution of the problem (1.7.2), corresponding to $\xi \in X$ and $f \in L^1(a,b;X)$, is called *the mild solution operator* attached to the problem (1.7.2).

REMARK 1.7.2 The operator $\mathcal{M}$ is Lipschitz with constant $Me^{|\omega|(T-a)}$, where $M \geq 1$ and $\omega \in \mathbb{R}$ are given by Theorem 1.6.1 and therefore it maps bounded subsets in $X \times L^1(a,b;X)$ into bounded subsets in $C([a,b];X)$.

THEOREM 1.7.2 (Vrabie) *Let* $A : D(A) \subseteq X \to X$ *be the generator of a* C_0-*semigroup* $\{S(t) : X \to X;\ t \in \mathbb{R}_+\}$, *let* $\mathcal{D}$ *be a bounded subset in* X *and* $\mathcal{F}$ *a uniformly integrable subset in* $L^1(a,b;X)$. *Then* $\mathcal{M}(\mathcal{D}, \mathcal{F})$ *is relatively compact in* $C([\sigma,b];X)$ *for each* $\sigma \in (a,b)$, *if and only if there exists a dense subset* D *in* $[a,b]$ *such that, for each* $t \in D$, *the* t-*section of* $\mathcal{M}(\mathcal{D}, \mathcal{F})$, *i.e.*,

$$\mathcal{M}(\mathcal{D}, \mathcal{F})(t) = \{\mathcal{M}(\xi, f)(t);\ (\xi, f) \in \mathcal{D} \times \mathcal{F}\},$$

is relatively compact in X. *Moreover, if the latter condition is satisfied and* $a \in \mathcal{D}$, *then* $\mathcal{M}(\mathcal{D}, \mathcal{F})$ *is relatively compact even in* $C([a,b];X)$.

See Vrabie [250] or Vrabie [253, Theorem 8.4.1, p. 194].

THEOREM 1.7.3 *Let* $A : D(A) \subseteq X \to X$ *be the generator of a* C_0-*semigroup* $\{S(t) : X \to X;\ t \in \mathbb{R}_+\}$, *let* $\mathcal{D}$ *be a bounded subset in* X *and* $\mathcal{F}$ *a subset in* $L^1(a,b;X)$ *for which there exists a compact set* $K \subseteq X$ *such that* $f(t) \in K$ *for each* $f \in \mathcal{F}$ *and a.e. for* $t \in [a,b]$. *Then* $\mathcal{M}(\mathcal{D}, \mathcal{F})$ *is relatively compact in* $C([\sigma,b];X)$ *for each* $\sigma \in (a,b)$. *If, in addition,* $\mathcal{D}$ *is relatively compact, then* $\mathcal{M}(\mathcal{D}, \mathcal{F})$ *is relatively compact even in* $C([a,b];X)$.

The proof of Theorem 1.7.3 is based on Theorem 1.7.2, combined with the lemma below, which is interesting by itself.

LEMMA 1.7.1 (Becker) *Let* K *be a compact subset in* X *and let* $\mathcal{F}$ *be a family of continuous functions from* $[a,b]$ *to* K. *Let* $\{S(t) : X \to X;\ t \in \mathbb{R}_+\}$ *be a* C_0-*semigroup on* X. *Then, for each* $t \in [a,b]$, *the set*

$$\left\{ \int_a^t S(t-s)f(s)\,ds;\ f \in \mathcal{F} \right\}$$

is relatively compact in X.

Lemma 1.7.1 due to Becker [25] follows from Lemma 1.7.2 below.

28 *Delay Differential Evolutions Subjected to Nonlocal Initial Conditions*

DEFINITION 1.7.5 The C_0-semigroup $\{S(t):X \to X;\ t \in \mathbb{R}_+\}$ is called *compact* if for each $t > 0$, $S(t)$ is a compact operator.

A very useful consequence of Theorem 1.7.2 is the following sufficient condition of relative compactness of the set $\mathcal{M}(\mathcal{D}, \mathcal{F})$ in $C([a, b]; X)$.

THEOREM 1.7.4 (Baras–Hassan–Veron) *Let $A:D(A) \subseteq X \to X$ be the generator of a compact C_0-semigroup, let $\xi \in X$, $\mathcal{D} = \{\xi\}$, and let $\mathcal{F}$ be a uniformly integrable subset in $L^1(a, b; X)$. Then $\mathcal{M}(\mathcal{D}, \mathcal{F})$ is relatively compact in $C([a, b]; X)$.*

See Baras, Hassan and Veron [20] or Vrabie [253, Theorem 8.4.2, p. 196].

We conclude with a useful extension of Lemma 1.7.1.

LEMMA 1.7.2 *Let $K \subseteq X$ be compact, let $\mathcal{F}$ be a family of Lebesgue integrable functions from $[a, b]$ to K and let $\{S(t) : X \to X;\ t \in \mathbb{R}_+\}$ be a C_0-semigroup on X. Then, for every $t \in [a, b]$, the set*

$$\left\{ \int_a^t S(t - s)f(s)\, ds;\ f \in \mathcal{F} \right\}$$

is relatively compact in X.

Proof. Since $(s, x) \mapsto S(s)x$ is continuous from $\mathbb{R}_+ \times X$ to X and $[0, t - a] \times K$ is compact, it follows that $\{S(s)x;\ (s, x) \in [0, t - a] \times K\}$ is compact. The conclusion follows from the simple observation that

$$\left\{ \int_a^t S(t - s)f(s)ds;\ f \in \mathcal{F} \right\} \subseteq \frac{1}{t}\overline{\mathrm{conv}}\{S(s)x;\ (s, x) \in [0, t - a] \times K\}$$

while, by Theorem 1.4.2, the latter set is compact. $\qquad\square$

1.8 Evolutions governed by m-dissipative operators

1.8.1 Semi-inner products

Let X be a Banach space and let $\|\cdot\|$ be the norm on X. If $x, y \in X$, we denote by $[x, y]_+$ the *right directional derivative of the norm* calculated at x in the direction y, i.e.,

$$[x, y]_+ = \lim_{h\downarrow 0} \frac{\|x + hy\| - \|x\|}{h}$$

and by $(x, y)_+$ the *right directional derivative of $\frac{1}{2}\|\cdot\|^2$* calculated at x in the direction y, i.e.,

$$(x, y)_+ = \lim_{h\downarrow 0} \frac{\|x + hy\|^2 - \|x\|^2}{2h}.$$

Preliminaries

Analogously, we denote by $[x, y]_-$ the *left directional derivative of the norm* calculated at x in the direction y, i.e.,

$$[x, y]_- = \lim_{h \uparrow 0} \frac{\|x + hy\| - \|x\|}{h}$$

and by $(x, y)_-$ the *left directional derivative of* $\frac{1}{2}\| \cdot \|^2$ calculated at x in the direction y, i.e.,

$$(x, y)_- = \lim_{h \uparrow 0} \frac{\|x + hy\|^2 - \|x\|^2}{2h}.$$

The mappings $(\cdot, \cdot)_\pm$ are called the *semi-inner products on* X, while $[\cdot, \cdot]_\pm$ are called the *normalized semi-inner products on* X.

PROPOSITION 1.8.1 *The functions* $[\cdot, \cdot]_\pm$ *and* $(\cdot, \cdot)_\pm$ *satisfy*

(i) $(x, y)_\pm = \|x\| [x, y]_\pm$

(ii) $|[x, y]_\pm| \leq \|y\|$

(iii) $|[x, y]_\pm - [x, z]_\pm| \leq \|y - z\|$

(iv) $[x, y]_+ = -[-x, y]_- = -[x, -y]_-$

(v) $[ax, by]_\pm = b[x, y]_\pm$ *for* $a, b > 0$

(vi) $[x, y + z]_+ \leq [x, y]_+ + [x, z]_+$ *and* $[x, y + z]_- \geq [x, y]_- + [x, z]_-$

(vii) $[x, y + z]_+ \geq [x, y]_+ + [x, z]_-$ *and* $[x, y + z]_- \leq [x, y]_- + [x, z]_+$

(viii) $[x, y + \alpha x]_\pm = [x, y]_\pm + \alpha\|x\|$ *for* $\alpha \in \mathbb{R}$

(ix) *if* $u : [a, b] \to X$ *is differentiable from the right at* $t \in [a, b)$ *(differentiable from the left at* $t \in (a, b]$*), then both* $s \mapsto \|u(s)\|$ *and* $s \mapsto \|u(s)\|^2$ *are differentiable from the right (left) at* t *and*

$$\frac{d^\pm}{dt}(\|u(\cdot)\|)(t) = [u(t), u'_\pm(t)]_\pm$$

$$\frac{d^\pm}{dt}(\|u(\cdot)\|^2)(t) = 2(u(t), u'_\pm(t))_\pm,$$

where $u'_\pm(t) = \frac{d^\pm}{dt}(u(\cdot))(t)$

(x) $\| \cdot \|$ *is Gâteaux differentiable at each* $x \in X$, $x \neq 0$, *if and only if, for each* $x \in X \setminus \{0\}$ *and each* $y \in X$, *we have*

$$[x, y]_+ = -[-x, y]_+$$

(xi) *the normalized inner product* $[\cdot, \cdot]_+$ *is u.s.c., and the normalized inner product* $[\cdot, \cdot]_-$ *is l.s.c.*

30   Delay Differential Evolutions Subjected to Nonlocal Initial Conditions

See Barbu [22, Proposition 3.7, p. 101]. We recall that $J : X \rightsquigarrow X^*$ denotes the duality mapping on X.

PROPOSITION 1.8.2 *For each* $x, y \in X$, *we have*

(i) *there exists* $x_+^* \in J(x)$ *such that*

$$\|x\|[x, y]_+ = \sup\{(y, x^*); \ x^* \in J(x)\} = (y, x_+^*)$$

(ii) *there exists* $x_-^* \in J(x)$ *such that*

$$\|x\|[x, y]_- = \inf\{(y, x^*); \ x^* \in J(x)\} = (y, x_-^*).$$

See Miyadera [188, Theorem 2.5, p. 16] or Barbu [22, Proposition 3.7, p. 101].

EXAMPLE 1.8.1 Let Ω be a bounded subset in $\mathbb{R}^d$, $d \geq 1$, let $p \in (1, +\infty)$, and let $X = L^p(\Omega)$. Then, for each $f, g \in L^p(\Omega)$, we have

$$[f, g]_+ = [f, g]_- = \begin{cases} \|f\|_{L^p(\Omega)}^{1-p} \displaystyle\int_\Omega g(x)|f(x)|^{p-1}\mathrm{sgn}(f(x))dx & \text{if } f \neq 0, \\[2ex] \|g\|_{L^p(\Omega)} & \text{if } f = 0. \end{cases}$$

If $p = 1$ then, for each $f, g \in L^1(\Omega)$, we have

$$[f, g]_\pm = \int_{\{y \in \Omega;\, f(y) > 0\}} g(x)dx - \int_{\{y \in \Omega;\, f(y) < 0\}} g(x)dx \pm \int_{\{y \in \Omega;\, f(y) = 0\}} |g(x)|dx.$$

See Sato [231].

1.8.2 Dissipative operators and evolution equations

If $A : X \rightsquigarrow X$, we say that A is an *operator* or *multifunction*. We denote by $D(A)$ the set of all elements $x \in X$ for which $A(x) \neq \emptyset$ and we simply write $A : D(A) \subseteq X \rightsquigarrow X$. For each $x \in D(A)$, we denote $Ax = A(x)$. Whenever A is a single-valued operator on $D(A)$, we shall identify A with a function defined on $D(A)$, i.e., with its unique selection and we shall write $A : D(A) \subseteq X \to X$ and $Ax = y$ instead of $A : D(A) \subseteq X \rightsquigarrow X$ and of $Ax = \{y\}$. Obviously, each function $f : D(f) \subseteq X \to X$ can be identified with a single-valued operator whose domain is $D(f)$. If $A : D(A) \subseteq X \rightsquigarrow X$ is an operator, then $R(A) = \cup_{x \in D(A)} Ax$. If A and B are operators and $\lambda \in \mathbb{R}$, then $R(A)$, A^{-1}, $A + B$, AB and λA are defined in the usual sense of relations in $X \times X$.

We say that the operator $A : D(A) \subseteq X \rightsquigarrow X$ is *dissipative* if

$$(x_1 - x_2, y_1 - y_2)_- \leq 0$$

for each $x_i \in D(A)$ and $y_i \in Ax_i$, $i = 1, 2$, and *m-dissipative* if it is dissipative and, for each $\lambda > 0$, or equivalently for some $\lambda > 0$, $R(I - \lambda A) = X$. If $\omega \in \mathbb{R}$,

Preliminaries 31

we say that the operator $A : D(A) \subseteq X \rightsquigarrow X$ is ω-*dissipative* if $A + \omega I$ is dissipative. We say that A is ω-m-*dissipative* if A is both ω-dissipative and m-dissipative. We emphasize that the concept of the ω-m-dissipative operator is different from that of the *dissipative operator of type* ω, as defined in Miyadera [187, §2, p. 130], which is an operator A such that $A - \omega I$ is dissipative.

Let $A : D(A) \subseteq X \rightsquigarrow X$ be an ω-m-dissipative operator, let $\xi \in \overline{D(A)}$, let $f \in L^1(a, b; X)$ and let us consider the differential equation

$$u'(t) \in Au(t) + f(t). \tag{1.8.1}$$

DEFINITION 1.8.1 A function $u : [a, b] \to X$ is called a *strong solution* of (1.8.1) on $[a, b]$ if

(S_1) $u(t) \in D(A)$ a.e. for $t \in (a, b)$

(S_2) $u(t) \in W_{\text{loc}}^{1,1}((a, b]; X)$ and there exists $g \in L_{\text{loc}}^1(a, b; X)$, such that

$$\begin{cases} g(t) \in Au(t) & \text{a.e. for } t \in (a, b), \\ u'(t) = g(t) + f(t) & \text{a.e. for } t \in (a, b). \end{cases}$$

A *strong solution* of (1.8.1) on $[a, b)$ is a function, $u : [a, b) \to X$, whose restriction to each interval of the form $[a, c]$, with $a < c < b$, is a strong solution of (1.8.1) on $[a, c]$.

Because when X is infinite dimensional, the problem (1.8.1) may have no strong solution, another general concept was introduced. Namely:

DEFINITION 1.8.2 A C^0-*solution* of the problem (1.8.1) on $[a, b]$ is a function u in $C([a, b]; X)$ satisfying: for each $a < c < b$ and $\varepsilon > 0$ there exist

(i) $a = t_0 < t_1 < \cdots < c \leq t_n < b$, $\quad t_k - t_{k-1} \leq \varepsilon$ $\quad$ for $k = 1, 2, \ldots, n$

(ii) $f_1, \ldots, f_n \in X$ $\quad$ with $\quad \displaystyle\sum_{k=1}^{n} \int_{t_{k-1}}^{t_k} \|f(t) - f_k\| \, dt \leq \varepsilon$

(iii) $v_0, \ldots, v_n \in X$ $\quad$ satisfying

$$\frac{v_k - v_{k-1}}{t_k - t_{k-1}} \in Av_k + f_k \text{ for } k = 1, 2, \ldots, n \text{ and such that}$$

$$\|u(t) - v_k\| \leq \varepsilon \text{ for } t \in [t_{k-1}, t_k), \quad k = 1, 2, \ldots, n.$$

A function $v_\varepsilon : [a, t_n] \to D(A)$, defined by $v_\varepsilon(t) = v_k$ for $t \in [t_{k-1}, t_k)$, $k = 1, 2, \ldots, n$, where t_k, v_k and f_k, for $k = 1, 2, \ldots, n$, are as above, is called an ε-*difference scheme-solution* or briefly, ε-*DS-solution*.

THEOREM 1.8.1 (Benilan) *Let X be a Banach space, let $\omega \in \mathbb{R}$ and let $A : D(A) \subseteq X \rightsquigarrow X$ be an ω-m-dissipative operator. Then, for each $\xi \in \overline{D(A)}$ and $f \in L^1(a, b; X)$, there exists a unique C^0-solution of (1.8.1) on $[a, b]$ that*

32 *Delay Differential Evolutions Subjected to Nonlocal Initial Conditions*

satisfies $u(a) = \xi$. *If* $f, g \in L^1(a, b; X)$ *and* u, v *are two* C^0-*solutions of* (1.8.1) *corresponding to* f *and* g, *respectively, then*

$$\|u(t) - v(t)\| \le e^{-\omega(t-s)}\|u(s) - v(s)\| + \int_s^t e^{-\omega(t-\theta)}\|f(\theta) - g(\theta)\|d\theta \quad (1.8.2)$$

for each $a \le s \le t \le b$.

In particular, if $x \in D(A)$ and $y \in Ax$, we have

$$\|u(t) - x\| \le e^{-\omega(t-s)}\|u(s) - x\| + \int_s^t e^{-\omega(t-\theta)}\|f(\theta) + y\|d\theta \quad (1.8.3)$$

for each $a \le s \le t \le b$.

See Benilan [32], Barbu [22, Theorem 4.1, p. 128] or Lakshmikantham and Leela [163, Theorem 3.6.1, p. 116].

DEFINITION 1.8.3 Let C be a nonempty and closed subset in X and let $\gamma \in \mathbb{R}$. A family $\{S(t) : C \to C; \ t \in \mathbb{R}_+\}$ is a *semigroup of type* γ on C, if

(i) $S(0) = I$

(ii) $S(t + s) = S(t)S(s)$ for all $t, s \in \mathbb{R}_+$

(iii) for each $\xi \in C$ the function $s \mapsto S(s)\xi$ is continuous at $s = 0$

(iv) $\|S(t)\xi - S(t)\eta\| \le e^{\gamma t}\|\xi - \eta\|$ for all $t \in \mathbb{R}_+$ and $\xi, \eta \in C$.

If $\gamma = 0$ the semigroup is called a *semigroup of nonexpansive mappings* or a *semigroup of contractions*.

PROPOSITION 1.8.3 *If* $\{S(t) : C \to C; \ t \in \mathbb{R}_+\}$ *is a semigroup of type* γ, *then the mapping* $(t, \xi) \mapsto S(t)\xi$ *is continuous from* $\mathbb{R}_+ \times C$ *to* C.

REMARK 1.8.1 In the case in which A is single-valued, linear, and of course, m-dissipative, u is a C^0-solution of (1.8.1) on $[a, b]$ in the sense of Definition 1.8.2 if and only if u is a mild solution on $[a, b]$ in the sense of Definition 1.7.3. This is an easy consequence of the fact that in the linear case, each mild solution can be approximated uniformly with strong solutions of some suitably chosen approximate problems.

See Vrabie [252, Theorem 1.8.2, p. 29].

We denote by $u(\cdot, a, \xi, f) : [a, b] \to \overline{D(A)}$ the unique C^0-solution of (1.8.1) satisfying $u(a, a, \xi, f) = \xi$.

THEOREM 1.8.2 (Benilan) *Let* X *be a real Banach space, let* $\omega \in \mathbb{R}$ *and let* $A : D(A) \subseteq X \rightsquigarrow X$ *be an* ω-m-*dissipative operator. Let* $\xi, \eta \in \overline{D(A)}$, *let*

Preliminaries   33

$f, g \in L^1(a, b; X)$ and let us denote by $\widetilde{u} = u(\cdot, a, \xi, f)$ and $\widetilde{v} = u(\cdot, a, \eta, g)$. We have

$$\|\widetilde{u}(t) - \widetilde{v}(t)\| \le e^{-\omega(t-a)}\|\xi - \eta\| + \int_a^t e^{-\omega(t-s)}[\,\widetilde{u}(s) - \widetilde{v}(s), f(s) - g(s)\,]_+ \, ds,$$

(1.8.4)

for each $t \in [a, b]$. Moreover, for each $a \le \nu \le t \le b$, we have

$$u(t, a, \xi, f) = u(t, \nu, u(\nu, a, \xi, f), f|_{[\nu, b]}). \tag{1.8.5}$$

See Benilan [32] or Barbu [22, Theorem 4.1, p. 128].

The relation (1.8.5) is known as the *evolution property*.

PROPOSITION 1.8.4 *Let* $\omega \in \mathbb{R}$ *be such that* $A : D(A) \subseteq X \rightsquigarrow X$ *is* ω-*m*-*dissipative. Let* $t \in \mathbb{R}_+$ *and let* $S(t) : \overline{D(A)} \to \overline{D(A)}$ *be defined by*

$$S(t)\xi = u(t, 0, \xi, 0)$$

for each $\xi \in \overline{D(A)}$. *Then the family* $\{S(t) : \overline{D(A)} \to \overline{D(A)};\ t \in \mathbb{R}_+\}$ *is a semigroup of type* $-\omega$ *(called the semigroup of type* $-\omega$ *generated by* A *on* $\overline{D(A)}$*).*

According to Definition 1.8.3, if $\omega = 0$ the semigroup in Proposition 1.8.4 is called the *semigroups of contractions*, or the *semigroup of nonexpansive mappings* generated by A.

We notice that the semigroup $\{S(t) : \overline{D(A)} \to \overline{D(A)};\ t \in \mathbb{R}_+\}$ is given by the Crandall–Liggett Exponential Formula

$$S(t)x = \lim_n \left(I - \frac{t}{n}A\right)^{-n} x, \tag{1.8.6}$$

for each $x \in \overline{D(A)}$, uniformly for t in bounded subsets in $\mathbb{R}_+$. See Crandall and Liggett [90].

From (1.8.4) and (ii) in Proposition 1.8.1, we easily deduce Proposition 1.8.5 below, which is both a continuous dependence result with respect to the data, and a uniqueness result as well.

PROPOSITION 1.8.5 *If* $A : D(A) \subseteq X \rightsquigarrow X$ *is m-dissipative,* $\xi, \eta \in \overline{D(A)}$ *and* $f, g \in L^1(a, b; X)$, *then* $\widetilde{u} = u(\cdot, a, \xi, f)$ *and* $\widetilde{v} = u(\cdot, a, \eta, g)$ *satisfy*

$$\|\widetilde{u}(t) - \widetilde{v}(t)\| \le \|\xi - \eta\| + \int_a^b \|f(s) - g(s)\| \, ds,$$

for each $t \in [a, b]$.

THEOREM 1.8.3 (Benilan) *Let* X *be a Banach space, let* $A : D(A) \subseteq X \rightsquigarrow X$ *be* ω-*m*-*dissipative for some* $\omega \in \mathbb{R}$, *let* $\xi \in \overline{D(A)}$ *and* $f \in L^1(a, b; X)$. *Then* $\widetilde{u} : [a, b] \to \overline{D(A)}$ *coincides with* $u(\cdot, a, \xi, f)$ *if and only if it is continuous and*

$$\|\widetilde{u}(t) - x\|^2 \le e^{-2\omega(t-s)}\|\widetilde{u}(s) - x\|^2 + 2\int_s^t e^{-2\omega(t-\theta)}(\widetilde{u}(\theta) - x, f(\theta) + y)_+ \, d\theta$$

34   Delay Differential Evolutions Subjected to Nonlocal Initial Conditions

for each $x \in D(A)$, each $y \in Ax$ and each $a \leq s \leq t \leq b$. Equivalently, the function $\widetilde{u} : [a, b] \to \overline{D(A)}$ coincides with $u(\cdot, a, \xi, f)$ if and only if it is continuous and

$$\|\widetilde{u}(t) - x\| \leq e^{-\omega(t-s)}\|\widetilde{u}(s) - x\| + \int_s^t e^{-\omega(t-\theta)}[\widetilde{u}(\theta) - x, f(\theta) + y]_+ \, d\theta$$

for each $x \in D(A)$, each $y \in Ax$ and each $a \leq s \leq t \leq b$.

See Benilan [32] or Lakshmikantham and Leela [163, Theorem 3.5.1, p. 104] or Miyadera [187, Theorem 5.18, p. 157].

1.8.3   Compactness of the solution operator

DEFINITION 1.8.4 The semigroup $\{S(t) : \overline{D(A)} \to \overline{D(A)}; \ t \in \mathbb{R}_+\}$ is *compact* if, for each $t > 0$, $S(t)$ is a compact operator.

LEMMA 1.8.1 Let $A : D(A) \subseteq X \rightsquigarrow X$ be an m-dissipative operator and let $\{S(t) : \overline{D(A)} \to \overline{D(A)}; \ t \in \mathbb{R}_+\}$ be the semigroup of contractions generated by A on $\overline{D(A)}$ via the Crandall–Liggett Exponential Formula (1.8.6). Let $f \in L^1_{\mathrm{loc}}([a, b); X)$ be arbitrary, $a < b \leq +\infty$ and let u be a C^0-solution of (1.8.1). Then, for each $t \in [a, b)$ and each $s > 0$ with $t + s < b$, we have

$$\|S(s)u(t) - u(t + s)\| \leq \int_t^{t+s} \|f(\theta)\| \, d\theta.$$

We also need the following specific form of a general result due to Brezis [51].

LEMMA 1.8.2 (Brezis) Let $A : D(A) \subseteq X \rightsquigarrow X$ be an m-dissipative operator, let $\lambda > 0$ and $J_\lambda = (I - \lambda A)^{-1}$ and let $\{S(t) : \overline{D(A)} \to \overline{D(A)}; \ t \in \mathbb{R}_+\}$ be the semigroup of contractions generated by A on $\overline{D(A)}$ via the Crandall–Liggett Exponential Formula (1.8.6). Then, for each $\lambda > 0$ and $x \in \overline{D(A)}$, we have

$$\|J_\lambda x - x\| \leq \frac{4}{\lambda} \int_0^\lambda \|S(s)x - x\| \, ds.$$

See Brezis [50] or Vrabie [252, Theorem 2.1.1, p. 33].

We have the following two compactness results.

THEOREM 1.8.4 (Vrabie) Let X be a Banach space, let $A : D(A) \subseteq X \rightsquigarrow X$ be an m-dissipative operator, $\xi \in \overline{D(A)}$ and $\mathcal{F}$ a uniformly integrable subset in $L^1(a, b; X)$. Then the following conditions are equivalent:

(i) the set $\{u(\cdot, a, \xi, f); \ f \in \mathcal{F}\}$ is relatively compact in $C([a, b]; X)$

(ii) there exists a dense subset E in $[a, b]$ such that, for each $t \in E$, the cross section $\{u(t, a, \xi, f); \ f \in \mathcal{F}\}$ is relatively compact in X.

Preliminaries   35

See Vrabie [250] or Vrabie [252, Theorem 2.3.1, p. 45].

THEOREM 1.8.5 (Baras) *Let X be a Banach space, let $A: D(A) \subseteq X \rightsquigarrow X$ be an m-dissipative operator, and let us assume that A generates a compact semigroup. Let ξ in $\overline{D(A)}$ be arbitrary but fixed and let $\mathcal{F}$ be uniformly integrable in $L^1(a, b; X)$. Then the set $\{u(\cdot, a, \xi, f); f \in \mathcal{F}\}$ is relatively compact in $C([a, b]; X)$.*

See Baras [19] or Vrabie [252, Theorem 2.3.3, p. 47].

THEOREM 1.8.6 *Let X be a Banach space, let $A: D(A) \subseteq X \rightsquigarrow X$ be an m-dissipative operator and let us assume that A generates a compact semigroup. Let $B \subseteq \overline{D(A)}$ be bounded and let $\mathcal{F}$ be uniformly integrable in $L^1(a, b; X)$. Then, for each $\sigma \in (a, b)$, the set $\{u(\cdot, a, \xi, f); (\xi, f) \in B \times \mathcal{F}\}$ is relatively compact in $C([\sigma, b]; X)$. If, in addition, B is relatively compact, then the C^0-solutions set $\{u(\cdot, a, \xi, f); (\xi, f) \in B \times \mathcal{F}\}$ is relatively compact even in $C([a, b]; X)$.*

See Vrabie [252, Theorems 2.3.2 and 2.3.3, pp. 46–47].

THEOREM 1.8.7 (Mitidieri–Vrabie) *Let $A: D(A) \subseteq X \rightsquigarrow X$ be m-dissipative and let us assume that $(I - A)^{-1}$ is compact. Let $\mathcal{F} \subseteq L^1(a, b; X)$ be a uniformly integrable set and let $\xi \in X$ be fixed. Then, the following conditions are equivalent:*

(*i*) *the set $\{u(\cdot, a, \xi, f); f \in \mathcal{F}\}$ is equicontinuous from the right on $[a, b)$*

(*ii*) *the set $\{u(\cdot, a, \xi, f); f \in \mathcal{F}\}$ is relatively compact in $C([a, b]; X)$.*

See Mitidieri and Vrabie [185] or Vrabie [252, Theorem 2.3.2, p. 46].

We also need a general existence result in Vrabie [247]. Let us consider the nonlinear perturbed evolution equation

$$\begin{cases} u'(t) \in Au(t) + f(t, u(t)), & t \in [a, b], \\ u(a) = \xi. \end{cases} \tag{1.8.7}$$

THEOREM 1.8.8 (Vrabie) *Let X be a Banach space, let $A: D(A) \subseteq X \rightsquigarrow X$ be an m-dissipative operator and let us assume that A generates a compact semigroup. Let $U \subseteq X$ be nonempty and open, let $a < b \leq +\infty$ and let $f: [a, b) \times U \to X$ be a continuous function. Then, for each $\xi \in U \cap \overline{D(A)}$, there exists $c \in (a, b]$ such that (1.8.7) has at least one C^0-solution $u: [a, c] \to U \cap \overline{D(A)}$. If, in addition, $U = X$ and f has affine growth, i.e., there exist $\ell > 0$ and $m \geq 0$ such that $\|f(t, u)\| \leq \ell\|u\| + m$ for each $t \in [a, b)$ and $u \in \overline{D(A)}$, then each C^0-solution of (1.8.7) can be continued to the whole interval $[a, b)$.*

For the proof see Vrabie [247] or Vrabie [252, Theorem 3.8.1, p. 131].

36 *Delay Differential Evolutions Subjected to Nonlocal Initial Conditions*

DEFINITION 1.8.5 We say that the m-dissipative operator A is of *complete continuous type* if for each $a < b$ and every sequence $(f_n)_n$ in $L^1(a, b; X)$ and $(u_n)_n$ in $C([a, b]; X)$, with u_n a C^0-solution on $[a, b]$ of the problem

$$u_n'(t) \in Au_n(t) + f_n(t), \quad n = 1, 2, \ldots,$$

$$\lim_n f_n = f \quad \text{weakly in} \quad L^1(a, b; X)$$

and

$$\lim_n u_n = u \quad \text{strongly in} \quad C([a, b]; X),$$

it follows that u is a C^0-solution on $[a, b]$ of the limit problem

$$u'(t) \in Au(t) + f(t).$$

REMARK 1.8.2 If the topological dual of X is uniformly convex and the operator A generates a compact semigroup, then A is of complete continuous type. See Vrabie [252, Corollary 2.3.1, p. 49]. An m-dissipative operator of complete continuous type in a nonreflexive Banach space (and, by consequence, whose dual is not uniformly convex) is the nonlinear diffusion operator $\Delta\varphi$ in $L^1(\Omega)$. See Theorem 1.9.6 in Section 1.9.

1.9 Examples of m-dissipative operators

1.9.1 Sobolev spaces

To fix the idea, let us first recall some notations. If Ω is a nonempty and open subset in $\mathbb{R}^d$, $d \geq 1$, with boundary Σ, we denote by $C_0^\infty(\Omega)$ the space of C^∞-real functions with compact support in Ω. Further, if $1 \leq p < +\infty$ and $m \in \mathbb{N}$, $W^{m,p}(\Omega)$ denotes the space of all functions $u : \Omega \to \mathbb{R}$ which, together with their partial derivatives up to the order m, in the sense of distributions over Ω, belong to $L^p(\Omega)$. Endowed with the norm

$$\|u\|_{W^{m,p}(\Omega)} = \left(\sum_{0 \leq |\kappa| \leq m} \|\mathcal{D}^\kappa u\|_{L^p(\Omega)}^p \right)^{1/p},$$

$W^{m,p}(\Omega)$ is a separable real Banach space, densely and continuously imbedded in $L^p(\Omega)$. Here, as usual, if $\kappa = (\kappa_1, \kappa_2, \ldots, \kappa_n)$ is a multi-index, we denote by

$$\mathcal{D}^\kappa u = \frac{\partial^{\kappa_1 + \kappa_2 + \cdots + \kappa_n} u}{\partial x_1^{\kappa_1} \partial x_2^{\kappa_2} \ldots \partial x_n^{\kappa_n}},$$

where the partial derivatives are in the sense of distributions over Ω.

Preliminaries 37

We denote by $W_0^{m,p}(\Omega)$ the closure of $C_0^\infty(\Omega)$ in $W^{m,p}(\Omega)$, by $H^1(\Omega) = W^{1,2}(\Omega)$, $H_0^1(\Omega) = W_0^{1,2}(\Omega)$, $H^{-1}(\Omega) = [\,H_0^1(\Omega)\,]^*$ and $H^2(\Omega) = W^{2,2}(\Omega)$.

Finally, we make the conventional notation $W^{0,p}(\Omega) = L^p(\Omega)$.

For details on Sobolev spaces, see Adams [1], Evans [111] and Maz'ja [180].

THEOREM 1.9.1 (Sobolev, Rellich, Kondrašov, Gagliardo) *Let Ω be a nonempty, open and bounded subset in $\mathbb{R}^d$, $d \geq 1$, whose boundary Σ is of class C^1. Let $m \in \mathbb{N}$ and let $p, q \in [\,1, +\infty)$.*

(i) *If $mp < d$ and $q < \frac{dp}{d-mp}$, then $W^{m,p}(\Omega)$ is compactly imbedded in $L^q(\Omega)$.*

(ii) *If $mp = d$ and $q \in [\,1, +\infty)$, then $W^{m,p}(\Omega)$ is compactly imbedded in $L^q(\Omega)$.*

(iii) *If $mp > d$, then $W^{m,p}(\Omega)$ is compactly imbedded in $C(\overline{\Omega})$.*

See Sobolev [239], Rellich [223], Kondrašov [157] and Gagliardo [121]. For a complete proof, see either Adams [1, Theorem 5.4, p. 97 and Theorem 6.2, p. 144] or Maz'ja [180, Theorem 1.4.5, p. 60 and Theorem 1.4.6/2, p. 64].

1.9.2 The Laplace operator

EXAMPLE 1.9.1 *The Laplace operator in $L^2(\Omega)$.* Let Ω be a nonempty and open subset in $\mathbb{R}^d$, $d \geq 1$, with C^1 boundary, let $X = L^2(\Omega)$, and let us consider the operator A on X defined by

$$\begin{cases} D(A) = \{u \in H_0^1(\Omega);\ \Delta u \in L^2(\Omega)\}, \\ Au = \Delta u, \ \text{for each}\ u \in D(A). \end{cases}$$

THEOREM 1.9.2 *The Laplace operator Δ with homogeneous Dirichlet boundary conditions on $L^2(\Omega)$, i.e., the linear operator A defined above, is the infinitesimal generator of a compact C_0-semigroup of contractions.*

See Vrabie [253, Theorem 4.1.2, p. 79].

Before passing to another example, we recall for easy reference a specific form of the Poincaré Inequality. For the general case, see Evans [111, Theorem 1, p. 290].

LEMMA 1.9.1 *If Ω is a bounded domain with C^1-boundary in $\mathbb{R}^d$, $d \geq 1$, then for each $u \in H_0^1(\Omega)$, we have*

$$\|u\|_{L^2(\Omega)} \leq \lambda_1^{-1}\|\nabla u\|_{L^2(\Omega)},$$

where λ_1 is the first eigenvalue of $-\Delta$, with Δ defined as in Example 1.9.1.

38   *Delay Differential Evolutions Subjected to Nonlocal Initial Conditions*

EXAMPLE 1.9.2 *The Laplace operator in* $L^1(\Omega)$. Let Ω be a nonempty, bounded and open subset in $\mathbb{R}^d$, $d \geq 1$, with C^2 boundary Σ, let $X = L^1(\Omega)$ and let us consider the operator A on X, defined by

$$\begin{cases} D(A) = \{u \in W_0^{1,1}(\Omega); \, \Delta u \in L^1(\Omega)\}, \\ Au = \Delta u, \text{ for each } u \in D(A). \end{cases}$$

THEOREM 1.9.3 *The Laplace operator* Δ *with homogeneous Dirichlet boundary conditions on* $L^1(\Omega)$, *i.e., the linear operator* A *defined above, is the infinitesimal generator of a compact* C_0-*semigroup of contractions on* $L^1(\Omega)$.

See Vrabie [253, Theorem 7.2.7, p. 160 and Remark 4.1.3, p. 82].

1.9.3 C_0-groups generators

EXAMPLE 1.9.3 (*The Wave Operator*). Let Ω be a nonempty and open subset in $\mathbb{R}^d$, $d \geq 1$, with C^1 boundary, let $H = H_0^1(\Omega) \times L^2(\Omega)$ and let $A \colon D(A) \subset H \to H$ be defined by

$$\begin{cases} D(A) = [H_0^1(\Omega) \cap H^2(\Omega)] \times H_0^1(\Omega), \\ A(u, v) = (v, \Delta u), \end{cases}$$

for each $(u, v) \in D(A)$.

In this example, the space H is endowed with the inner product $\langle \cdot, \cdot \rangle$, defined by

$$\langle (u, v), (f, g) \rangle = \langle u, f \rangle_{H_0^1(\Omega)} + \langle v, g \rangle_{L^2(\Omega)},$$

where $\langle \cdot, \cdot \rangle_{H_0^1(\Omega)}$ is the inner product in $H_0^1(\Omega)$ defined by

$$\langle u, v \rangle_{H_0^1(\Omega)} = \langle \nabla u, \nabla v \rangle_{L^2(\Omega)}$$

for each $u, v \in H_0^1(\Omega)$.

THEOREM 1.9.4 *The operator* A, *defined as above, is the generator of a* C_0-*group of unitary operators on* H.

See Vrabie [253, Theorem 4.6.2, p. 93].

EXAMPLE 1.9.4 Let $X = L^p(\mathbb{R}^d)$, $d \geq 1$, with $1 \leq p < +\infty$ and let $v \in \mathbb{R}^d$. Let us define the operator $A \colon D(A) \subseteq X \to X$ by

$$\begin{cases} D(A) = \{u \in X; \, v \cdot \nabla u \in X\}, \\ Au = v \cdot \nabla u = \sum_{i=1}^d v_i \dfrac{\partial u}{\partial x_i}, \text{ for } u \in D(A), \end{cases}$$

where the partial derivatives are in the sense of distributions over $\mathbb{R}^d$.

Preliminaries 39

THEOREM 1.9.5 *The operator A defined as above is the infinitesimal generator of the C_0-group of isometries $\{G(t) : X \to X;\ t \in \mathbb{R}\}$, given by*

$$[\,G(t)f\,](x) = f(x + tv)$$

for each $f \in X$, $t \in \mathbb{R}$ and $x \in \mathbb{R}^d$.

See Vrabie [253, Theorem 4.4.1, p. 88]. For additional details, see Evans [111, 2.1. Transport Equation, p. 18].

1.9.4 The nonlinear diffusion operator

EXAMPLE 1.9.5 As before, the operator Δ is the Laplace operator in the sense of distributions over Ω. If $\varphi : D(\varphi) \subseteq \mathbb{R} \rightsquigarrow \mathbb{R}$ and $u : \Omega \to D(\varphi)$, we denote by

$$\mathcal{S}_\varphi(u) = \{v \in L^1(\Omega);\ v(x) \in \varphi(u(x)) \text{ a.e. for } x \in \Omega\}.$$

We say that $\varphi : D(\varphi) \subseteq \mathbb{R} \rightsquigarrow \mathbb{R}$ is *maximal-monotone*[5] if $-\varphi$ is m-dissipative.

The (i) part in Theorem 1.9.6 below is essentially due to Brezis and Strauss [53], the (ii) part to Badii, Díaz and Tesei [16], and the (iii) part to Cârjă, Necula and Vrabie [74].

THEOREM 1.9.6 *Let Ω be a nonempty, bounded and open subset in $\mathbb{R}^d$, $d \geq 1$, with C^1 boundary Σ and let $\varphi : D(\varphi) \subseteq \mathbb{R} \rightsquigarrow \mathbb{R}$ be maximal-monotone with $0 \in D(\varphi)$ and $0 \in \varphi(0)$.*

(i) *Then the operator $\Delta\varphi : D(\Delta\varphi) \subseteq L^1(\Omega) \rightsquigarrow L^1(\Omega)$, defined by*

$$\begin{cases} D(\Delta\varphi) = \{u \in L^1(\Omega);\ \exists v \in \mathcal{S}_\varphi(u) \cap W_0^{1,1}(\Omega),\ \Delta v \in L^1(\Omega)\}, \\ \Delta\varphi(u) = \{\Delta v;\ v \in \mathcal{S}_\varphi(u) \cap W_0^{1,1}(\Omega)\} \cap L^1(\Omega) \text{ for } u \in D(\Delta\varphi) \end{cases}$$

is m-dissipative on $L^1(\Omega)$.

(ii) *If, in addition, $\varphi : \mathbb{R} \to \mathbb{R}$ is continuous on $\mathbb{R}$ and C^1 on $\mathbb{R} \setminus \{0\}$ and there exist two constants $C > 0$ and $\alpha > 0$ if $d \leq 2$ and $\alpha > (d-2)/d$ if $d \geq 3$ such that*

$$\varphi'(r) \geq C|r|^{\alpha-1}$$

for each $r \in \mathbb{R} \setminus \{0\}$, then $\Delta\varphi$ generates a compact semigroup.

(iii) *In the hypotheses of (ii), $\Delta\varphi$ is of complete continuous type.*

---

[5]The name comes from the property that, in the case of a Hilbert space H, an operator $A : D(A) \subseteq H \rightsquigarrow H$, with $-A$ dissipative, is called *monotone* and an operator B is m-dissipative if and only if it is *maximal dissipative*, i.e., if its graph is not strictly contained in the graph of another dissipative operator.

40   *Delay Differential Evolutions Subjected to Nonlocal Initial Conditions*

For the proof of (i), see Barbu [22, Theorem 3.5, p. 115], and for the proof of (ii) see Vrabie [252, Theorem 2.7.1, p. 70]. The proof of (iii) is a consequence of Theorem 1.9.7 below due to Díaz and Vrabie [97].

THEOREM 1.9.7 *In the hypotheses of (i) and (ii) in Theorem 1.9.6, if $q > 1$ is such that $L^q(\Omega) \subseteq H^{-1}(\Omega)$, then, for each arbitrary but fixed $\xi \in L^q(\Omega)$, the mapping $f \mapsto u(\cdot, a, \xi, f)$ is weakly-strongly sequentially continuous from $L^1(a, b; L^q(\Omega))$ to $C([a, b]; L^1(\Omega))$.*

We can now pass to the proof of (iii) in Theorem 1.9.6 due to Cârjă, Necula and Vrabie [74, Theorem 1.7.9, p. 22], showing that the m-dissipative operator $\Delta \varphi$ is of complete continuous type in $L^1(\Omega)$.

Proof. Let $\xi \in L^1(\Omega)$, let $f \in L^1(a, b; L^1(\Omega))$, and let $(f_n)_n$ be a sequence in $L^1(a, b; L^1(\Omega))$ such that $\lim_n f_n = f$ weakly in $L^1(a, b; L^1(\Omega))$. As, by Fubini Theorem 1.2.5, $L^1([a, b] \times \Omega) = L^1(a, b; L^1(\Omega))$, we have that $\lim_n f_n = f$ weakly in $L^1([a, b] \times \Omega)$. Let $k \in \mathbb{N}$ be arbitrary but fixed and let us define

$$P_k : L^1([a, b] \times \Omega) \to L^1([a, b] \times \Omega)$$

by

$$P_k(g)(t, x) = \begin{cases} g(t, x) & \text{if } |g(t, x)| \le k, \\ 0 & \text{if } |g(t, x)| > k \end{cases}$$

for each $g \in L^1([a, b] \times \Omega)$. Clearly, $(f_n)_n$ is bounded in $L^1([a, b] \times \Omega)$, say by $M > 0$. Throughout this proof, we denote by $\| \cdot \|_{L^1}$ the norm of $L^1([a, b] \times \Omega)$ and by $\| \cdot \|_{L^\infty}$ the norm of $L^\infty([a, b] \times \Omega)$. Since

$$k\mu(\{(s, y) \in [a, b] \times \Omega; \ |f_n(s, y)| > k\}) \le \int_{|f_n(t,x)|>k} |f_n(s, y)| ds \, dy \le M,$$

we get

$$\mu(\{(s, y) \in [a, b] \times \Omega; \ |f_n(s, y)| > k\}) \le \frac{M}{k}$$

for $n, k = 1, 2, \dots$. Further, since

$$\|P_k f_n - f_n\|_{L^1} = \int_{|f_n(t,x)|>k} |f_n(s, y)| ds \, dy \tag{1.9.1}$$

for each $k, n \in \mathbb{N}$ and, by Theorem 1.4.11, $\{f_n; \ n \in \mathbb{N}\}$ is uniformly integrable, from (1.9.1) we deduce

$$\lim_k P_k f_n = f_n \tag{1.9.2}$$

strongly in $L^1([a, b] \times \Omega)$, uniformly for $n = 1, 2, \dots$. Since $\lim_n f_n = f$, weakly in $L^1([a, b] \times \Omega)$, from (1.9.2), it follows that, for each arbitrary but fixed element g in the dual of $L^1([a, b] \times \Omega)$, i.e., $g \in L^\infty([a, b] \times \Omega)$, we have

$$\lim_{n,k} |(P_k f_n - P_k f, g)| = 0. \tag{1.9.3}$$

Preliminaries 41

Indeed, let us observe that

$$|(P_k f_n - P_k f, g)| \le |(P_k f_n - f_n, g)| + |(f_n - f, g)| + |(f - P_k f, g)|$$

$$\le [\,\|P_k f_n - f_n\|_{L^1} + \|f - P_k f\|_{L^1}\,]\|g\|_{L^\infty} + |(f_n - f, g)|,$$

and thus (1.9.2) and $\lim_n f_n = f$, weakly in $L^1([a,b] \times \Omega)$, imply (1.9.3). Next, let $q > 1$ be arbitrary but fixed and take $(\xi_p)_p$ in $L^q(\Omega)$ with $\lim_p \xi_p = \xi$ strongly in $L^1(\Omega)$. We have

$$\|u(t,a,\xi,f_n) - u(t,a,\xi,f)\| \le \|u(t,a,\xi,f_n) - u(t,a,\xi_p,f_n)\|$$

$$+\|u(t,a,\xi_p,f_n) - u(t,a,\xi_p,P_k f_n)\| + \|u(t,a,\xi_p,P_k f_n) - u(t,a,\xi_p,P_k f)\|$$

$$+\|u(t,a,\xi_p,P_k f) - u(t,a,\xi_p,f)\| + \|u(t,a,\xi_p,f) - u(t,a,\xi,f)\|$$

$$\le 2\|\xi - \xi_p\| + \|u(t,a,\xi_p,f_n) - u(t,a,\xi_p,P_k f_n)\|$$

$$+\|u(t,a,\xi_p,P_k f_n) - u(t,a,\xi_p,P_k f)\| + \|u(t,a,\xi_p,P_k f) - u(t,a,\xi_p,f)\|,$$

where $\|\cdot\|$ stands for the norm in $L^1(\Omega)$. Let $\varepsilon > 0$. Fix $p = p(\varepsilon)$ such that

$$\|\xi - \xi_p\| \le \varepsilon.$$

In view of (1.9.3) and Theorem 1.9.7, for this fixed p, we can find $n_1(\varepsilon) \in \mathbb{N}$ such that

$$\|u(t,a,\xi_p,P_k f_n) - u(t,a,\xi_p,P_k f)\| \le \varepsilon$$

for each $n, k \in \mathbb{N}$, $n \ge n_1(\varepsilon)$ and $k \ge n_1(\varepsilon)$. Furthermore, in view of (1.9.2), for the very same $\varepsilon > 0$ and $p = p(\varepsilon)$, there exists $n_2(\varepsilon) \in \mathbb{N}$, such that we have both

$$\|u(t,a,\xi_p,f_n) - u(t,a,\xi_p,P_k f_n)\| \le \|f_n - P_k f_n\|_{L^1} \le \varepsilon,$$

$$\|u(t,a,\xi_p,P_k f) - u(t,a,\xi_p,f)\| \le \|f - P_k f\|_{L^1} \le \varepsilon$$

for each $k \in \mathbb{N}$, $k \ge n_2(\varepsilon)$ and each $n \in \mathbb{N}$. Set $n(\varepsilon) = \max\{n_1(\varepsilon), n_2(\varepsilon)\}$. We have

$$\|u(t,a,\xi,f_n) - u(t,a,\xi,f)\| \le 5\varepsilon$$

for each $n \in \mathbb{N}$, $n \ge n(\varepsilon)$ and this completes the proof. $\qquad\square$

The following extension of Lemma 11.1 in Cârjă, Necula and Vrabie [75] is also needed. See also Cârjă, Necula and Vrabie [74, Lemma 13.5.11, p. 273].

LEMMA 1.9.2 *Let Ω be a bounded domain in $\mathbb{R}^d$, $d \ge 1$, with C^2 boundary Γ, let $\omega \ge 0$, let $\psi : \mathbb{R} \to \mathbb{R}$ be strictly increasing with $\psi(0) = 0$, let $u_0, \widetilde{u}_0 \in L^1(\Omega)$, $f_0, \widetilde{f}_0 \in L^1(a,b;L^1(\Omega))$ and let $u : [a,b] \to L^1(\Omega)$ be the unique C^0-solution of the Cauchy problem*

$$\begin{cases} \dfrac{\partial u}{\partial t} = \Delta \psi(u) - \omega u + f_0(t,x) & \text{in } [a,b] \times \Omega, \\ u = 0 & \text{on } [a,b] \times \Gamma, \\ u(a,x) = u_0(x) & \text{in } \Omega \end{cases}$$

42   *Delay Differential Evolutions Subjected to Nonlocal Initial Conditions*

and let $\widetilde{u} : [a, b] \to L^1(\Omega)$ be the unique C^0-solution of the same Cauchy problem but with f_0 replaced by $\widetilde{f}_0$ and u_0 replaced by $\widetilde{u}_0$. If $u_0(x) \le \widetilde{u}_0(x)$ a.e. for $x \in \Omega$ and $f_0(s, x) \le \widetilde{f}_0(s, x)$ for each $s \in [a, b]$ and a.e. for $x \in \Omega$, then

$$u(s, x) \le \widetilde{u}(s, x)$$

for each $s \in [a, b]$ and a.e. for $x \in \Omega$.

1.9.5   The p-Laplace operator

EXAMPLE 1.9.6 Let Ω be a nonempty, bounded and open subset in $\mathbb{R}^d$, $d \ge 1$, with C^2 boundary Σ, let $p \in [2, +\infty)$ and $\lambda > 0$, let $\beta : D(\beta) \subseteq \mathbb{R} \rightsquigarrow \mathbb{R}$ a maximal-monotone operator, i.e., an operator such that $-\beta$ is m-dissipative and let $\Delta_p^\lambda : D(\Delta_p^\lambda) \subseteq L^2(\Omega) \to L^2(\Omega)$ be defined as

$$\Delta_p^\lambda u = \sum_{i=1}^d \frac{\partial}{\partial x_i} \left(\left| \frac{\partial u}{\partial x_i} \right|^{p-2} \frac{\partial u}{\partial x_i} \right) - \lambda |u|^{p-2} u,$$

$$D(\Delta_p^\lambda) = \left\{ u \in W^{1,p}(\Omega); \ \Delta_p^\lambda u \in L^2(\Omega), \ -\frac{\partial u}{\partial \nu_p}(x) \in \beta(u(x)) \text{ a.e. for } x \in \Sigma \right\},$$

the p-conormal derivative of u, u_{ν_p}, being defined as

$$\frac{\partial u}{\partial \nu_p} = \sum_{i=1}^d \left| \frac{\partial u}{\partial x_i} \right|^{p-2} \frac{\partial u}{\partial x_i} \cos(\overrightarrow{n}, \overrightarrow{e_i}).$$

In the above formula, $\overrightarrow{n}$ is the unitary exterior normal to the boundary Σ, and $\{\overrightarrow{e_1}, \overrightarrow{e_2}, \dots, \overrightarrow{e_d}\}$ is the canonical base in $\mathbb{R}^d$.

THEOREM 1.9.8 *Let Ω be a nonempty, bounded and open subset in $\mathbb{R}^d$, $d \ge 1$, with C^2 boundary Σ, $p \in [2, +\infty)$ and $\lambda > 0$ and let $\Delta_p^\lambda : D(\Delta_p^\lambda) \subseteq L^2(\Omega) \to L^2(\Omega)$ be defined as above. If $-\beta : D(\beta) \subseteq \mathbb{R} \rightsquigarrow \mathbb{R}$ is m-dissipative, $0 \in D(\beta)$ and $0 \in \beta(0)$, then Δ_p^λ is m-dissipative on $L^2(\Omega)$ and generates a compact semigroup.*

See Vrabie [252, Theorem 2.8.2, p. 77].

DEFINITION 1.9.1 The nonlinear partial differential operator Δ_p^λ, defined as above, is called the *p-Laplace operator*.

For other results concerning partial differential nonlinear operators similar to the p-Laplace operator, see Lions [172, 2.3 Exemples, pp. 177–179].

Preliminaries 43

1.10 Strong solutions

DEFINITION 1.10.1 Let H be a real Hilbert space with the inner product $\langle \cdot, \cdot \rangle$ and $\varphi : H \to \mathbb{R} \cup \{+\infty\}$ a proper, l.s.c., convex function. The set

$$\partial\varphi(x) = \{z \in H; \ \varphi(x) \le \varphi(y) + \langle x - y, z \rangle \text{ for each } y \in H\}$$

is called the *subdifferential* of φ calculated at x.

DEFINITION 1.10.2 Let $\varphi : H \to \mathbb{R} \cup \{+\infty\}$ be a proper, l.s.c., convex function, and let $D(\partial\varphi) = \{x \in H; \ \partial\varphi(x) \neq \emptyset\}$. The operator $\partial\varphi : D(\partial\varphi) \subseteq H \rightsquigarrow H$, which assigns to each $x \in H$ the subset $\partial\varphi(x)$ in H, is called the *subdifferential* of the convex function φ.

THEOREM 1.10.1 *Let $\varphi : H \to \mathbb{R} \cup \{+\infty\}$ be a proper, l.s.c., convex function. Then $D(\partial\varphi)$ is dense in $D(\varphi)$.*

See Barbu [21, Corollary 2.1, p. 55].

THEOREM 1.10.2 (Minty, Moreau) *If $\varphi : H \to \mathbb{R} \cup \{+\infty\}$ is a proper, l.s.c., convex function, then $-\partial\varphi : D(\partial\varphi) \subseteq H \rightsquigarrow 2^H$ is an m-dissipative operator.*

See Barbu [21, Theorem 2.1, p. 54]. In what follows, $AC([a,b])$ denotes the space of absolutely continuous functions from $[a,b]$ to $\mathbb{R}$.

THEOREM 1.10.3 (Brezis) *Let $\varphi : H \to \mathbb{R}_+ \cup \{+\infty\}$ be a proper, l.s.c., convex function. Then for each $\xi \in \overline{D(A)}$ and $f \in L^2([a,b]; H)$ the problem*

$$\begin{cases} u'(t) \in -\partial\varphi(u(t)) + f(t), & t \in [a,b], \\ u(a) = \xi \end{cases}$$

has a unique strong solution u – see Definition 1.8.1 – satisfying

$$u(t) \in D(A) \quad a.e \text{ for } t \in (a,b)$$

$$t \mapsto (t-a)^{1/2} u'(t) \quad \text{belongs to} \quad L^2([a,b]; H)$$

$$\left(\int_a^b (t-a) \|u'(t)\|^2 \, dt \right)^{1/2} \le \left(\int_a^b \|f(t)\|^2 dt \right)^{1/2} + \frac{1}{\sqrt{2}} \int_a^b \|f(t)\| \, dt + \frac{1}{\sqrt{2}} \|\xi - \eta\|$$

for each $\eta \in H$ with $\varphi(\eta) = 0$

$$t \mapsto \varphi(u(t)) \text{ belongs to } L^1([a,b]) \cap AC([\sigma,b]) \text{ for each } a < \sigma < b$$

$$\|u'(t)\|^2 + \frac{d}{dt}\, \varphi(u(t)) = \langle f(t), u'(t) \rangle \text{ a.e. for } t \in (a,b).$$

If, in addition, $\xi \in D(\varphi)$, then

$$t \mapsto u'(t) \quad belongs \ to \ \ L^2([\,a,b\,]); H),$$

$$\left(\int_a^b \|u'(t)\|^2 \, dt \right)^{1/2} \leq \left(\int_a^b \|f(t)\|^2 dt \right)^{1/2} + \sqrt{\varphi(\xi)},$$

and $t \mapsto \varphi(u(t))$ belongs to $AC([\,a,b\,])$.

See Brezis [49, Proposition 3.6, p. 72].

1.11 Nonautonomous evolution equations

Here we consider the autonomous evolution equation

$$\mathcal{U}'(t) \in \mathcal{A}(t)\mathcal{U}(t), \quad t \in \mathbb{R}_+,$$

where $\mathcal{X}$ is a Banach space and $\{\mathcal{A}(t) : D(\mathcal{A}(t)) \to D(\mathcal{A}(t)); \ t \in \mathbb{R}_+\}$ is a family of possible nonlinear operators. We notice that in this section, for the sake of simplicity, we are using exactly the notations in the forthcoming Section 1.12, to whom the main result here presented is especially intended.

In order to introduce a concept of generalized solution for this problem, some preliminaries are needed.

DEFINITION 1.11.1 Let $\{\mathcal{C}(t) \subseteq \mathcal{X}; \ t \in \mathbb{R}_+\}$ and let $\gamma \in \mathbb{R}$ be fixed. A family of operators

$$\{\mathcal{U}(t,s) : \mathcal{C}(s) \to \mathcal{C}(t); \ 0 \leq s \leq t < +\infty\}$$

is called an *evolution system of type γ* in $\{\mathcal{C}(t) \subseteq \mathcal{X}; \ t \in \mathbb{R}_+\}$ if it satisfies

(E_1) $\mathcal{U}(t,\theta)\mathcal{U}(\theta,s) = \mathcal{U}(t,s)$ for each $0 \leq s \leq \theta \leq t < +\infty$

(E_2) $\lim_{t \downarrow s} \mathcal{U}(t,s)\varphi = \varphi$ for each $0 \leq s < +\infty$ and each $\varphi \in \mathcal{C}(s)$

(E_3) $\|\mathcal{U}(t,s)\varphi - \mathcal{U}(t,s)\psi\|_{\mathcal{X}} \leq e^{\gamma(t-s)}\|\varphi - \psi\|_{\mathcal{X}}$

for each $s \in \mathbb{R}_+$, each $t \in [\,s,+\infty)$ and each $\varphi, \psi \in \mathcal{C}(s)$.

We say that the family of operators $\{\mathcal{A}(t) : D(\mathcal{A}(t)) \to D(\mathcal{A}(t)); \ t \in \mathbb{R}_+\}$ is the infinitesimal generator of the evolution system

$$\{\mathcal{U}(t,s) : \mathcal{C}(s) \to \mathcal{C}(t); \ 0 \leq s \leq t < +\infty\}$$

if

$$\mathcal{U}(t,s)\varphi = \lim_{n \to +\infty} \prod_{i=1}^{n} \left(I - \frac{t-s}{n} \mathcal{A} \left(s + i\frac{t-s}{n} \right) \right)^{-1} \varphi$$

for each $0 \leq s \leq t < +\infty$ and each $\varphi \in \mathcal{C}(s)$.

Preliminaries 45

The theorem below is a specific form of a more general result. See Pavel [208, Theorem 3.6, p. 35].

THEOREM 1.11.1 *Let $\gamma \in \mathbb{R}_+$ and let $\{\mathcal{A}(t) : D(\mathcal{A}(t)) \to D(\mathcal{A}(t)); \ t \in \mathbb{R}_+\}$ be a family of operators satisfying*

(i) *there exist $\lambda_0 \in \left(0, \frac{1}{\gamma}\right)$, a continuous function $h : \mathbb{R}_+ \to X$, and a nondecreasing function $L : \mathbb{R}_+ \to \mathbb{R}_+$ such that, for each $s \in \mathbb{R}_+$, each $t \in [s, +\infty)$, each $v \in D(\mathcal{A}(s))$, $w \in D(\mathcal{A}(t))$, each $\widetilde{v} \in \mathcal{A}(s)v$ and $\widetilde{w} \in \mathcal{A}(t)w$, we have*

$$(1 - \lambda\gamma)\|v - w\|_X \le \|v - w - \lambda(\widetilde{v} - \widetilde{w})\|_X + \lambda\|h(s) - h(t)\|L(\|w\|_X)$$

for each $\lambda \in (0, \lambda_0\,]$

(ii) *for each $t \in \mathbb{R}_+$, $\mathcal{A}(t)$ is γ-m-dissipative*

(iii) *for each $s \in \mathbb{R}_+$, each $t \in (s, +\infty)$, each $(t_n)_n$ with $t_n \uparrow t$ and each $(v_n)_n$ with $v_n \in D(\mathcal{A}(t_n))$ and $\lim_n v_n = v$, we have $v \in D(\mathcal{A}(t))$.*

Then, the family of operators $\{\mathcal{A}(t) : D(\mathcal{A}(t)) \to D(\mathcal{A}(t)); \ t \in \mathbb{R}_+\}$ is the infinitesimal generator of an evolution system

$$\{\mathcal{U}(t, s) : \overline{D(\mathcal{A}(s))} \to \overline{D(\mathcal{A}(t))}; \ 0 \le s \le t < +\infty\}.$$

REMARK 1.11.1 Condition (i) is in fact a γ-m-dissipativity condition, coupled with a stability hypotheses with respect to t. Condition (iii) is satisfied whenever (i) and (ii) are satisfied and $D(\mathcal{A}(t))$ is independent of $t \in \mathbb{R}_+$.

1.12 Delay evolution equations

Our aim here is to show how the abstract theory of evolution equations driven by ω-m-dissipative operators can be applied to get information concerning delay evolution equations.

Let X be a Banach space, let $A : D(A) \subseteq X \rightsquigarrow X$ be an ω-m-dissipative operator for some $\omega \in \mathbb{R}$, let $\tau \ge 0$, let $\mathcal{X} = C([-\tau, 0]; X)$, and let $f : \mathcal{X} \to X$ be a given function. If $u \in C([-\tau, +\infty); X)$ and $t \in \mathbb{R}_+$, $u_t : [-\tau, 0] \to X$ denotes the function defined by

$$u_t(s) = u(t + s)$$

for each $s \in [-\tau, 0]$. Obviously, $u_t \in \mathcal{X}$.

46   *Delay Differential Evolutions Subjected to Nonlocal Initial Conditions*

1.12.1   The autonomous case

Let us now consider the autonomous delay evolution equation

$$u'(t) \in Au(t) + f(u_t), \quad t \in \mathbb{R}_+. \tag{1.12.1}$$

As we will see later, under some usual global Lipschitz condition on f, for each $\varphi \in X$ with $\varphi(0) \in \overline{D(A)}$, the problem (1.12.1) has a unique C^0-solution $u(\cdot, 0, \varphi) : [-\tau, +\infty) \to X$ satisfying $u(t, 0, \varphi) = \varphi(t)$ for each $t \in [-\tau, 0]$. See Theorem 2.6.3. Of course, $u(t, 0, \varphi) \in \overline{D(A)}$ for each $t \in \mathbb{R}_+$.

Our next goal is to show that this delay evolution problem in X can be rewritten as a non-delay evolution equation governed by a γ-m-dissipative operator defined in an appropriate larger Banach space. The main idea is that, instead of $u : [-\tau, +\infty) \to X$, we consider as a new unknown function the translate u_t of u, i.e., $\mathcal{U} : \mathbb{R}_+ \to \mathfrak{X}$, defined as

$$\mathcal{U}(t)(s) = u_t(s) = u(t+s)$$

for each $t \in \mathbb{R}_+$ and $s \in [-\tau, 0]$, and show that $\mathcal{U}$ satisfies a suitably defined evolution equation which can be directly handled by the Crandall and Liggett [90] nonlinear semigroup theory in the space $\mathfrak{X}$. If $t \mapsto \mathcal{S}(t)\varphi$ is the solution of the new equation in the larger space, then the unique C^0-solution $t \mapsto u(t, 0, \varphi)$ of (1.12.1) satisfying $u(s, 0, \varphi) = \varphi(s)$ for each $s \in [-\tau, 0]$ is given by $u(t, 0, \varphi) = [\mathcal{S}(t)\varphi](0)$ for $t \in \mathbb{R}_+$[6].

So, let $\mathcal{A} : D(\mathcal{A}) \subseteq \mathfrak{X} \rightsquigarrow \mathfrak{X}$ be defined by

$$\begin{cases} D(\mathcal{A}) = \{\varphi \in \mathfrak{X}; \ \varphi' \in \mathfrak{X}, \ \varphi(0) \in D(A), \ \varphi'(0) \in A\varphi(0) + f(\varphi)\}, \\ \mathcal{A}(\varphi) = \varphi', \ \text{for } \varphi \in D(\mathcal{A}). \end{cases} \tag{1.12.2}$$

THEOREM 1.12.1 *Let X be a Banach space, let $A : D(A) \subseteq X \rightsquigarrow X$ and let $f : \mathfrak{X} \to X$. Let us assume that there exist $\omega \in \mathbb{R}$ and $\ell > 0$ such that*

(i) A is ω-m-dissipative

(ii) f is Lipschitz with constant ℓ, i.e.,

$$\|f(\varphi) - f(\psi)\| \le \ell \|\varphi - \psi\|_{\mathfrak{X}}$$

for each $\varphi, \psi \in \mathfrak{X}$.

Then, for $\gamma = \max\{0, \ell - \omega\}$, the operator $\mathcal{A}$ defined by (1.12.2) is γ-m-dissipative in $\mathfrak{X}$. Accordingly, $\mathcal{A}$ is the infinitesimal generator of a nonlinear semigroup of type γ, $\{\mathcal{S}(t) : \overline{D(\mathcal{A})} \to \overline{D(\mathcal{A})}; \ t \in \mathbb{R}_+\}$, i.e., a nonlinear semigroup satisfying

$$\|\mathcal{S}(t)\varphi - \mathcal{S}(t)\psi\|_{\mathfrak{X}} \le e^{\gamma t} \|\varphi - \psi\|_{\mathfrak{X}} \tag{1.12.3}$$

for each $t \in \mathbb{R}_+$ and each $\varphi, \psi \in \overline{D(\mathcal{A})}$. In addition,

$$\overline{D(\mathcal{A})} = \{\varphi \in \mathfrak{X}; \ \varphi(0) \in \overline{D(A)}\}. \tag{1.12.4}$$

---

[6]We emphasize that the proof of this statement, although not surprising, is not simple.

Preliminaries 47

See Webb [266, Proposition 1].

We denote by

$$\mathcal{D} = \{\varphi \in \mathcal{X}; \ \varphi(0) \in \overline{D(A)}\}.$$

The problem is to determine when the C^0-solution of a Cauchy problem associated with (1.12.1) is given by the semigroup $\{\mathcal{S}(t) : \overline{D(A)} \to \overline{D(A)}; \ t \in \mathbb{R}_+\}$ generated by $\mathcal{A}$. The answer to this question was given by Plant [219, Proposition 2.1]. Namely, we have

THEOREM 1.12.2 *Under the hypotheses of Theorem 1.12.1, if $\varphi \in \mathcal{D}$, then the function $u(\cdot, 0, \varphi) : [-\tau, +\infty) \to X$, defined by*

$$u(t, 0, \varphi) = \begin{cases} \varphi(t), & t \in [-\tau, 0], \\ [\mathcal{S}(t)\varphi](0), & t \in (0, +\infty), \end{cases} \tag{1.12.5}$$

is a C^0-solution of (1.12.1) satisfying $u_0(\cdot, 0, \varphi) = \varphi$. In addition,

$$u_t(\cdot, 0, \varphi)(s) = u(t + s, 0, \varphi) = [\mathcal{S}(t)\varphi](s)$$

for each $t \in \mathbb{R}_+$ and $s \in [-\tau, 0]$.

Conversely, if $\varphi \in \mathcal{D}$, the unique C^0-solution, $u(\cdot, 0, \varphi) : [-\tau, +\infty) \to X$, of the delay evolution equation (1.12.1) satisfying $u(t, 0, \varphi) = \varphi(t)$ for each $t \in [-\tau, 0]$, is given by (1.12.5).

DEFINITION 1.12.1 The C^0-solution, $u(\cdot, 0, \varphi) : [-\tau, +\infty) \to X$ of the delay evolution equation (1.12.1) satisfying $u(t, 0, \varphi) = \varphi(t)$ for each $t \in [-\tau, 0]$, is called *stable* if for each $\varepsilon > 0$ there exists $\delta(\varepsilon, \varphi) > 0$ such that, for each $\psi \in \mathcal{D}$ satisfying $\|\varphi - \psi\|_X \le \delta(\varepsilon, \varphi)$, we have

$$\|u(t, 0, \varphi) - u(t, 0, \psi)\| \le \varepsilon$$

for each $t \in \mathbb{R}_+$.

The C^0-solution $u(\cdot, 0, \varphi) : [-\tau, +\infty) \to X$ of the delay evolution equation (1.12.1) satisfying $u(t, 0, \varphi) = \varphi(t)$ for each $t \in [-\tau, 0]$, is called *globally uniformly asymptotically stable* if it is stable and, for each $\psi \in \mathcal{D}$, the unique C^0-solution, $u(\cdot, 0, \psi)$, of (1.12.1) satisfying $u(t, 0, \psi) = \psi(t)$ for each $t \in [-\tau, 0]$, satisfies

$$\lim_{t \to +\infty} \|u(t, 0, \varphi) - u(t, 0, \psi)\| = 0.$$

If $\ell - \omega < 0$, then it is not evident that we can take $\gamma = \ell - \omega$ in (1.12.3) to get global uniform asymptotic stability. However, we describe below how to define an equivalent norm on the space $\mathcal{X}$ in order to show that the condition $\ell < \omega$ implies global uniform asymptotic stability, as in the nondelayed case. More precisely, let $\sigma \in \mathbb{R}$ and let us define the norm $\| \cdot \|_\sigma$ on $\mathcal{X}$, by

$$\|\varphi\|_\sigma = \sup_{\theta \in [-\tau, 0]} e^{-\sigma\theta} \|\varphi(\theta)\|.$$

48 *Delay Differential Evolutions Subjected to Nonlocal Initial Conditions*

One may easily verify that, for each $\sigma \le \gamma$ and $\theta \in [-\tau, 0]$, we successively have

$$1 \le e^{(\gamma-\sigma)(\tau+\theta)} = e^{(\gamma-\sigma)\tau} \cdot e^{(\gamma-\sigma)\theta}$$

and

$$e^{-\gamma\theta} \le e^{(\gamma-\sigma)\tau} e^{-\sigma\theta}.$$

Hence

$$\|\varphi\|_\sigma \le \|\varphi\|_\gamma \le e^{(\gamma-\sigma)\tau} \|\varphi\|_\sigma$$

and

$$\|\varphi\|_0 = \|\varphi\|_x.$$

So, all the norms $\|\varphi\|_\sigma$ are equivalent with $\|\varphi\|_0 = \|\varphi\|_x$.

If $\ell < \omega$, the equation

$$\alpha = -\omega + \ell e^{-\alpha\tau} \tag{1.12.6}$$

has a unique solution $\alpha \in (-\infty, 0)$.

The result below is due to Plant [219].

THEOREM 1.12.3 *Under the hypotheses of Theorem 1.12.1, if $\ell < \omega$ and α is given by (1.12.6), then*

$$\|\mathbb{S}(t)\varphi - \mathbb{S}(t)\psi\|_\alpha \le e^{\alpha t} \|\varphi - \psi\|_\alpha$$

for each $\varphi, \psi \in \mathcal{D}$ and $t \in \mathbb{R}_+$. Hence, each global C^0-solution of the problem (1.12.1) is globally uniformly asymptotically stable in the sense of Definition 1.12.1.

We conclude with a consequence of Theorems 1.12.1, 1.12.2 and 1.12.3.

THEOREM 1.12.4 *Let us assume that all the hypotheses of Theorem 1.12.1 are satisfied and, in addition, $0 \in D(A)$, $0 \in A0$ and $\ell \le \omega$. Then, for each $\varphi \in \mathcal{D}$, the unique C^0-solution $u(\cdot, 0, \varphi)$ of the problem (1.12.1) satisfying $u_0(\cdot, 0, \varphi) = \varphi$ is stable in the sense of Definition 1.12.1. If $\ell < \omega$, then $u(\cdot, 0, \varphi)$ is bounded on $[-\tau, +\infty)$.*

Proof. The stability follows from (1.12.3). To prove that $u(\cdot, 0, \varphi)$ is bounded on $[-\tau, +\infty)$ it suffices to show that it is bounded on $\mathbb{R}_+$. To this end, let us observe first that

$$\|u(t, 0, \varphi) - u(t, 0, 0)\| \le \|\varphi\|_x \tag{1.12.7}$$

for each $t \in \mathbb{R}_+$. So, if $u(\cdot, 0, 0)$ is bounded on $\mathbb{R}_+$, we have nothing to prove. Then let us assume by contradiction that $u(\cdot, 0, 0)$ is unbounded on $\mathbb{R}_+$. Then there exists $(t_n)_n$ with $\lim_n t_n = +\infty$ such that

$$\|u(t_n, 0, 0)\| = \|u(\cdot, 0, 0)\|_{C([0, t_n]; X)}$$

Preliminaries 49

and
$$\lim_n \|u(t_n, 0, 0)\| = \sup_{t \in \mathbb{R}_+} \|u(t, 0, 0)\| = +\infty.$$

At this point, let us observe that, for each $t \in \mathbb{R}_+$, we have

$$\|u(t, 0, 0)\| \le (1 - e^{-\omega t}) \frac{\ell}{\omega} \left(\|u(\cdot, 0, 0)\|_{C([-\tau, t]; X)} + \frac{1}{\ell} \|f(0)\| \right).$$

As $u(s, 0, 0) = 0$ for each $s \in [-\tau, 0]$, this yields

$$\|u(t, 0, 0)\| \le (1 - e^{-\omega t}) \frac{\ell}{\omega} \left(\|u(\cdot, 0, 0)\|_{C([0, t]; X)} + \frac{1}{\ell} \|f(0)\| \right). \qquad (1.12.8)$$

Setting $t = t_n$ in the inequality (1.12.8) and dividing by $\|u(t_n, 0, 0)\|$ we get

$$1 \le \frac{\ell}{\omega} (1 - e^{-\omega t_n}) \left(1 + \frac{\|f(0)\|}{\ell \|u(t_n, 0, 0)\|} \right)$$

for each $n \in \mathbb{N}$. Passing to the limit for $n \to +\infty$ on both sides of this inequality, we get

$$1 \le \frac{\ell}{\omega} < 1.$$

This contradiction can be eliminated only if $u(\cdot, 0, 0)$ is bounded on $\mathbb{R}_+$. From (1.12.7) we deduce that $u(\cdot, 0, \varphi)$ is bounded on $\mathbb{R}_+$ and since $u(\cdot, 0, \varphi)$ is a fortiori bounded on $[-\tau, 0]$, this completes the proof. $\qquad \square$

1.12.2 The quasi-autonomous case

For the sake of completeness, we will say now a few words about the quasi-autonomous delay evolution equation

$$u'(t) \in Au(t) + f(t, u_t), \qquad t \in \mathbb{R}_+, \qquad (1.12.9)$$

where X is a Banach space, $A : D(A) \subseteq X \rightsquigarrow X$ is an ω-m-dissipative operator for some $\omega \in \mathbb{R}$, and $f : \mathbb{R}_+ \times \mathfrak{X} \to X$ is a function satisfying the Lipschitz condition with respect the second argument, i.e., there exists $\ell > 0$ such that

$$\|f(t, \varphi) - f(t, \psi)\| \le \ell \|\varphi - \psi\|_{\mathfrak{X}}$$

for each $t \in \mathbb{R}_+$ and $\varphi, \psi \in \mathfrak{X}$.

As in the autonomous case, we consider the family of nonlinear operators $\{\mathcal{A}(t) : D(\mathcal{A}(t)) \subseteq \mathfrak{X} \rightsquigarrow \mathfrak{X}; \ t \in \mathbb{R}_+\}$ defined by

$$\begin{cases} D(\mathcal{A}(t)) = \{\varphi \in \mathfrak{X}; \ \varphi' \in \mathfrak{X}, \ \varphi(0) \in D(A), \ \varphi'(0) \in A\varphi(0) + f(t, \varphi)\}, \\ \mathcal{A}(t)(\varphi) = \varphi', \ \text{for } \varphi \in D(\mathcal{A}(t)). \end{cases}$$

$$(1.12.10)$$

50 Delay Differential Evolutions Subjected to Nonlocal Initial Conditions

From (1.12.4), we easily deduce that

$$\overline{D(\mathcal{A}(t))} = \{\varphi \in \mathfrak{X}; \ \varphi(0) \in \overline{D(A)}\}$$

which means that $\overline{D(\mathcal{A}(t))}$ is independent of $t \in \mathbb{R}_+$. As in the autonomous case, we denote this set by

$$\mathcal{D} = \{\varphi \in \mathfrak{X}; \ \varphi(0) \in \overline{D(A)}\}. \tag{1.12.11}$$

The next result, essentially based on Theorem 1.11.1, is a very particular instance of a generation theorem valid in the fully nonautonomous setting due to Ghavidel [126, Theorem 3.3].

THEOREM 1.12.5 *Let X be a Banach space, let $A : D(A) \subseteq X \rightsquigarrow X$ and let $f : \mathbb{R}_+ \times \mathfrak{X} \to X$. Let $\omega \in \mathbb{R}$ and let us assume that*

(i) the operator A is ω-m-dissipative

(ii) the function f is globally Lipschitz with respect to the second argument, i.e., there exists $\ell > 0$ such that

$$\|f(t, \varphi) - f(t, \psi)\| \le \ell \|\varphi - \psi\|_{\mathfrak{X}}$$

for each $t \in \mathbb{R}_+$ and each $\varphi, \psi \in \mathfrak{X}$

(iii) for each $T > 0$ there exist a continuous function $h : [0, T] \to X$ and a monotone increasing function $\mu : \mathbb{R}_+ \to \mathbb{R}_+$ such that

$$\|f(t, \varphi) - f(s, \varphi)\| \le \|h(t) - h(s)\| \mu(\|\varphi\|_{\mathfrak{X}})$$

for each $s, t \in [0, T]$ and each $\varphi \in \mathfrak{X}$.

Let $\gamma = \max\{0, \ell - \omega\}$ and let $\mathcal{D}$ be given by (1.12.11). Then, we have the following:

(c_1) For each $t \in \mathbb{R}_+$, the operator $\mathcal{A}(t)$, given by (1.12.10), is γ-m-dissipative in $\mathfrak{X}$ and so, the family $\{\mathcal{A}(t) : D(\mathcal{A}(t)) \to D(\mathcal{A}(t)); \ t \in \mathbb{R}_+\}$ in (1.12.10), generates an evolution system

$$\{\mathcal{U}(t, s) : \mathcal{D} \to \mathcal{D}; \ 0 \le s \le t < +\infty\}$$

of type γ in $\mathcal{D}$.

(c_2) For each $(s, \varphi) \in \mathbb{R}_+ \times \mathcal{D}$, the function $u(\cdot, s, \varphi) : [s - \tau, +\infty) \to X$, defined by

$$u(t, s, \varphi) = \begin{cases} \varphi(t - s), & t \in [s - \tau, s], \\ [\mathcal{U}(t, s)\varphi](0), & t \in (s, +\infty), \end{cases} \tag{1.12.12}$$

satisfies

$$\mathcal{U}(t, s)\varphi = u_t(\cdot, s, \varphi) \tag{1.12.13}$$

for each $t \in [s, +\infty)$. Thus $u(\cdot, s, \varphi) : [s - \tau, +\infty) \to X$ is a C^0-solution of the problem (1.12.9) satisfying $u(t, s, \varphi) = \varphi(t - s)$ for $t \in [s - \tau, s]$.

Preliminaries 51

(c_3) *Conversely, if $s \in \mathbb{R}_+$, $\varphi \in \mathcal{D}$ and $u(\cdot, s, \varphi) : [s - \tau, +\infty) \to X$ is the C^0-solution of the problem (1.12.9) satisfying $u(t, s, \varphi) = \varphi(t - s)$ for each $t \in [s - \tau, s]$, then u is given by (1.12.12) and (1.12.13) holds true.*

REMARK 1.12.1 Clearly, whenever f is jointly globally Lipschitz on $\mathbb{R}_+ \times \mathfrak{X}$, it satisfies the hypothesis (iii).

As in the autonomous case, we introduce:

DEFINITION 1.12.2 The C^0-solution $u(\cdot, 0, \varphi) : [-\tau, +\infty) \to X$ of the problem (1.12.9) is called *stable* if for each $\varepsilon > 0$ there exists $\delta(\varepsilon, \varphi) > 0$ such that, for each $\psi \in \mathcal{D}$ satisfying $\|\varphi - \psi\|_{\mathfrak{X}} \le \delta(\varepsilon, \varphi)$, we have

$$\|u(t, 0, \varphi) - u(t, 0, \psi)\| \le \varepsilon$$

for each $t \in \mathbb{R}_+$.

The C^0-solution $u(\cdot, 0, \varphi) : [-\tau, +\infty) \to$ of (1.12.9) is called *globally uniformly asymptotically stable* if it is stable and, for each $\psi \in \mathcal{D}$, the unique C^0-solution, $u(\cdot, 0, \psi)$, of (1.12.9) with $u(t, 0, \psi) = \psi(t)$ for each $t \in [-\tau, 0]$, satisfies

$$\lim_{t \to +\infty} \|u(t, 0, \varphi) - u(t, 0, \psi)\| = 0.$$

Clearly, if $\ell \le \omega$, (c_1) in Theorem 1.12.5 implies the stability of C^0-solutions of the problem (1.12.9). We conclude this section with some direct consequences of Theorem 1.12.5.

The next asymptotic behavior result, inspired by Plant [219], is from Ghavidel [126, Proposition 4.3].

THEOREM 1.12.6 *Let $\omega > 0$ and let us assume that the conditions $(i) \sim (iii)$ in Theorem 1.12.5 are satisfied and, in addition, that $\ell < \omega$. Then there exists $K \ge 1$ and $\beta > 0$ such that, for each $\varphi, \psi \in \mathcal{D}$, we have*

$$\|\mathcal{U}(t, s)\varphi - \mathcal{U}(t, s)\psi\|_{\mathfrak{X}} \le K e^{-\beta(t-s)} \|\varphi - \psi\|_{\mathfrak{X}}$$

for each $0 \le s \le t < +\infty$.

THEOREM 1.12.7 *If the hypotheses of Theorem 1.12.6 are satisfied, then all global C^0-solutions of the problem (1.12.9) are bounded on $[-\tau, +\infty)$ and globally uniformly asymptotically stable in the sense of Definition 1.12.2.*

The proof of Theorem 1.12.7 is similar to that of Theorem 1.12.4.

1.13 Integral inequalities

The following lemma is a slight extension of the well-known Gronwall Lemma – see Vrabie [254, Lemma 1.5.2, p. 44] – allowing x and k to be only integrable.

52   *Delay Differential Evolutions Subjected to Nonlocal Initial Conditions*

LEMMA 1.13.1 *Let $m \in \mathbb{R}$ and let $z, k : [a, b) \to \mathbb{R}$ be measurable, $k(s) \geq 0$ a.e. for $s \in [a, b)$ and $k \in L^1(a, b)$. Let us assume that $s \mapsto k(s)z(s)$ is locally integrable on $[a, b)$ and*

$$z(t) \leq m + \int_a^t k(s)z(s)\, ds$$

for every $t \in [a, b)$. Then

$$z(t) \leq m e^{\int_a^t k(s)\, ds}$$

for every $t \in [a, t)$.

Below, we state a variant of Bellman's Inequality.

LEMMA 1.13.2 *Let $\alpha, z : [a, b) \to \mathbb{R}$ be two continuous functions, and let $\beta : [a, b) \to \mathbb{R}$ be measurable, $\beta(s) \geq 0$ a.e. for $s \in [a, b)$ and $\beta \in L^1(0, T)$. If*

$$z(t) \leq \alpha(t) + \int_a^t \beta(s)z(s)\, ds$$

for every $t \in [a, b)$, then

$$z(t) \leq \alpha(t) + \int_a^t \alpha(s)\beta(s)e^{\int_\tau^s \beta(\theta)\, d\theta}\, ds$$

for every $t \in [a, b)$.

See Vrabie [254, Problem 1.16, p. 48].

LEMMA 1.13.3 *Let $\mathcal{R} = C([-\tau, 0]; \mathbb{R})$ and let $y : [-\tau, +\infty) \to \mathbb{R}_+$ and $\alpha_0 : \mathbb{R}_+ \to \mathbb{R}_+$ be continuous functions with α_0 nondecreasing and let $\beta : \mathbb{R}_+ \to \mathbb{R}_+$ be locally integrable. If*

$$y(t) \leq \alpha_0(t) + \int_0^t \beta(s)\|y_s\|_{\mathcal{R}}\, ds \tag{1.13.1}$$

for each $t \in \mathbb{R}_+$, then

$$y(t) \leq \alpha(t) + \int_0^t \alpha(s)\beta(s)e^{\int_s^t \beta(\sigma)\, d\sigma}\, ds, \tag{1.13.2}$$

for each $t \in \mathbb{R}_+$, where
$$\alpha(t) = \|y_0\|_{\mathcal{R}} + \alpha_0(t)$$

for each $t \in \mathbb{R}_+$.

Preliminaries 53

Proof. Let $z : [-\tau, +\infty) \to \mathbb{R}_+$ be the function defined as

$$z(t) = \sup\{\|y_s\|_{\mathcal{R}};\ s \in [0,t]\} = \sup\{y(\theta);\ \theta \in [-\tau, t]\}.$$

Clearly

$$y(t) \le z(t)$$

for each $t \in \mathbb{R}_+$. Since y is continuous, for each $t \in \mathbb{R}_+$, there exists $s_t \in [-\tau, t]$ such that

$$z(t) = y(s_t).$$

We distinguish between two cases.

Case 1. If $s_t \in [-\tau, 0]$ and $y(s) < z(s)$ for each $s \in (0, t]$, then

$$z(t) = y(s_t) \le \|y_0\|_{\mathcal{R}} \le \alpha(t) + \int_0^t \beta(s)z(s)\,ds. \qquad (1.13.3)$$

Case 2. If $s_t \in (0, t]$, by (1.13.1), we have

$$z(t) = y(s_t) \le \alpha_0(s_t) + \int_0^{s_t} \beta(\sigma)\|y_\sigma\|_{\mathcal{R}}\,d\sigma$$

$$\le \alpha_0(t) + \|y_0\|_{\mathcal{R}} + \int_0^t \beta(\sigma)z(\sigma)\,d\sigma = \alpha(t) + \int_0^t \beta(\sigma)z(\sigma)\,d\sigma.$$

So, (1.13.3) holds true for each $t \in \mathbb{R}_+$. From Lemma 1.13.2, we get

$$z(t) \le \alpha(t) + \int_0^t \alpha(s)\beta(s)e^{\int_s^t \beta(\sigma)\,d\sigma}\,ds,$$

for each $t \in \mathbb{R}_+$. Since $y(t) \le z(t)$ for each $t \in \mathbb{R}_+$, the inequality above implies (1.13.2). The proof is complete. $\qquad\square$

1.14 Brezis–Browder ordering principle

In this section we include, without proof, a very general ordering principle, less restrictive than Zorn's Lemma, but extremely useful in applications because it can be applied in almost all situations encountered when dealing with "maximal elements."

Let $\mathcal{S}$ be a nonempty set. We recall that a *preorder on* $\mathcal{S}$ is a binary relation $\preceq \subseteq \mathcal{S} \times \mathcal{S}$ which is reflexive, i.e., $\xi \preceq \xi$ for each $\xi \in \mathcal{S}$, and transitive, i.e., $\xi \preceq \eta$ and $\eta \preceq \zeta$ imply $\xi \preceq \zeta$.

DEFINITION 1.14.1 Let $\mathcal{S}$ be a nonempty set, $\preceq \subseteq \mathcal{S} \times \mathcal{S}$ a preorder on $\mathcal{S}$, and let $\mathcal{N} : \mathcal{S} \to \mathbb{R} \cup \{+\infty\}$ be an increasing function. We say that $\overline{\xi} \in \mathcal{S}$ is an $\mathcal{N}$-*maximal element* if $\mathcal{N}(\xi) = \mathcal{N}(\overline{\xi})$, for all $\xi \in \mathcal{S}$ with $\overline{\xi} \preceq \xi$.

54   *Delay Differential Evolutions Subjected to Nonlocal Initial Conditions*

We may now formulate the Brezis–Browder Ordering Principle.

THEOREM 1.14.1 *Let $\mathcal{S}$ be a nonempty set, $\preceq \subseteq \mathcal{S} \times \mathcal{S}$ a preorder on $\mathcal{S}$ and let $\mathcal{N} : \mathcal{S} \to \mathbb{R} \cup \{+\infty\}$ be a given function. Suppose that*

(i) *each increasing sequence in $\mathcal{S}$ is bounded from above*

(ii) *the function $\mathcal{N}$ is increasing.*

Then, for each $\xi_0 \in \mathcal{S}$, there exists an $\mathcal{N}$-maximal element $\overline{\xi} \in \mathcal{S}$ satisfying $\xi_0 \preceq \overline{\xi}$.

For the proof of this fundamental result, see Brezis and Browder [52] or Cârjă, Necula and Vrabie [74, Theorem 2.1.1, p. 30].

1.15 Bibliographical notes and comments

Section 1.1. Theorem 1.1.1 and Corollary 1.1.1 are due to Mazur [181]. Theorem 1.1.2 was proved by Pettis [215]. The notion of uniformly convex space was introduced by Clarkson [83]. Theorem 1.1.3 is from Milman [183]. Theorem 1.1.4 is due to Kato [152].

Section 1.2. The extension of the Lebesque integral to vector-valued functions is due to Bochner [37]. The strong measurability necessary and sufficient condition in Theorem 1.2.1 was implicitly proved by Pettis [214]. Theorem 1.2.2 is due to Bochner [37]. Theorem 1.2.3 is a simple extension of the Lebesgue Dominated Convergence Theorem. See Lebesgue [167] and Dinculeanu [102]. Lemma 1.2.1 is essentially due to Fatou [114, pp. 375–376] and its proof can be found in Dunford and Schwartz [105, 19 Theorem, p. 152]. Theorem 1.2.4 is due to Bochner and Taylor [38]. It should be noted that, in the very same fundamental paper, they introduced the notion known nowadays as Banach space having the Radon–Nicodým property. Theorem 1.2.5 is due to Fubini [120]. For the proof, see Dunford and Schwartz [105, 9 Theorem, p. 190]. Theorem 1.2.6 was proved by Clarkson [83].

Section 1.3. Lemma 1.3.1, attributed to Krasnoselskii, is essentially based on Vitali's Theorem. See Dunford and Schwartz [105, Theorem 6, p. 122].

Section 1.4. Theorem 1.4.1 was established by Hausdorff [138, II, p. 312 and III, p. 313], Theorem 1.4.2 is due to Mazur [181], Theorem 1.4.3 to Krein and Šmulian [158]. Theorem 1.4.4 was proved by Schauder [232], Theorem 1.4.5 is due to Tychonoff [245] and Theorem 1.4.6 to Schaefer [233], the latter being a specific form of the famous Leray–Schauder Alternative. See Granas and Dugundji [132, Theorem 5.4, p. 124]. Theorem 1.4.7 is a consequence of Mazur Theorem 1.4.2; Theorem 1.4.8 is due to Eberlein [109] and

Preliminaries 55

Šmulian [241], while Theorem 1.4.9 follows from Theorem 1.4.8 combined with the following:

THEOREM. (Kakutani) *A Banach space is reflexive if and only if its closed unit ball is weakly compact.*

See Kakutani [150] or Hille and Philips [143, Theorem 2.10.2, p. 38]. For $X = \mathbb{R}$ and μ finite, Theorem 1.4.11 is due to Dunford [104], Theorem 1.4.12 to Diestel [99]. For other characterizations of weak compactness in $L^1(\Sigma; \mu, X)$, see Benabdellah and Castaing [27] and Diestel, Ruess and Schachermayer [100]. Theorem 1.4.13 is from Vrabie [255].

Section 1.5. As far as multifunctions are concerned, we begin by noting the pioneering works of Moore [189] and Vasilesco [246]. The concept of the u.s.c. multivalued function – Definition 1.5.2 – was introduced independently by Bouligand [45] and Kuratowski [162]. Lemma 1.5.1, due to Berge [34, Théorème 3, p. 116], is a Weierstrass–type result whose proof is closely related to its classical counterpart. Theorem 1.5.1 is a variant of a closed graph–type result in Castaing and Valadier [72, Theorem VI-4, p. 170]. Lemma 1.5.2 is perhaps known but, in this form, we did not find it in the literature. The property of a multifunction of being almost strongly-weakly u.s.c. – see Definition 1.5.3 – is a Scorza Dragoni–type property. See Scorza Dragoni [234], Zygmunt [271] and the references therein. Proposition 1.5.1 is a specific case of a general result due to Berge [34, Théorème 7, p. 117]. Lemma 1.5.3 is from Vrabie [255] and extends a related result in Paicu [201]. Lemma 1.5.4, inspired by Cârjă, Necula and Vrabie [74, Problem 2.6.1, p. 46], is a simplified version of Lemma 5.1 in Vrabie [256], while Lemma 1.5.5 is new. For related results, see the detailed survey monograph of Appell, De Pascale, Thái and Zabreĭko [11]. Theorem 1.5.2 is essentially due to Ky Fan [113] and Glicksberg [127] and extends the classical Kakutani Fixed-Point Theorem [151]. Theorem 1.5.3 is a simple consequence of Theorem 1.5.2, while Theorem 1.5.4 extends to the multivalued case a result due to Arino, Gautier and Penot [12]. For a systematic study of fixed points for multivalued mappings, see Górniewicz [131].

Section 1.6. Theorems 1.6.1, 1.6.2 and 1.6.3 are classical and may be found in many monographs, for instance, Hille and Phillips [143], Pazy [212] and Vrabie [253]. Theorem 1.6.5 is due to Hille [142] and to Yosida [267]. Theorem 1.6.6 was proved independently by Feller [115], Miyadera [187] and Phillips [217] and is an extension of Theorem 1.6.5. The notion of m-dissipative operator was introduced by Lumer and Phillips [177] who also proved Theorems 1.6.7 and 1.6.8.

Section 1.7. Theorem 1.7.1 is classical and extends to infinite dimensional Banach spaces its finite-dimensional counterpart due to Peano [213]. The notion of a mild solution was introduced by Browder [54]. Theorem 1.7.2 is a linear version of a general compactness result established by Vrabie [250]. Theorem 1.7.4 is a consequence of Theorem 1.7.2, while Lemma 1.7.1 is essentially due to Becker [25]. Theorem 1.7.4 is due to Baras, Hassan and Verron [20]

56 *Delay Differential Evolutions Subjected to Nonlocal Initial Conditions*

and uses some arguments developed by Pazy [211]. Lemma 1.7.2 is a slight extension of Lemma 1.7.1.

Section 1.8. The semi-inner products were introduced by Lumer [176], while, as we already have mentioned, the notion of the linear dissipative operator was defined by Lumer and Phillips [177]. The concept of DS-*solution* – C^0-*solution* in our terminology – was introduced by Takahashi [242] in the case of autonomous equations with $\omega = 0$. Benilan [32] introduced the notion of the *integral solution* as a continuous function satisfying the inequality in Theorem 1.8.3 and showed that, in the case of dissipative operators, DS-solutions and integral solutions are one and the same mathematical object, called a *mild solution* by Crandall [88], as in the linear case. It should be noted that this concept was defined earlier in a Hilbert space frame by Benilan and Brezis [33]. Theorem 1.8.1 is due to Benilan [32], Crandall and Evans [89] and Kobayashi [156]. Proposition 1.8.3 is a simple copy of its linear counterpart. Theorem 1.8.2, Proposition 1.8.5 and Theorem 1.8.5 are from Benilan [32]. The semigroup in Proposition 1.8.4 is simply the semigroup generated by A via the Crandall and Liggett Exponential Formula (1.8.6). See Crandall and Liggett [90]. For other related results, see Kobayashi [156] and Takahashi [242]. Lemma 1.8.1 is a consequence of Proposition 1.8.5 and Lemma 1.8.2 is due to Brezis [50]. Theorems 1.8.4 and 1.8.5 were proved by Vrabie [250], while Theorem 1.8.6 is due to Baras [19]. A specific form was established independently by Vrabie [247]. Theorem 1.8.7 is from Mitidieri and Vrabie [185] and is an extension of a previous compactness result in Vrabie [249]. Theorem 1.8.8 is from Vrabie [247]. For details on evolution equations governed by m-dissipative operators, see the monographs Barbu [21], [22], Miyadera [187], Lakshmikantham and Leela [163] and Showalter [238].

Section 1.9. The spaces $W^{m,p}(\Omega)$ were introduced by Sobolev [239] who also proved the continuity of the imbeddings in Theorem 1.9.1. See also Sobolev [240]. The compactness of the imbeddings was established by Rellich [223] for $p = 2$ and by Kondrašov [157] for $p > 1$. The case $p = 1$ was closed by Gagliardo [121]. Theorems 1.9.2 and 1.9.3, which are consequences of Theorem 1.6.5 combined with a compactness argument due to Pazy [210], are classical. Theorems 1.9.4 and 1.9.5 are also classical and can be obtained also from Theorem 1.6.5. Theorem 1.9.6 is due (i) to Brezis and Strauss [53], (ii) to Badii, Díaz and Tesei [16] and (iii) to Cârjă, Necula and Vrabie [74, Theorem 1.7.9, p. 22]. Theorem 1.9.7 is due to Díaz and Vrabie [96]. Lemma 1.9.2 is simply an extension of Lemma 11.1 in Cârjă, Necula and Vrabie [75]. See also Cârjă, Necula and Vrabie [74, Lemma 13.5.11, p. 273].

Section 1.10. Theorem 1.10.2 is due to Minty [184] and Moreau [190], while Theorem 1.10.3, which is a Hilbert space nonlinear version of a regularity result referring to analytic C_0-semigroups, was established by Brezis [51].

Section 1.11. Theorem 1.11.1 is a particular case of a generation theorem in Pavel [208, Theorem 3.6, p. 35]. For other basic results concerning

Preliminaries 57

generators of evolution systems, see the fundamental papers of Crandall and Pazy [91], Pavel [207] and K. Kobayashi, Y. Kobayashi and Oharu [155].

Section 1.12. Initial-value problems for delay equations in finite dimensional spaces were considered by Hale [136] who showed that a linear differential equation with delay of the form

$$\begin{cases} u'(t) = Au_t, & t \in \mathbb{R}_+, \\ u(t) = \varphi(t), & t \in [-\tau, 0], \end{cases}$$

in $\mathbb{R}^d$, $d \geq 1$, gives birth to a C_0-semigroup defined in a suitably defined function space, i.e., in the history space $C([-\tau, 0]; \mathbb{R}^d)$. This remark proved of great importance in the study of delay linear, or even nonlinear, evolution equations in infinite dimensional Banach spaces and was the starting point of an elegant and powerful abstract theory on this subject. For further details on this kind of linear problems, see Hale [136, Section 7.1, pp. 166–167] or Vrabie [253, Section 6.5, pp. 147–149].

The infinite dimensional case was considered successively by Webb [265], [266] and Fitzgibbon [116]. More precisely, Webb [265] considers the initial-value problem for the delay differential equation

$$\begin{cases} u'(t) = f(u_t), & t \in \mathbb{R}_+, \\ u(t) = \varphi(t), & t \in [-\tau, 0], \end{cases}$$

where f is Lipschitz continuous and shows that it can be transformed into a non-delay evolution Cauchy problem in a function space, i.e., in $\mathcal{X}$. Subsequently, Webb [266] analyzes the more general case

$$\begin{cases} u'(t) = Au(t) + f(u_t), & t \in \mathbb{R}_+, \\ u(t) = \varphi(t), & t \in [-\tau, 0], \end{cases}$$

with $A: D(A) \subseteq X \rightsquigarrow X$ m-dissipative and f Lipschitz continuous and shows that, also in this general case, it can be transformed into a non-delay evolution Cauchy problem in $\mathcal{X}$. He also proved that the operator $\mathcal{A}$ is m-dissipative and the delay Cauchy problem above can be rewritten as

$$\begin{cases} U'(t) = \mathcal{A}U(t), & t \in \mathbb{R}_+, \\ U(0) = \varphi. \end{cases}$$

Moreover, he showed that, if either A is everywhere defined and continuous or X is a Hilbert space, then each strong solution of the problem above – if any – defines a strong solution of the former delay equation. As one can easily see, this is the converse implication of that observed by Hale [136] and proves extremely useful in applications.

Further extensions are due to Ruess [227] who obtained the solution via a suitably defined nonlinear evolution system. His results cover the more general case of nonlinear time-dependent operators A and Lipschitz time-dependent

58   *Delay Differential Evolutions Subjected to Nonlocal Initial Conditions*

perturbations F. For more details on this topic in the case in which A is linear, see the monograph of Bátkai and Piazzera [24].

Theorem 1.12.1 is essentially due to Webb [266]. For related results, the reader is referred to Plant [218], Ruess [227], Ruess and Summers [230] and Ghavidel [125], [126].

Results referring to the asymptotic behavior of solutions are due to Brewer [48], Plant [219], Ruess [228] and Webb [266]. Theorem 1.12.2 is from Plant [219, Proposition 2.1]. Previous results referring to classical or strong solutions are obtained by Flashka and Leitman [117] and Plant [218] in the specific case $A \equiv 0$. For the general case, see Ruess [229]. Theorem 1.12.3 is also from Plant [219], but the idea of using an equivalent norm on $\mathcal{X}$ goes back to Bielecki [35]. A similar stability result was obtained by Webb [265], under the additional assumption that X is a Hilbert space. Theorem 1.12.5 is a particular instance of a general result of Ghavidel [126] who considered the general case in which A is replaced with a family of $\omega(t)$-m-dissipative operators $\{A(t); \ t \in \mathbb{R}_+\}$, where $\omega : \mathbb{R}_+ \to \mathbb{R}$ and the Lipschitz constant of f with respect to its second argument depends on $t \in \mathbb{R}_+$. For this general case, Dyson and Bressan [106] have proved that $\overline{D(\mathcal{A}(t))}$ is independent of $t \in \mathbb{R}_+$. Theorems 1.12.6 and 1.12.6 are from Ghavidel [125, Proposition 3.4, p. 73] and [126]. It should be emphasized that the t-continuity condition (iii) in Theorem 1.12.5, as the proof of Ghavidel [125, Proposition 3.4, p. 73] shows, is essential in obtaining the global exponential stability result in Theorems 1.12.6.

Section 1.13. Lemma 1.13.1 is a slight extension of the Gronwall Lemma, see Vrabie [254, Lemma 1.5.2, p. 44], and covers the more general case in which the functions involved are only measurable. Lemma 1.13.2 is an extension of the Bellman Integral Inequality [26]. See also Vrabie [254, Problem 1.16, p. 48]. Lemma 1.13.3 is from Burlică and Roşu [58]. A differential version of the integral inequality in Lemma 1.13.3, in the constant coefficients case, was proved earlier by Halanay [134, Lemma, p. 378]. For an extension of the latter result to vector-valued functions, see Jia, Erbe and Mert [148].

Section 1.14. Theorem 1.14.1, due Brezis and Browder [52], is essentially based on a less restrictive axiom than the Axiom of Choice, i.e., the Axiom of Dependent Choice stated below:

Let $\mathcal{S}$ be a nonempty set and let $\mathcal{R} \subseteq \mathcal{S} \times \mathcal{S}$ be a binary relation with the property that, for each $\xi \in \mathcal{S}$, the set $\{\eta \in \mathcal{S}; \ \xi \mathcal{R} \eta\}$ is nonempty. Then, for each $\xi \in \mathcal{S}$, there exists a sequence $(\xi_k)_k$ in $\mathcal{S}$ such that $\xi_0 = \xi$ and $\xi_k \mathcal{R} \xi_{k+1}$ for each $k \in \mathbb{N}$.

This fundamental ordering principle of Brezis and Browder [52] is a strictly less restrictive substitute of Zorn's Lemma. Therefore, we hope that it will find its right place in any textbook or monograph on Ordinary Differential Equations. For more details and comments on this subject, see Cârjă, Necula and Vrabie [74, Section 2.1, pp. 310–311].

Chapter 2

Local Initial Conditions

Overview

In this chapter, we establish some sufficient conditions for the local and global existence, uniqueness and uniform asymptotic stability of a C^0-solution for the nonlinear delay differential evolution equation with local initial data

$$\begin{cases} u'(t) \in Au(t) + f(t, u_t), & t \in [\sigma, T], \\ u(t) = \psi(t - \sigma), & t \in [\sigma - \tau, \sigma], \end{cases}$$

where $\tau \geq 0$, $\sigma \in \mathbb{R}$, X is a Banach space, $\mathcal{X} = C([-\tau, 0]; X)$, the operator $A: D(A) \subseteq X \rightsquigarrow X$ is the infinitesimal generator of a nonlinear semigroup of contractions, the function $f : [\sigma, +\infty) \times \mathcal{X} \to X$ is continuous, and the initial history $\psi \in \mathcal{X}$ satisfies $\psi(0) \in \overline{D(A)}$. The case $A \equiv 0$ is also considered.

2.1 An existence result for ODEs with delay

We begin with some notations needed later. If X is a Banach space, D is a closed subset in X and I is an interval, $C_b(I; X)$ denotes the space of all bounded and continuous functions from I to X, equipped with the sup-norm $\| \cdot \|_{C_b(I;X)}$, while $C_b(I; D)$ denotes the closed subset in $C_b(I; X)$ consisting of all elements $u \in C_b(I; X)$ satisfying $u(t) \in D$ for each $t \in I$. Further, $C([a,b]; X)$ stands for the space of all continuous functions from $[a,b]$ to X endowed with the sup-norm $\| \cdot \|_{C([a,b];X)}$ and $C([a,b]; D)$ is the closed subset of $C([a,b]; X)$ containing all $u \in C([a,b]; X)$ with $u(t) \in D$ for each $t \in [a,b]$. If $\tau \geq 0$, $\mathcal{X} = C([-\tau, 0]; X)$. If $\sigma \in \mathbb{R}$, $u \in C([\sigma - \tau, +\infty); X)$ and $t \in [\sigma, +\infty)$, then $u_t \in \mathcal{X}$ is defined by

$$u_t(s) = u(t + s)$$

for each $s \in [-\tau, 0]$. If $\tau = 0$, $\mathcal{X}$ identifies with X and u_t identifies with $u(t)$.

59

60   *Delay Differential Evolutions Subjected to Nonlocal Initial Conditions*

Let X be a Banach space, let $\tau \geq 0$, let $\sigma \in \mathbb{R}$, and let $f \colon [\,\sigma, +\infty)\times \mathfrak{X} \to X$ be a continuous function and $\psi \in \mathfrak{X}$.

We consider the nonlinear delay differential evolution equation with local initial data

$$\begin{cases} u'(t) = f(t, u_t), & t \in [\,\sigma, T\,], \\ u(t) = \psi(t - \sigma), & t \in [\,\sigma - \tau, \sigma\,]. \end{cases} \tag{2.1.1}$$

DEFINITION 2.1.1 By a *classical solution* of the problem (2.1.1) on $[\,\sigma - \tau, T\,]$ we mean a function $u \colon [\,\sigma - \tau, T\,] \to X$ which is of class C^1 on $[\,\sigma, T\,]$, satisfies $u'(t) = f(t, u_t)$ for each $t \in [\,\sigma, T\,]$ and $u(t) = \psi(t - \sigma)$ for each $t \in [\,\sigma - \tau, \sigma\,]$. By a *classical solution* of the problem (2.1.1) on $[\,\sigma, +\infty)$ we mean a function $u \colon [\,\sigma - \tau, +\infty) \to X$ which is a classical solution of (2.1.1) on $[\,\sigma - \tau, T\,]$ for each $T > \sigma$.

2.1.1 A Weierstrass-type boundedness result

We need the following simple boundedness lemma which slightly extends the celebrated Weierstrass Theorem on continuous real functions defined on compact sets.

LEMMA 2.1.1 *Let X and Y be two Banach spaces, let $C \subseteq Y$ be a compact set and let $f \colon [\,\sigma, T\,] \times Y \to X$ be a function which is continuous on $[\,\sigma, T\,] \times C$. Then there exist $r > 0$ and $M > 0$ such that*

$$\|f(t, y)\| \leq M$$

for each $t \in [\,\sigma, T\,]$ and $y \in \cup_{\eta \in C} D(\eta, r)$.

Proof. We proceed by contradiction. So let us assume that even though C is compact and f is continuous on $[\,\sigma, T\,] \times C$, there exist two sequences $(r_n)_n$ and $(M_n)_n$ in $(0, +\infty)$ with $\lim_n r_n = 0$ and $\lim_n M_n = +\infty$ and three sequences $(t_n)_n$ in $[\,\sigma, T\,]$, $(\eta_n)_n$ in C and $(y_n)_n$ in Y satisfying

$$\|\eta_n - y_n\| \leq r_n$$

and

$$\|f(t_n, y_n)\| > M_n$$

for each $n \in \mathbb{N}$. Since both $[\,\sigma, T\,]$ and C are compact, we can assume without loss of generality that there exist $t \in [\,\sigma, T\,]$ and $\eta \in C$ such that $\lim_n t_n = t$ and $\lim_n \eta_n = \eta$. From the preceding inequalities and the continuity of f at (t, η), it follows that $\lim_n y_n = \eta$ and $\|f(t, \eta)\| \geq +\infty$, which is a contradiction. This contradiction can be eliminated only if the conclusion of the lemma holds true. $\qquad\square$

Local Initial Conditions 61

2.1.2 The main local existence theorem

THEOREM 2.1.1 *If* $f : [\sigma, +\infty) \times \mathfrak{X} \to X$ *is continuous and compact, then for each* $\psi \in \mathfrak{X}$ *there exists* $T = T(\psi) > \sigma$ *such that the delay initial-value problem* (2.1.1) *has at least one classical solution* $u : [\sigma - \tau, T] \to X$.

Proof. Let $\psi \in \mathfrak{X}$, and let $T > \sigma$ and $r > 0$ be arbitrary but fixed. Let

$$C_\psi([\sigma - \tau, T]; X) = \{v \in C([\sigma - \tau, T]; X);\ v(t) = \psi(t - \sigma),\ t \in [\sigma - \tau, \sigma]\}.$$

Let us define the function $\widetilde{\psi} \in C([\sigma - \tau, T]; X)$ by

$$\widetilde{\psi}(t) = \begin{cases} \psi(t - \sigma), & t \in [\sigma - \tau, \sigma], \\ \psi(0), & t \in (\sigma, T], \end{cases} \tag{2.1.2}$$

which clearly belongs to $C_\psi([\sigma - \tau, T]; X)$ and let

$$K = \{v \in C_\psi([\sigma - \tau, T]; X);\ \|v(t) - \widetilde{\psi}(t)\| \le r,\ \forall t \in [\sigma - \tau, T]\}.$$

Obviously, for each $v \in K$, we have $\|v_t - \widetilde{\psi}_t\|_{\mathfrak{X}} \le r$, for all $t \in [\sigma, T]$.

Now, let us remark that $\{\widetilde{\psi}_t;\ t \in [\sigma, T]\}$ is compact in $\mathfrak{X}$. Indeed, since $\widetilde{\psi}([\sigma - \tau, T])$ is compact, it follows that $\{\widetilde{\psi}_t;\ t \in [\sigma, T]\}$ has relatively compact cross sections. As $\widetilde{\psi}$ is uniformly continuous being continuous on a compact interval, we deduce that the family $\{\widetilde{\psi}_t;\ t \in [\sigma, T]\}$ is equicontinuous. Thanks to Arzelà–Ascoli Theorem 1.4.10, we conclude that $\{\widetilde{\psi}_t;\ t \in [\sigma, T]\}$ is relatively compact in $\mathfrak{X}$. Since it is also closed, we conclude that it is compact in $\mathfrak{X}$. Thus Lemma 2.1.1 applies and so, diminishing $r > 0$ if necessary, we may assume that there exists $M > 0$ such that

$$\|f(t, v_t)\| \le M \tag{2.1.3}$$

for each $v \in K$ and each $t \in [\sigma, T]$.

Since $\widetilde{\psi}$ given by (2.1.2) is in K, the latter is nonempty. Furthermore, K is closed and convex in $C([\sigma - \tau, T]; X)$.

Now, let us define the operator $Q : K \to C([\sigma - \tau, T]; X)$ by

$$Q(v) = u,$$

for each $v \in K$, where u is given by

$$\begin{cases} u(t) = \widetilde{\psi}(\sigma) + \displaystyle\int_\sigma^t f(s, v_s)\, ds, & t \in [\sigma, T], \\ u(t) = \psi(t - \sigma), & t \in [\sigma - \tau, \sigma]. \end{cases}$$

Clearly, u is a classical solution of the problem (2.1.1) if and only if u is a fixed point for the operator Q. So, to complete the proof, it suffices to show that, if $T > \sigma$ is "small enough," the operator Q maps K into itself, is continuous and compact, and then to apply the Schauder Fixed-Point Theorem 1.4.4.

One may easily see that we can diminish $T > \sigma$, if necessary, in order to have
$$(T - \sigma)M \leq r,$$
where $M > 0$ satisfies (2.1.3). With $T > \sigma$ fixed as above, from (2.1.3), we get
$$\|Q(v)(t) - \widetilde{\psi}(t)\| = \|Q(v)(t) - \psi(0)\| \leq \int_\sigma^T \|f(s, v_s)\|\, ds \leq (T - \sigma)M \leq r$$
for each $v \in K$ and each $t \in [\sigma, T]$. On the other hand, by (2.1.2), we get
$$\|Q(v)(t) - \widetilde{\psi}(t)\| = \|\psi(t - \sigma) - \psi(t - \sigma)\| = 0$$
for each $t \in [\sigma - \tau, \sigma]$. Thus $Q(K) \subseteq K$. Moreover
$$\|Q(v)(t) - Q(w)(t)\| \leq \int_\sigma^t \|f(s, v_s) - f(s, w_s)\|\, ds$$
for each $v, w \in K$ and each $t \in [\sigma, T]$. On the other hand $Q(v)(t) = Q(w)(t)$ for each $t \in [\sigma - \tau, \sigma]$ and f is continuous. So, the above inequality proves that Q is continuous from K to K in the norm topology of $C([\sigma - \tau, T]; X)$. Next, since by hypothesis f is compact and K is bounded in $C_\psi([\sigma - \tau, T]; X)$, by Lemma 1.7.1 it follows that, for each $t \in [\sigma - \tau, T]$, the cross sections of $Q(K)$ at t, i.e., $Q(K)(t) = \{Q(v)(t);\ v \in K\}$ is relatively compact in X. On the other hand, in view of (2.1.3), we have
$$\|Q(v)(t) - Q(v)(s)\| \leq M|t - s|$$
for each $v \in K$ and each $t, s \in [\sigma - \tau, T]$. Thus $Q(K)$ is equicontinuous on $[\sigma - \tau, T]$. By Arzelà–Ascoli Theorem 1.4.10, it follows that $Q(K)$ is relatively compact in $C([\sigma - \tau, T]; X)$. So, we are in the hypotheses of the Schauder Fixed-Point Theorem 1.4.4, from which it follows that Q has at least one fixed point $u \in K$. But, as we already have mentioned, u is a classical solution of the problem (2.1.1) and this completes the proof. $\qquad\square$

2.2 An application to abstract hyperbolic problems

We will derive next an abstract result regarding delay initial-value problems governed by compact perturbations of infinitesimal generators of C_0-groups. So, let us consider the delay initial-value problem in the Banach space X

$$\begin{cases} u'(t) = Au(t) + F\left(t, \int_{-\tau}^0 u(t + s)\, ds\right), & t \in [0, T], \\ u(t) = \varphi(t), & t \in [-\tau, 0]. \end{cases} \tag{2.2.1}$$

Local Initial Conditions 63

2.2.1 The main abstract result

THEOREM 2.2.1 *Let $A : D(A) \subseteq X \to X$ be the infinitesimal generator of a C_0-group of linear operators $\{G(t) : X \to X; \ t \in \mathbb{R}\}$, let $\mathcal{X} = C([-\tau, 0]; X)$ and let $F : \mathbb{R}_+ \times \mathcal{X} \to X$ be a continuous and compact function. Then, for each $\varphi \in \mathcal{X}$, there exists $T = T(\varphi) > 0$ such that the delay evolution initial-value problem (2.2.1) has at least one mild solution defined on $[0, T]$.*

Proof. We will show that the problem (2.2.1) can be transformed into an ordinary delay differential equation of the form

$$\begin{cases} u'(t) = f(t, u_t), & t \in [0, T], \\ u(t) = \psi(t), & t \in [-\tau, 0], \end{cases} \tag{2.2.2}$$

where $f : \mathbb{R}_+ \times \mathcal{X} \to X$ is continuous and compact, and then we will make use of the infinite-dimensional variant of the Peano Local Existence Theorem 2.1.1. First, let us recall that $u : [-\tau, T] \to X$ is a mild solution of the problem (2.2.1) if and only if u satisfies

$$u(t) = \begin{cases} \varphi(t), & t \in [-\tau, 0), \\ G(t)\varphi(0) + \displaystyle\int_0^t G(t-s) F\left(s, \int_{-\tau}^0 u(s+\theta)\,d\theta\right) ds, & t \in [0, T]. \end{cases}$$

But $G(t)$ is invertible and so we get

$$G^{-1}(t)u(t) = G^{-1}(t)\varphi(t), \quad t \in [-\tau, 0),$$

$$G^{-1}(t)u(t) = \varphi(0) + \int_0^t G^{-1}(s) F\left(s, \int_{-\tau}^0 u(s+\theta)\,d\theta\right) ds,$$

for $t \in [0, T]$. For $t \in [-\tau, T]$, we denote $G^{-1}(t)u(t)$ by $U(t)$. Then, the integral equation above takes the equivalent form

$$U(t) = \varphi(0) + \int_0^t G^{-1}(s) F\left(s, \int_{-\tau}^0 G(s+\theta)U(s+\theta)\,d\theta\right) ds$$

for $t \in [0, T]$. Now, let us define the function $f : \mathbb{R}_+ \times \mathcal{X} \to X$ by

$$f(t, v) = G^{-1}(t) F\left(t, \int_{-\tau}^0 G(t+\theta)v(\theta)\,d\theta\right)$$

each $(t, v) \in \mathbb{R}_+ \times \mathcal{X}$. So, u is a mild solution of the problem (2.2.1) if and only if U is a classical solution of the initial-value problem (2.2.2), where f is defined as above and $\psi(t) = G^{-1}(t)\varphi(t)$ for each $t \in [-\tau, 0]$. Since F is continuous and compact, it readily follows that f enjoys the very same properties. So, (2.2.2) can be handled by Theorem 2.1.1. $\qquad\square$

64 Delay Differential Evolutions Subjected to Nonlocal Initial Conditions

2.2.2 A semilinear wave equation with delay

Here we will analyze an example showing that Theorem 2.1.1 is very suitable in the study of second-order semilinear hyperbolic problems on bounded domains in $\mathbb{R}^d$, $d \geq 1$. It is interesting to note that this example proves that such problems are, in some sense, of "finite-dimensional" nature.

So, let Ω be a nonempty bounded and open subset in $\mathbb{R}^d$, $d \geq 1$, with C^1 boundary Σ, let $Q_+ = \mathbb{R}_+ \times \Omega$, $Q_\tau = [-\tau, 0] \times \Omega$, $\Sigma_+ = \mathbb{R}_+ \times \Sigma$, and let us consider the initial-value problem for the wave equation with delay:

$$\begin{cases} \dfrac{\partial^2 u}{\partial t^2}(t, x) = \Delta u(t, x) + h\left(t, \displaystyle\int_{-\tau}^0 u(t + s, x)\, ds\right), & \text{in } Q_+, \\[2ex] u(t, x) = 0, & \text{on } \Sigma_+, \\[2ex] u(t, x) = \psi_1(t)(x), & \text{in } Q_\tau, \\[2ex] \dfrac{\partial u}{\partial t}(t, x) = \psi_2(t)(x), & \text{in } Q_\tau. \end{cases} \tag{2.2.3}$$

The auxiliary result below will prove useful in the sequel.

LEMMA 2.2.1 *Let $h : \mathbb{R}_+ \times \mathbb{R} \to \mathbb{R}$ be a continuous function for which there exist $\widetilde{\ell} > 0$, $\widetilde{m} \geq 0$ such that*

$$|h(t, y)| \leq \widetilde{\ell}|y| + \widetilde{m}$$

for all $(t, y) \in \mathbb{R}_+ \times \mathbb{R}$. Then the superposition operator h_2 associated with h is well-defined on $\mathbb{R}_+ \times L^2(\Omega)$ in the sense of Definition 1.3.1, is both continuous from $\mathbb{R}_+ \times L^2(\Omega)$ to $L^2(\Omega)$, and continuous and compact from $\mathbb{R}_+ \times H_0^1(\Omega)$ to $L^2(\Omega)$.

Proof. The fact that h_2 is well-defined on $\mathbb{R}_+ \times L^2(\Omega)$ and is continuous from $\mathbb{R}_+ \times L^2(\Omega)$ to $L^2(\Omega)$, follows from *(iii)* in Lemma 1.3.1. The continuity from $\mathbb{R}_+ \times H_0^1(\Omega)$ to $L^2(\Omega)$ follows from the continuity from $\mathbb{R}_+ \times L^2(\Omega)$ to $L^2(\Omega)$ and the continuity of the embedding $H_0^1(\Omega) \subseteq L^2(\Omega)$. We will prove next that h_2 is compact from $\mathbb{R}_+ \times H_0^1(\Omega)$ to $L^2(\Omega)$. So, let $((t_n, u_n))_n$ be an arbitrary bounded sequence in $\mathbb{R}_+ \times H_0^1(\Omega)$. Clearly $\{t_n; \ n \in \mathbb{N}\}$ is relatively compact in $\mathbb{R}_+$. By Theorem 1.9.1, we know that $H_0^1(\Omega)$ is compactly embedded in $L^2(\Omega)$, and so the set $\{u_n; \ n \in \mathbb{N}\}$ is relatively compact in $L^2(\Omega)$. Recalling that the operator h_2 is continuous from $\mathbb{R}_+ \times H_0^1(\Omega)$ to $L^2(\Omega)$, it follows that the set

$$\{h_2(t_n, u_n); \ n \in \mathbb{N}\}$$

is relatively compact in $L^2(\Omega)$. Thus h_2 is a compact operator from $\mathbb{R}_+ \times H_0^1(\Omega)$ to $L^2(\Omega)$ and this completes the proof. $\qquad\square$

Let us denote both $\mathcal{H}_0^1 = C([-\tau, 0]; H_0^1(\Omega))$ and $\mathcal{X}_2 = C([-\tau, 0]; L^2(\Omega))$.

Local Initial Conditions   65

THEOREM 2.2.2 *Let Ω be a nonempty bounded and open subset in $\mathbb{R}^d$, $d \geq 1$, with C^1 boundary Σ, let $\tau \geq 0$, let $h : \mathbb{R}_+ \times \mathbb{R} \to \mathbb{R}$, and let $\psi_1 \in \mathcal{H}_0^1$ and $\psi_2 \in X_2$. In addition, let us assume that*

(h_1) *h is continuous and there exist $\widetilde{\ell} > 0$, $\widetilde{m} \geq 0$ such that*

$$|h(t, y)| \leq \widetilde{\ell}|y| + \widetilde{m}$$

for all $(t, y) \in \mathbb{R}_+ \times \mathbb{R}$.

Then there exists $T = T(\psi_1, \psi_2) > 0$ such that the problem (2.2.3) has at least one mild solution $u \in \mathcal{H}_0^1$ with $\dfrac{\partial u}{\partial t} \in X_2$.

REMARK 2.2.1 We note that the notion of a mild solution in this context refers to the fact that, as we will see later in the proof of Theorem 2.2.2, the problem (2.2.3) rewrites as an abstract semilinear evolution equation in the product space

$$X = \begin{pmatrix} H_0^1(\Omega) \\ \times \\ L^2(\Omega) \end{pmatrix}. \tag{2.2.4}$$

So, by a mild solution of (2.2.3) we mean a mild solution of the corresponding semilinear evolution equation in the product space X.

Proof. It is easy to see that (2.2.3) can be rewritten as a first-order system of partial differential equations of the form

$$\begin{cases} \dfrac{\partial u}{\partial t}(t, x) = v(t, x), & \text{in } Q_+, \\[2mm] \dfrac{\partial v}{\partial t}(t, x) = \Delta u(t, x) + h\left(t, \displaystyle\int_{-\tau}^{0} u(t + s, x)\, ds\right), & \text{in } Q_+, \\[2mm] u(t, x) = 0, & \text{on } \Sigma_+, \\[2mm] u(t, x) = \psi_1(t)(x), & \text{in } Q_\tau, \\[2mm] v(t, x) = \psi_2(t)(x), & \text{in } Q_\tau. \end{cases} \tag{2.2.5}$$

The product space X, defined by (2.2.4), endowed with the natural inner product

$$\left\langle \begin{pmatrix} u \\ v \end{pmatrix}, \begin{pmatrix} \widetilde{u} \\ \widetilde{v} \end{pmatrix} \right\rangle = \int_\Omega \nabla u(x) \cdot \nabla \widetilde{u}(x)\, dx + \int_\Omega v(x)\widetilde{v}(x)\, dx$$

for each $\begin{pmatrix} u \\ v \end{pmatrix}, \begin{pmatrix} \widetilde{u} \\ \widetilde{v} \end{pmatrix} \in X$, is a real Hilbert space. Clearly, (2.2.5) can be rewritten as an abstract evolution equation subjected to nonlocal initial

66 *Delay Differential Evolutions Subjected to Nonlocal Initial Conditions*

conditions of the form (2.2.1) in the space X, where A, F and φ are defined as follows.

First, let us define the linear operator $A : D(A) \subseteq X \to X$ by

$$D(A) = \begin{pmatrix} H_0^1(\Omega) \cap H^2(\Omega) \\ \times \\ H_0^1(\Omega) \end{pmatrix}, \quad A\begin{pmatrix} u \\ v \end{pmatrix} = \begin{pmatrix} v \\ \Delta u \end{pmatrix}$$

for each $\begin{pmatrix} u \\ v \end{pmatrix} \in D(A)$. By Theorem 1.9.4, we know that the linear operator A, previously defined, is the infinitesimal generator of a C_0-group of unitary operators $\{G(t) : X \to X; \ t \in \mathbb{R}\}$. Now, let us define $F : \mathbb{R}_+ \times X \to X$ by

$$F\left(t, \begin{pmatrix} u \\ v \end{pmatrix}\right)(x) = \begin{pmatrix} 0 \\ h(t, u(x)) \end{pmatrix}$$

for each $t \in \mathbb{R}_+$, each $\begin{pmatrix} u \\ v \end{pmatrix} \in X$ and a.e. for $x \in \Omega$. So, the problem (2.2.5) rewrites in the space X as

$$\begin{cases} \begin{pmatrix} u \\ v \end{pmatrix}'(t) = A\begin{pmatrix} u \\ v \end{pmatrix}(t) + F\left(t, \int_{-\tau}^{0} \begin{pmatrix} u(t+s) \\ v(t+s) \end{pmatrix} ds\right), & t \in \mathbb{R}_+, \\ \begin{pmatrix} u(t) \\ v(t) \end{pmatrix} = \begin{pmatrix} \psi_1(t) \\ \psi_2(t) \end{pmatrix}, & t \in [-\tau, 0]. \end{cases}$$

Thus (2.2.5) rewrites in the form (2.2.1), where A, F and $\varphi = \begin{pmatrix} \psi_1 \\ \psi_2 \end{pmatrix}$.

By virtue of Lemma 2.2.1, it follows that F is continuous and compact and thus satisfies the hypotheses of Theorem 2.2.1, from which the conclusion follows. $\qquad\square$

2.3 Local existence: The case f Lipschitz

Let $A : D(A) \subseteq X \rightsquigarrow X$ be the infinitesimal generator of a semigroup of nonlinear contractions, and let $\tau \geq 0$, $\sigma \in \mathbb{R}$,

$$\begin{cases} \mathfrak{X} = C([-\tau, 0]; X), \\ \mathcal{D} = \{\varphi \in \mathfrak{X}; \ \varphi(0) \in \overline{D(A)}\}. \end{cases} \tag{2.3.1}$$

Let $f : [\sigma, +\infty) \times \mathfrak{X} \to X$ be a continuous function and $\psi \in \mathcal{D}$. We consider the nonlinear delay differential evolution equation with local initial data

$$\begin{cases} u'(t) \in Au(t) + f(t, u_t), & t \in [\sigma, T], \\ u(t) = \psi(t - \sigma), & t \in [\sigma - \tau, \sigma]. \end{cases} \tag{2.3.2}$$

Local Initial Conditions

In this section we will give a direct and rather elementary approach to the existence and uniqueness problem for the problem (2.3.2) under some hypotheses on f more general than those used in Section 1.12. Indeed, if f is merely continuous or even locally but not globally Lipschitz, it is not possible to apply the abstract theory developed in Section 1.12. More than this, even f is globally Lipschitz with respect to its second argument, in order to apply Theorem 1.12.5, besides the t continuity of f, one has to assume an additional hypothesis. More precisely, the absence of this condition, labeled as (iii) in Theorem 1.12.5, rules out the possibility of making use of the latter abstract existence result. Of course, there is a price to pay for this generality. Namely, in this setting, we cannot expect to get bounded or even global solutions in the absence of some extra-conditions. Anyhow, from these simple remarks, a good lesson is to be learned. It is extremely important to keep a good balance between the various ways and sets of hypotheses used to get a certain expected conclusion in order to choose the most appropriate and, at the same time elegant, pairing of a set of hypotheses and methods of proof.

We begin with the concept of a solution, to which we will refer later on.

DEFINITION 2.3.1 Let $T_0 \in (\sigma, T]$. By a C^0-*solution* of the problem (2.3.2) on $[\sigma - \tau, T_0]$, we mean a continuous function $u : [\sigma - \tau, T_0] \to X$ satisfying $u(t) = \psi(t - \sigma)$ for each $t \in [\sigma - \tau, \sigma]$ and which is a C^0-solution on $[\sigma, T_0]$ in the sense of Definition 1.8.2 for the equation

$$u'(t) \in Au(t) + h(t),$$

where $h(t) = f(t, u_t)$ for $t \in [\sigma, T_0]$.

If $u : [\sigma - \tau, T_0] \to X$ is a C^0-solution for (2.3.2), from Definition 1.8.2, we deduce that $u(t) \in \overline{D(A)}$ for each $t \in [\sigma, T_0]$.

Our aim here is to prove a local existence and uniqueness result for C^0-solutions by assuming a usual local Lipschitz condition on f with respect to its second argument. To be more specific, let us introduce:

DEFINITION 2.3.2 The function $f : [\sigma, +\infty) \times X \to X$ is called *Lipschitz on bounded sets with respect to its second argument* if for each $T > \sigma$ and $r > 0$ there exists $\ell = \ell(T, r) > 0$ such that

$$\|f(t, v) - f(t, w)\| \le \ell \|v - w\|_X$$

for each $t \in [\sigma, T]$ and each $v, w \in X$ satisfying

$$\begin{cases} \|v\|_X \le r, \\ \|w\|_X \le r. \end{cases}$$

A typical existence and uniqueness result, which is simply a copy of its ODEs counterpart, is the following:

68   *Delay Differential Evolutions Subjected to Nonlocal Initial Conditions*

THEOREM 2.3.1 *Let $A: D(A) \subseteq X \rightsquigarrow X$ be an m-dissipative operator with $0 \in D(A)$ and $0 \in A0$, let $\tau \geq 0$, $\sigma \in \mathbb{R}$, and let X and D be defined as in (2.3.1). Let $f: [\sigma, +\infty) \times X \to X$ be a continuous function which is Lipschitz on bounded sets with respect to its second argument. Then, for each $\psi \in D$, there exist $T = T(\psi) > \sigma$ and a unique C^0-solution $u: [\sigma - \tau, T] \to X$ of the problem (2.3.2).*

REMARK 2.3.1 We can easily get rid of the assumptions $0 \in D(A)$ and $0 \in A0$ by setting $\widetilde{A}x = A(x+x_0)-y_0$ and $\widetilde{f}(t, v) = f(t, v+x_0)+y_0$, where $x_0 \in D(A)$ and $y_0 \in Ax_0$ are arbitrary but fixed, and $D(\widetilde{A}) = D(A) + \{x_0\}$. Clearly $\widetilde{A}$ and $\widetilde{f}$ satisfy all the assumptions of Theorem 2.3.1 and each C^0-solution u of (2.3.2) corresponds to a C^0-solution $\widetilde{u} = u - x_0$ of the same problem with A replaced by $\widetilde{A}$ and f replaced by $\widetilde{f}$.

Proof. Let $\psi \in D$, let $T > \sigma$ be arbitrary but fixed, and let us define

$$r = \|\psi\|_X + 1. \tag{2.3.3}$$

Let $\ell = \ell(T, r)$ be given by Definition 2.3.2 and let

$$K = \{v \in C([\sigma - \tau, T]; X); \ \|v\|_{C([\sigma-\tau,T];X)} \leq r, \ v(t) \in \overline{D(A)}, \ t \in [\sigma, T]\}.$$

Since $0 \in D(A)$, the set K is nonempty. Moreover, one may easily verify that it is closed in $C([\sigma - \tau, T]; X)$.

Let us define the operator $Q : K \to C([\sigma - \tau, T]; X)$ by

$$Q(v) = u,$$

for each $v \in K$, where u is the unique C^0-solution of the problem

$$\begin{cases} u'(t) \in Au(t) + f(t, v_t), & t \in [\sigma, T], \\ u(t) = \psi(t - \sigma), & t \in [\sigma - \tau, \sigma], \end{cases}$$

whose existence and uniqueness is ensured by Theorem 1.8.1. To complete the proof, it suffices to show that, if $T > \sigma$ is "small enough" the operator Q maps K into itself and it is a strict contraction. Then, by the Banach Fixed-Point Theorem, Q has a unique fixed point $u \in K$. Since $u \in K$ is a fixed point of Q if and only if u is a C^0-solution of the problem (2.3.2), this would complete the proof.

Set

$$m = \|f(\cdot, 0)\|_{C([\sigma,T];X)}$$

and let us observe that we can diminish $T > \sigma$ if necessary in order to have both

$$(T - \sigma)\ell < 1$$

and

$$(T - \sigma)(\ell r + m) \leq 1.$$

Local Initial Conditions

We note that this is always possible because $\ell(T_1, r) \leq \ell(T_2, r)$ for each $r > 0$ and each $\sigma < T_1 \leq T_2$. With $T > \sigma$ fixed as above, from (1.8.3) with $\omega = 0$, $x = 0$, $y = 0$ and using (2.3.3), we get

$$\|Q(v)(t)\| \leq \|\psi(0)\| + \int_\sigma^t \|f(s, v_s)\| \, ds \leq r - 1 + (T - \sigma)(\ell r + m) \leq r$$

for each $v \in K$ and each $t \in [\sigma, T]$. Since $Q(v)(t) = \psi(t - \sigma)$ for each $t \in [\sigma - \tau, \sigma]$, from (2.3.3) and the above inequality, it readily follows that

$$\|Q(v)\|_{C([\sigma - \tau, T]; X)} \leq r$$

and thus $Q(K) \subseteq K$. Moreover, from (1.8.2) with $\omega = 0$, we deduce that

$$\|Q(v)(t) - Q(w)(t)\| \leq \int_\sigma^t \ell \|v_s - w_s\|_X \, ds$$

$$\leq (T - \sigma)\ell \|v - w\|_{C([\sigma - \tau, T]; X)}$$

for each $v, w \in K$ and each $t \in [\sigma, T]$. But $\|Q(v)(t) - Q(w)(t)\| = 0$ for each $t \in [\sigma - \tau, \sigma]$ and so it follows that

$$\|Q(v) - Q(w)\|_{C([\sigma - \tau, T]; X)} \leq (T - \sigma)\ell \|v - w\|_{C([\sigma - \tau, T]; X)}$$

for each $v, w \in K$. Since $(T - \sigma)\ell < 1$, this shows that Q is a strict contraction from K to K, and this completes the proof. $\qquad\square$

2.4 Local existence: The case f continuous

As in the preceding section, $\tau \geq 0$, $\sigma \in \mathbb{R}$, $A : D(A) \subseteq X \rightsquigarrow X$ is m-dissipative, $f : [\sigma, +\infty) \times \mathcal{X} \to X$ is continuous and the initial history $\psi \in \mathcal{D}$. Here, instead of the assumption that f is Lipschitz on bounded sets with respect to its second argument, we assume that the semigroup generated by A is compact. As expected, the price paid in this case is the possible lack of uniqueness.

THEOREM 2.4.1 *Let $A : D(A) \subseteq X \rightsquigarrow X$ be an m-dissipative operator with $0 \in D(A)$, $0 \in A0$ and let $\mathcal{X}$ and $\mathcal{D}$ be as in (2.3.1). Let $f : [\sigma, +\infty) \times \mathcal{X} \to X$ be a continuous function. If A generates a compact semigroup and $\overline{D(A)}$ is convex, then, for each $\psi \in \mathcal{D}$, there exists $T = T(\psi) > \sigma$ such that the problem (2.3.2) has at least one C^0-solution $u : [\sigma - \tau, T] \to X$.*

Proof. Let $\psi \in \mathcal{D}$, and let $T > \sigma$ and $r > 0$ be arbitrary but fixed. Let

$$C_\psi([\sigma - \tau, T]; X) = \{v \in C([\sigma - \tau, T]; X); \ v(t) = \psi(t - \sigma), \ t \in [\sigma - \tau, \sigma]\}.$$

70   *Delay Differential Evolutions Subjected to Nonlocal Initial Conditions*

Let us define the function $\widetilde{\psi} \in C([\sigma - \tau, T]; X)$ by

$$\widetilde{\psi}(t) = \begin{cases} \psi(t - \sigma), & t \in [\sigma - \tau, \sigma], \\ \psi(0), & t \in (\sigma, T], \end{cases} \tag{2.4.1}$$

which clearly belongs to $C_\psi([\sigma - \tau, T]; X)$ and let $K = K_1 \cap K_2$, where

$$K_1 = \{v \in C_\psi([\sigma - \tau, T]; X); \ \|v(t) - \widetilde{\psi}(t)\| \le r, \ \forall t \in [\sigma - \tau, T]\}$$

and

$$K_2 = \{v \in C_\psi([\sigma - \tau, T]; X); \ v(t) \in \overline{D(A)}, \forall t \in [\sigma, T]\}.$$

Obviously, for each $v \in K$, we have $\|v_t - \widetilde{\psi}_t\|_X \le r$, for all $t \in [\sigma, T]$.

Now, let us remark that $\{\widetilde{\psi}_t; \ t \in [\sigma, T]\}$ is compact in X. Indeed, since $\widetilde{\psi}([\sigma - \tau, T])$ is compact, it follows that $\{\widetilde{\psi}_t; \ t \in [\sigma, T]\}$ has relatively compact cross sections. As $\widetilde{\psi}$ is uniformly continuous, we deduce that $\{\widetilde{\psi}_t; \ t \in [\sigma, T]\}$ is equicontinuous. By Arzelà–Ascoli Theorem 1.4.10, we conclude that the set $\{\widetilde{\psi}_t; \ t \in [\sigma, T]\}$ is relatively compact in X. Since this family is obviously closed, it is even compact in X, as claimed. From this remark and Lemma 2.1.1, recalling that f is continuous and diminishing $r > 0$ if necessary, we may assume that there exists $M > 0$ such that

$$\|f(t, v_t)\| \le M \tag{2.4.2}$$

for each $v \in K$ and each $t \in [\sigma, T]$.

Since $\widetilde{\psi}$ given by (2.4.1) is in K, the latter is nonempty. In addition it is closed in $C([\sigma - \tau, T]; X)$ and, due to the fact that $\overline{D(A)}$ is convex, it is convex too. Let us define the operator $Q : K \to C([\sigma - \tau, T]; X)$ by

$$Q(v) = u,$$

for each $v \in K$, where u is the unique C^0-solution of the problem

$$\begin{cases} u'(t) \in Au(t) + f(t, v_t), & t \in [\sigma, T], \\ u(t) = \psi(t - \sigma), & t \in [\sigma - \tau, \sigma], \end{cases}$$

whose existence and uniqueness is ensured by Theorem 1.8.1. To complete the proof, it suffices to show that, if $T > \sigma$ is "small enough" the operator Q maps K into itself, and is continuous and compact. Then, by the Schauder Fixed-Point Theorem, it would follow that Q has at least one fixed point $u \in K$ and this would complete the proof.

We can diminish $T > \sigma$ if necessary in order to have

$$(T - \sigma)M \le r,$$

where $M > 0$ satisfies (2.4.2). With $T > \sigma$ fixed as above, from (2.4.2), we get

$$\|Q(v)(t) - \widetilde{\psi}(t)\| = \|Q(v)(t) - \psi(0)\| \le \int_\sigma^T \|f(s, v_s)\| \, ds \le (T - \sigma)M \le r$$

Local Initial Conditions 71

for each $v \in K$ and each $t \in [\sigma, T]$. Clearly,

$$\|Q(v)(t) - \widetilde{\psi}(t)\| = \|\psi(t - \sigma) - \psi(t - \sigma)\| = 0 \le r$$

for each $t \in [\sigma - \tau, \sigma]$. Since $Q(v)(t) \in \overline{D(A)}$ for each $v \in K$ and each $t \in [\sigma, T]$, we conclude that $Q(K) \subseteq K$. Moreover,

$$\|Q(v)(t) - Q(w)(t)\| \le \int_\sigma^t \|f(s, v_s) - f(s, w_s)\| \, ds$$

for each $v, w \in K$ and each $t \in [\sigma, T]$. On the other hand, $Q(v)(t) = Q(w)(t)$ for each $t \in [\sigma - \tau, \sigma]$ and f is continuous. So, the above inequality proves that Q is continuous from K to K in the norm topology of $C([\sigma - \tau, T]; X)$. Next, since by (2.4.2), the family of functions $\{t \mapsto f(t, v_t); \ v \in K\}$ is uniformly bounded on $[\sigma, T]$ by M, it is uniformly integrable. Recalling that the semigroup generated by A is compact, by Theorem 1.8.5, it follows that $Q(K)$ is relatively compact in $C([\sigma, T]; X)$. Taking into account the initial condition, we conclude that $Q(K)$ is relatively compact in $C([\sigma - \tau, T]; X)$. So, we are in the hypotheses of the Schauder Fixed-Point Theorem 1.4.4 from which it follows that Q has at least one fixed point $u \in K$. Obviously u is a C^0-solution of the problem (2.3.2) and this completes the proof. $\qquad \square$

2.5 Local existence: The case f compact

Here we will show that, in the case in which A is linear, one can get rid of the compactness of the generated C^0-semigroup if, instead, we assume that the function f is compact. More precisely, we have the existence result below, which will prove very useful for second-order semilinear hyperbolic equations, as we will see later. We note that if A is linear, $\overline{D(A)} = X$ and then $\mathcal{D} = \mathfrak{X}$.

THEOREM 2.5.1 *Let $A : D(A) \subseteq X \to X$ be a linear m-dissipative operator and let $f : [\sigma, +\infty) \times \mathfrak{X} \to X$ be continuous and compact. Then, for each $\psi \in \mathfrak{X}$, there exists $T = T(\psi) > \sigma$ such that the problem (2.3.2) has at least one mild solution $u : [\sigma - \tau, T] \to X$.*

Proof. Let $\psi \in \mathfrak{X}$, let $T > \sigma$ and $r > 0$ be chosen as in the proof of Theorem 2.4.1. Furthermore, let us consider the set $C_\psi([\sigma - \tau, T]; X)$, the function $\widetilde{\psi} \in C([\sigma - \tau, T]; X)$ and the set $K \subseteq C_\psi([\sigma - \tau, T]; X)$ as there defined. Next, let $Q : K \to C([\sigma - \tau, T]; X)$ be given by

$$Q(v) = u,$$

72 *Delay Differential Evolutions Subjected to Nonlocal Initial Conditions*

for each $v \in K$, where u is the unique mild solution of the problem

$$\begin{cases} u'(t) \in Au(t) + f(t, v_t), & t \in [\sigma, T], \\ u(t) = \psi(t - \sigma), & t \in [\sigma - \tau, \sigma], \end{cases}$$

in the sense of Definition 1.7.3. Then, reasoning exactly as in the proof of Theorem 2.4.1, we can show that the set K is nonempty, closed and convex in $C([\sigma - \tau, T]; X)$, and Q maps K into itself and is continuous. To verify the compactness of Q, we proceed as follows.

Since f is compact and K is bounded, by Lemma 1.7.1, it follows that $Q(K)$ is relatively compact in $C([\sigma, T]; X)$. Clearly, the restriction of $Q(K)$ to $[\sigma - \tau, \sigma]$ is relatively compact in $C([\sigma - \tau, \sigma]; X)$ and thus $Q(K)$ is relatively compact in $C([\sigma - \tau, T]; X)$. So, the Schauder Fixed-Point Theorem 1.4.4 applies, from which it follows that Q has at least one fixed point $u \in K$. But u is a mild solution of the problem (2.3.2) and this completes the proof. $\square$

2.6 Global existence

Here we will prove some sufficient conditions in order to show that the C^0-solution whose local existence is ensured either by Theorem 2.3.1, Theorem 2.4.1 or by Theorem 2.5.1 can be continued up to a global one. For the sake of simplicity, we confine ourselves only to the case $\sigma = 0$. More precisely, we consider

$$\begin{cases} u'(t) \in Au(t) + f(t, u_t), & t \in [0, T], \\ u(t) = \psi(t), & t \in [-\tau, 0], \end{cases} \tag{2.6.1}$$

where $A : D(A) \subseteq X \rightsquigarrow X$ is m-dissipative, X and $\mathcal{D}$ are defined as in (2.3.1), and $f : X \to X$ is continuous and $\psi \in \mathcal{D}$.

DEFINITION 2.6.1 A C^0-solution $u : [-\tau, T) \to X$ of the problem (2.6.1) is called *continuable* if there exist $\widetilde{T} > T$ and a C^0-solution $\widetilde{u} : [-\tau, \widetilde{T}) \to X$ of (2.6.1) such that $u(t) = \widetilde{u}(t)$ for each $t \in [-\tau, T)$.

A C^0-solution $u : [-\tau, T) \to X$ of the problem (2.6.1) is called *non-continuable* or *saturated* if it is not continuable.

We begin with a simple but fundamental lemma.

LEMMA 2.6.1 *Let $A : D(A) \subseteq X \rightsquigarrow X$ be an m-dissipative operator with $0 \in D(A)$, $0 \in A0$, let $f : \mathbb{R}_+ \times X \to X$ be continuous, and let $\psi \in \mathcal{D}$. Assume that either the semigroup generated by A is compact and $\overline{D(A)}$ is convex, or f is Lipschitz on bounded sets with respect to its last argument, or even A is linear and f is compact. Then a C^0-solution $u : [-\tau, T) \to X$ of (2.6.1) is continuable if and only if there exists*

$$u^* = \lim_{t \uparrow T} u(t). \tag{2.6.2}$$

Local Initial Conditions 73

Proof. The necessity is obvious, while for the sufficiency we have to observe that $u^* \in \overline{D(A)}$. Hence, for the initial history

$$\eta(t) = \begin{cases} u(T+t), & t \in [-\tau, 0), \\ u^*, & t = 0, \end{cases}$$

there exists $\delta > 0$ such that the problem

$$\begin{cases} w'(t) \in Aw(t) + f(t, w_t), & t \in [T, T+\delta], \\ w(t) = \eta(t-T), & t \in [T-\tau, T], \end{cases}$$

has either a unique C^0-solution if f is Lipschitz on bounded sets with respect to its last argument, or at least one C^0-solution either if A generates a compact semigroup and $\overline{D(A)}$ is convex, or if A is linear and f is compact. For simplicity, we denote this C^0-solution in all three cases by $w : [T-\tau, T+\delta] \to X$. Finally, we observe that $\widetilde{u} : [-\tau, T+\delta] \to X$, defined by

$$\widetilde{u}(t) = \begin{cases} u(t), & t \in [-\tau, T), \\ w(t), & t \in [T, T+\delta], \end{cases}$$

is a C^0-solution for the problem (2.6.1) which coincides with u on $[-\tau, T)$. The proof is complete. $\qquad\square$

REMARK 2.6.1 Thanks to Lemma 2.6.1, we easily conclude that each non-continuable C^0-solution of the problem (2.6.1) is necessarily defined on an open at the right interval, i.e., of the form $[-\tau, T)$.

A sufficient condition for the existence of the limit (2.6.2) is stated below.

PROPOSITION 2.6.1 *Let $A : D(A) \subseteq X \rightsquigarrow X$ be an m-dissipative operator with $0 \in D(A)$, $0 \in A0$, let $f : \mathbb{R}_+ \times X \to X$ be continuous and let $\psi \in \mathcal{D}$. Furthermore, let $u : [-\tau, T) \to X$ be a C^0-solution of (2.6.1) and let us assume that $T < +\infty$ and $t \mapsto f(t, u_t)$ belongs to $L^1(0, T; X)$. Then there exists*

$$u^* = \lim_{t \uparrow T} u(t).$$

Proof. Since $t \mapsto g(t) = f(t, u_t)$ is integrable on $[0, T]$, it follows that the Cauchy problem

$$\begin{cases} v'(t) \in Av(t) + g(t), & t \in [0, T], \\ v(0) = u(0) \end{cases}$$

has a unique C^0-solution $v : [0, T] \to X$ which, by Proposition 1.8.5, must coincide with u on $[0, T)$. As v is continuous, it follows that there exists

$$\lim_{t \uparrow T} u(t) = \lim_{t \uparrow T} v(t) = v(T)$$

74   *Delay Differential Evolutions Subjected to Nonlocal Initial Conditions*

and this completes the proof.   □

A useful characterization of continuable C^0-solutions of (2.6.1) is the following:

THEOREM 2.6.1 *Let $A : D(A) \subseteq X \rightsquigarrow X$ be an m-dissipative operator with $0 \in D(A)$, $0 \in A0$, let $f : \mathbb{R}_+ \times \mathcal{X} \to X$ be a continuous function which maps bounded sets in $\mathbb{R}_+ \times \mathcal{X}$ into bounded sets in X. Let $\psi \in \mathcal{D}$, and let us assume that either the semigroup generated by A is compact and $\overline{D(A)}$ is convex or f is Lipschitz on bounded sets with respect to its last argument, or even A is linear and f is compact. A necessary and sufficient condition for a C^0-solution $u : [-\tau, T) \to X$ of (2.6.1) to be continuable is that the graph of u, i.e.,*

$$\text{graph } u = \{(t, u(t)) \in [-\tau, +\infty) \times X; \ t \in [-\tau, T)\}$$

is included in a compact subset in $\mathbb{R} \times X$.

Proof. To prove the necessity, we begin by observing that if the C^0-solution $u : [-\tau, T) \to X$ of (2.6.1) is continuable, there exist $\widetilde{T} > T$ and at least one C^0-solution $\widetilde{u} : [-\tau, \widetilde{T}) \to X$ of (2.6.1) which coincides with u on $[-\tau, T)$. So, graph $u \subseteq$ graph $\widetilde{u}$, which, due to the continuity of the mapping $t \mapsto (t, \widetilde{u}(t))$ on $[-\tau, T]$, is obviously compact and included in $\mathbb{R} \times X$.

For the sufficiency, let us remark that if $K_0 \subseteq \mathbb{R} \times X$ is a compact set and the graph of u is included in K_0, it follows that $\{u(t); \ t \in [-\tau, T)\}$ is compact in X. Thus, the family $\{u_t; \ t \in [0, T)\}$ is bounded in $\mathcal{X}$. Since f maps bounded sets in $\mathbb{R}_+ \times \mathcal{X}$ into bounded sets in X, it follows that the function $t \mapsto f(t, u_t)$ is bounded on $[0, T)$. So, by virtue of Proposition 2.6.1, there exists $\lim_{t \uparrow T} u(t) = u^*$. Since $(T, u^*) \in K_0$ and $u^* \in \overline{D(A)}$, we can apply either Theorem 2.3.1, Theorem 2.4.1 or Theorem 2.5.1 to conclude that there exists $\widetilde{T} > T$ such that the problem

$$\begin{cases} v'(t) \in Av(t) + f(t, v_t), & t \in [T, \widetilde{T}], \\ v(t) = u(t), & t \in [T - \tau, T] \end{cases}$$

has a C^0-solution $v : [T - \tau, \widetilde{T}] \to X$ which, in the case in which f is Lipschitz on bounded sets with respect to its last argument, is unique.

Finally, we have only to observe that the function $\widetilde{u} : [-\tau, \widetilde{T}) \to X$, defined by

$$\widetilde{u}(t) = \begin{cases} u(t), & t \in [-\tau, T), \\ v(t), & t \in [T, \widetilde{T}], \end{cases}$$

is a C^0-solution of (2.6.1). The proof is complete.   □

THEOREM 2.6.2 *Let $A : D(A) \subseteq X \rightsquigarrow X$ be an m-dissipative operator with $0 \in D(A)$, $0 \in A0$, let $f : \mathbb{R}_+ \times \mathcal{X} \to X$ be continuous and let $\psi \in \mathcal{D}$. Let $u : [-\tau, T) \to X$ be a C^0-solution of (2.6.1). Then, either u is non-continuable, or u can be continued up to a non-continuable one.*

Local Initial Conditions 75

Proof. We shall apply Brezis–Browder Ordering Principle Theorem 1.14.1. Let $u : [\,-\tau, T) \to X$ be a C^0-solution of (2.6.1). If u is non-continuable, we have nothing to prove. If u is continuable, it follows that the set $\mathcal{S}$ of all C^0-solutions of (2.6.1) extending u strictly to the right of T is nonempty. On this set, we define the partial order "$\preceq$" as follows: if $u_1 : [\,-\tau, T_1) \to X$ and $u_2 : [\,-\tau, T_2) \to X$ are in $\mathcal{S}$, we say that $u_1 \preceq u_2$ if $T_1 \leq T_2$ and $u_1(t) = u_2(t)$ for each $t \in [\,-\tau, T_1)$. Further, we define the mapping $\mathcal{N} : \mathcal{S} \to \mathbb{R} \cup \{+\infty\}$ by $\mathcal{N}(v) = T_v$, where $v : [\,-\tau, T_v) \to X$. One may easily check that $(\mathcal{S}, \preceq)$ and $\mathcal{N}$ satisfy all the hypotheses of Theorem 1.14.1. Accordingly there exists at least one $\mathcal{N}$-maximal element $\widetilde{u}$ in $\mathcal{S}$. Finally, we have merely to observe that $\widetilde{u}$ is a non-continuable C^0-solution of (2.6.1) extending u. The proof is complete. $\square$

From Theorems 2.3.1, 2.4.1, 2.5.1 and 2.6.2 it follows:

COROLLARY 2.6.1 *Let* $A : D(A) \subseteq X \rightsquigarrow X$ *be an m-dissipative operator with* $0 \in D(A)$, $0 \in A0$ *and let* $f : \mathbb{R}_+ \times \mathcal{X} \to X$ *be continuous. Assume that either the semigroup generated by A is compact and $\overline{D(A)}$ is convex, or f is Lipschitz on bounded sets with respect to its last argument or even A is linear and f is compact. Then for each initial data $\psi \in \mathcal{D}$, the problem (2.6.1) has at least one non-continuable C^0-solution. If f is Lipschitz on bounded sets with respect to its last argument, then this C^0-solution is unique.*

We conclude this section with a sufficient condition on f ensuring the existence of global C^0-solutions of (2.6.1).

THEOREM 2.6.3 *Let* $A : D(A) \subseteq X \rightsquigarrow X$ *be an m-dissipative operator with* $0 \in D(A)$, $0 \in A0$ *and let* $f : \mathbb{R}_+ \times \mathcal{X} \to X$ *be continuous. Assume that either the semigroup generated by A is compact and $\overline{D(A)}$ is convex, or f is Lipschitz on bounded sets with respect to its last argument, or even that A is linear and f is compact. Let us assume further that there exist two continuous functions $h, k : \mathbb{R}_+ \to \mathbb{R}_+$ such that*

$$\| f(t, v) \| \leq k(t) \| v \|_{\mathcal{X}} + h(t), \tag{2.6.3}$$

for each $(t, v) \in \mathbb{R}_+ \times \mathcal{X}$. Then, for each $\psi \in \mathcal{D}$, the problem (2.6.1) has at least one global C^0-solution, i.e., defined on $[\,-\tau, +\infty)$. If f is Lipschitz on bounded sets with respect to its last argument, then this C^0-solution is unique.

Proof. Let $\psi \in \mathcal{D}$ and let $u : [\,-\tau, T) \to X$ be a non-continuable C^0-solution of (2.6.1) whose existence is ensured by Corollary 2.6.1. We will show that $T = +\infty$. Indeed, if we assume the contrary, by taking $x = y = 0$ in (1.8.3) and using (2.6.3), we get

$$\| u(t) \| \leq \| u(0) \| + \int_0^t k(s) \| u_s \|_{\mathcal{X}} \, ds + \int_0^t h(s) \, ds$$

76   *Delay Differential Evolutions Subjected to Nonlocal Initial Conditions*

for each $t \in [0, T)$. Set

$$
\begin{cases}
\alpha_0(t) = \|u(0)\| + \displaystyle\int_0^t h(s)\,ds, & t \in \mathbb{R}_+, \\[2mm]
\beta(t) = k(t), & t \in \mathbb{R}_+, \\[2mm]
y(t) = \|u(t)\|, & t \in [-\tau, T).
\end{cases}
$$

Taking into account that both k and h are bounded on $[0, T]$, from Lemma 1.13.3, we conclude that $\|u\|$ is bounded on $[0, T)$ and thus it is bounded on $[-\tau, T)$. Again by (2.6.3), we deduce that the mapping $t \mapsto f(t, u_t)$ is bounded on $[0, T)$. Hence, by Proposition 2.6.1, there exists $\lim_{t \uparrow T} u(t) = u^*$. From Lemma 2.6.1, it follows that u is continuable – a contradiction. This contradiction can be eliminated only if $T = +\infty$ and this completes the proof. $\qquad\square$

2.7   Examples

The aim of this section is to give two illustrative examples of ODEs coming from mathematical modeling that can be analyzed by the previously developed abstract existence theory. We would like to add a comment here. Some practitioners,[1] who are beneficiaries of Mathematical Modeling, are convinced that it is completely useless to prove existence and uniqueness referring to a given ODE or PDE describing the evolution of a certain phenomenon. Their argument, although deeply incorrect, is apparently somehow convincing: Why spend time to prove the existence of a solution to a given model as long as we can see that the real phenomenon evolves and thus it, a priori, has a solution? The main objection to this argument, which we have encountered several times, is that there is no perfect mathematical model simply because, in the process of obtaining it, one deliberately ignores some parameters that are heuristically considered to have less importance. Due to this fact, it could happen that a model thus obtained is contradictory by itself. At this point, mathematics comes into play and shows whether or not the model in question is non-contradictory and consistent. But this can be done, only by proving an existence result.

More than this, in order for a model to be reliable, one has either to prove uniqueness or find a clear selection procedure of "good solutions."

Let $0 < \sigma \leq \tau$. Let $\mathcal{R} = C([-\tau, 0]; \mathbb{R})$ and $\mathcal{R}_\sigma = C([-\sigma, 0]; \mathbb{R})$. If $\psi_1 \in \mathcal{R}$

---

[1] Fortunately, not too many.

Local Initial Conditions 77

and $\psi_2 \in \mathcal{R}_\sigma$, we define $\psi \in \begin{pmatrix} \mathcal{R} \\ \times \\ \mathcal{R} \end{pmatrix}$, by

$$\psi(t) = \begin{pmatrix} \widetilde{\psi}_1(t), \\ \widetilde{\psi}_2(t), \end{pmatrix}, \quad t \in [-\tau, 0],$$

where

$$\widetilde{\psi}_1(t) = \psi_1(t), \qquad t \in [-\tau, 0],$$

$$\widetilde{\psi}_2(t) = \begin{cases} \psi_2(t), & t \in [-\sigma, 0], \\ \psi_2(-\sigma), & t \in [-\tau, -\sigma). \end{cases}$$

2.7.1   Spring mass system with delay

The next example is inspired by McKibben [182, Model III.4, p. 333].

EXAMPLE 2.7.1 Let $\beta > 0$, $\omega > 0$ and let us consider the system

$$\begin{cases} x''(t) + \beta x'(t) + \omega^2 x(t) = h(t, x(t-\tau), x'(t-\sigma)), & t \in [0, T], \\ x(t) = \psi_1(t), & t \in [-\tau, 0], \quad (2.7.1) \\ x'(t) = \psi_2(t), & t \in [-\sigma, 0], \end{cases}$$

where $h : [0, T] \times \begin{pmatrix} \mathbb{R} \\ \times \\ \mathbb{R} \end{pmatrix} \to \mathbb{R}$ is continuous. One may easily see that (2.7.1) can be rewritten as a first-order differential system with delay of the form

$$\begin{cases} x'(t) = y(t), & t \in [0, T], \\ y'(t) = -\omega^2 x(t) - \beta y(t) + h(t, x(t-\tau), y(t-\sigma)), & t \in [0, T], \\ x(t) = \psi_1(t), & t \in [-\tau, 0], \\ y(t) = \psi_2(t), & t \in [-\sigma, 0]. \end{cases} \quad (2.7.2)$$

In order to rewrite this system in a more convenient abstract form, let us observe that

$$\begin{cases} x(s-\tau) = \delta(-\tau)x_s, & s \in [0, T], \\ y(s-\sigma) = \delta(-\sigma)y_s, & s \in [0, T], \end{cases}$$

where $\delta(\theta)$ is the Dirac delta concentrated at $\theta \in \mathbb{R}$. Now, let us define the operator $A : \begin{pmatrix} \mathbb{R} \\ \times \\ \mathbb{R} \end{pmatrix} \to \begin{pmatrix} \mathbb{R} \\ \times \\ \mathbb{R} \end{pmatrix}$ by

$$A \begin{pmatrix} \widetilde{x} \\ \widetilde{y} \end{pmatrix} = \begin{pmatrix} \widetilde{y} \\ -\widetilde{x} \end{pmatrix} \quad \text{for each} \quad \begin{pmatrix} \widetilde{x} \\ \widetilde{y} \end{pmatrix} \in \begin{pmatrix} \mathbb{R} \\ \times \\ \mathbb{R} \end{pmatrix}.$$

78   *Delay Differential Evolutions Subjected to Nonlocal Initial Conditions*

Clearly A is linear, m-dissipative and the C_0-semigroup (in fact the C_0-group) generated by A on $X = \begin{pmatrix} \mathbb{R} \\ \times \\ \mathbb{R} \end{pmatrix}$ is compact. Let $f : [0,T] \times \begin{pmatrix} \mathcal{R} \\ \times \\ \mathcal{R} \end{pmatrix} \to \begin{pmatrix} \mathbb{R} \\ \times \\ \mathbb{R} \end{pmatrix}$ be given by

$$
f\left(t, \begin{pmatrix} z \\ w \end{pmatrix}\right) = \begin{pmatrix} 0 \\ (1-\omega^2)\delta(0)z - \beta\delta(0)w + h(t, \delta(-\tau)z, \delta(-\sigma)w) \end{pmatrix}
$$

for each $t \in [0,T]$ and each $\begin{pmatrix} z \\ w \end{pmatrix} \in \begin{pmatrix} \mathcal{R} \\ \times \\ \mathcal{R} \end{pmatrix}$. Setting $u = \begin{pmatrix} \widetilde{x} \\ \widetilde{y} \end{pmatrix}$, the problem (2.7.2) and equivalently (2.7.1) can be rewritten as

$$
\begin{cases} u'(t) = Au(t) + f(t, u_t), & t \in [0,T], \\ u(t) = \psi(t), & t \in [-\tau, 0]. \end{cases} \tag{2.7.3}
$$

From Corollary 2.6.1 and Theorem 2.6.3, we deduce:

THEOREM 2.7.1 *If the function* $h : [0,T] \times \begin{pmatrix} \mathbb{R} \\ \times \\ \mathbb{R} \end{pmatrix} \to \mathbb{R}$ *is continuous, then for each* $\begin{pmatrix} \psi_1 \\ \psi_2 \end{pmatrix} \in \begin{pmatrix} \mathcal{R} \\ \times \\ \mathcal{R}_\sigma \end{pmatrix}$ *the problem (2.7.3) has at least one non-continuable solution* $u : [-\tau, T_m) \to \begin{pmatrix} \mathbb{R} \\ \times \\ \mathbb{R} \end{pmatrix}$, $u = \begin{pmatrix} \widetilde{x} \\ \widetilde{y} \end{pmatrix}$, *where* $x = \widetilde{x}$ *is a non-continuable solution of (2.7.1). If, in addition, h is Lipschitz with respect to its last two arguments, then the solution u is unique and global.*

2.7.2   A delayed glucose level–dependent dosage

Also inspired by a model described by McKibben [182, Model II.3, p. 191], the example below shows how the previously developed abstract theory applies in various specific cases.

EXAMPLE 2.7.2 Let us consider the system describing the evolution of the glucose level in the blood stream as well as in the gastrointestinal system in the presence of a feedback – called dosage – taking into account the histories of both levels, i.e.,

$$
\begin{cases} y'(t) = -ay(t) + D_1(t, y(t-\tau), z(t-\sigma)), & t \in [0,T], \\ z'(t) = ay(t) - bz(t) + D_2(t, y(t-\tau), z(t-\sigma)), & t \in [0,T], \\ y(t) = \psi_1(t), & t \in [-\tau, 0], \\ z(t) = \psi_2(t), & t \in [-\sigma, 0]. \end{cases} \tag{2.7.4}
$$

Local Initial Conditions

Here $0 < a \leq b$ and the dosages D_1, D_2 depend on the retarded value of the glucose level in the blood stream, $y(t - \tau)$, as well as on the glucose level in the gastrointestinal system, $z(t - \tau)$. Let $0 \leq \sigma \leq \tau$ and let us define

$$\psi(t) = \begin{pmatrix} \widetilde{\psi}_1(t) \\ \widetilde{\psi}_2(t) \end{pmatrix}, \quad t \in [-\tau, 0],$$

as at the beginning of this section.

For each $y \in \mathcal{R}$ and $z \in \mathcal{R}_\sigma$, we have

$$\begin{cases} y(s - \tau) = \delta(-\tau)y_s, & s \in [0, T], \\ z(s - \sigma) = \delta(-\sigma)z_s, & s \in [0, T]. \end{cases}$$

Now, let

$$A : \begin{pmatrix} \mathbb{R} \\ \times \\ \mathbb{R} \end{pmatrix} \to \begin{pmatrix} \mathbb{R} \\ \times \\ \mathbb{R} \end{pmatrix}$$

be defined by

$$A \begin{pmatrix} \widetilde{y} \\ \widetilde{z} \end{pmatrix} = \begin{pmatrix} -a\widetilde{y} \\ -b\widetilde{z} \end{pmatrix} \quad \text{for each } \begin{pmatrix} \widetilde{y} \\ \widetilde{z} \end{pmatrix} \in \begin{pmatrix} \mathbb{R} \\ \times \\ \mathbb{R} \end{pmatrix}.$$

Clearly

$$\left\langle A \begin{pmatrix} \widetilde{y} \\ \widetilde{z} \end{pmatrix}, \begin{pmatrix} \widetilde{y} \\ \widetilde{z} \end{pmatrix} \right\rangle = \left\langle \begin{pmatrix} -a\widetilde{y} \\ -b\widetilde{z} \end{pmatrix}, \begin{pmatrix} \widetilde{y} \\ \widetilde{z} \end{pmatrix} \right\rangle = -a\widetilde{y}^2 - b\widetilde{z}^2 \leq 0.$$

So, A is linear, m-dissipative and since $X = \begin{pmatrix} \mathbb{R} \\ \times \\ \mathbb{R} \end{pmatrix}$ is finite dimensional, A

generates a compact C_0-semigroup. Finally, let $f : [0, T] \times \begin{pmatrix} \mathcal{R} \\ \times \\ \mathcal{R} \end{pmatrix} \to \begin{pmatrix} \mathbb{R} \\ \times \\ \mathbb{R} \end{pmatrix}$

be given by

$$f \left(t, \begin{pmatrix} v \\ w \end{pmatrix} \right) = \begin{pmatrix} D_1(t, \delta(-\tau)v, \delta(-\sigma)w) \\ a\delta(0)v + D_2(t, \delta(-\tau)v, \delta(-\sigma)w) \end{pmatrix}$$

for each $t \in [0, T]$ and each $\begin{pmatrix} v \\ w \end{pmatrix} \in \begin{pmatrix} \mathcal{R} \\ \times \\ \mathcal{R} \end{pmatrix}$. With the notations above,

recalling that $u = \begin{pmatrix} \widetilde{y} \\ \widetilde{z} \end{pmatrix} \in \begin{pmatrix} \mathbb{R} \\ \times \\ \mathbb{R} \end{pmatrix}$, the problem (2.7.4) can be rewritten

under the form

$$\begin{cases} u'(t) = Au(t) + f(t, u_t), & t \in [0, T], \\ u(t) = \psi(t), & t \in [-\tau, 0]. \end{cases} \tag{2.7.5}$$

From Corollary 2.6.1 and Theorem 2.6.3, we get:

80   *Delay Differential Evolutions Subjected to Nonlocal Initial Conditions*

THEOREM 2.7.2 *If $D_i : [0, T] \times \begin{pmatrix} \mathbb{R} \\ \times \\ \mathbb{R} \end{pmatrix} \to \mathbb{R}$, $i = 1, 2$, are continuous,*

then for each $\begin{pmatrix} \psi_1 \\ \psi_2 \end{pmatrix} \in \begin{pmatrix} \mathcal{R} \\ \times \\ \mathcal{R}_\sigma \end{pmatrix}$ the problem (2.7.5) has at least one non-

continuable solution $u : [-\tau, T_m) \to \begin{pmatrix} \mathbb{R} \\ \times \\ \mathbb{R} \end{pmatrix}$, $u = \begin{pmatrix} \widetilde{y} \\ \widetilde{z} \end{pmatrix}$, where $\begin{pmatrix} \widetilde{y} \\ z \end{pmatrix}$, with

$z = \widetilde{z}_{|[-\sigma, T_m)}$, *is a non-continuable solution of (2.7.4). If, in addition, D_i, $i = 1, 2$, are Lipschitz with respect to the last two arguments, then the solution is unique and global, i.e., $\widetilde{y} : [-\tau, T] \to \mathbb{R}$ and $z : [-\sigma, T] \to \mathbb{R}$.*

2.8   Global existence of bounded C^0-solutions

Let us consider the evolution equation subjected to a nonlocal initial condition

$$\begin{cases} u'(t) \in Au(t) + f(t, u_t), & t \in \mathbb{R}_+, \\ u(t) = \psi(t), & t \in [-\tau, 0]. \end{cases} \tag{2.8.1}$$

Here, $A : D(A) \subseteq X \rightsquigarrow X$ is an m-dissipative operator in the (infinite dimensional) Banach space X, $\tau \geq 0$, $f : \mathbb{R}_+ \times \mathcal{X} \to X$ is continuous and $\psi \in \mathcal{D}$. In the limiting case $\tau = 0$, i.e., when the delay is absent, $\mathcal{X} = X$ and $\mathcal{D} = \overline{D(A)}$. Then, in this case, $f : \mathbb{R}_+ \times X \to X$ and $\psi \in \overline{D(A)}$.

As we have already seen in Theorem 2.6.3, under rather general assumptions on A and on f, the problem (2.8.1) has global C^0-solutions. Now, we are interested in getting sufficient conditions so that the global C^0-solutions of (2.8.1) be bounded.

We begin with the assumptions we need in what follows.

(H_A) The operator $A : D(A) \subseteq X \rightsquigarrow X$ satisfies

(A_1) $0 \in D(A)$, $0 \in A0$ and A is ω-m-dissipative for some $\omega > 0$

(A_2) A generates a compact semigroup

(A_3) $\overline{D(A)}$ is convex

(A_4) A is linear and ω-m-dissipative for some $\omega > 0$.

(H_f) The function $f : \mathbb{R}_+ \times \mathcal{X} \to X$ is continuous and satisfies

(f_1) f is globally Lipschitz with respect to its second argument, i.e., there exists $\ell > 0$ such that

$$\|f(t, v) - f(t, \widetilde{v})\| \leq \ell \|v - \widetilde{v}\|_{\mathcal{X}}$$

for each $t \in \mathbb{R}_+$ and $v, \widetilde{v} \in \mathcal{X}$

Local Initial Conditions

(f_2) there exists $m \geq 0$ such that

$$\|f(t,0)\| \leq m$$

for each $t \in \mathbb{R}_+$

(f_3) there exist ℓ and m such that

$$\|f(t,v)\| \leq \ell\|v\|_X + m$$

for each $t \in \mathbb{R}_+$ and $v \in X$

(f_4) f is compact.

(H_c) The constants ℓ and ω satisfy the nonresonance condition

$$\ell < \omega.$$

(H_ψ) The initial history $\psi \in \mathcal{D}$.

The first boundedness result concerning (2.8.1) is:

THEOREM 2.8.1 *If (A_1) in (H_A), (f_1), (f_2) in (H_f), (H_c) and (H_ψ) are satisfied, then the unique C^0-solution, u, of the problem (2.8.1) whose existence is ensured by Theorem 2.6.3 is bounded on $[-\tau, +\infty)$, i.e., $u \in C_b([-\tau, +\infty); X)$. More precisely, u satisfies*

$$\|u\|_{C_b([-\tau,+\infty);X)} \leq \max\left\{ m_0, \frac{m}{\omega - \ell} \right\}, \tag{2.8.2}$$

where m is given by (f_2) and $m_0 = \|\psi\|_X$.

Proof. The fact that u is bounded follows exactly as in Theorem 1.12.4 and therefore we do not give details.

To prove the estimate (2.8.2), let $u \in C([-\tau, +\infty); X)$ be the unique C^0-solution of (2.8.1). Since $\|u\|_X = \|\psi\|_X = m_0$, by (1.8.3), we get

$$\|u(t)\| \leq e^{-\omega t}\|u(0)\| + \int_0^t e^{-\omega(t-s)} \left(\ell\|u_s\|_X + m \right) ds$$

$$\leq e^{-\omega t}\|u\|_{C([0,T];X)} + \int_0^t e^{-\omega(t-s)} \left(\ell\|u\|_{C([-\tau,T];X)} + m \right) ds$$

$$\leq e^{-\omega t}\|u\|_{C([0,T];X)} + \int_0^t e^{-\omega(t-s)} \left(\ell \max\{\|u\|_{C([0,T];X)}, \|u\|_X\} + m \right) ds,$$

for each $T > 0$ and $t \in [0,T]$. Hence

$$\|u(t)\| \leq e^{-\omega t}\|u\|_{C([0,T];X)} + \frac{\ell}{\omega}\left(1 - e^{-\omega t}\right)\left(\max\{\|u\|_{C([0,T];X)}, m_0\} + \frac{m}{\ell} \right) \tag{2.8.3}$$

82   *Delay Differential Evolutions Subjected to Nonlocal Initial Conditions*

for each $T > 0$ and $t \in [0, T]$.

If for each $T > 0$, $\|u\|_{C([0,T];X)} \leq \|\psi\|_X = m_0$, we have nothing to prove. If there exists $T_0 > 0$ such that

$$\|u\|_{C([0,T_0];X)} > m_0,$$

then, for each $T \geq T_0$, from (2.8.3), we get

$$\|u(t)\| \leq e^{-\omega t}\|u\|_{C([0,T];X)} + \frac{\ell}{\omega}\left(1 - e^{-\omega t}\right)\left(\|u\|_{C([0,T];X)} + \frac{m}{\ell}\right) \quad (2.8.4)$$

for each $t \in [0, T]$. Now, let $\widetilde{t} > 0$ be such that

$$\|u(\widetilde{t})\| = \|u\|_{C([0,T];X)}.$$

Setting $t = \widetilde{t}$ in (2.8.4) and observing that

$$\|u\|_{C([0,T];X)} = \|u\|_{C([-\tau,T];X)},$$

we deduce

$$\|u\|_{C([-\tau,T];X)} \leq \frac{m}{\omega - \ell}.$$

As $T \geq T_0$ is arbitrary, we get

$$\|u\|_{C_b([-\tau,+\infty);X)} \leq \frac{m}{\omega - \ell}.$$

Thus u satisfies (2.8.2) and this completes the proof. $\qquad\square$

THEOREM 2.8.2 *If (A_1), (A_2) and (A_3) in (H_A), (f_3) in (H_f), (H_c) and (H_ψ) are satisfied, then each C^0-solution, u, of the problem (2.8.1) whose existence is ensured by Theorem 2.6.3 is bounded on $[-\tau, +\infty)$, i.e., $u \in C_b([-\tau, +\infty); X)$. More precisely, u satisfies (2.8.2).*

THEOREM 2.8.3 *If (A_4) in (H_A), (f_3), (f_4) in (H_f), (H_c) and (H_ψ) are satisfied, then each mild solution, u, of the problem (2.8.1) whose existence is ensured by Theorem 2.6.3 is bounded on $[-\tau, +\infty)$, i.e., $u \in C_b([-\tau, +\infty); X)$. More precisely, u satisfies (2.8.2).*

The proofs of both Theorems 2.8.2 and 2.8.3 are very similar to that of Theorem 2.8.1 and therefore we do not provide details.

2.9   Three more examples

We now present the next three completely different examples of both parabolic and hyperbolic delay equations which can be analyzed by the abstract theory previously presented.

Local Initial Conditions   83

2.9.1 The nonlinear diffusion equation with delay

Let Ω be a nonempty, bounded domain in $\mathbb{R}^d$, $d \geq 1$, with C^1 boundary Σ and let $\tau \geq 0$ and $\omega > 0$. In the following, Δ denotes the Laplace operator in the sense of distributions over the domain Ω. Let $\varphi : D(\varphi) \subseteq \mathbb{R} \rightsquigarrow \mathbb{R}$ be a maximal-monotone operator, let $\mathcal{X}_1 = C([-\tau, 0]; L^1(\Omega))$, let $f : \mathbb{R}_+ \times \mathcal{X}_1 \to L^1(\Omega)$, and let $\psi \in \mathcal{X}_1$. We denote by $Q_+ = \mathbb{R}_+ \times \Omega$, $\Sigma_+ = \mathbb{R}_+ \times \Sigma$, $Q_\tau = [-\tau, 0] \times \Omega$ and we consider the following delay nonlinear diffusion equation:

$$\begin{cases} \dfrac{\partial u}{\partial t}(t, x) = \Delta\varphi(u(t, x)) - \omega u(t, x) + f(t, u_t)(x), & \text{in } Q_+, \\[2mm] \varphi(u(t, x)) = 0, & \text{on } \Sigma_+, \\[2mm] u(t, x) = \psi(t)(x), & \text{in } Q_\tau. \end{cases} \quad (2.9.1)$$

This problem describes *the controlled porous media diffusion equation*. See Barbu [22, Section 5.3, p. 222]. Here, Ω is a porous medium filled by a fluid, which may be a gas or a liquid. Roughly speaking, if the medium Ω is homogeneous, i.e., the permeability and porosity do not depend on $x \in \Omega$, $u(t, x)$ represents the rescaled density of the fluid at the time t and point $x \in \Omega$, while f is a feedback operator which controls the evolution of the density by reacting at each moment t taking into account the delayed values of u over the interval $[t - \tau, t]$. Usually, the function φ is given by

$$\varphi(u) = u|u|^{m-1}$$

for $u \in \mathbb{R}$. The case $m \in (0, 1)$ corresponds to the *fast diffusion* and the case $m > 1$ to the *slow diffusion*. For details, see Showalter [238, A2, pp. 253–254]. We finally note that a feedback depending on the history of the states is more realistic than an instantaneous one. An example of such a feedback function f is $f(t, v)(x) = h\left(t, x, \displaystyle\int_{-\tau}^0 v(s)(x)\, ds\right)$ for each $(t, v) \in \mathbb{R}_+ \times \mathcal{X}_1$ and a.e. for $x \in \Omega$, where $h : \mathbb{R}_+ \times \Omega \times \mathbb{R} \to \mathbb{R}$. For other examples, see the monographs of Curtain and Zwart [92] and of Lasiecka and Triggiani [166].

THEOREM 2.9.1 *Let Ω be a nonempty and bounded domain in $\mathbb{R}^d$, $d \geq 1$, with C^1 boundary Σ, let $\tau > 0$, $\omega > 0$ and let $\varphi : D(\varphi) \subseteq \mathbb{R} \rightsquigarrow \mathbb{R}$ be a maximal-monotone operator with $0 \in D(\varphi)$ and $0 \in \varphi(0)$. Let $f : \mathbb{R}_+ \times \mathcal{X}_1 \to L^1(\Omega)$ be a continuous function, and let $\psi \in \mathcal{X}_1$. Let us assume that the following hypotheses are satisfied*

(i) *$\varphi : \mathbb{R} \to \mathbb{R}$ is continuous on $\mathbb{R}$ and C^1 on $\mathbb{R} \setminus \{0\}$ and there exist two constants $C > 0$ and $\alpha > 0$ if $d \leq 2$ and $\alpha > (d - 2)/d$ if $d \geq 3$ such that*

$$\varphi'(s) \geq C|s|^{\alpha-1}$$

for each $s \in \mathbb{R} \setminus \{0\}$

84   *Delay Differential Evolutions Subjected to Nonlocal Initial Conditions*

(ii) there exist $\ell > 0$ and $m > 0$ such that

$$\|f(t,v)\|_{L^1(\Omega)} \le \ell\|v\|_{\mathfrak{X}_1} + m$$

for each $(t,v) \in \mathbb{R}_+ \times \mathfrak{X}_1$.

Then (2.9.1) has at least one C^0-solution, $u \in C([-\tau,+\infty); L^1(\Omega))$. If, in addition, $\ell < \omega$, then each global C^0-solution of (2.9.1) satisfies

$$\|u\|_{C_b([-\tau,+\infty);L^1(\Omega))} \le \max\left\{m_0, \frac{m}{\omega - \ell}\right\},$$

where $m_0 = \|\psi\|_{\mathfrak{X}_1}$.

Proof. Take $X = L^1(\Omega)$ and let the operator $A : D(A) \subseteq L^1(\Omega) \rightsquigarrow L^1(\Omega)$ be given by $D(A) = D(\Delta\varphi)$, $Au = \Delta\varphi(u) - \omega u$, for each $u \in D(A)$, where $\Delta\varphi$ is defined as in Theorem 1.9.6. We apply Theorem 2.6.3. Namely, thanks to (i), we are in the hypotheses of Theorem 1.9.6. So, the operator A is m-dissipative, $0 \in D(A)$, $A0 = 0$, and the semigroup generated by A on $L^1(\Omega)$ is compact. From (ii), we deduce that f satisfies (2.6.3). The additional conclusion in the case $\ell < \omega$ follows from Theorem 2.8.2 and this completes the proof. $\qquad\square$

2.9.2   A singular transport equation

Let us consider the singular transport equation

$$\begin{cases} \dfrac{\partial u}{\partial t}(t,x) + \dfrac{\partial(xu)}{\partial x}(t,x) = \mu u(t-\tau,\alpha x)(1-u(t-\tau,\alpha x)), & (t,x) \in Q_+, \\ u(t,x) = \psi(t)(x), & (t,x) \in Q_\tau, \end{cases} \quad (2.9.2)$$

where $\mu > 0$, $\alpha \in (0,1)$, $\tau > 0$, $\psi \in C([-\tau,0]; C([0,1];\mathbb{R}))$, $Q_+ = \mathbb{R}_+ \times (0,1)$ and $Q_\tau = [-\tau,0] \times (0,1)$. Up to some normalized constants, this equation is a model, proposed by Rey and Mackey [222], describing the production of proliferative stem and precursor cells in the bone marrow, $u(t,x)$ being the population density of cells with respect to maturity $x \in [0,1]$, at time $t \in \mathbb{R}_+$. It is interesting to emphasize that the authors conclude that it is more convenient to consider a delay evolution equation rather a non-delay one because the former offers more accurate information on the evolution of the process than the latter.

Let us observe that the problem (2.9.2) can be rewritten, in the function space

$$X = C([0,1];\mathbb{R}),$$

as an abstract semilinear delay evolution equation of the form

$$\begin{cases} u'(t) = Au(t) + f(u_t), & t \in \mathbb{R}_+, \\ u(t) = \psi(t), & t \in [-\tau,0], \end{cases} \quad (2.9.3)$$

Local Initial Conditions

where the operator $A : D(A) \subseteq X \to X$ is defined by

$$\begin{cases} D(A) = \{w \in X; \ x \mapsto xw(x) \in C^1([0,1]; \mathbb{R})\}, \\ (Aw)(x) = -\dfrac{d}{dx}[xw(x)], \quad w \in D(A), \ x \in (0,1], \\ (Aw)(0) = 0 \end{cases}$$

and $f : X \to X$ is given by[2]

$$f(v)(x) = \mu v(-\tau)(\alpha x)(1 - v(-\tau)(\alpha x))$$

for each $v \in X$ and each $x \in [0,1]$. Then A generates a C_0-semigroup of contractions

$$\{S(t) : X \to X; \ t \in \mathbb{R}_+\}.$$

To find the analytical expression of this semigroup, let us consider the semigroup $\{T(t) : X \to X; \ t \in \mathbb{R}_+\}$ given by

$$[T(t)\varphi](x) = \varphi(e^{-t}x)$$

for each $\varphi \in X$, each $t \in \mathbb{R}_+$ and $x \in [0,1]$. One may easily verify that the infinitesimal generator of the C_0-semigroup $\{T(t) : X \to X; \ t \in \mathbb{R}_+\}$ is the operator $B : D(B) \subseteq X \to X$ defined by

$$\begin{cases} D(B) = \{w \in X; \ x \mapsto xw(x) \in C^1([0,1]; \mathbb{R})\}, \\ (Bw)(x) = -xw'(x), \quad w \in D(B), \ x \in (0,1], \\ (Bw)(0) = 0. \end{cases}$$

So, $D(A) = D(B)$ and the operator $Aw = Bw - w$ for each $w \in D(A)$. Therefore

$$[S(t)\varphi](x) = e^{-t}[T(t)\varphi](x) = e^{-t}\varphi(e^{-t}x)$$

for each $\varphi \in X$, each $t \in \mathbb{R}_+$ and $x \in [0,1]$.

Putting all these together, we get:

THEOREM 2.9.2 *Let $\mu > 0$, $\alpha \in (0,1)$ and $\tau > 0$. Then, for each $\psi \in X$, the problem (2.9.3) and thus (2.9.2) has a unique saturated mild solution $u : [-\tau, T_\psi) \to X$*

$$u(t,x) = \begin{cases} \psi(t)(x), & (t,x) \in Q_\tau, \\ e^{-t}\psi(0)(e^{-t}x) + \displaystyle\int_0^t e^{-(t-s)}g(t,s,x)\,ds, & (t,x) \in Q_+^\psi, \end{cases}$$

g being given by

$$g(t,s,x) = \mu u(s - \tau, e^{-(t-s)}\alpha x)(1 - u(s - \tau, e^{-(t-s)}\alpha x))$$

for $(t,x) \in Q_+^\psi$ and $0 \le s \le t$, while $Q_+^\psi = [0, T_\psi) \times (0,1)$.

---

[2]We recall that $X = C([-\tau, 0]; X)$.

86 *Delay Differential Evolutions Subjected to Nonlocal Initial Conditions*

Proof. Clearly A is 1-m-dissipative and f, defined as above, is Lipschitz on bounded sets. So, we are in the hypotheses of Corollary 2.6.1, from which the conclusion follows. $\qquad\square$

2.9.3 A semilinear damped wave equation with delay

Here we will show how Theorem 2.5.1 can be applied in the study of second-order semilinear hyperbolic problems on bounded domains in $\mathbb{R}^d$, $d \geq 1$. Let Ω be a nonempty bounded and open subset in $\mathbb{R}^d$ with C^1 boundary Σ, let $Q_+ = \mathbb{R}_+ \times \Omega$, $Q_\tau = [-\tau, 0] \times \Omega$, $\Sigma_+ = \mathbb{R}_+ \times \Sigma$, let $\omega > 0$, and let us consider the initial-value problem for the damped wave equation with delay:

$$\begin{cases} \dfrac{\partial^2 u}{\partial t^2}(t,x) = \mathcal{L}u(t,x) + h\left(t, \displaystyle\int_{-\tau}^0 u(t+s,x)\,ds\right), & \text{in } Q_+, \\[2mm] u(t,x) = 0, & \text{on } \Sigma_+, \\[2mm] u(t,x) = \psi_1(t)(x), & \text{in } Q_\tau, \\[2mm] \dfrac{\partial u}{\partial t}(t,x) = \psi_2(t)(x), & \text{in } Q_\tau, \end{cases} \qquad (2.9.4)$$

where $\mathcal{L}$ is defined by

$$\mathcal{L}u(t,x) = \Delta u(t,x) - 2\omega\frac{\partial u}{\partial t}(t,x) - \omega^2 u(t,x),$$

for $(t,x) \in Q_+$.

Let $\mathcal{H}_0^1 = C([-\tau, 0]; H_0^1(\Omega))$ and $\mathcal{X}_2 = C([-\tau, 0]; L^2(\Omega))$.

THEOREM 2.9.3 *Let Ω be a nonempty bounded and open subset in $\mathbb{R}^d$, $d \geq 1$, with C^1 boundary Σ, let $\tau \geq 0$, $\omega > 0$, let $h : \mathbb{R}_+ \times \mathbb{R} \to \mathbb{R}$, and let $\psi_1 \in \mathcal{H}_0^1$ and $\psi_2 \in \mathcal{X}_2$. In addition, let us assume that*

(h_1) *h is continuous and there exists $\widetilde{\ell} > 0$, $\widetilde{m} \geq 0$ such that*

$$|h(t,y)| \leq \widetilde{\ell}|y| + \widetilde{m}$$

for all $(t,y) \in \mathbb{R}_+ \times \mathbb{R}$

(c_1) *The constants[3] $\ell = \widetilde{\ell}\sqrt{|\Omega|}$ and ω satisfy $\ell < \omega$.*

Then the problem (2.9.4) has at least one mild solution $u \in C_b([-\tau, +\infty); H_0^1(\Omega))$ with $\dfrac{\partial u}{\partial t} \in C_b([-\tau, +\infty); L^2(\Omega))$, that satisfies

$$\|u\|_{C_b([-\tau,+\infty);H_0^1(\Omega))} + \left\|\frac{\partial u}{\partial t}\right\|_{C_b([-\tau,+\infty);L^2(\Omega))} \leq \max\left\{m_0, \frac{m}{\omega - \ell}\right\},$$

---

[3]Hereinafter, $|\Omega|$ denotes the Lebesgue measure of Ω.

Local Initial Conditions 87

where $m_0 = \|\psi_1\|_{\mathcal{H}_0^1} + \|\psi_2\|_{X_2}$ and $m = |\widetilde{m}|\sqrt{|\Omega|}$.

If, in addition, h is jointly globally Lipschitz of constant $\widetilde{\ell}$, then the mild solution is unique and globally asymptotically stable.

Proof. Setting

$$\psi_3 = \omega\psi_1 + \psi_2,$$

it is easy to see that (2.9.4) can be rewritten as a first-order system of partial differential equations of the form

$$\begin{cases} \dfrac{\partial u}{\partial t}(t,x) = v(t,x) - \omega u(t,x), & \text{in } Q_+, \\[2mm] \dfrac{\partial v}{\partial t}(t,x) = \Delta u(t,x) - \omega v(t,x) + h\left(t, \displaystyle\int_{-\tau}^{0} u(t+s,x)\,ds\right), & \text{in } Q_+, \\[2mm] u(t,x) = 0, & \text{on } \Sigma_+, \\[2mm] u(t,x) = \psi_1(t)(x), & \text{in } Q_\tau, \\[2mm] v(t,x) = \psi_3(t)(x), & \text{in } Q_\tau. \end{cases} \qquad (2.9.5)$$

Let us define the product space

$$X = \begin{pmatrix} H_0^1(\Omega) \\ \times \\ L^2(\Omega) \end{pmatrix},$$

which, endowed with the natural inner product

$$\left\langle \begin{pmatrix} u \\ v \end{pmatrix}, \begin{pmatrix} \widetilde{u} \\ \widetilde{v} \end{pmatrix} \right\rangle = \int_\Omega \nabla u(x) \cdot \nabla \widetilde{u}(x)\,dx + \int_\Omega v(x)\widetilde{v}(x)\,dx$$

for each $\begin{pmatrix} u \\ v \end{pmatrix}, \begin{pmatrix} \widetilde{u} \\ \widetilde{v} \end{pmatrix} \in X$, is a real Hilbert space. In turn, (2.9.5) can be rewritten as an abstract evolution equation subjected to nonlocal initial conditions of the form (2.8.1) in X, where A, f and ψ are defined as follows. First, let us define the linear operator $A : D(A) \subseteq X \to X$ by

$$D(A) = \begin{pmatrix} H_0^1(\Omega) \cap H^2(\Omega) \\ \times \\ H_0^1(\Omega) \end{pmatrix}, \qquad A\begin{pmatrix} u \\ v \end{pmatrix} = \begin{pmatrix} -\omega u + v \\ \Delta u - \omega v \end{pmatrix}$$

for each $\begin{pmatrix} u \\ v \end{pmatrix} \in D(A)$. By Theorem 1.9.4, we know that the linear operator $B : D(B) \subseteq X \to X$, with $D(B) = D(A)$ and $B = A + \omega I$, where A is defined as above and I is the identity on X, is the infinitesimal generator of a C_0-group of unitary operators $\{G(t) : X \to X;\ t \in \mathbb{R}\}$. Consequently, A

88   *Delay Differential Evolutions Subjected to Nonlocal Initial Conditions*

generates a C_0-semigroup of contractions $\{S(t) : X \to X; \ t \in \mathbb{R}_+\}$, defined by

$$S(t)\xi = e^{-\omega t}G(t)\xi$$

for each $t \in \mathbb{R}_+$ and each $\xi \in X$. Thus A is ω-m-dissipative and therefore (A_4) in (H_A) is satisfied.

Now, let $\mathfrak{X} = C([-\tau, 0]; X)$ and let us define $f : \mathbb{R}_+ \times \mathfrak{X} \to X$ by

$$f\left(t, \begin{pmatrix} u \\ v \end{pmatrix}\right)(x) = \begin{pmatrix} 0 \\ h\left(t, \int_{-\tau}^0 u(s)(x)ds\right) \end{pmatrix}$$

for each $t \in \mathbb{R}_+$, each $\begin{pmatrix} u \\ v \end{pmatrix} \in \mathfrak{X}$ and a.e. for $x \in \Omega$.

By virtue of (h_1) it follows that f satisfies (f_3) in (H_f). From Lemma 2.2.1 combined with Lemma 1.7.1, we conclude that f is continuous and compact and thus it satisfies (f_4) in (H_f). Since (H_c) is equivalent in our case to (c_1), we are in the hypotheses of Theorems 2.6.3 and 2.8.3, from which the conclusion follows. Finally, if h is jointly globally Lipschitz of constant $\widehat{\ell}$, the uniqueness is a consequence of Theorem 2.6.3, while the global asymptotic stability follows from Remark 1.12.1 and Theorem 1.12.7. $\square$

2.10   Bibliographical notes and comments

We begin with a few comments on delay evolution equations subjected to a local initial condition.

The theory of delay evolution equations in finite dimensional spaces is well developed since many years ago. See the books of Halanay [134], Driver [103] and the references therein. Both the finite delay, i.e., when the values of the unknown function are given on $[-\tau, 0]$, and the infinite or unbounded delay, i.e., when $[-\tau, 0]$ is replaced by $(-\infty, 0]$, were analyzed. Although this monograph is focused merely on the case of finite delay problems, we emphasize that the infinite delay case is equally important. For a detailed overview of differential equations with unbounded delay, the reader is referred to the survey article of Corduneanu and Lakshmikantham [87], discussing the state-of-the-art of the 1980s, and to the monograph of Hino, Murakami and Naito [144]. See also Corduneanu [85]. Starting with the monographs of Hale [135], a new stage of the theory emerged. The semigoups of linear or even nonlinear operators as well as the evolution systems enter into play to furnish new information on the existence and asymptotic behavior of solutions. For the linear (mostly) autonomous case, see Hale [136] and Batkai and Piazzera [24]. For the semilinear integro-differential case with delay, see the pioneering paper of Travis and Webb [244] and that of Fitzgibbon [116]. The identification problem of

Local Initial Conditions 89

a source term in an evolution semilinear equation with delay was considered by Lorenzi and Vrabie [175], [174]. The most important contributions to the nonlinear case, very briefly presented in Section 1.12, are due Webb [265], [266], Flaschka and Leitman [117], Dyson and Villella Bressan [106], [107], Plant [218], [219], Ruess [227], [229], and Ruess and Summers [230]. For a classical approach in the study of asymptotic stability of solutions for the neutral delay logistic equation see Gopalsami and Zhang [128]. A good show of applications in population biology and epidemiology involving delay differential equations can be found in Brauer and Castillo–Chavez [46]. For models in population dynamics, see Kuang [161].

Section 2.1. Lemma 2.1.1 is a simple but useful extension of the well-known Weierstrass Theorem in Real Analysis saying that continuous real functions on compact spaces are bounded. Theorem 2.1.1 is an extension of the Peano Local Existence Theorem to delay initial-value problems in infinite-dimensional Banach spaces. For the latter result, see Vrabie [253, Theorem 10.1.1, p. 231].

Section 2.2. Theorem 2.2.1 is a particular instance of a general result in Vrabie [262]. Surprisingly, it shows that semilinear delay equations governed by compact perturbations of infinitesimal generators of C_0-groups can be handled by the infinite dimensional delay version of the celebrated Peano Local Existence Theorem 2.1.1. Theorem 2.2.2 is also from Vrabie [262] where some other abstract results referring to pseudoparabolic problems are proved.

Section 2.3. Although very important, Theorem 2.3.1 is simply an immediate application of the Banach Fixed-Point Theorem. It is interesting to note that the first result of this kind, in the infinite dimensional case, for a non-delay semilinear evolution equation, was obtained by Segal [235].

Section 2.4. The idea to compensate the lack of compactness of bounded and closed subsets in infinite dimensional Banach spaces by the compactness of the semigroup goes back to Pazy [211]. He had proved the specific form of Theorem 2.4.1 when A is linear and $\tau = 0$, i.e., the delay is absent. The main argument used by Pazy [211], i.e., the compactness of the mild solution operator – see Theorem 1.7.4 – was observed independently by Baras, Hassan and Veron [20]. As far as the nonlinear case is concerned, a particular form of Theorem 2.4.1, i.e., for $\tau = 0$ and A fully nonlinear, is due to Vrabie [247]. For some details on the compactness machinery subtleties entering into the proof of all these results, the interested reader is referred to Vrabie [250]. For more examples and details, see Vrabie [252, Theorem 2.3.1, p. 45]. Theorem 2.4.1 is from Mitidieri and Vrabie [185, Theorem 7]. We emphasize that the compactness of the semigroup is a parabolicity condition, and so Theorem 2.4.1 cannot give information on hyperbolic problems.

Section 2.5. Theorem 2.5.1 is a replica of Theorem 2.4.1 that proves useful in the study of second-order linear hyperbolic problems. In fact, in the case of second-order hyperbolic problems on bounded domains, the lack of the

90   *Delay Differential Evolutions Subjected to Nonlocal Initial Conditions*

compactness of the generated C_0-group is compensated by the compactness of the reaction term, as long as the latter depends only on u but not on the t-partial derivative of u. This is a simple consequence of the fact that the key point concerning the contribution of $u(t, \cdot) \in H_0^1(\Omega)$ in the reaction term f is that only its behavior with respect to the $L^2(\Omega)$-norm really counts. As $H_0^1(\Omega)$ is compactly embedded in $L^2(\Omega)$, this implies the compactness of the reaction term f.

Section 2.6. The results in this section are simple extensions of some propositions referring to the semilinear case proved in Vrabie [253, Section 10.1, pp. 236–242], to the case of fully nonlinear delay evolution equations.

Section 2.7. Examples 2.7.1 and 2.7.2 are adapted from the monograph of McKibben [182, Model III.4, p. 333, Model II.3, p. 191]. Theorems 2.7.1 and 2.7.2 are rather obvious consequences of the abstract theory developed in the preceding sections.

Section 2.8. Theorem 2.8.1 cannot be obtained from Theorem 1.12.5, simply because f does not satisfy condition (iii). Of course, if we assume, in addition, that f is jointly globally Lipschitz, then we can use Theorem 1.12.5 in order to get both existence and asymptotic behavior. The estimate (2.8.2), which probably is known, was proved in Vrabie [259, Lemma 4.3].

Section 2.9. In the form here presented, Theorem 2.9.1 is new. We note that by using sharper compactness arguments than those developed in Díaz and Vrabie [96], we can prove that the conclusion of Theorem 2.9.1 still holds true if, instead of the condition (i), we merely assume that φ is strictly increasing and $L^1(\Omega)$ is replaced by $L^\infty(\Omega)$ endowed with the $L^1(\Omega)$-topology. In fact, the key point in this setting is that the semigroup generated by the nonlinear diffusion operator $\Delta\varphi$, although not compact, maps weakly compact sets in $L^1(\Omega)$ into norm compact sets in $L^1(\Omega)$. The example referring to the singular transport equation is an adaptation from the paper of Dyson, Villella Bressan and Web [108], where several other interesting properties were studied. We mean regularity, uniqueness, invariance, global existence and asymptotic behavior of solutions, existence of equilibrium solutions, as well as continuous dependence on the data. Theorem 2.9.3 is inspired by its nondelayed counterpart in Vrabie [253, Theorem 10.4.2, p. 243].

Chapter 3

Nonlocal Initial Conditions: The Autonomous Case

Overview

In this chapter, we establish some sufficient conditions for the existence, uniqueness, and global uniform asymptotic stability of a C^0-solution for the nonlinear delay autonomous differential evolution equation

$$\begin{cases} u'(t) \in Au(t) + f(u_t), & t \in \mathbb{R}_+, \\ u(t) = g(u)(t), & t \in [-\tau, 0], \end{cases}$$

where $\tau \geq 0$, X is a real Banach space, the operator $A: D(A) \subseteq X \rightsquigarrow X$ is the infinitesimal generator of a nonlinear semigroup of contractions, $\mathcal{X} = C([-\tau, 0]; X)$, $\mathcal{D} = \{\varphi \in \mathcal{X}; \ \varphi(0) \in \overline{D(A)}\}$, the function $f: \mathcal{X} \to X$ is continuous, and the mapping $g : C_b(\mathbb{R}_+; \overline{D(A)}) \to \mathcal{D}$ is continuous and has affine growth.

3.1 The problem to be studied

In this chapter, we prove some existence, uniqueness, stability and even uniform asymptotic stability results regarding C^0-solutions for a large class of nonlinear delay evolution equations with nonlocal initial data. For the sake of simplicity here, we confine ourselves to the study of the autonomous case which, under fairly reasonable hypotheses, can be handled with the results from Section 1.12. Moreover, from now on, we will focus our attention mainly on bounded C^0-solutions. The bias of this preference is explained by the fact that many phenomena exhibit a long-time bounded behavior. Accordingly, if the corresponding mathematical models do not have bounded solutions, they are usually considered inappropriate and rejected. This is the case of the well-known demographic model proposed by Malthus – see Vrabie [254, Section 1.4.6, pp. 3-34] – which was severely criticized even by his contemporaries.

91

92 *Delay Differential Evolutions Subjected to Nonlocal Initial Conditions*

So, let us consider the autonomous delay evolution equation subjected to a nonlocal initial condition

$$\begin{cases} u'(t) \in Au(t) + f(u_t), & t \in \mathbb{R}_+, \\ u(t) = g(u)(t), & t \in [-\tau, 0]. \end{cases} \tag{3.1.1}$$

In the problem (3.1.1), $A : D(A) \subseteq X \rightsquigarrow X$ is an m-dissipative operator in the real (infinite dimensional) Banach space X, $\tau \geq 0$ and let

$$\mathcal{X} = C([-\tau, 0]; X) \quad \text{and} \quad \mathcal{D} = \{\varphi \in \mathcal{X}; \; \varphi(0) \in \overline{D(A)}\}. \tag{3.1.2}$$

The function $f : \mathcal{X} \to X$ is continuous and $g : C_b(\mathbb{R}_+; \overline{D(A)}) \to \mathcal{D}$ is a non-expansive or merely continuous mapping. In the limiting case when the delay is absent, i.e., when $\tau = 0$, $\mathcal{X}$ identifies with X, $\mathcal{D}$ identifies with $\overline{D(A)}$, and therefore $f : X \to X$ and $g : C_b(\mathbb{R}_+; \overline{D(A)}) \to \overline{D(A)}$. Hereinafter, for the sake of simplicity, if $u \in C_b([-\tau, +\infty); X)$ and $u(t) \in \overline{D(A)}$ for each $t \in \mathbb{R}_+$, we denote by

$$g(u) = g(u_{|\mathbb{R}_+}).$$

Customarily, if $u \in C([-\tau, +\infty); X)$ and $t \in \mathbb{R}_+$, $u_t \in \mathcal{X}$ is defined by

$$u_t(s) = u(t + s)$$

for each $s \in [-\tau, 0]$.

We emphasize that, in many practical circumstances, A acts as a partial differential operator in a function space X. Namely, A is the mathematical expression of an evolution law established partly empirically but verified in concrete situations which, in the absence of any external force or reaction f, drives the system according to a set of physical, biological, economical, and other constraints. The perturbation term f is suitably chosen in order to make the system modify its own natural evolution in order to behave according to some final objectives. Its role is to produce an appropriate response establishing the manner in which the system reacts at any time t in order to correct its instantaneous rate of change. Clearly, the type of reaction is usually dictated by the cumulative history of the states over the time interval $[t - \tau, t]$, i.e., by u_t.

DEFINITION 3.1.1 By a C^0-*solution* of (3.1.1) we mean a continuous function $u : [-\tau, +\infty) \to X$ satisfying $u(t) = g(u)(t)$ for each $t \in [-\tau, 0]$ and which, for each $T > 0$, is a C^0-solution on $[0, T]$ in the sense of Definition 1.8.2 for the equation

$$u'(t) \in Au(t) + h(t),$$

where $h(t) = f(u_t)$ for $t \in [0, T]$.

REMARK 3.1.1 Clearly, if u is a C^0-solution of (3.1.1), then necessarily we have $g(u)(0) \in \overline{D(A)}$. This explains why, in what follows, we will only consider functions $g : C_b(\mathbb{R}_+; \overline{D(A)}) \to \mathcal{D}$, where $\mathcal{D}$ is defined as before.

Nonlocal Initial Conditions: The Autonomous Case 93

REMARK 3.1.2 If $\tau = 0$, (3.1.1) reduces to a nonlocal nondelayed evolution problem, i.e.,

$$\begin{cases} u'(t) \in Au(t) + f(u(t)), & t \in \mathbb{R}_+, \\ u(0) = g(u). \end{cases}$$

If g does not depend on u, i.e., if there exists an element $\psi \in \mathcal{D}$ such that $g(u)(t) = \psi(t)$ for each $t \in [-\tau, 0]$, then (3.1.1) is simply the specific initial-value nonlocal delay evolution problem

$$\begin{cases} u'(t) \in Au(t) + f(u_t), & t \in \mathbb{R}_+, \\ u(t) = \psi(t), & t \in [-\tau, 0], \end{cases}$$

considered in Chapter 2. If, in addition, $\tau = 0$, we obtain a classical initial-value or Cauchy problem, i.e.,

$$\begin{cases} u'(t) \in Au(t) + f(u(t)), & t \in \mathbb{R}_+, \\ u(0) = \xi, \end{cases}$$

where $\xi \in \overline{D(A)}$.

So, (3.1.1) is general enough to encompass various classes of evolution problems.

REMARK 3.1.3 Likewise in the case in which $A = 0$ and $X = \mathbb{R}^d$ – see for instance Hale [135, Section 2, p. 11] – in the case here considered, the problem (3.1.1) includes specific classes of evolution functional equations such as differential-difference problems of the form

$$\begin{cases} u'(t) \in Au(t) + F(u(t), u(t - \tau_1), \dots, u(t - \tau_n)), & t \in \mathbb{R}_+, \\ u(t) = g(u)(t), & t \in [-\tau, 0], \end{cases} \quad (3.1.3)$$

where $0 < \tau_1 < \tau_2 < \cdots < \tau_n \le \tau$ are some intermediate delays and the forcing term $F : X^{n+1} \to X$ is continuous or Lipschitz. Indeed, if $\delta(-\theta) : \mathcal{X} \to X$ denotes the vector-valued Dirac delta concentrated at $-\theta$, i.e.,

$$\delta(-\theta)v = v(-\theta),$$

for each $v \in \mathcal{X}$, we have

$$F(u(t), u(t - \tau_1), \dots, u(t - \tau_n)) = F(\delta(-0)u_t, \delta(-\tau_1)u_t, \dots, \delta(-\tau_n)u_t)$$

and

$$F(\delta(-0)u_t, \delta(-\tau_1)u_t, \dots, \delta(-\tau_n)u_t) = f(u_t)$$

for each $u \in C_b([-\tau, +\infty); X)$ and each $t \in \mathbb{R}_+$. Since $\delta(-\theta)$ is nonexpansive from $\mathcal{X}$ to X, we conclude that $f : \mathcal{X} \to X$, defined as above, preserves the continuity properties of F. More precisely, f is continuous whenever F is continuous and f is Lipschitz whenever F is Lipschitz. So, our main result applies also to problems of the form (3.1.3), but not only that form.

DEFINITION 3.1.2 The initial condition $u(t) = g(u)(t)$ for each $t \in [-\tau, 0]$ is called *purely nonlocal* if $g(0) = 0$.

3.2 The case f and g Lipschitz

The assumptions we need in what follows are listed below.

(H_A) The operator $A: D(A) \subseteq X \rightsquigarrow X$ satisfies

$\quad$ (A_1) $0 \in D(A)$, $0 \in A0$ and A is ω-m-dissipative for some $\omega > 0$.

(H_f) The function $f: X \to X$ is continuous and satisfies

$\quad$ (f_1) there exists $\ell > 0$ such that

$$\|f(v) - f(\widetilde{v})\| \le \ell \|v - \widetilde{v}\|_X$$

$\quad$ for each $v, \widetilde{v} \in X$.

(H_c) The constants ℓ and ω satisfy $\ell < \omega$.

(H_g) The function $g : C_b(\mathbb{R}_+; \overline{D(A)}) \to \mathcal{D}$ satisfies

$\quad$ (g_1) there exists $a > 0$ such that for each $u, v \in C_b(\mathbb{R}_+; \overline{D(A)})$, we have

$$\|g(u) - g(v)\|_X \le \|u - v\|_{C_b([a,+\infty);X)}.$$

COMMENT 3.2.1 A main difficulty encountered in the case of nonlocal initial conditions is due to the fact that the C^0-solutions do not obey semigroup properties. In fact, the nonlocal problems are closer to "multi-point" boundary problems rather than to initial-value problems as those analyzed in Sections 2.6~2.9. Actually, the main difference between these two classes of problems is that the solutions in the nonlocal initial condition case cannot be concatenated as it always happens in the case of initial-value problems. The lack of this property creates a lot of trouble, especially when trying to prove global existence. To be clearer, in this general case, we cannot proceed as in the case of local initial conditions. In the latter, as we have already seen in the preceding chapter, we prove first a local existence result and then, under some appropriate conditions, by using a Zorn Lemma maximality argument, for instance the Brezis–Browder Ordering Principle, i.e., Theorem 1.14.1, combined with a concatenation procedure as described in Vrabie [254, Proposition 2.1.2, p. 53], we can pass to non-continuable or even global solutions. See Sections 2.3~2.9. In contrast, in this frame, we are forced to solve the problem of global existence directly from the very beginning. So, one may easily realize why this requires some stronger hypotheses than in the classical case of local initial conditions. We mean here the assumption (A_1) and the fact that A dominates f, i.e., the assumption (H_c). Compared with the general frame used in Section 1.12, as well as in Sections 2.3~2.9, the condition (H_c) seems rather restrictive. However, it should be emphasized that exactly this condition ensures both the existence and the global asymptotic stability of bounded global C^0-solutions.

Nonlocal Initial Conditions: The Autonomous Case

REMARK 3.2.1 The hypothesis (g_1), introduced by Vrabie [257], is very easy to verify in practical situations. See Remark 3.2.4 below. Namely, (g_1) in conjunction with (H_A) and (H_f), ensures that a suitably defined operator, whose fixed points produce C^0-solutions for the problem (3.1.1), is a strict contraction. It should be emphasized that (g_1) is the key assumption for getting uniqueness, even for problems without delay, as the example below shows.

EXAMPLE 3.2.1 Let us consider the problem

$$\begin{cases} u'(t) = -2u(t) + u(t), & t \in \mathbb{R}_+, \\ u(0) = g(u). \end{cases}$$

Then, $\tau = 0$, $X = \mathbb{R}$, $A : \mathbb{R} \to \mathbb{R}$, $f : \mathbb{R} \to \mathbb{R}$ and $g : \mathbb{R} \to \mathbb{R}$ are defined by $Au = -2u$, $f(u) = u$ and $g(u) = u(0)$ for each $u \in \mathbb{R}$. Clearly, A, f and g satisfy (H_A), (H_f) and (H_c) with $\omega = 2$, $\ell = 1$, but (H_g) does not hold true. On the other hand, the problem above has infinitely many solutions, i.e., all functions in the family $\{t \mapsto \xi e^{-t};\ \xi \in \mathbb{R}\}$.

REMARK 3.2.2 If g satisfies (g_1), then it also satisfies

(g_2) g has affine growth, i.e., there exists $m_0 \geq 0$ such that, for each $u \in C_b(\mathbb{R}_+; \overline{D(A)})$, we have

$$\|g(u)\|_X \leq \|u\|_{C_b([a,+\infty);X)} + m_0,$$

where a is given by (g_1) and m_0 can be taken $\|g(0)\|_X$

(g_3) for each $u, v \in C_b(\mathbb{R}_+; \overline{D(A)})$ with $u(t) = v(t)$ for each $t \in [a, +\infty)$, where a is given by (g_1), we have $g(u) = g(v)$.

Indeed, from (g_1), we easily deduce

$$\|g(u)\|_X \leq \|g(u) - g(0)\|_X + \|g(0)\|_X$$

$$\leq \|u\|_{C_b([a,+\infty);X)} + m_0$$

for each $u \in C_b(\mathbb{R}_+; \overline{D(A)})$, which proves (g_2).

To prove (g_3) let us remark that if $u, v \in C_b(\mathbb{R}_+; \overline{D(A)})$ are such that $u(t) = v(t)$ for each $t \in [a, +\infty)$, then, from (g_1), it follows that

$$\|g(u) - g(v)\|_X = 0.$$

REMARK 3.2.3 If g satisfies (g_3), then it depends only on the values of u on $[a, +\infty)$, i.e., $g(u)(t) = g(\widetilde{u})(t)$ for each $t \in [-\tau, 0]$, whenever u and $\widetilde{u}$ are solutions of (3.1.1) and $u(s) = \widetilde{u}(s)$ for each $s \in [a, +\infty)$. By Remark 3.2.2, the same conclusion holds true if g satisfies (g_1). As a consequence, in the frame here considered, g can be defined only on $C_b([a, +\infty); \overline{D(A)})$ and not on the larger domain $C_b(\mathbb{R}_+; \overline{D(A)})$. We note that this observation will play a very important role in several places within this book.

96 *Delay Differential Evolutions Subjected to Nonlocal Initial Conditions*

REMARK 3.2.4 If the function g is defined as

(i) $g(u)(t) = u(T + t)$, $t \in [-\tau, 0]$ and $T > \tau$ (T-periodicity condition)

(ii) $g(u)(t) = -u(T + t)$, $t \in [-\tau, 0]$ and $T > \tau$ (T-anti-periodicity condition)[1]

(iii) $g(u)(t) = \int_{\tau}^{+\infty} k(\theta) u(t + \theta)\, d\theta$, $t \in [-\tau, 0]$, where $k \in L^1([\tau, +\infty); \mathbb{R})$ and $\int_{\tau}^{+\infty} |k(\theta)|\, d\theta = 1$ (mean condition)

(iv) $g(u)(t) = \sum_{i=1}^{n} \alpha_i u(t + t_i)$ for each $t \in [-\tau, 0]$, where $\sum_{i=1}^{n} |\alpha_i| \leq 1$ and $\tau < t_1 < t_2 < \cdots < t_n$ (multi-point discrete mean condition)

then g satisfies (g_2) with $m_0 = 0$ and (g_1) (with $a = t_1 - \tau > 0$ in the case of (iv)).

A more general case is that in which the function g is given by

$$g(u)(t) = \int_{\tau}^{+\infty} \mathcal{N}(u(t + \theta))\, d\mu(\theta) + \psi(t), \qquad (3.2.1)$$

for each $u \in C_b(\mathbb{R}_+; \overline{D(A)})$ and $t \in [-\tau, 0]$, where $\mathcal{N} : X \to X$ is a (possibly nonlinear) nonexpansive operator with $\mathcal{N}(0) = 0$ and μ is a σ-finite and complete measure on $[\tau, +\infty)$, satisfying $\mathrm{supp}\, \mu = [\delta, +\infty)$, $\delta \geq \tau$, $\mu([\tau, +\infty)) = 1$ and $\psi \in \mathcal{X}$ is such that $g(u)(0) \in \overline{D(A)}$.

To see that $(i) \sim (iv)$ correspond to particular choices of both $\mathcal{N}$ and μ in (3.2.1), let $s \in \mathbb{R}_+$ and let us denote by μ_s the atomic measure defined by

$$\mu_s(E) = \begin{cases} 1, & \text{if } s \in E, \\ 0, & \text{if } s \notin E, \end{cases}$$

for each $E \subseteq \mathbb{R}_+$. Then (i) corresponds to $\mathcal{N} = I$ and $\mu = \mu_T$, (ii) to $\mathcal{N} = -I$ and $\mu = \mu_T$, (iii) to $\mathcal{N} = I$ and $\mu = k(\theta)d\theta$, and (iv) to $\mathcal{N} = I$ and $\mu = \sum_{i=1}^{n} \alpha_i \mu_{t_i}$.

Finally, if g_0 satisfies (g_2), $g_0(0) = 0$ and $\psi \in \mathcal{X}$ is such that

$(H_{\psi,g})$ $g_0(u)(0) + \psi(0) \in \overline{D(A)}$ for each $u \in C_b(\mathbb{R}_+; \overline{D(A)})$,

then the function $g : C_b(\mathbb{R}_+; \overline{D(A)}) \to \mathcal{D}$, defined by

$$g(u) = g_0(u) + \psi$$

for each $u \in C_b(\mathbb{R}_+; \overline{D(A)})$, satisfies in turn (g_2) with $m_0 = \|\psi\|_{\mathcal{X}}$. This is the case of the function g defined by (3.2.1).

---

[1] Both conditions (i) and (ii) will be analyzed in the next chapter because they are really relevant in the quasi-autonomous case when f depends on t as well and is either T-periodic or T-anti-periodic when the C^0-solution is T-periodic or T-anti-periodic.

Nonlocal Initial Conditions: The Autonomous Case 97

REMARK 3.2.5 Concerning the condition $(H_{\psi,g})$, there are several simple but significant instances in which this is automatically satisfied

(i) if $g_0 \equiv 0$ and $\psi \in \mathfrak{X}$ satisfies $\psi(0) \in \overline{D(A)}$

(ii) if $g_0 : C_b(\mathbb{R}_+; \overline{D(A)}) \to C([-\tau, 0]; \overline{D(A)})$ satisfies (g_2), $g_0(0) = 0$ and $\psi \equiv 0$

(iii) if $g_0 : C_b(\mathbb{R}_+; \overline{D(A)}) \to C([-\tau, 0]; \overline{D(A)})$ satisfies (g_2), $g_0(0) = 0$, $\overline{D(A)}$ is a linear subspace in X and $\psi \in C([-\tau, 0]; \overline{D(A)})$

(iv) if the operator A is linear, $g_0 : C_b(\mathbb{R}_+; X) \to \mathfrak{X}$ satisfies (g_2), $g_0(0) = 0$ and $\psi \in \mathfrak{X}$.

We assume that (H_A), (H_f) are satisfied.

DEFINITION 3.2.1 We say that the C^0-solution u of (3.1.1) is *stable* if, for each $\eta \in \mathcal{D}$, the unique C^0-solution $v = v(\cdot, \eta)$ of the Cauchy problem

$$\begin{cases} v'(t) \in Av(t) + f(v_t), & t \in \mathbb{R}_+, \\ v(t) = \eta(t), & t \in [-\tau, 0], \end{cases}$$

whose existence and uniqueness is ensured by Theorem 2.8.1, satisfies

$$\|u(t) - v(t)\| \le \|g(u) - \eta\|_{\mathfrak{X}}$$

for each $t \in \mathbb{R}_+$.

We say that the C^0-solution u of (3.1.1) is *globally asymptotically stable* if, for each initial datum $\eta \in \mathcal{D}$, the unique C^0-solution $v = v(\cdot, \eta)$ of the Cauchy problem above satisfies

$$\lim_{t \to +\infty} \|u(t) - v(t)\| = 0.$$

3.2.1 The main result

The main global existence, uniqueness, stability and even global uniform stability results of a bounded C^0-solution for the problem (3.1.1) are:

THEOREM 3.2.1 *If (H_A), (H_f), (H_c) and (H_g) are satisfied, then (3.1.1) has a unique C^0-solution $u \in C_b([-\tau, +\infty); X)$. Moreover, u is globally asymptotically stable.*

THEOREM 3.2.2 *If (H_A), (H_f), (H_c) and (H_g) are satisfied, then the unique C^0-solution u of the problem (3.1.1), given by Theorem 3.2.1, satisfies*

$$\|u\|_{C_b([-\tau,+\infty);X)} \le \frac{m}{\omega - \ell} + k \cdot m_0, \tag{3.2.2}$$

where

$$k = k(a, \omega, \ell) := \frac{\omega}{\omega - \ell} \cdot \left(\frac{1}{e^{\omega a} - 1} + \frac{\ell}{\omega} \right) + 1, \tag{3.2.3}$$

$m = \|f(0)\|$ *and* $m_0 = \|g(0)\|_{\mathfrak{X}}$.

98 *Delay Differential Evolutions Subjected to Nonlocal Initial Conditions*

As concerns the regularity of the C^0-solution whose existence is ensured by Theorem 3.2.1, from Theorem 1.10.3, we deduce:

THEOREM 3.2.3 *If, in addition to the hypotheses of Theorem 3.2.1, we assume that X is a Hilbert space and $A = -\partial\varphi$ is the subdifferential of an l.s.c., proper and convex function $\varphi : X \to [0, +\infty]$, then the unique C^0-solution u of the problem* (3.1.1) *satisfies*

(i) $u(t) \in D(A)$ *a.e. for* $t \in \mathbb{R}_+$

(ii) $t \mapsto t^{1/2}u'(t)$ *belongs to* $L^2(0, T; X)$ *for each $T > 0$*

(iii) $t \mapsto \varphi(u(t))$ *belongs to* $L^1(0, T) \cap AC([\delta, T])$ *for each $0 < \delta < T$*

(iv) *if, in addition, $g(u)(0) \in D(\varphi)$, then $t \mapsto u'(t)$ belongs to $L^2(0, T; X)$ and $t \mapsto \varphi(u(t))$ belongs to $AC([0, T])$ for each $T > 0$.*

REMARK 3.2.6 *If $g(u)(t) = u(t + T)$ for each $u \in C_b(\mathbb{R}_+; \overline{D(A)})$ and each $t \in [-\tau, 0]$, then we have $g(u)(0) \in D(\varphi)$ and thus* (iv) *holds. Similarly, if $g(u)(t) = -u(t + T)$ for each $u \in C_b(\mathbb{R}_+; \overline{D(A)})$ and each $t \in [-\tau, 0]$ and $D(A) = -D(A)$, then we have $g(u)(0) \in D(\varphi)$ and thus* (iv) *holds.*

3.2.2   The nondelayed case

Taking $\tau = 0$, we get some existence results for evolution equations without delay subjected to nonlocal initial conditions. We recall first that, in this case, $\mathfrak{X}$ reduces to X and $\mathcal{D}$ reduces to $\overline{D(A)}$ and therefore $f : X \to X$ and $g : C_b(\mathbb{R}_+; \overline{D(A)}) \to \overline{D(A)}$. Namely, let us consider

$$\begin{cases} u'(t) \in Au(t) + f(u(t)), & t \in \mathbb{R}_+, \\ u(0) = g(u), \end{cases} \tag{3.2.4}$$

where A is as above and f and g satisfy the hypotheses below, which are copies of the hypotheses imposed previously in this section, adapted for the specific case $\tau = 0$. More precisely, we assume the following:

$(H_f^{[\tau=0]})$ The function $f : X \to X$ is continuous and satisfies

$(f_1^{[\tau=0]})$ there exists $\ell > 0$ such that

$$\|f(v) - f(\widetilde{v})\| \le \ell\|v - \widetilde{v}\|$$

for each $v, \widetilde{v} \in X$.

$(H_g^{[\tau=0]})$ The function $g : C_b(\mathbb{R}_+; \overline{D(A)}) \to \overline{D(A)}$ satisfies

$(g_1^{[\tau=0]})$ there exists $a > 0$ such that for each $u, v \in C_b(\mathbb{R}_+; \overline{D(A)})$, we have

$$\|g(u) - g(v)\| \le \|u - v\|_{C_b([a, +\infty); X)}.$$

Nonlocal Initial Conditions: The Autonomous Case 99

REMARK 3.2.7 In the nondelayed case, the condition $(f_1^{[\tau=0]})$ in $(H_f^{[\tau=0]})$ in conjunction with the assumption that $A + \omega I$ is dissipative and (H_c) shows that, for each $t \in \mathbb{R}_+$, $u \mapsto Au + f(u)$ is dissipative in X. Indeed, from (ii), (vii) and $(viii)$ in Proposition 1.8.1, we have

$$[u - v, Au + f(u) - Av - f(v)]_-$$

$$= [u - v, Au + \omega u - Av - \omega v - \omega(u - v) + (f(u) - f(v))]_-$$

$$\le -\omega\|u - v\| + \ell\|u - v\| = (\ell - \omega)\|u - v\| \le 0$$

for each $u, v \in D(A)$, which proves that $u \mapsto Au + f(u)$ is dissipative.

An immediate consequence of Theorems 3.2.1 and 3.2.2 is stated below.

THEOREM 3.2.4 *If* (H_A), $(H_f^{[\tau=0]})$, (H_c), $(H_g^{[\tau=0]})$ *are satisfied, then the problem* (3.2.4) *has a unique* C^0-*solution* $u \in C_b(\mathbb{R}_+; \overline{D(A)})$. *Moreover,* u *is globally asymptotically stable and satisfies*

$$\|u\|_{C_b(\mathbb{R}_+;X)} \le \frac{m}{\omega - \ell} + k \cdot m_0,$$

where k *is given by* (3.2.3), $m = \|f(0)\|$ *and* $m_0 = \|g(0)\|$.

We can pass next to the proofs of Theorems 3.2.1 and 3.2.2, which rest heavily on the abstract theory presented in Section 1.12. More precisely, the idea is to rewrite the problem (3.1.1) as an abstract differential equation subjected to a nonlocal initial condition, in the space $\mathcal{X} = C([-\tau, 0]; X)$ and to take advantage of the Banach Fixed-Point Theorem in order to conclude the existence of C^0-solutions stated in Theorem 3.2.1. We emphasize, however, that there is an alternate proof, using less specialized arguments. We mean here the direct approach we chose in Chapter 4 of obtaining a similar result in the quasi-autonomous case. We note that in this more general case, an appeal to the results in Section 1.12 is possible only if f satisfies a stronger t-continuity assumption which, in the autonomous case, is always fulfilled.

3.3 Proofs of the main theorems

3.3.1 Proof of Theorem 3.2.1

We shall use a fixed-point argument.

Proof. Let $\mathcal{D}$ be given by (3.1.2) and let us define the operator $Q : \mathcal{D} \to \mathcal{D}$ by

$$Qv = g(u_{|\mathbb{R}_+})$$

100 *Delay Differential Evolutions Subjected to Nonlocal Initial Conditions*

for each $v \in \mathcal{D}$, where $u \in C_b([-\tau, +\infty); X)$ is the unique C^0-solution of the auxiliary problem

$$\begin{cases} u'(t) \in Au(t) + f(u_t), & t \in \mathbb{R}_+, \\ u(t) = v(t), & t \in [-\tau, 0]. \end{cases} \qquad (3.3.1)$$

By virtue of Theorems 1.12.1 and 1.12.4, Q is everywhere defined. We will complete the proof by showing that Q has a unique fixed point v which, by means of $v \mapsto u$ via (3.3.1), produces a unique C^0-solution of the problem (3.1.1).

To begin with, let us observe that, in view of Theorem 1.12.1, the problem (3.3.1) rewrites in the space X as

$$\begin{cases} \mathcal{U}'(t) \in \mathcal{A}\mathcal{U}(t), & t \in \mathbb{R}_+, \\ \mathcal{U}(0) = v, \end{cases}$$

where $\mathcal{A}$ is the m-dissipative operator defined by (1.12.2), i.e.,

$$\begin{cases} D(\mathcal{A}) = \{\varphi \in \mathcal{X};\ \varphi' \in \mathcal{X},\ \varphi(0) \in D(A),\ \varphi'(0) \in A\varphi(0) + f(\varphi)\}, \\ \mathcal{A}(\varphi) = \varphi', \ \text{ for } \varphi \in D(\mathcal{A}). \end{cases}$$

Then, the C^0-solution u of (3.3.1) is defined as in Theorem 1.12.2, i.e.,

$$u(t) = \begin{cases} v(t), & t \in [-\tau, 0], \\ [\mathcal{S}(t)v](0), & t \in \mathbb{R}_+, \end{cases}$$

where $\{\mathcal{S}(t) : \mathcal{D} \to \mathcal{D};\ t \in \mathbb{R}_+\}$ is the semigroup generated by $\mathcal{A}$. By (1.12.4), we know that $\mathcal{D} = \overline{D(\mathcal{A})}$.

Let us denote the C^0-solution of the problem (3.3.1) in which v is replaced by $\widetilde{v}$ by $\widetilde{u}$. Since by (H_c) we have $\ell < \omega$, from Theorem 1.12.3, we know that

$$\|\mathcal{S}(t)v - \mathcal{S}(t)\widetilde{v}\|_\alpha = \|u_t - \widetilde{u}_t\|_\alpha \le e^{\alpha t}\|v - \widetilde{v}\|_\alpha$$

for each $t \in \mathbb{R}_+$, where $\alpha \in (-\infty, 0)$ is the unique solution of the equation

$$\alpha = -\omega + \ell e^{-\alpha \tau},$$

while $\|\cdot\|_\alpha$ is an equivalent norm on $\mathcal{X}$, defined by

$$\|\varphi\|_\alpha = \sup_{\theta \in [-\tau, 0]} e^{-\alpha \theta}\|\varphi(\theta)\|,$$

for each $\varphi \in \mathcal{X}$.

Therefore, we have

$$\|Qv - Q\widetilde{v}\|_\alpha = \|g(u) - g(\widetilde{u})\|_\alpha = \sup_{\theta \in [-\tau, 0]} e^{-\alpha \theta}\|g(u)(\theta) - g(\widetilde{u})(\theta)\|$$

$$\leq \sup_{\theta \in [-\tau, 0]} e^{-\alpha\theta} \sup_{t \in [a, +\infty)} \|u(t) - \tilde{u}(t)\| = \sup_{t \in [a, +\infty)} \|u_t(0) - \tilde{u}_t(0)\|$$

$$\leq \sup_{t \in [a, +\infty)} \|u_t - \tilde{u}_t\|_\alpha \leq \sup_{t \in [a, +\infty)} e^{\alpha t} \|v - \tilde{v}\|_\alpha = e^{\alpha a} \|v - \tilde{v}\|_\alpha$$

for each $v, \tilde{v} \in \mathcal{D}$. As $e^{\alpha a} < 1$, it follows that Q is a strict contraction. Since $\mathcal{D}$ is complete being closed in $\mathcal{X}$, Q has a unique fixed point $v \in \mathcal{D}$. Clearly, the C^0-solution u of (3.3.1) corresponding to this fixed point v is a C^0-solution of (3.1.1). Finally, the global asymptotic stability of the C^0-solution follows from Theorem 1.12.3 and this completes the proof of Theorem 3.2.1. $\qquad \square$

3.3.2 Proof of Theorem 3.2.2

The conclusion of Theorem 3.2.2 is an easy consequence of the technical lemma below.

LEMMA 3.3.1 *If* (H_g) *and* (H_c) *are satisfied and* $u \in C_b([-\tau, +\infty); X)$ *is such that*

$$\|u(t)\| \leq e^{-\omega t}\|u(0)\| + (1 - e^{-\omega t})\frac{\ell}{\omega}\left[\|u\|_{C_b([-\tau,+\infty);X)} + \frac{m}{\ell}\right]$$

for each $t \in \mathbb{R}_+$ *and*

$$u(t) = g(u)(t)$$

for $t \in [-\tau, 0]$, *then*

$$\|u\|_{C_b([-\tau,+\infty);X)} \leq \frac{m}{\omega - \ell} + \left[\frac{\omega}{\omega - \ell} \cdot \left(\frac{1}{e^{\omega a} - 1} + \frac{\ell}{\omega}\right) + 1\right] m_0. \qquad (3.3.2)$$

Proof. If there exists $t \in (0, +\infty)$ such that

$$\|u(t)\| = \|u\|_{C_b([-\tau,+\infty);X)}, \qquad (3.3.3)$$

we deduce

$$\left(1 - \frac{\ell}{\omega}\right)\|u\|_{C_b([-\tau,+\infty);X)} \leq \frac{m}{\omega}$$

and thus

$$\|u\|_{C_b([-\tau,+\infty);X)} \leq \frac{m}{\omega - \ell}. \qquad (3.3.4)$$

So, in this case, from (3.3.4), we get (3.3.2). Next, if there exists $t \in [-\tau, 0]$ such that (3.3.3) holds true, from (H_g) and (g_2) in Remark 3.2.2, it follows that

$$\|u\|_{C_b([-\tau,+\infty);X)} \leq \|u\|_{C_b([a,+\infty);X)} + m_0.$$

102 Delay Differential Evolutions Subjected to Nonlocal Initial Conditions

From the last inequality, we deduce that, for each $t \in [a, +\infty)$,

$$\|u(t)\| \le e^{-\omega t}\|u(0)\| + (1 - e^{-\omega t})\frac{\ell}{\omega}\left(\|u\|_{C_b([-\tau,+\infty);X)} + \frac{m}{\ell}\right)$$

$$\le e^{-\omega t}[\|u\|_{C_b([a,+\infty);X)} + m_0] + (1 - e^{-\omega t})\frac{\ell}{\omega}\left(\|u\|_{C_b([a,+\infty);X)} + \frac{m}{\ell} + m_0\right)$$

$$\le e^{-\omega t}\|u\|_{C_b([a,+\infty);X)} + (1 - e^{-\omega t})\frac{\ell}{\omega}\left(\|u\|_{C_b([a,+\infty);X)} + \frac{m}{\ell}\right)$$

$$+ \left(e^{-\omega t} + (1 - e^{-\omega t})\frac{\ell}{\omega}\right)m_0.$$

Let $\varepsilon > 0$ and let $t_\varepsilon \in [a, +\infty)$ such that

$$\|u\|_{C_b([a,+\infty);X)} \le \|u(t_\varepsilon)\| + \varepsilon.$$

Then it follows that

$$\|u\|_{C_b([a,+\infty);X)} \le e^{-\omega t_\varepsilon}\|u\|_{C_b([a,+\infty);X)}$$

$$+(1 - e^{-\omega t_\varepsilon})\frac{\ell}{\omega}\left(\|u\|_{C_b([a,+\infty);X)} + \frac{m}{\ell}\right) + \left(e^{-\omega t_\varepsilon} + (1 - e^{-\omega t_\varepsilon})\frac{\ell}{\omega}\right)m_0 + \varepsilon.$$

Hence we get

$$\left(1 - \frac{\ell}{\omega}\right)\|u\|_{C_b([a,+\infty);X)} \le \frac{m}{\omega} + \left(\frac{1}{e^{\omega a} - 1} + \frac{\ell}{\omega}\right)m_0 + \frac{\varepsilon}{1 - e^{-\omega a}}.$$

Since $\varepsilon > 0$ is arbitrary and $\ell < \omega$, we conclude that

$$\|u\|_{C_b([a,+\infty);X)} \le \frac{m}{\omega - \ell} + \frac{\omega}{\omega - \ell} \cdot \left(\frac{1}{e^{\omega a} - 1} + \frac{\ell}{\omega}\right)m_0.$$

As we already mentioned, in this case, i.e., when (3.3.3) holds true for some t in $[-\tau, 0]$, we have $\|u\|_{C_b([-\tau,+\infty);X)} \le \|u\|_{C_b([a,+\infty);X)} + m_0$. Hence, from the last inequality, we get

$$\|u\|_{C_b([-\tau,+\infty);X)} \le \frac{m}{\omega - \ell} + \left[\frac{\omega}{\omega - \ell} \cdot \left(\frac{1}{e^{\omega a} - 1} + \frac{\ell}{\omega}\right) + 1\right]m_0.$$

If, for each $t \in [-\tau, +\infty)$, we have $\|u(t)\| < \|u\|_{C_b([-\tau,+\infty);X)}$, then there exists at least one sequence $(t_n)_n$ in $\mathbb{R}_+$, with $\lim_{n \to +\infty} t_n = +\infty$, such that

$$\lim_{n \to +\infty}\|u(t_n)\| = \|u\|_{C_b([-\tau,+\infty);X)}.$$

Setting $t = t_n$ in the first inequality in the hypothesis and passing to the limit of both sides for $n \to +\infty$, we get (3.3.4). This completes the proof of Lemma 3.3.1. $\qquad\square$

We are now ready to prove Theorem 3.2.2.

Nonlocal Initial Conditions: The Autonomous Case 103

Proof. From (H_f), we have

$$\|u(t)\| \le e^{-\omega t}\|u(0)\| + e^{-\omega t} \int_0^t e^{\omega s} \left(\|f(u_s) - f(0)\| + \|f(0)\|\right) ds$$

$$\le e^{-\omega t}\|u(0)\| + (1 - e^{-\omega t})\frac{\ell}{\omega} \left(\|u\|_{C_b([-\tau,+\infty);X)} + \frac{m}{\ell}\right),$$

for each $t \in \mathbb{R}_+$. This shows that we are in the hypotheses of Lemma 3.3.1, which implies that (3.2.2) holds true with k given by (3.2.3). $\square$

3.4 The transport equation in $\mathbb{R}^d$

Our aim here is to show how the abstract theory previously developed can be applied in order to get information about the consistency of a mathematical model that describes the flowing of a fluid in the whole space, with velocity v. Usually, the mathematical model of the time×space evolution of the fluid is described by a transport equation subjected to some initial data. As we already mentioned in the "Motivation" section, if we intend to describe the evolution of the fluid under the action of a feedback law containing information on some recent history of the positions of the particles, we have to use a delay evolution transport equation. This heuristic principle is similar to the one that led to the formulation of the model described by the singular transport equation considered in Section 2.9. Furthermore, taking into account that, in various mathematical models, empiric considerations have shown that it is more convenient to consider, instead of an initial history, the average of several measured values at certain points near the initial one, we have to accept, as an alternative, a delay transport equation subjected to a nonlocal initial condition.

So, let us consider the transport equation with delay, subjected to nonlocal initial conditions

$$\begin{cases} \dfrac{\partial u}{\partial t}(t,x) = v \cdot \nabla u(t,x) - \omega u(t,x) + h(x, u_t), & \text{in } \mathbb{R}_+ \times \mathbb{R}^d, \\[2mm] u(t,x) = \displaystyle\sum_{i=1}^{\infty} \alpha_i u(t_i + t, x) + \psi(t)(x), & \text{in } [-\tau, 0] \times \mathbb{R}^d. \end{cases} \tag{3.4.1}$$

Here, $v = (v_1, v_2, \dots v_d)$ is a fixed element in $\mathbb{R}^d$, $d \ge 1$, and $v \cdot \nabla u(t,x)$ is the v-directional distributional derivative of $u(t, \cdot)$, i.e.,

$$v \cdot \nabla u(t,x) = \sum_{i=1}^{d} v_i \frac{\partial u}{\partial x_i}(t,x)$$

104   *Delay Differential Evolutions Subjected to Nonlocal Initial Conditions*

for each $u(t, \cdot) \in L^p(\mathbb{R}^d)$ with $v \cdot \nabla u(t, \cdot) \in L^p(\mathbb{R}^d)$, $1 \le p < +\infty$ and a.e. for $x \in \mathbb{R}^d$. Moreover, $\omega > 0$, $h : \mathbb{R}^d \times C([-\tau, 0]; L^p(\mathbb{R}^d)) \to \mathbb{R}$, $\alpha_i \in [-1, 1]$ for $i = 1, 2, \ldots$, $0 < t_1 < t_2 < \ldots$ and $\psi \in C([-\tau, 0]; L^p(\mathbb{R}^d))$.

THEOREM 3.4.1 *Let $\tau \ge 0$, $\omega > 0$, $1 \le p < +\infty$ and $v \in \mathbb{R}^d$, where $d \ge 1$, let $\psi \in C([-\tau, 0]; L^p(\mathbb{R}^d))$, $h : \mathbb{R}^d \times C([-\tau, 0]; L^p(\mathbb{R}^d)) \to \mathbb{R}$ and let $\alpha_i \in [-1, 1]$, for $i = 1, 2, \ldots$ and $0 < t_1 < t_2 < \ldots$. We assume that*

(h_1) *there exists a nonnegative function $\ell_1 \in L^p(\mathbb{R}^d)$ such that*

$$|h(x, w) - h(x, \widetilde{w})| \le \ell_1(x) \|w - \widetilde{w}\|_{C([-\tau, 0]; L^p(\mathbb{R}^d))}$$

a.e. for $x \in \mathbb{R}^d$ and for each $w, \widetilde{w} \in C([-\tau, 0]; L^p(\mathbb{R}^d))$

(α) $\displaystyle\sum_{i=1}^{\infty} |\alpha_i| \le 1$ *and $\tau < t_1 < t_2 < \ldots$*

(c) $\|\ell_1\|_{L^p(\mathbb{R}^d)} = \ell < \omega$.

Then the problem (3.4.1) has a unique mild solution $u \in C_b([-\tau, +\infty); L^p(\mathbb{R}^d))$. Moreover, u is globally asymptotically stable and

$$\|u\|_{C_b([-\tau, +\infty); L^p(\mathbb{R}^d))} \le \frac{m}{\omega - \ell} + k\|\psi\|_{C([-\tau, 0]; L^p(\mathbb{R}^d))},$$

where $m = \|h(\cdot, 0)\|_{L^p(\mathbb{R}^d)}$ and k is given by (3.2.3).

REMARK 3.4.1 As we shall see in the proof of Theorem 3.4.1, the problem (3.4.1) rewrites as an abstract autonomous semilinear evolution equation subjected to a nonlocal initial condition in the Banach space $X = L^p(\mathbb{R}^d)$. Accordingly, by a mild solution of the problem (3.4.1), we mean a mild solution of the corresponding evolution equation in X in the sense of Definition 1.7.3 on $[0, T]$ for each $T > 0$. Namely, in our case, a mild solution of (3.4.1) is a function $u \in C_b([-\tau, +\infty); L^p(\mathbb{R}^d))$ satisfying

$$\begin{cases} u(t, x) = e^{-\omega t} u(0, x) + \displaystyle\int_0^t e^{-\omega(t-s)} h(x + (t-s)v, u_s) \, ds, & \text{in } \mathbb{R}_+ \times \mathbb{R}^d, \\ u(t, x) = \displaystyle\sum_{i=1}^{n} \alpha_i u(t_i + t, x) + \psi(t)(x), & \text{in } [-\tau, 0] \times \mathbb{R}^d. \end{cases}$$

See Vrabie [254, Problem 6.15, p. 237], Theorem 1.9.5 and Definition 1.7.3.

We can now proceed to the proof of Theorem 3.4.1.
Proof. Let $X = L^p(\mathbb{R}^d)$ and $A : D(A) \subseteq X \to X$ defined by

$$D(A) = \{ u \in L^p(\mathbb{R}^d); \ v \cdot \nabla u \in L^p(\mathbb{R}^d) \},$$

$$Au = v \cdot \nabla u - \omega u$$

Nonlocal Initial Conditions: The Autonomous Case

for each $u \in D(A)$. By Theorem 1.9.5, the operator $B : D(B) \subseteq X \to X$ with $D(B) = D(A)$ and $Bu = Au + \omega u$ for each $u \in D(A)$, where A is defined as above, is the infinitesimal generator of the C_0-group of isometries $\{G(t) : X \to X;\ t \in \mathbb{R}\}$, given by

$$[G(t)\xi](x) = \xi(x + tv)$$

for each $\xi \in X$, $t \in \mathbb{R}$ and a.e. for $x \in \mathbb{R}^d$. Thus A generates a C_0-semigroup of contractions $\{S(t) : X \to X;\ t \in \mathbb{R}_+\}$, defined by

$$[S(t)\xi](x) = e^{-\omega t}\xi(x + tv)$$

for each $t \in \mathbb{R}_+$, each $\xi \in X$ and a.e. for $x \in \mathbb{R}^d$.

Since h is Lipschitz continuous, it follows that we can define the superposition operator $f : \mathcal{X} \to \mathcal{X}$ by

$$f(w)(x) = h(x, w)$$

for each $w \in \mathcal{X}$ and a.e. for $x \in \mathbb{R}^d$. Moreover, let us define the function $g : C_b(\mathbb{R}_+; X) \to \mathcal{X}$ by

$$[g(u)(t)](x) = \sum_{i=1}^{\infty} \alpha_i u(t_i + t, x) + \psi(t)(x)$$

for each $u \in C_b([-\tau, +\infty); X)$, each $t \in [-\tau, 0]$ and $x \in \mathbb{R}^d$. So, the problem (3.4.1) can be rewritten in the abstract form (3.1.1), where X, A, f and g are as above. Clearly A, which is linear, satisfies (H_A). Furthermore, in view of (h_1), it follows that the function f satisfies (H_F) while from (α), we deduce that g satisfies (H_g) with $a = t_1 - \tau > 0$ and $m_0 = \|\psi\|_{\mathcal{X}}$, where $\mathcal{X} = C([-\tau, 0]; X)$ for $X = L^p(\mathbb{R}^d)$. Hence, the conclusion follows from Theorem 3.2.1 combined with Remark 1.8.1. $\qquad\square$

REMARK 3.4.2 It is not difficult to realize that, in (3.4.1), it would be more convenient – in fact more practical and more realistic – if, instead of the feedback law h, we would consider a localized one, i.e., of the form

$$h(x, w) = h_0(\chi_\Omega(x), w)$$

for each $w \in C([-\tau, 0]; \mathbb{R}^d)$ and a.e. for $x \in \mathbb{R}^d$, where $\Omega \subseteq \mathbb{R}^d$ is a bounded domain and χ_Ω is its characteristic function. We note that Theorem 3.4.1 extends, with no major modifications, to this case.

REMARK 3.4.3 We indicate some examples of forcing terms h, satisfying the hypothesis (h_1) and defined as partial superposition operators, with respect to the second argument, of a real-valued function $\widetilde{h} : \mathbb{R}^d \times \mathbb{R} \to \mathbb{R}$.

106 *Delay Differential Evolutions Subjected to Nonlocal Initial Conditions*

(i)
$$h(x, u) = \widetilde{h}(x, u(-\tau)(x))$$

for $u \in C([-\tau, 0]; L^p(\mathbb{R}^d))$ and a.e. for $x \in \mathbb{R}^d$. Then,

$$h(x, u_t) = \widetilde{h}(x, u(t-\tau)(x)),$$

for $t \in \mathbb{R}_+$, $u \in C_b([-\tau, +\infty); L^p(\mathbb{R}^d))$ and a.e. for $x \in \mathbb{R}^d$, is the usual delay difference forcing term.

(ii)
$$h(x, u) = \widetilde{h}\left(x, \int_{-\tau}^{0} k(s)u(s)(x)\, ds\right)$$

for each $u \in C([-\tau, 0]; L^p(\mathbb{R}^d))$ and a.e. for $x \in \mathbb{R}^d$, is a neutral type forcing term, i.e.,

$$h(x, u_t) = \widetilde{h}\left(x, \int_{-\tau}^{0} k(s)u(t+s)(x)\, ds\right),$$

for $u \in C_b([-\tau, +\infty); L^p(\mathbb{R}^d))$, $t \in \mathbb{R}_+$ and a.e. for $x \in \mathbb{R}^d$.

It is easy to verify that whenever $\widetilde{h}$ is continuous and there exists the non-negative function $\ell_2 \in L^1(\mathbb{R}^d)$ such that $|\widetilde{h}(x, u) - \widetilde{h}(x, v)| \leq \ell_2(x)|u - v|$ a.e. for $x \in \mathbb{R}^d$ and for each $u, v \in \mathbb{R}$, then h, defined as in (i), satisfies (h_1) with $\ell_1(x) = \ell_2(x)$ a.e. for $x \in \mathbb{R}^d$. If $\widetilde{h}$ is as above and $k \in L^1(-\tau, 0; \mathbb{R})$, then h defined as in (ii) satisfies (h_1) with $\ell_1(x) = \ell_2(x)\|k\|_{L^1(-\tau,0;\mathbb{R})}^{-1}$ a.e. for $x \in \mathbb{R}^d$.

REMARK 3.4.4 Particularizing the systems of points α_i, t_i, for $i = 1, 2, \ldots$ and $\psi \in C([-\tau, 0]; L^p(\mathbb{R}^d))$, we obtain two important nonlocal initial conditions. More precisely,

(j) for $n = 1$, $\alpha_1 = 1$, $\tau < t_1 = T$ and $\psi = 0$, we get the T-periodicity condition, i.e.,
$$u(t, x) = u(T + t, x),$$
for each $(t, x) \in [-\tau, 0] \times \mathbb{R}^d$

(jj) for $n = 1$, $\alpha_1 = -1$, $\tau < t_1 = T$ and $\psi = 0$, we get the T-anti-periodicity condition, i.e.,
$$u(t, x) = -u(T + t, x),$$
for each $(t, x) \in [-\tau, 0] \times \mathbb{R}^d$.

REMARK 3.4.5 With essentially the same proof, we can get an existence, uniqueness, and global asymptotic stability result referring to a similar transport equation with delay subjected to more general nonlocal initial conditions of the form

$$u(t, x) = \int_{\tau}^{+\infty} \mathcal{N}(x, u(t+s, x))\, d\mu(s) + \psi(t)(x), \quad \text{in } [-\tau, 0] \times \mathbb{R}^d,$$

where $\mathcal{N} : \mathbb{R}^d \times \mathbb{R} \to \mathbb{R}$ is continuous and μ is a σ-finite and complete measure on $[\tau, +\infty)$, both satisfying the following:

Nonlocal Initial Conditions: The Autonomous Case 107

(k_1) there exists a nonnegative function $\eta \in L^p(\mathbb{R}^d)$ such that

$$|\mathcal{N}(x, u) - \mathcal{N}(x, \widetilde{u})| \leq \eta(x)|u - \widetilde{u}|$$

for each $u, \widetilde{u} \in \mathbb{R}$ and a.e for $x \in \mathbb{R}^d$

(k_2) $\|\eta\|_{L^p(\mathbb{R}^d)} \leq 1$

(k_3) μ is continuous with respect to the Lebesgue measure at $t = \tau$, i.e., $\lim_{h \downarrow \tau} \mu([\tau, \tau + h]) = 0$.

3.5 The damped wave equation with nonlocal initial conditions

Here we consider a situation more general than in Example 2.7.1. Roughly speaking, we have a domain in which each point obeys a delay Spring Mass System, i.e., a second-order equation derived from the system in Example 2.7.1, and which is subjected to a nonlocal initial condition. So, let Ω be a nonempty bounded and open subset in $\mathbb{R}^d$, $d \geq 1$, with C^1 boundary Σ, let $Q_+ = \mathbb{R}_+ \times \Omega$, $Q_\tau = [-\tau, 0] \times \Omega$, $\Sigma_+ = \mathbb{R}_+ \times \Sigma$, let $\omega > 0$, and let us consider the damped wave equation with delay, subjected to nonlocal initial conditions:

$$\begin{cases} \dfrac{\partial^2 u}{\partial t^2} = \Delta u - 2\omega \dfrac{\partial u}{\partial t} - \omega^2 u + h\left(u_t, \left(\dfrac{\partial u}{\partial t}\right)_t\right), & \text{in } Q_+, \\[2ex] u(t, x) = 0, & \text{on } \Sigma_+, \\[2ex] u(t, x) = \displaystyle\int_\tau^{+\infty} \alpha(s)u(t + s, x)\, ds + \psi_1(t)(x), & \text{in } Q_\tau, \\[2ex] \dfrac{\partial u}{\partial t}(t, x) = \displaystyle\int_\tau^{+\infty} \mathcal{N}\left(s, u(t + s, x), \dfrac{\partial u}{\partial t}(t + s, x)\right) ds + \psi_2(t)(x), & \text{in } Q_\tau. \end{cases} \quad (3.5.1)$$

The next result is a nontrivial application of Theorem 3.2.1. Let us recall first some notations, i.e.,

$$\mathcal{H}_0^1 = C([-\tau, 0]; H_0^1(\Omega)) \text{ and } \mathcal{X}_2 = C([-\tau, 0]; L^2(\Omega)).$$

THEOREM 3.5.1 *Let Ω be a nonempty bounded and open subset in $\mathbb{R}^d$, $d \geq 1$, with C^1 boundary Σ, let $\tau \geq 0$, $\omega > 0$ and let $\psi_1 \in \mathcal{H}_0^1$ and $\psi_2 \in \mathcal{X}_2$. Finally, let $h : \mathcal{H}_0^1 \times \mathcal{X}_2 \to L^2(\Omega)$, $\alpha \in L^2(\mathbb{R}_+)$ and $\mathcal{N} : \mathbb{R}_+ \times \mathbb{R} \times \mathbb{R} \to \mathbb{R}$ be continuous functions. In addition, let us assume that*

(h_1) *there exists $\widetilde{\ell} > 0$ such that*

$$\|h(w, y) - h(\widetilde{w}, \widetilde{y})\|_{L^2(\Omega)} \leq \widetilde{\ell}\left(\|w - \widetilde{w}\|_{\mathcal{H}_0^1} + \|y - \widetilde{y}\|_{\mathcal{X}_2}\right)$$

for all $w, \widetilde{w} \in \mathcal{H}_0^1$ and $y, \widetilde{y} \in \mathcal{X}_2$

108   Delay Differential Evolutions Subjected to Nonlocal Initial Conditions

(n_1) *there exists a nonnegative continuous function $\eta \in L^2(\mathbb{R}_+)$ such that*

$$|\mathcal{N}(t, u, v)| \le \eta(t)(|u| + |v|),$$

for each $t \in \mathbb{R}_+$ and $u, v \in \mathbb{R}$

(n_2) $|\mathcal{N}(t, u, v) - \mathcal{N}(t, \widetilde{u}, \widetilde{v})| \le \eta(t)(|u - \widetilde{u}| + |v - \widetilde{v}|)$, *for each $t \in \mathbb{R}_+$ and $u, \widetilde{u}, v, \widetilde{v} \in \mathbb{R}$, where η is given by (n_1)*

(n_3) *with λ_1 the first eigenvalue of $-\Delta$, we have*

$$\begin{cases} \|\eta\|_{L^2(\mathbb{R}_+)} \le 1, \\ (1 + \lambda_1^{-1}\omega)\|\alpha\|_{L^2(\mathbb{R}_+)} + \lambda_1^{-1}(1 + \omega)\|\eta\|_{L^2(\mathbb{R}_+)} \le 1 \end{cases}$$

(n_4) *there exists $b > \tau$ such that $\alpha(t) = \eta(t) = 0$ for each $t \in [0, b]$*

(c_1) $\ell = \widetilde{\ell}(1 + \omega\lambda_1^{-1}) < \omega.$

Then the problem (3.5.1) has a unique mild solution $u \in C_b([-\tau, +\infty); H_0^1(\Omega))$, with $\dfrac{\partial u}{\partial t} \in C_b([-\tau, +\infty); L^2(\Omega))$. Moreover, u is globally asymptotically stable and

$$\|u\|_{C_b([-\tau, +\infty); H_0^1(\Omega))} + \left\| \frac{\partial u}{\partial t} \right\|_{C_b([-\tau, +\infty); L^2(\Omega))} \le \frac{m}{\omega - \ell} + k \cdot m_0,$$

where $m = \|h(0, 0)\|_{L^2(\Omega)}$, k is given by (3.2.3) and

$$m_0 = \|\psi_1\|_{\mathcal{H}_0^1} + \omega\|\psi_1\|_{\mathcal{X}_2} + \|\psi_2\|_{\mathcal{X}_2}.$$

For the meaning of mild solution in this context see Remark 2.2.1.

Proof. First, let us observe that (3.5.1) can be equivalently rewritten as a first-order system of partial differential equations of the form

$$\begin{cases} \dfrac{\partial u}{\partial t}(t, x) = v(t, x) - \omega u(t, x), & \text{in } Q_+, \\[2ex] \dfrac{\partial v}{\partial t}(t, x) = \Delta u(t, x) - \omega v(t, x) + h(u_t, w_t), & \text{in } Q_+, \\[2ex] u(t, x) = 0, & \text{on } \Sigma_+, \quad (3.5.2) \\[2ex] u(t, x) = \displaystyle\int_\tau^{+\infty} \alpha(s)u(t + s, x)\, ds + \psi_1(t)(x), & \text{in } Q_\tau, \\[2ex] v(t, x) = \displaystyle\int_\tau^{+\infty} \mathcal{R}(s, u(t + s, x), w(t + s, x))\, ds + \psi_3(t)(x), & \text{in } Q_\tau, \end{cases}$$

where

$$\begin{cases} w = v - \omega u, \\ \psi_3 = \omega\psi_1 + \psi_2, \\ \mathcal{R}(t, u, w) = \omega\alpha(t)u + \mathcal{N}(t, u, w), \end{cases} \quad (3.5.3)$$

Nonlocal Initial Conditions: The Autonomous Case 109

in the product space $X = \begin{pmatrix} H_0^1(\Omega) \\ \times \\ L^2(\Omega) \end{pmatrix}$. Endowed with the usual inner product

$$\left\langle \begin{pmatrix} u \\ v \end{pmatrix}, \begin{pmatrix} \widetilde{u} \\ \widetilde{v} \end{pmatrix} \right\rangle = \int_\Omega \nabla u(x) \cdot \nabla \widetilde{u}(x)\, dx + \int_\Omega v(x)\widetilde{v}(x)\, dx$$

for each $\begin{pmatrix} u \\ v \end{pmatrix}, \begin{pmatrix} \widetilde{u} \\ \widetilde{v} \end{pmatrix} \in X$, X is a real Hilbert space. In turn, (3.5.2) can be rewritten as an abstract evolution equation subjected to nonlocal initial conditions of the form (3.1.1), where A, f and g are defined as follows.

First, let us define the linear operator $A : D(A) \subseteq X \to X$ by

$$D(A) = \begin{pmatrix} H_0^1(\Omega) \cap H^2(\Omega) \\ \times \\ H_0^1(\Omega) \end{pmatrix}, \qquad A\begin{pmatrix} u \\ v \end{pmatrix} = \begin{pmatrix} -\omega u + v \\ \Delta u - \omega v \end{pmatrix}$$

for each $\begin{pmatrix} u \\ v \end{pmatrix} \in D(A)$.

Second, let us define $f : \mathcal{X} \to X$ by

$$f\left(\begin{pmatrix} z \\ y \end{pmatrix} \right) = \begin{pmatrix} 0 \\ h(z, \zeta) \end{pmatrix}$$

for each $\begin{pmatrix} z \\ y \end{pmatrix} \in \mathcal{X}$, where $\zeta = y - \omega z$ and $\mathcal{X} = C([-\tau, 0]; X)$.

Third, the nonlocal constraint $g : C_b(\mathbb{R}_+; X) \to \mathcal{X}$ is given by

$$\left[g\begin{pmatrix} u \\ v \end{pmatrix}(t) \right](x) = \begin{pmatrix} \displaystyle\int_\tau^{+\infty} \alpha(s)u(t+s, x)\, ds + \psi_1(t)(x) \\ \displaystyle\int_\tau^{+\infty} \mathcal{R}\left(s, u(t+s, x), w(t+s, x)\right) ds + \psi_3(t)(x) \end{pmatrix}$$

for each $\begin{pmatrix} u \\ v \end{pmatrix} \in C_b(\mathbb{R}_+; X)$, each $t \in [-\tau, 0]$, and a.e. for $x \in \Omega$, where w, ψ_3 and $\mathcal{R}$ are defined in (3.5.3).

By Theorem 1.9.4, it follows that the linear operator $B : D(B) \subseteq X \to X$, with $D(B) = D(A)$ and $B = A + \omega I$, where A is defined as above and I is the identity on X, is the infinitesimal generator of a C_0-group of unitary operators $\{G(t) : X \to X; \ t \in \mathbb{R}\}$. Consequently, A generates a C_0-semigroup of contractions $\{S(t) : X \to X; \ t \in \mathbb{R}_+\}$, defined by

$$S(t)\xi = e^{-\omega t} G(t)\xi$$

for each $t \in \mathbb{R}_+$ and each $\xi \in X$ and thus it is m-dissipative. So, A satisfies

110   *Delay Differential Evolutions Subjected to Nonlocal Initial Conditions*

(H_A). Furthermore, in view of (h_1), it follows that the function f satisfies (H_f). Indeed, for each $\begin{pmatrix} z \\ y \end{pmatrix}, \begin{pmatrix} \widetilde{z} \\ \widetilde{y} \end{pmatrix} \in \mathfrak{X}$, we have

$$\left\| f\left(\begin{pmatrix} z \\ y \end{pmatrix} \right) - f\left(\begin{pmatrix} \widetilde{z} \\ \widetilde{y} \end{pmatrix} \right) \right\| = \| h(z, y - \omega z) - h(\widetilde{z}, \widetilde{y} - \omega \widetilde{z}) \|_{L^2(\Omega)}$$

$$\leq \widetilde{\ell} \left(\| z - \widetilde{z} \|_{\mathcal{H}_0^1} + \| (y - \omega z) - (\widetilde{y} - \omega \widetilde{z}) \|_{\mathfrak{X}_2} \right)$$

$$\leq \widetilde{\ell} \| z - \widetilde{z} \|_{\mathcal{H}_0^1} + \widetilde{\ell} \| y - \widetilde{y} \|_{\mathfrak{X}_2} + \widetilde{\ell} \omega \| z - \widetilde{z} \|_{\mathfrak{X}_2}.$$

By the Poincaré Inequality – see Lemma 1.9.1 – we have

$$\| z - \widetilde{z} \|_{\mathfrak{X}_2} \leq \lambda_1^{-1} \| z - \widetilde{z} \|_{\mathcal{H}_0^1},$$

where λ_1 is the first eigenvalue of $-\Delta$. Consequently, we have

$$\left\| f\left(\begin{pmatrix} z \\ y \end{pmatrix} \right) - f\left(\begin{pmatrix} \widetilde{z} \\ \widetilde{y} \end{pmatrix} \right) \right\| \leq \widetilde{\ell}(1 + \omega \lambda_1^{-1}) \left\| \begin{pmatrix} z \\ y \end{pmatrix} - \begin{pmatrix} \widetilde{z} \\ \widetilde{y} \end{pmatrix} \right\|_{\mathfrak{X}}$$

for each $\begin{pmatrix} z \\ y \end{pmatrix}, \begin{pmatrix} \widetilde{z} \\ \widetilde{y} \end{pmatrix} \in \mathfrak{X}$, which in view of (c_1), shows that f satisfies (f_1) in (H_f) in Theorem 3.2.1 with the Lipschitz constant $\widetilde{\ell}(1 + \omega \lambda_1^{-1})$.

We show next that g satisfies (g_1) in (H_g) in Theorem 3.2.1. To this end, let $\begin{pmatrix} u \\ v \end{pmatrix}, \begin{pmatrix} \widetilde{u} \\ \widetilde{v} \end{pmatrix} \in C_b(\mathbb{R}_+; X)$. Then

$$\left\| g\left(\begin{pmatrix} u \\ v \end{pmatrix} \right)(t) - g\left(\begin{pmatrix} \widetilde{u} \\ \widetilde{v} \end{pmatrix} \right)(t) \right\| \leq \int_\tau^{+\infty} \alpha(s) \| \nabla u(t+s, \cdot) - \nabla \widetilde{u}(t+s, \cdot) \|_{L^2(\Omega)} \, ds$$

$$+ \int_\tau^{+\infty} \| \mathcal{R}\left(s, u(t+s, \cdot), w(t+s, \cdot) \right) - \mathcal{R}\left(s, \widetilde{u}(t+s, \cdot), \widetilde{w}(t+s, \cdot) \right) \|_{L^2(\Omega)} \, ds$$

for each $t \in [-\tau, 0]$, where $w = v - \omega u$ and $\widetilde{w} = \widetilde{v} - \omega \widetilde{u}$. Using $(n_2) \sim (n_4)$, we finally get

$$\left\| g\left(\begin{pmatrix} u \\ v \end{pmatrix} \right) - g\left(\begin{pmatrix} \widetilde{u} \\ \widetilde{v} \end{pmatrix} \right) \right\|_{\mathfrak{X}} \leq \| \alpha \|_{L^2(\mathbb{R}_+)} \| u - \widetilde{u} \|_{C_b([a,+\infty); H_0^1(\Omega))}$$

$$+ \left(\omega \| \alpha \|_{L^2(\mathbb{R}_+)} + (1 + \omega) \| \eta \|_{L^2(\mathbb{R}_+)} \right) \| u - \widetilde{u} \|_{C_b([a,+\infty); L^2(\Omega))}$$

$$+ \| \eta \|_{L^2(\mathbb{R}_+)} \| v - \widetilde{v} \|_{C_b([a,+\infty); L^2(\Omega))}$$

$$\leq \left[\| \alpha \|_{L^2(\mathbb{R}_+)} + \lambda_1^{-1} \left(\omega \| \alpha \|_{L^2(\mathbb{R}_+)} + (1 + \omega) \| \eta \|_{L^2(\mathbb{R}_+)} \right) \right] \| u - \widetilde{u} \|_{C_b([a,+\infty); H_0^1(\Omega))}$$

$$+\|\eta\|_{L^2(\mathbb{R}_+)}\|v - \widetilde{v}\|_{C_b([a,+\infty);L^2(\Omega))} \leq \widetilde{m}\left\|\begin{pmatrix} u \\ v \end{pmatrix} - \begin{pmatrix} \widetilde{u} \\ \widetilde{v} \end{pmatrix}\right\|_{C_b([a,+\infty);X)},$$

where $a = b - \tau$ and $\widetilde{m}$ is defined by

$$\widetilde{m} = \max\left\{(1 + \lambda_1^{-1}\omega)\|\alpha\|_{L^2(\mathbb{R}_+)} + \lambda_1^{-1}(1 + \omega)\|\eta\|_{L^2(\mathbb{R}_+)}, \|\eta\|_{L^2(\mathbb{R}_+)}\right\}.$$

From (n_3), we deduce that $\widetilde{m} \leq 1$ and accordingly g satisfies (g_1) in (H_g). The conclusion follows from Theorem 3.2.1 and Remark 1.8.1. $\qquad\square$

3.6 The case f Lipschitz and g continuous

Let $a \in \mathbb{R}$. We endow now the space $C_b([a,+\infty);X)$ with a locally convex topology needed in our next analysis. So, let us consider the family of seminorms $\{\|\cdot\|_k;\ k \in \mathbb{N},\ k \geq a\}$, defined by $\|u\|_k = \sup\{\|u(t)\|;\ t \in [a,k]\}$ for each $k \in \mathbb{N},\ k \geq a$. Endowed with this family of seminorms, $C_b([a,+\infty);X)$ is a separated locally convex space, denoted by $\widetilde{C}_b([a,+\infty);X)$ and whose topology is strictly weaker than the norm topology. In fact, $\widetilde{C}_b([a,+\infty);X)$ is a Fréchet space and the usual metric d on $\widetilde{C}_b([a,+\infty);X)$ defining a topology which coincides with the uniform convergence topology on compacta defined by the above family of seminorms, is given by

$$d_X(u,\widetilde{u}) = \sum_{k=1}^{\infty} \frac{1}{2^k} \frac{\|u - \widetilde{u}\|_k}{1 + \|u - \widetilde{u}\|_k}$$

for each $u, \widetilde{u} \in C_b([a,+\infty);X)$.

REMARK 3.6.1 One may ask why we are not working with the space $C_b([a,+\infty);X)$ endowed with its own Banach space topology. The reason for considering a new topology on $C_b([a,+\infty);X)$ is that some very important compactness arguments needed in the forthcoming proofs, although valid in $\widetilde{C}_b([a,+\infty);X)$, do not work in $C_b([a,+\infty);X)$. Namely, as in the case of $C([a,b];X)$, where a relatively compact set should be equicontinuous on $[a,b]$ and, in particular, at $t = b$, similarly, in $C_b([a,+\infty);X)$, a relatively compact set has to be "equicontinuous at $t = +\infty$." As the latter property is almost impossible to verify, instead of working with compact sets in the norm topology of $C_b([a,+\infty);X)$, we must work with compact sets in the convergence on compacta topology, i.e., in the space $\widetilde{C}_b([a,+\infty);X)$.

We recall that $\mathcal{X} = C([-\tau,0];X)$, $\mathcal{D} = \{\varphi \in \mathcal{X};\ \varphi(0) \in \overline{D(A)}\}$ and we reconsider next the problem (3.1.1) under the assumptions below.

112   *Delay Differential Evolutions Subjected to Nonlocal Initial Conditions*

(H_A) The operator $A : D(A) \subseteq X \rightsquigarrow X$ satisfies

> (A_1) $0 \in D(A)$, $0 \in A0$ and A is ω-m-dissipative for some $\omega > 0$
>
> (A_2) the semigroup generated by A is compact
>
> (A_3) $\overline{D(A)}$ is convex.

(H_f) The function $f : \mathfrak{X} \to X$ is continuous and satisfies

> (f_1) there exists $\ell > 0$ such that
>
> $$\|f(v) - f(\widetilde{v})\| \le \ell \|v - \widetilde{v}\|_{\mathfrak{X}}$$
>
> for each $v, \widetilde{v} \in \mathfrak{X}$.

(H_c) The constants ℓ and ω satisfy $\ell < \omega$.

(H_g) The function $g : C_b(\mathbb{R}_+; \overline{D(A)}) \to \mathcal{D}$ is continuous and satisfies

> (g_2) g has affine growth, i.e., there exist $a > 0$ and $m_0 \ge 0$ such that
>
> $$\|g(u)\|_{\mathfrak{X}} \le \|u\|_{C_b([a,+\infty);X)} + m_0$$
>
> for each $u \in C_b(\mathbb{R}_+; \overline{D(A)})$
>
> (g_3) with a given by (g_2), for each $u, v \in C_b(\mathbb{R}_+; \overline{D(A)})$ satisfying $u(t) = v(t)$ for each $t \in [a, +\infty)$, we have $g(u) = g(v)$
>
> (g_4) g is continuous from $C_b([a,+\infty); \overline{D(A)})$ endowed with the $\widetilde{C}_b([a,+\infty); X)$ topology to $\mathfrak{X}$, where a is given by (g_2).

REMARK 3.6.2 If g is defined as in Remark 3.2.4, then it satisfies the conditions (g_2), (g_3), and (g_4) in (H_g).

THEOREM 3.6.1 *If (H_A), (H_f), (H_c) and (H_g) are satisfied, then the problem (3.1.1) has at least one C^0-solution, $u \in C_b([-\tau, +\infty); X)$. Moreover, each C^0-solution of (3.1.1) satisfies*

$$\|u\|_{C_b([-\tau,+\infty);X)} \le \frac{m}{\omega - \ell} + k \cdot m_0, \tag{3.6.1}$$

where $m = \|f(0)\|$ and k is given by (3.2.3), and is globally asymptotically stable.

Proof. To prove the existence part, we will use a fixed-point argument based on the Schauder Fixed-Point Theorem 1.4.4. The idea is to consider the same auxiliary problem as we already did in Section 3.3, i.e.,

$$\begin{cases} u'(t) \in Au(t) + f(u_t), & t \in \mathbb{R}_+, \\ u(t) = v(t), & t \in [-\tau, 0], \end{cases} \tag{3.6.2}$$

Nonlocal Initial Conditions: The Autonomous Case 113

which, for a fixed $v \in \mathcal{D}$, due to Theorems 1.12.1 and 1.12.4, has a unique global C^0-solution u in $C_b([-\tau, +\infty); X)$ and to define $Q : \mathcal{D} \to \mathcal{D}$, by

$$Qv = g(u_{|\mathbb{R}_+})$$

for each $v \in \mathcal{D}$. Then, it suffices to show that Q has at least one fixed point. First, we will prove that for a suitably chosen $r > 0$, Q maps

$$B_\alpha = D_\alpha(0, r) \cap \mathcal{D}$$

into itself, where $D_\alpha(0, r)$ denotes the closed ball with radius r and centered at 0 in X endowed with the norm $\|\cdot\|_\alpha$. Here, $\alpha \in (-\infty, 0)$ is the unique solution of the equation $\alpha = -\omega + \ell e^{-\alpha\tau}$. For the definition and the main properties of the norm $\|\cdot\|_\alpha$, see Section 1.12. Then, we will prove that Q is continuous and compact and thus, in view of the Schauder Fixed-Point Theorem, it has at least one fixed point $v \in B_\alpha$. Clearly, the unique global C^0-solution $u \in C_b([-\tau, +\infty); X)$ of the problem (3.6.2) corresponding to this fixed point v is a C^0-solution of (3.1.1) and this will complete the proof of the existence part in Theorem 3.6.1.

To begin with, let us observe that, in view (A_1) and (A_3) in (H_A), B_α is nonempty, closed, and convex. Moreover, thanks to Theorem 1.12.1, the problem (3.6.2) rewrites as

$$\begin{cases} \mathcal{U}'(t) \in \mathcal{A}\mathcal{U}(t), & t \in \mathbb{R}_+, \\ \mathcal{U}(0) = v, \end{cases}$$

where $\mathcal{A}$ is the m-dissipative operator defined by (1.12.2), i.e.,

$$\begin{cases} D(\mathcal{A}) = \{\varphi \in \mathcal{X}; \ \varphi' \in \mathcal{X}, \ \varphi(0) \in D(A), \ \varphi'(0) \in A\varphi(0) + f(\varphi)\}, \\ \mathcal{A}(\varphi) = \varphi', \ \text{for } \varphi \in D(\mathcal{A}). \end{cases}$$

Then, the C^0-solution u of (3.6.2) is defined by (1.12.5), i.e.,

$$u(t) = \begin{cases} v(t), & t \in [-\tau, 0], \\ [S(t)v](0), & t \in \mathbb{R}_+. \end{cases}$$

Here $\{S(t) : \mathcal{D} \to \mathcal{D}; t \in \mathbb{R}_+\}$ is the semigroup generated by $\mathcal{A}$ on $\mathcal{D} = \overline{D(\mathcal{A})}$. Let $r > 0$ and let $v \in D_\alpha(0, r)$ be arbitrary. Then, we have

$$\|Qv\|_\alpha = \|g(u)\|_\alpha = \sup_{\theta \in [-\tau, 0]} e^{-\alpha\theta} \|g(u)(\theta)\|$$

$$\leq \sup_{t \in [a, +\infty)} \|u(t)\| + m_0 \leq \sup_{t \in [a, +\infty)} \|u_t(0)\| + m_0$$

$$\leq \sup_{t \in [a, +\infty)} \|u_t\|_\alpha + m_0.$$

114 *Delay Differential Evolutions Subjected to Nonlocal Initial Conditions*

From Theorem 1.12.1, recalling that, for each $t \in \mathbb{R}_+$, $u_t = \mathcal{S}(t)v$, we conclude that

$$\|Qv\|_\alpha \leq \sup_{t \in [a,+\infty)} [\|\mathcal{S}(t)v - \mathcal{S}(t)0\|_\alpha + \|\mathcal{S}(t)0\|_\alpha] + m_0$$

$$\leq \sup_{t \in [a,+\infty)} [e^{\alpha t}\|v\|_\alpha + \|\mathcal{S}(t)0\|_\alpha] + m_0.$$

Taking into account that $\|v\|_\alpha \leq r$ and that, by Theorem 1.12.4, there exists $m_1 \geq 0$ such that

$$\|\mathcal{S}(t)0\|_\alpha \leq m_1$$

for each $t \in \mathbb{R}_+$, we deduce

$$\|Qv\|_\alpha \leq e^{\alpha a}r + m_1 + m_0.$$

So, if we choose a sufficiently large r, i.e.,

$$\frac{m_1 + m_0}{1 - e^{\alpha a}} \leq r,$$

we get $\|Qv\|_\alpha \leq r$. Thus, with $r > 0$ fixed as above, Q maps B_α into itself.

To prove the compactness of $Q(B_\alpha)$, we observe that Q can be decomposed as

$$Q = g \circ S$$

where $S : \mathcal{D} \to C_b([-\tau, +\infty); X)$ is defined by $S(v) = u_{|\mathbb{R}_+}$ for each $v \in \mathcal{D}$, where u is the unique global C^0-solution of (3.6.2). Since, by (g_4), g is continuous from $\widetilde{C}_b([a, +\infty); X)$ to X, it would be sufficient to show that $S(B_\alpha)$ is relatively compact in $\widetilde{C}_b([-\tau, +\infty); X)$. Clearly, by 1.12.3 in Theorem 1.12.1, it follows that $S(B_\alpha)$ is bounded in $C_b(\mathbb{R}_+; X)$. Thanks to (H_f), we conclude that the family of functions $\{t \mapsto f(S(v)_t); \ v \in B_\alpha\}$ is bounded in $C_b(\mathbb{R}_+; X)$. From Theorem 1.8.6, we deduce that $S(B_\alpha)$ is relatively compact in $\widetilde{C}_b([a, +\infty); X)$. As a consequence, $g(S(B_\alpha)) = Q(B_\alpha)$ is relatively compact in X.

Finally, thanks to the compactness of $S(B_\alpha)$, it follows that Q is continuous from B_α into B_α. Indeed, if $(v_n)_n$ is a sequence in B_α that converges to some function $v \in B_\alpha$, it follows that the corresponding sequence $(u_n)_n$ of C^0-solutions for the problem (3.6.2) is in $S(B_\alpha)$ and converges to the unique C^0-solution for the same problem, corresponding to the function v. Indeed, this simply follows because, for each $n \in \mathbb{N}$, we have

$$u_n(t) = \begin{cases} v_n(t), & t \in [-\tau, 0], \\ [\mathcal{S}(t)v_n](0), & t \in \mathbb{R}_+. \end{cases}$$

We conclude that $(Q(v_n))_n$ converges in X to $Q(v)$. Thus Q satisfies the hypotheses of the Schauder Fixed-Point Theorem 1.4.4. So, it has at least one fixed point $v \in B_\alpha$. Obviously, $u = S(v)$ is a C^0-solution of the problem (3.1.1). Since (3.6.1) follows from Lemma 3.3.1 and the global asymptotic stability from Theorem 1.12.3, this completes the proof. $\square$

3.7 Parabolic problems governed by the p-Laplacian

Let Ω be a nonempty, bounded and open subset in $\mathbb{R}^d$, $d \geq 1$, with C^2 boundary Σ, $p \in [2, +\infty)$, $\omega > 0$ and $\lambda > 0$, let $Q_+ = \mathbb{R}_+ \times \Omega$, $\Sigma_+ = \mathbb{R}_+ \times \Sigma$, $Q_\tau = [-\tau, 0] \times \Omega$ and let us consider the nonlinear problem with the nonlocal retarded initial condition

$$
\begin{cases}
\dfrac{\partial u}{\partial t}(t, x) = \Delta_p^\lambda u(t, x) - \omega u(t, x) + f(u_t)(x), & \text{in } Q_+, \\[2mm]
-\dfrac{\partial u}{\partial \nu_p}(t, x) \in \beta(u(t, x)), & \text{on } \Sigma_+, \\[2mm]
u(t, x) = \displaystyle\int_\tau^{+\infty} \mathcal{N}(u(t + s, x)) \, d\mu(s) + \psi(t)(x), & \text{in } Q_\tau.
\end{cases}
\tag{3.7.1}
$$

Here Δ_p^λ is the p-Laplace operator in the sense of Definition 1.9.1. Since in our case, $\overline{D(\Delta_p^\lambda)} = L^2(\Omega)$, we have $\mathcal{X}_2 = \mathcal{D}_2 = C([-\tau, 0]; L^2(\Omega))$. Further, $\mathcal{N} : \mathbb{R} \to \mathbb{R}$, $f \colon \mathcal{X}_2 \to L^2(\Omega)$, μ is a σ-finite and complete measure on $[\tau, +\infty)$ with $\mu([\tau, +\infty)) = 1$ and $\psi \in \mathcal{X}_2$.

From Theorem 3.6.1, we deduce:

THEOREM 3.7.1 *Let Ω be a nonempty bounded and open subset in $\mathbb{R}^d$, $d \geq 1$, with C^2 boundary Σ, let $\tau \geq 0$, $\omega > 0$, $p \in [2, +\infty)$ and $\lambda > 0$. Let $\beta : D(\beta) \subseteq \mathbb{R} \rightsquigarrow \mathbb{R}$ be m-accretive with $0 \in D(\beta)$ and $0 \in \beta(0)$, let $\mathcal{N} : \mathbb{R} \to \mathbb{R}$, let $\psi \in \mathcal{X}_2$ and let $f \colon \mathcal{X}_2 \to L^2(\Omega)$ be a continuous function. Let us assume that*

(h_1) *there exists $\ell > 0$ such that*

$$
\|f(v) - f(w)\|_{L^2(\Omega)} \leq \ell \|v - w\|_{\mathcal{X}_2}
$$

for each $v, w \in \mathcal{X}_2$

(h_2) *there exists $m > 0$ such that*

$$
\|f(0)\|_{L^2(\Omega)} \leq m
$$

(c_1) $\ell < \omega$

$(\widetilde{N}_1)$ *$\mathcal{N}$ is continuous and $|\mathcal{N}(u)| \leq |u|$ for each $u \in \mathbb{R}$.*

Let μ be a σ-finite and complete measure on $[\tau, +\infty)$ which is continuous with respect to the Lebesgue measure at $t = \tau$, i.e., $\lim_{\delta \downarrow \tau} \mu([\tau, \tau + \delta]) = 0$. Then, the problem (3.7.1) has at least one C^0-solution $u \in C_b([-\tau, \infty); L^2(\Omega))$ which satisfies

$$
\|u\|_{C_b([-\tau, +\infty); L^2(\Omega))} \leq \frac{m}{\omega - \ell}
\tag{3.7.2}
$$

$$
\Delta_p^\lambda u \in L_{\mathrm{loc}}^2(\mathbb{R}_+; L^2(\Omega))
\tag{3.7.3}
$$

116   *Delay Differential Evolutions Subjected to Nonlocal Initial Conditions*

$$-\frac{\partial u}{\partial \nu_p}(t,x) \in \beta(u(t,x)) \text{ for each } t \in \mathbb{R}_+ \text{ and a.e. for } x \in \Sigma \qquad (3.7.4)$$

$$u \in W^{1,2}_{\text{loc}}(\mathbb{R}_+; L^2(\Omega)) \cap AC(\mathbb{R}_+; W^{1,p}(\Omega)). \qquad (3.7.5)$$

In addition, each C^0-solution of (3.7.1) is globally asymptotically stable.

REMARK 3.7.1 In the case of the problem (3.7.1), the notion of the C^0-solution should also be understood by identification, i.e., by observing that (3.7.1) can be reformulated as an abstract nonlinear evolution equation of the form (3.1.1), in $X = L^2(\Omega)$, subjected to a nonlocal initial condition. Therefore, by a C^0-solution of (3.7.1) we mean a C^0-solution of the corresponding nonlinear evolution equation on $[0,T]$ for each $T > 0$ in the sense of Definition 3.1.1.

Proof. On $L^2(\Omega)$, we define $A:D(A) \subseteq L^2(\Omega) \to L^2(\Omega)$, by $A = \Delta_p^\lambda - \omega I$. From Theorem 1.9.8, we know that A is m-dissipative on $L^2(\Omega)$ and generates a compact semigroup. Moreover, $0 \in D(A)$ and $A0 = 0$ and hence A satisfies (H_A) in Theorem 3.6.1. Further, let $g : C_b(\mathbb{R}_+; L^2(\Omega)) \to \mathcal{X}_2$ be given by

$$[g(u)(t)](x) = \int_\tau^{+\infty} \mathcal{N}(u(t+s)(x))\, d\mu(s) + \psi(t)(x)$$

for each $u \in C_b([-\tau, +\infty); L^2(\Omega))$ and each $(t,x) \in [-\tau, 0] \times \Omega$. Clearly, (3.7.1) can be rewritten in the abstract form (3.1.1), where A, f and g satisfy all the hypotheses of Theorem 3.6.1. Hence, the problem (3.7.1) has a unique C^0-solution satisfying (3.7.2). Since (3.7.3), (3.7.4) and (3.7.5) follow from the fact that $-A$ is a subdifferential (see Vrabie [252, Example 1.6.3, p. 21]) combined with Theorem 1.10.3, the proof is complete. $\qquad \square$

3.8   Bibliographical notes and comments

We begin with some historical comments referring to the two cases below.

3.8.1   The nondelayed case

A detailed presentation of nonlocal initial value problems in finite dimensional spaces can be found in Ntouyas [198]. General existence results for periodic problems without delay were obtained by Aizicovici, Papageorgiou and Staicu [4], Caşcaval and Vrabie [73], Hirano [146], Hirano and Shioji [147], Paicu [202] and Vrabie [251]. As concerns the more general case of differential equations subjected to nonlocal initial data without delay, we mention the pioneering work of Byszewski [63], [64] and Byszewski and Lakshmikantham [68].

Nonlocal Initial Conditions: The Autonomous Case

More precisely, Byszewski [63] was the first to consider an abstract semilinear nonlocal problem of the form

$$\begin{cases} u'(t) = Au(t) + f(t, u(t)), & t \in (t_0, t_0 + a], \\ u(t_0) + g(u(t_1), u(t_2), \ldots, u(t_n)) = u_0, \end{cases}$$

where A is the infinitesimal generator of a C_0-semigroup in a Banach space X, $a > 0$, $r > 0$, while $f \colon [t_0, t_0 + a] \times D(0, r) \to X$, $g : X^n \to X$ are Lipschitz continuous functions, and $u_0 \in X$. We will use the following notations:

(i) $M = \sup_{t \in [0, a]} \|S(t)\|$, where $\{S(t) : X \to X; \ t \in \mathbb{R}_+\}$ is the C_0-semigroup generated by A

(ii) $L > 0$ is the Lipschitz constant of f, i.e., $\|f(s, v) - f(s, w)\| \le L\|v - w\|$ for each $t \in [t_0, t_0 + a]$ and each $v, w \in D(0, r)$

(iii) $N = \sup_{t \in [t_0, t_0 + a]} \|f(t, 0)\|$

(iv) K is the Lipschitz constant of g, i.e.,

$$\|g(u(t_1), u(t_2), \ldots, u(t_n)) - g(v(t_1), v(t_2), \ldots, v(t_n))\| \le K\|u - v\|$$

for each $u, v \in C([t_0, t_0 + a]; X)$

(v) $G = \sup_{u \in C([t_0, t_0 + a]; X)} \|u(t_0) + g(u(t_1), u(t_2), \ldots, u(t_n))\|$.

Byszewski assumes that

(H_1) $M[\|u_0\| + G + raL + aN] \le r$

(H_2) $M[K + aL] < 1$

and he proves three existence results concerning the existence and uniqueness of mild, strong and classical solutions to the above problem.

The assumption (H_1) is an invariance condition, while (H_2) implies that a suitably defined operator satisfies the hypotheses of the Banach Fixed-Point Theorem. All these results are very interesting because they have opened a new area of study. However, we have to note that (H_2), combined with the fact that $M \ge 1$, precludes their applicability to periodic or anti-periodic problems when g should be nonexpansive but not a strict contraction, i.e., K must be 1.

Notable results on this subject, relaxing the above hypotheses on g, were subsequently obtained by Aizicovici and Lee [2], Aizicovici and McKibben [3], Benedetti, Malaguti and Taddei [29], Byszewski [66], Z. Fan, Q. Dong and G. Li [112], García–Falset [122], and García–Falset and Reich [123]. Some other general results referring to semilinear non-autonomous functional-differential equations subjected to nonlocal initial conditions can be found in the papers of Al-Omair, Ibrahim [7] and Cardinali, Precup and Rubbioni [69]. Variational inequalities subjected to nonlocal initial conditions were considered

118   *Delay Differential Evolutions Subjected to Nonlocal Initial Conditions*

by Jin-Mun Jeong, Dong-Hwa Kim and Jong-Yeoul [149], while semilinear integro-differential equations in Banach spaces subjected to nonlocal initial conditions were studied, among others, by Tran Dinh Ke, Obukhovskii, Ngai-Ching Wong and Jen-Chih Yao [154]. These studies are essentially motivated by the fact that these kinds of problems represent mathematical models for the evolution of various phenomena. A model of the gas flow through a thin transparent tube, expressed as a problem with nonlocal initial conditions, was analyzed in Deng [95]. Parabolic equations with discrete nonlocal in time initial conditions were considered by Gordeziani [129]. Some models in pharmacokinetics were discussed in McKibben [182, Section 10.2, pp. 394–398]. Several models from physics were studied by Olmstead, Roberts [200] and Shelukhin [236], [237]. A class of linear second-order evolution equations subjected to linear nonlocal initial conditions in Hilbert triples, motivated by mathematical models for long-term weather forecasting as mentioned in Rabier, Courtier and Ehrendorfer [221], was considered by G. Avalishvili and M. Avalishvili [15]. For Navier-Stokes equations subjected to initial nonlocal conditions, see Gordeziani [130]. Reaction–diffusion subjected to similar initial conditions were studied by Pao [205]. See also Liu and Chang [173].

We would like to add that we have decided to consider separately this specific case referring to autonomous problems because on one hand, it can be easily approached by the elegant abstract theory developed in Section 1.12 under fairly general assumptions and on the other hand it is important enough for applications. As far as the pseudo-autonomous case is concerned, the situation is different. Namely, in that case, we can obtain existence and uniqueness results by using a classical direct approach avoiding the abstract theory in Section 1.12. It should be emphasized that, for a pseudo-autonomous problem to satisfy the hypotheses in Theorem 1.12.5, besides the t-continuity, we must impose some additional t-regularity assumptions on the function f, which are unnecessary for a direct approach. For details, see Chapter 4.

3.8.2   The delayed case

Section 3.1 Particular quasi-autonomous versions of (3.1.1) were intensively studied in the last years. We begin by mentioning the paper of Y. Li [171] which extends some previous results in Y. Li [170], by proving some existence, uniqueness, global uniform asymptotic stability as well as regularity results for a particular problem of type (3.1.1). Namely, Y. Li [171] assumes that A is the infinitesimal generator of an analytic compact semigroup in a Hilbert space H, while the history function $g : C_b(\mathbb{R}_+; H) \to C([-\tau, 0]; H)$ has the form $g(u)(s) = u(\omega + s)$ for some $\omega > 0$ and each $s \in [-\tau, 0]$, which corresponds to an ω-periodicity condition. Nonlocal initial value problems for delay semilinear evolution equations were studied by Byszewski [65], Byszewski and Akca [67], Balachandran and Park [17]. An existence in the large result for a class of fully nonautonomous semilinear equations subjected to nonlocal initial condi-

Nonlocal Initial Conditions: The Autonomous Case 119

tions was obtained, via compactness arguments, by Wang and Zhu [263]. For a systematic study of semilinear functional evolution equations with delay and nonlocal initial conditions, see the recent book of Benchora and Abbas [28]. Integro-differential equations with variable delay were considered by Chandrasekaran [79], as well as by Balachandran and Samuel [18]. Neutral partial differential equations subjected to nonlocal initial conditions were studied by Chang and Liu [80].

Section 3.2 Theorem 3.2.1 is an extension, in the autonomous case, of the main result of Vrabie [261] which, in turn, generalizes the main result of Burlică and Roşu [58] in the simplest case in which the history function, instead of linear growth, has affine growth. Both this theorem and its proof are new and appear for the first time here. It is interesting to note that we allow g to be of the form $g(u)(t) = \psi(t)$ for $t \in [-\tau, 0]$ with $\psi \in \mathfrak{X}$, a situation which is ruled out by the assumption that g has linear growth used by Burlică and Roşu [58]. So, Theorem 3.2.1 handles not only purely nonlocal problems – see Definition 3.1.2 – but Cauchy problems too, i.e., problems of the form

$$\begin{cases} u'(t) = Au(t) + f(u_t), & t \in \mathbb{R}_+, \\ u(t) = \psi(t), & t \in [-\tau, 0]. \end{cases}$$

In turn, the main result of Burlică and Roşu [58] generalizes an existence result in Vrabie [257] referring to nonlinear purely nonlocal initial-value problems of the form

$$\begin{cases} u'(t) = Au(t) + f(t, u(t), u(t - \tau_1), \dots, u(t - \tau_n)), & t \in \mathbb{R}_+, \\ u(t) = g(u)(t), & t \in [-\tau, 0]. \end{cases}$$

The main result of Vrabie [257] extends to the fully nonlinear case and to general nonlocal initial conditions, i.e., for initial history functions satisfying the additional condition $g(0) = 0$, in a general Banach space frame, the main results of Y. Li [171].

Section 3.3 The proof of Theorem 3.2.1 is essentially based on the semigroup approach of delay evolution equations initiated by Hale [135]. Another good source of information for the linear case is the monograph of Batkay and Piazzera [24]. The general theory in the nonlinear autonomous case was developed subsequently by Brewer [47], Webb [265], [266], Plant [219], [218] to cite only a few. The idea to reduce the study of a delay evolution equation to a non-delay one in a larger Banach space proved extremely useful, at least in the autonomous case. Lemma 3.3.1 is from Vrabie [261].

Section 3.4 The example referring to the transport equation in $\mathbb{R}^d$, as well as Theorem 3.4.1, are new and appear for the first time here. It should be noted that, by using similar arguments, we can obtain existence, uniqueness

120   *Delay Differential Evolutions Subjected to Nonlocal Initial Conditions*

and global asymptotic stability results for many other transport equations in bounded domains in $\mathbb{R}^d$, subjected to more general nonlocal initial conditions.

Section 3.5 Also, Theorem 3.5.1 concerning the damped wave equation is new and was not published until now. Here we have preferred to consider a rather general nonlocal initial condition, because the presentation of the particular cases, which are really of interest in practice, is not based on simpler arguments. We mean here some variants of mean initial conditions imposed both on u and on its time derivative.

Section 3.6 The main result in this section is an autonomous version of an existence theorem due to Mitidieri and Vrabie [185]. Its proof, essentially based on some compactness arguments previously developed by Pazy [210], Baras, Hassan and Veron [20], in the linear case, and by Baras [19], Vrabie [247], [250] and Mitidieri and Vrabie [185], in the fully nonlinear case, is completely new. Of course, it seems much simpler than the original proof given in Mitidieri and Vrabie [185], but we cannot forget that behind this apparent shortcut are the deep arguments developed in Section 1.12 relating delay evolution equations with non-delay evolution problems in the space of histories, $\mathfrak{X}$.

Section 3.7 The example concerning the nonlinear diffusion driven by a delay perturbation of the p-Laplace operator, although new, is closely related to its non-delay quasi-autonomous counterpart considered in Paicu and Vrabie [204]. The main point in Theorem 3.7.1 is that the function appearing in the nonlocal condition is merely continuous but not Lipschitz. A similar result referring to a quasi-autonomous differential inclusion driven by the p-Laplace operator and subjected to a nonexpansive nonlocal initial condition was obtained by Vrabie [258].

Chapter 4

Nonlocal Initial Conditions: The Quasi-Autonomous Case

Overview

In this chapter, we establish some sufficient conditions for the existence, uniqueness, and global uniform asymptotic stability of a C^0-solution for the nonlinear delay quasi-autonomous differential evolution equation

$$\begin{cases} u'(t) \in Au(t) + f(t, u_t), & t \in \mathbb{R}_+, \\ u(t) = g(u)(t), & t \in [-\tau, 0], \end{cases}$$

where $\tau \geq 0$, X is a real Banach space, the operator $A : D(A) \subseteq X \rightsquigarrow X$ is the infinitesimal generator of a nonlinear semigroup of contractions, the function $f : \mathbb{R}_+ \times \mathcal{X} \to X$ is continuous, and $g : C_b(\mathbb{R}_+; \overline{D(A)}) \to \mathcal{D}$ is a continuous mapping having affine growth, where $\mathcal{X} = C([-\tau, 0]; X)$ and $\mathcal{D} = \{\varphi \in \mathcal{X}; \ \varphi(0) \in \overline{D(A)}\}$.

4.1 The quasi-autonomous case with f and g Lipschitz

This chapter is concerned with the quasi-autonomous case, i.e., the case in which f also depends on t. So, let us consider the delay evolution equation subjected to a nonlocal initial condition

$$\begin{cases} u'(t) \in Au(t) + f(t, u_t), & t \in \mathbb{R}_+, \\ u(t) = g(u)(t), & t \in [-\tau, 0]. \end{cases} \tag{4.1.1}$$

Here $A : D(A) \subseteq X \rightsquigarrow X$ is an m-dissipative operator in the real (infinite dimensional) Banach space X and $\tau \geq 0$. As in the preceding chapters, we denote by

$$\mathcal{X} = C([-\tau, 0]; X) \ \text{ and } \ \mathcal{D} = \{\varphi \in \mathcal{X}; \ \varphi(0) \in \overline{D(A)}\}.$$

121

122   Delay Differential Evolutions Subjected to Nonlocal Initial Conditions

So, in (4.1.1), $f : \mathbb{R}_+ \times \mathcal{X} \to X$ is continuous, and $g : C_b(\mathbb{R}_+; \overline{D(A)}) \to \mathcal{D}$ is nonexpansive or merely continuous.

In this more general setting, we are not going to make use of the existence theory in Section 1.12, simply because it would require stronger hypotheses on f than the simple continuity with respect to the t-variable. Just look at the condition (iii) in Theorem 1.12.5, to see that the continuity of f alone would be not enough to reduce the study of this problem to an abstract one involving an evolution system.

In contrast, by means of some usual classical methods involving, of course, the Benilan, Crandall and Liggett theory, although technical but rather elementary, we will see that the main results in Section 3.2 extend naturally to the quasi-autonomous case. It is important to notice that the proofs of these results can be easily adapted to handle the more general case in which f is strongly measurable with respect to its first variable and Lipschitz with respect to its second one and satisfies the Carathéodory condition, a case that is ruled out by the hypotheses on f in Theorem 1.12.5.

The assumptions we need here, less restrictive than those in Section 3.2, are listed below.

(H_A) The operator $A : D(A) \subseteq X \rightsquigarrow X$ satisfies

(A_1) $0 \in D(A)$, $0 \in A0$ and A is ω-m-dissipative for some $\omega > 0$

(H_f) The function $f : \mathbb{R}_+ \times \mathcal{X} \to X$ is continuous and satisfies

(f_1) there exists $\ell > 0$ such that

$$\|f(t, v) - f(t, \widetilde{v})\| \le \ell \|v - \widetilde{v}\|_{\mathcal{X}}$$

for each $t \in \mathbb{R}_+$ and $v, \widetilde{v} \in \mathcal{X}$

(f_2) there exists $m \ge 0$ such that

$$\|f(t, 0)\| \le m$$

for each $t \in \mathbb{R}_+$.

(H_c) The constants ℓ and ω satisfy $\ell < \omega$.

(H_g) The function $g : C_b(\mathbb{R}_+; \overline{D(A)}) \to \mathcal{D}$ satisfies

(g_1) there exists $a > 0$ such that, for each $u, v \in C_b(\mathbb{R}_+; \overline{D(A)})$, we have

$$\|g(u) - g(v)\|_{\mathcal{X}} \le \|u - v\|_{C_b([a, +\infty); X)}.$$

Let us assume that (H_A) is satisfied, $f : \mathbb{R}_+ \times \mathcal{X} \to X$ is continuous, and $g : C_b(\mathbb{R}_+; \overline{D(A)}) \to \mathcal{D}$. As in the autonomous case, we introduce:

Nonlocal Initial Conditions: The Quasi-Autonomous Case

DEFINITION 4.1.1 By a C^0-*solution* of (4.1.1) we mean a continuous function $u : [-\tau, +\infty) \to X$ satisfying $u(t) = g(u)(t)$ for each $t \in [-\tau, 0]$ and which, for each $T > 0$, is a C^0-solution on $[0, T]$ in the sense of Definition 1.8.2 for the equation

$$u'(t) \in Au(t) + h(t),$$

where $h(t) = f(t, u_t)$ for $t \in [0, T]$.

REMARK 4.1.1 If u is a C^0-solution of (4.1.1), then $u(t) \in \overline{D(A)}$ for each $t \in \mathbb{R}_+$.

The first existence, uniqueness and stability result, concerning the problem (4.1.1) and extending Theorems 3.2.1 and 3.2.2, is:

THEOREM 4.1.1 *If* (H_A), (H_f), (H_c) *and* (H_g) *are satisfied, then the problem* (4.1.1) *has a unique* C^0-*solution in the sense of Definition 4.1.1 satisfying*

$$\|u\|_{C_b([-\tau, +\infty); X)} \leq \frac{m}{\omega - \ell} + k \cdot m_0, \tag{4.1.2}$$

where m *is given by* (f_2) *in* (H_f), $m_0 = \|g(0)\|_X$ *and*

$$k = k(a, \omega, \ell) := \frac{\omega}{\omega - \ell} \cdot \left(\frac{1}{e^{\omega a} - 1} + \frac{\ell}{\omega} \right) + 1. \tag{4.1.3}$$

If, instead of (f_1) *in* (H_f), *the stronger condition*

(f_5) *there exists* $\ell > 0$ *such that*

$$\|f(t, v) - f(s, \widetilde{v})\| \leq \ell \left(|t - s| + \|v - \widetilde{v}\|_X \right)$$

for each $t, s \in \mathbb{R}_+$ *and* $v, \widetilde{v} \in \mathfrak{X}$,

is satisfied, then the C^0-*solution of* (4.1.1) *is globally asymptotically stable.*

Also in the quasi-autonomous case, there is a regularity result very similar to Theorem 3.2.3, in fact, a simple copy of the latter.

4.1.1 Periodic solutions

If we choose $\tau \in (0, T)$ and $g(u)(t) = u(t + T)$ for each $t \in [-\tau, 0]$, from Theorem 4.1.1, we obtain the next existence and uniqueness result concerning the periodic problem

$$\begin{cases} u'(t) \in Au(t) + f(t, u_t), & t \in \mathbb{R}_+, \\ u(t) = u(t + T), & t \in [-\tau, 0]. \end{cases} \tag{4.1.4}$$

124 *Delay Differential Evolutions Subjected to Nonlocal Initial Conditions*

THEOREM 4.1.2 *Let us assume that* (H_A), (H_f) *and* (H_c) *are satisfied and let* $f : \mathbb{R}_+ \times X \to X$ *be* T-*periodic with respect to its first argument. Then* (4.1.4) *has a unique* T-*periodic* C^0-*solution in the sense of Definition 4.1.1 satisfying*

$$\|u\|_{C([0,T];X)} \le \frac{m}{\omega - \ell}, \tag{4.1.5}$$

where m *is given by* (f_2) *in* (H_f). *If, instead of* (f_1) *in* (H_f), *the stronger condition* (f_5) *in Theorem 4.1.1 is satisfied, then the* T-*periodic* C^0-*solution of* (4.1.4) *is globally asymptotically stable.*

The existence and uniqueness part are consequences of Theorem 4.1.1 and Lemma 4.1.1 below. As long as the global asymptotic stability is concerned, this follows from Theorem 1.12.7.

LEMMA 4.1.1 *Let us assume that* (H_A), (H_f) *and* (H_c) *are satisfied and the function* $f : \mathbb{R}_+ \times X \to X$ *is* T-*periodic with respect to its first argument. Then the unique* C^0-*solution of the problem* (4.1.4), *whose existence is ensured by Theorem 4.1.1, is* T-*periodic.*

Proof. Let $u : [-\tau, +\infty) \to X$ be the unique C^0-solution of the problem (4.1.4). Then, from the periodicity of f and (1.8.2) in Theorem 1.8.1, we have

$$\|u(t+T) - u(t)\| \le e^{-\omega t} \|u(T) - u(0)\| + \int_0^t e^{-\omega(t-s)} \|f(s+T, u_{s+T}) - f(s, u_s)\| \, ds$$

$$= \int_0^t e^{-\omega(t-s)} \|f(s, u_{s+T}) - f(s, u_s)\| \, ds \le \ell \int_0^t \|u_{s+T} - u_s\|_X \, ds$$

for each $t \in \mathbb{R}_+$. Now, let us observe that we can apply Lemma 1.13.3 for

$$\begin{cases} y(t) = \|u(t+T) - u(t)\|, & t \in [-\tau, +\infty), \\ \alpha_0(t) = 0, & t \in \mathbb{R}_+, \\ \beta(t) = \ell, & t \in \mathbb{R}_+. \end{cases}$$

By the nonlocal initial condition we deduce that

$$\|y_0\|_{C([-\tau,0];\mathbb{R})} = \sup\{\|u(s+T) - u(s)\|; \ s \in [-\tau, 0]\} = 0,$$

and so the function $\alpha(t) = \|y_0\|_{C([-\tau,0];\mathbb{R})} + \alpha_0(t) \equiv 0$. From Lemma 1.13.3, we conclude that u is T-periodic on $[-\tau, +\infty)$. This completes the proof. $\square$

4.1.2 Anti-periodic solutions

If $\tau \in (0, T)$ and $g(u)(t) = -u(t+T)$ for each $t \in [-\tau, 0]$, from Theorem 4.1.1, we obtain an existence and uniqueness result for T-anti-periodic C^0-solutions

Nonlocal Initial Conditions: The Quasi-Autonomous Case 125

which are at the same time $2T$-periodic. More precisely, let us consider the problem

$$\begin{cases} u'(t) \in Au(t) + f(t, u_t), & t \in \mathbb{R}_+, \\ u(t) = -u(t + T), & t \in [-\tau, 0]. \end{cases} \tag{4.1.6}$$

Namely, we have:

THEOREM 4.1.3 *Let us assume that (H_A), (H_f) and (H_c) are satisfied, $D(A)$ is symmetric with respect to the origin, i.e., $D(A) = -D(A)$, $A(-\xi) = -A\xi$ for each $\xi \in D(A)$, and f satisfies*

$$-f(t + T, -v) = f(t, v), \quad t \in \mathbb{R}_+, \ v \in X. \tag{4.1.7}$$

Then (4.1.6) has a unique C^0-solution in the sense of Definition 4.1.1, which is $2T$-periodic and satisfies (4.1.5). If, instead of (f_1) in (H_f), the stronger condition (f_5) in Theorem 4.1.1 is satisfied, then the T-anti-periodic C^0-solution of (4.1.6) is globally asymptotically stable.

REMARK 4.1.2 From (4.1.7), we deduce that, for each $v \in X$, the function $t \mapsto f(t, v)$ is $2T$-periodic, i.e.,

$$f(t + 2T, v) = f(t, v)$$

for each $t \in \mathbb{R}_+$. Indeed, we have

$$f(t + 2T, v) = -f(t + T, -v) = f(t, v)$$

for each $t \in \mathbb{R}_+$ and $v \in X$, as claimed.

The conclusion of Theorem 4.1.3 follows from Theorems 4.1.1 and 1.12.7 in conjunction with:

LEMMA 4.1.2 *Let us assume that (H_A), (H_f) and (H_c) are satisfied, $D(A)$ is symmetric with respect to the origin, i.e., $D(A) = -D(A)$, $A(-\xi) = -A\xi$ for each $\xi \in D(A)$, and $f : \mathbb{R}_+ \times X \to X$ satisfies (4.1.7). Then the unique C^0-solution of the problem (4.1.6), whose existence is ensured by Theorem 4.1.1, is T-anti-periodic and $2T$-periodic.*

Proof. Let $u : [-\tau, +\infty) \to X$ be the unique C^0-solution of the problem

$$\begin{cases} u'(t) \in Au(t) + f(t, u_t), & t \in \mathbb{R}_+, \\ u(t) = -u(t + T), & t \in [-\tau, 0], \end{cases} \tag{4.1.8}$$

whose existence and uniqueness follow from Theorem 4.1.1 by particularizing the history constraint function g as

$$g(\widetilde{u})(t) = -\widetilde{u}(t + T)$$

126   *Delay Differential Evolutions Subjected to Nonlocal Initial Conditions*

for each $\tilde{u} \in C_b(\mathbb{R}_+; \overline{D(A)})$ and each $t \in [-\tau, 0]$.

From the condition $A(-\xi) = -A\xi$ and (4.1.7), we conclude that if u is a C^0-solution of the problem (4.1.8), then the function $v : [-\tau, +\infty) \to X$, defined by $v(t) = -u(t+T)$ for $t \in [-\tau, +\infty)$, is a C^0-solution of the problem

$$\begin{cases} v'(t) \in Av(t) + f(t, v_t), & t \in \mathbb{R}_+, \\ v(t) = u(t), & t \in [-\tau, 0]. \end{cases}$$

Indeed, if we assume that u is even a strong solution, we have

$$v'(t) = -u'(t+T) = -Au(t+T) - f(t+T, u_{t+T})$$
$$= -A(-v(t)) + f(t, -u_{t+T}) = Av(t) + f(t, v_t)$$

for each $t \in \mathbb{R}_+$ and

$$v(t) = -u(t+T) = u(t)$$

for each $t \in [-\tau, 0]$.

If u is merely a C^0-solution, the conclusion follows, via a purely "C^0-solution argument," from Theorem 1.8.3.[1]

Then, from (1.8.2) in Theorem 1.8.1 applied to u and v, we get

$$\|u(t) - v(t)\| \leq e^{-\omega t}\|u(0) - v(0)\| + \int_0^t e^{-\omega(t-s)}\|f(s, u_s) - f(s, v_s)\|\, ds$$

$$\leq \ell \int_0^t \|u_s - v_s\|_X\, ds$$

for each $t \in \mathbb{R}_+$. Now, let us observe that we can apply Lemma 1.13.3 for

$$\begin{cases} y(t) = \|u(t) - v(t)\|, & t \in [-\tau, +\infty), \\ \alpha_0(t) = 0, & t \in \mathbb{R}_+, \\ \beta(t) = \ell, & t \in \mathbb{R}_+. \end{cases}$$

By the nonlocal initial condition, we deduce that

$$\|y_0\|_{C([-\tau, 0]; \mathbb{R})} = \sup\{\|u(s) - v(s)\|;\ s \in [-\tau, 0]\} = 0,$$

and so the function $\alpha(t) = \|y_0\|_{C([-\tau, 0]; \mathbb{R})} + \alpha_0(t) \equiv 0$. From Lemma 1.13.3, we conclude that u is T-anti-periodic on $[-\tau, +\infty)$. So, $u(0) = -u(T) = u(2T)$ and thus, by the $2T$-periodicity of f – see Remark 4.1.2 – it follows that u is $2T$-periodic on $\mathbb{R}_+$, i.e., it satisfies

$$\begin{cases} u'(t) \in Au(t) + f(t, u_t), & t \in \mathbb{R}_+, \\ u(t) = u(t+2T), & t \in \mathbb{R}_+. \end{cases} \tag{4.1.9}$$

From (4.1.8) and (4.1.9) it follows that u is a C^0-solution of (4.1.6) and this completes the proof. $\square$

---

[1] As an almost general philosophy, if, for a certain C^0-solution u, we have to prove some estimates which do not involve the derivative of u, we can argue by simply assuming that the solution in question is even stronger.

Nonlocal Initial Conditions: The Quasi-Autonomous Case

4.1.3 The nondelayed case

Taking $\tau = 0$, we get some existence results for evolution equations, without delay, subjected to nonlocal initial conditions. More precisely, let us consider

$$\begin{cases} u'(t) \in Au(t) + f(t, u(t)), & t \in \mathbb{R}_+, \\ u(0) = g(u), \end{cases} \tag{4.1.10}$$

where A is as above and f and g satisfy the hypotheses below.

$(H_f^{[\tau=0]})$ The function $f : \mathbb{R}_+ \times X \to X$ is continuous and satisfies

$(f_1^{[\tau=0]})$ there exists $\ell > 0$ such that

$$\|f(t, v) - f(t, \widetilde{v})\| \le \ell \|v - \widetilde{v}\|$$

for each $t \in \mathbb{R}_+$ and $v, \widetilde{v} \in X$

$(f_2^{[\tau=0]})$ there exists $m \ge 0$ such that

$$\|f(t, 0)\| \le m$$

for each $t \in \mathbb{R}_+$.

$(H_g^{[\tau=0]})$ The function $g : C_b(\mathbb{R}_+; \overline{D(A)}) \to \overline{D(A)}$ satisfies

$(g_1^{[\tau=0]})$ there exists $a > 0$ such that for each $u, v \in C_b(\mathbb{R}_+; \overline{D(A)})$, we have

$$\|g(u) - g(v)\| \le \|u - v\|_{C_b([a, +\infty); X)}.$$

DEFINITION 4.1.2 By a C^0-*solution* of (4.1.10) we mean a continuous function $u : \mathbb{R}_+ \to X$ satisfying $u(0) = g(u)$ and which is, for each $T > 0$, a C^0-solution on $[0, T]$ in the sense of Definition 1.8.2 for the equation

$$u'(t) \in Au(t) + h(t),$$

where $h(t) = f(t, u(t))$ for $t \in [0, T]$.

REMARK 4.1.3 If u is a C^0-solution of (4.1.10), then $u(t) \in \overline{D(A)}$ for each $t \in \mathbb{R}_+$.

REMARK 4.1.4 In the nondelayed case, condition $(f_1^{[\tau=0]})$ in $(H_f^{[\tau=0]})$ in conjunction with the assumption that $A + \omega I$ is dissipative and (H_c) shows that, for each $t \in \mathbb{R}_+$, $u \mapsto Au + f(t, u)$ is dissipative in X. Indeed, from (ii), (vii) and $(viii)$ in Proposition 1.8.1, we have

$$[u - v, Au + f(t, u) - Av - f(t, v)]_-$$

$$= [u - v, Au + \omega u - Av - \omega v - \omega(u - v) + (f(t, u) - f(t, v))]_-$$

$$\le [u - v, Au + \omega u - Av - \omega v]_- - \omega\|u - v\| + [u - v, f(t, u) - f(t, v)]_+$$

$$\le -\omega\|u - v\| + \ell\|u - v\| = (\ell - \omega)\|u - v\| \le 0$$

for each $u, v \in D(A)$, which proves that $u \mapsto Au + f(t, u)$ is dissipative.

128   *Delay Differential Evolutions Subjected to Nonlocal Initial Conditions*

An immediate consequence of Theorem 4.1.1 is:

THEOREM 4.1.4 *If (H_A), $(H_f^{[\tau=0]})$, (H_c) and $(H_g^{[\tau=0]})$ are satisfied, then the problem (4.1.10) has a unique C^0-solution in the sense of Definition 4.1.2 satisfying*

$$\|u\|_{C_b(\mathbb{R}_+;X)} \le \frac{m}{\omega - \ell} + k \cdot m_0, \qquad (4.1.11)$$

where m is given by $(f_2^{[\tau=0]})$ in $(H_f^{[\tau=0]})$, $m_0 = \|g(0)\|$, and k is given by (4.1.3). If, instead of $(f_1^{[\tau=0]})$ in $(H_f^{[\tau=0]})$, the stronger condition

$(f_3^{[\tau=0]})$ *there exists $\ell > 0$ such that*

$$\|f(t,v) - f(s,\widetilde{v})\| \le \ell \left(|t - s| + \|v - \widetilde{v}\|\right)$$

for each $t, s \in \mathbb{R}_+$ and $v, \widetilde{v} \in X$,

is satisfied, then the C^0-solution of (4.1.10) is globally asymptotically stable.

As far as the periodic problem

$$\begin{cases} u'(t) \in Au(t) + f(t, u(t)), & t \in \mathbb{R}_+, \\ u(0) = u(T), \end{cases} \qquad (4.1.12)$$

is concerned, from Theorem 4.1.4, we deduce:

THEOREM 4.1.5 *Let us assume that (H_A), $(H_f^{[\tau=0]})$ and (H_c) are satisfied and $f : \mathbb{R}_+ \times X \to X$ is T-periodic with respect to its first argument. Then the periodic problem (4.1.12) has a unique C^0-solution in the sense of Definition 4.1.2 satisfying (4.1.11). If, instead of $(f_1^{[\tau=0]})$ in $(H_f^{[\tau=0]})$, the stronger condition $(f_3^{[\tau=0]})$ in Theorem 4.1.4 is satisfied, then the T-periodic C^0-solution of (4.1.12) is globally asymptotically stable.*

More generally, from Theorem 4.1.4, we deduce an existence result for two-point boundary nondelayed problems of the form

$$\begin{cases} u'(t) \in Au(t) + f(t, u(t)), & t \in \mathbb{R}_+, \\ u(0) = h(u(T)), \end{cases} \qquad (4.1.13)$$

where $h : \overline{D(A)} \to \overline{D(A)}$ is nonexpansive. More precisely, we have

THEOREM 4.1.6 *Let us assume that (H_A), $(H_f^{[\tau=0]})$ and (H_c) are satisfied, $h : \overline{D(A)} \to \overline{D(A)}$ is nonexpansive, and $h(0) = 0$. Then (4.1.13) has a unique C^0-solution in the sense of Definition 4.1.2 satisfying (4.1.11). If, instead of $(f_1^{[\tau=0]})$ in $(H_f^{[\tau=0]})$, the stronger condition $(f_3^{[\tau=0]})$ in Theorem 4.1.4 is satisfied, then the C^0-solution of (4.1.13) is globally asymptotically stable.*

Nonlocal Initial Conditions: The Quasi-Autonomous Case 129

Next, let us consider the T-anti-periodic problem

$$\begin{cases} u'(t) \in Au(t) + f(t, u(t)), & t \in \mathbb{R}_+, \\ u(0) = -u(T). \end{cases} \tag{4.1.14}$$

THEOREM 4.1.7 *Let us assume that* (H_A), $(H_f^{[\tau=0]})$ *and* (H_c) *are satisfied,* $D(A)$ *is symmetric with respect to the origin, i.e.,* $D(A) = -D(A)$, $A(-\xi) = -A\xi$ *for each* $\xi \in D(A)$, *and* f *satisfies*

$$-f(t + T, -v) = f(t, v)$$

for each $(t, v) \in \mathbb{R}_+ \times X$. *Then* (4.1.14) *has a unique* C^0-*solution,* $u \in C_b(\mathbb{R}_+; \overline{D(A)})$, *which is* $2T$-*periodic and satisfies* (4.1.11). *If, instead of* $(f_1^{[\tau=0]})$ *in* $(H_f^{[\tau=0]})$, *the stronger condition* $(f_3^{[\tau=0]})$ *in Theorem 4.1.4 is satisfied, then the* $2T$-*periodic* C^0-*solution of* (4.1.14) *is globally asymptotically stable.*

Obviously, from Theorem 4.1.6 we get both Theorem 4.1.5, by choosing h the identity on $\overline{D(A)}$, and Theorem 4.1.7, by choosing h the minus identity on $\overline{D(A)}$.

REMARK 4.1.5 *If* $X = \mathbb{R}$, *it is easy to see that the* T-*anti-periodic* C^0-*solution, given either by Theorem 4.1.3 or by Theorem 4.1.7, has infinitely many zeroes* $t_k \in [2kT, 2(k+1)T]$, $k \in \mathbb{Z}$.

4.2 Proofs of Theorems 4.1.1, 4.1.2

4.2.1 An auxiliary existence result

LEMMA 4.2.1 *Let us assume that* (H_A) *and* (H_g) *are satisfied. Then, for each* $h \in L^\infty(\mathbb{R}_+; X)$, *the problem*

$$\begin{cases} u'(t) \in Au(t) + h(t), & t \in \mathbb{R}_+, \\ u(t) = g(u)(t), & t \in [-\tau, 0], \end{cases} \tag{4.2.1}$$

has a unique C^0-*solution,* $u \in C_b([-\tau, +\infty); X)$, *satisfying*

$$\|u\|_{C_b([-\tau, +\infty); X)} \le \frac{e^{\omega a}}{e^{\omega a} - 1} m_0 + \frac{1}{\omega} \|h\|_{L^\infty(\mathbb{R}_+; X)}, \tag{4.2.2}$$

where $m_0 = \|g(0)\|_X$. *Moreover, the mapping* $h \mapsto u^h$ *is Lipschitz from* $L^\infty(\mathbb{R}_+; X)$ *to* $C_b([-\tau, +\infty); X)$ *with Lipschitz constant* $1/\omega$, *i.e., for each* $f, h \in L^\infty(\mathbb{R}_+; X)$, *we have*

$$\|u^f - u^h\|_{C_b([-\tau, +\infty); X)} \le \frac{1}{\omega} \|f - h\|_{L^\infty(\mathbb{R}_+; X)}. \tag{4.2.3}$$

130   *Delay Differential Evolutions Subjected to Nonlocal Initial Conditions*

Proof. Let $h \in L^\infty(\mathbb{R}_+; X)$ be arbitrary but fixed. Let $v \in C_b(\mathbb{R}_+; X)$ and let us consider the auxiliary problem

$$\begin{cases} u'(t) \in Au(t) + h(t), & t \in \mathbb{R}_+, \\ u(t) = g(v)(t), & t \in [-\tau, 0]. \end{cases} \tag{4.2.4}$$

From Theorem 1.8.1 it follows that the problem (4.2.4) has a unique C^0-solution $u \colon [-\tau, +\infty) \to X$, satisfying $u(t) \in \overline{D(A)}$ for each $t \in \mathbb{R}_+$ and

$$\|u(t)\| \le e^{-\omega t} \left[\|v\|_{C_b([a,+\infty); X)} + m_0 \right] + \left(1 - e^{-\omega t}\right) \frac{1}{\omega} \|h\|_{L^\infty(\mathbb{R}_+; X)} \tag{4.2.5}$$

for each $t \in \mathbb{R}_+$. So, $u \in C_b([-\tau, +\infty); X) \cap C_b(\mathbb{R}_+; \overline{D(A)})$ and thus, we can define the operator

$$T \colon C_b(\mathbb{R}_+; \overline{D(A)}) \to C_b(\mathbb{R}_+; \overline{D(A)})$$

by

$$T(v) = u_{|\mathbb{R}_+},$$

where $u_{|\mathbb{R}_+} \in C_b(\mathbb{R}_+; \overline{D(A)})$ is the restriction to $\mathbb{R}_+$ of the unique C^0-solution of the problem (4.2.4). Obviously, (4.2.1) has a unique C^0-solution if and only if T has a unique fixed point.

Then, to complete the proof, it suffices to show that T^2 is a strict contraction. So, let $w, z \in C_b(\mathbb{R}_+; \overline{D(A)})$. We have

$$\|Tw - Tz\|_{C_b([a,+\infty); X)} \le e^{-\omega a} \|w - z\|_{C_b([a,+\infty); X)}. \tag{4.2.6}$$

Indeed, if $t \ge a$, from (H_g), we get

$$\|(Tw)(t) - (Tz)(t)\| \le e^{-\omega t} \|g(w)(0) - g(z)(0)\|$$

$$\le e^{-\omega t} \|g(w) - g(z)\|_X \le e^{-\omega a} \|w - z\|_{C_b([a,+\infty); X)}.$$

Next, we prove that

$$\|T^2 w - T^2 z\|_{C_b(\mathbb{R}_+; X)} \le e^{-\omega a} \|w - z\|_{C_b(\mathbb{R}_+; X)}. \tag{4.2.7}$$

First, let $t \ge 0$. By (H_g) and (4.2.6), we have

$$\|(T^2 w)(t) - (T^2 z)(t)\| \le e^{-\omega t} \|(g(Tw))(0) - (g(Tz))(0)\|$$

$$\le e^{-\omega t} \|g(Tw) - g(Tz)\|_X \le \|Tw - Tz\|_{C_b([a,+\infty); X)}$$

$$\le e^{-\omega a} \|w - z\|_{C_b([a,+\infty); X)} \le e^{-\omega a} \|w - z\|_{C_b(\mathbb{R}_+; X)}.$$

Second, if $t \in [-\tau, 0)$, using the same inequalities, i.e., (H_g) and (4.2.6), we get

$$\|(T^2 w)(t) - (T^2 z)(t)\| = \|(g(Tw))(t) - (g(Tz))(t)\| \le \|g(Tw) - g(Tz)\|_X$$

Nonlocal Initial Conditions: The Quasi-Autonomous Case 131

$$\leq \|Tw - Tz\|_{C_b([a,+\infty);X)} \leq e^{-\omega a}\|w - z\|_{C_b(\mathbb{R}_+;X)}.$$

Thus, for each $t \in [-\tau, +\infty)$ we have

$$\|(T^2 w)(t) - (T^2 z)(t)\| \leq e^{-\omega a}\|w - z\|_{C_b(\mathbb{R}_+;X)},$$

which implies (4.2.7). But (4.2.7) shows that T^2 is a contraction of constant $e^{-\omega a}$. So, T has a unique fixed point u which is a C^0-solution of the problem (4.2.1).

This completes the proof of the existence and uniqueness part.

Finally, we will estimate $\|u(t)\|$ for $t \in [-\tau, +\infty)$. For $t \in [-\tau, 0]$, using (H_g), we deduce that

$$\|u(t)\| = \|g(u)(t)\| \leq \|u\|_{C_b([a,+\infty);X)} + m_0. \tag{4.2.8}$$

Next, setting $v = u$ in (4.2.5), we get

$$\|u(t)\| \leq e^{-\omega t}\|u\|_{C_b([a,+\infty);X)} + e^{-\omega t}m_0 + (1 - e^{-\omega t})\frac{1}{\omega}\|h\|_{L^\infty(\mathbb{R}_+;X)} \tag{4.2.9}$$

for each $t > 0$. If for some $t \geq a$

$$\|u(t)\| = \|u\|_{C_b([a,+\infty);X)},$$

from the last inequality, taking into account that the function $x \mapsto \dfrac{1}{e^{\omega x} - 1}$ is strictly decreasing on $[a, +\infty)$, we deduce

$$\|u\|_{C_b([a,+\infty);X)} \leq \frac{m_0}{e^{\omega a} - 1} + \frac{1}{\omega}\|h\|_{L^\infty(\mathbb{R}_+;X)}. \tag{4.2.10}$$

If for each $t \geq a$ we have $\|u(t)\| < \|u\|_{C_b([a,+\infty);X)}$, then there exists $t_k \to +\infty$ such that

$$\lim_k \|u(t_k)\| = \|u\|_{C_b([a,+\infty);X)}.$$

Setting $t = t_k$ in (4.2.9) and passing to the limit for $k \to +\infty$, we get

$$\|u\|_{C_b([a,+\infty);X)} \leq \frac{1}{\omega}\|h\|_{L^\infty(\mathbb{R}_+;X)}.$$

So, in any case, (4.2.10) holds true.

Taking $v = u$ in (4.2.5) and using (4.2.10), we obtain

$$\|u(t)\| \leq e^{-\omega t}\left(\frac{m_0}{e^{\omega a} - 1} + \frac{1}{\omega}\|h\|_{L^\infty(\mathbb{R}_+;X)} + m_0\right) + (1 - e^{-\omega t})\frac{1}{\omega}\|h\|_{L^\infty(\mathbb{R}_+;X)}$$

for each $t \in \mathbb{R}_+$. After some obvious rearrangements, we get

$$\|u\|_{C_b(\mathbb{R}_+;X)} \leq \frac{e^{\omega a}}{e^{\omega a} - 1}m_0 + \frac{1}{\omega}\|h\|_{L^\infty(\mathbb{R}_+;X)}.$$

132 *Delay Differential Evolutions Subjected to Nonlocal Initial Conditions*

From this inequality, taking into account (4.2.8) and (4.2.10), we deduce (4.2.2).

To establish (4.2.3), let us observe that there are only three possible cases.

Case 1. There exists a maximum point $\bar{t} \in [-\tau, 0]$ of the mapping

$$t \mapsto \|u^f(t) - u^h(t)\|,$$

i.e.,

$$\|u^f(\bar{t}) - u^h(\bar{t})\| = \|u^f - u^h\|_{C_b([-\tau,+\infty);X)}.$$

By the initial nonlocal condition and (g_1), it follows that

$$\|u^f(\bar{t}) - u^h(\bar{t})\| = \|g(u^f)(\bar{t}) - g(u^h)(\bar{t})\| \le \|u^f - u^h\|_{C_b([a,+\infty);X)}$$

and thus this case reduces to the one of the following two cases.

Case 2. There exists a maximum point $\bar{t} \in (0, +\infty)$ of $t \mapsto \|u^f(t) - u^h(t)\|$, i.e., $\|u^f(\bar{t}) - u^h(\bar{t})\| = \|u^f - u^h\|_{C_b([-\tau,+\infty);X)}$.

By (1.8.2), we get

$$\|u^f(\bar{t}) - u^h(\bar{t})\| = \|u^f - u^h\|_{C_b([-\tau,+\infty);X)}$$

$$\le e^{-\omega\bar{t}}\|u^f - u^h\|_{C_b([-\tau,+\infty);X)} + \frac{1 - e^{-\omega\bar{t}}}{\omega}\|f - h\|_{L^\infty(\mathbb{R}_+;X)}.$$

Hence

$$(1 - e^{-\omega\bar{t}})\|u^f - u^h\|_{C_b([-\tau,+\infty);X} \le \frac{1 - e^{-\omega\bar{t}}}{\omega}\|f - h\|_{L^\infty(\mathbb{R}_+;X)}$$

and so (4.2.3) holds true.

Case 3. There is no maximum point $t \in [-\tau, +\infty)$ of $t \mapsto \|u^f(t) - u^h(t)\|$. This means that there exists $(t_k)_k$ in $(0, +\infty)$, with $\lim_k t_k = +\infty$ and

$$\lim_k \|u^f(t_k) - u^h(t_k)\| = \|u^f - u^h\|_{C_b([-\tau,+\infty);X)}.$$

Passing to the limit for $k \to +\infty$ in the inequality (obtained as before)

$$\|u^f(t_k) - u^h(t_k)\| \le e^{-\omega t_k}\|u^f - u^h\|_{C_b([-\tau,+\infty);X)} + \frac{1 - e^{-\omega t_k}}{\omega}\|f - h\|_{L^\infty(\mathbb{R}_+;X)},$$

we get (4.2.3) and this completes the proof. $\qquad\square$

Nonlocal Initial Conditions: The Quasi-Autonomous Case 133

4.2.2   A boundedness lemma

LEMMA 4.2.2 *Let x, y be two nonnegative functions in $C_b([-\tau, +\infty); \mathbb{R})$ and let us assume that there exist $a > 0$, $\omega > 0$, $\ell > 0$, and $m > 0$ such that*

$$x(t) \leq \|x\|_{C_b([a,+\infty);\mathbb{R})} \qquad (4.2.11)$$

for each $t \in [-\tau, 0]$ and

$$x(t) \leq e^{-\omega t} \|x\|_{C_b([a,+\infty);\mathbb{R})} + \frac{\ell}{\omega}\left(1 - e^{-\omega t}\right)\left(\|y\|_{C_b([-\tau,+\infty);\mathbb{R})} + \frac{m}{\ell}\right) \quad (4.2.12)$$

for each $t \in \mathbb{R}_+$. Then

$$\|x\|_{C_b([-\tau,+\infty);\mathbb{R})} \leq \frac{\ell}{\omega}\left(\|y\|_{C_b([-\tau,+\infty);\mathbb{R})} + \frac{m}{\ell}\right). \qquad (4.2.13)$$

Proof. We prove first that

$$\|x\|_{C_b([a,+\infty);\mathbb{R})} \leq \frac{\ell}{\omega}\left(\|y\|_{C_b([-\tau,+\infty);\mathbb{R})} + \frac{m}{\ell}\right). \qquad (4.2.14)$$

We distinguish between two possible complementary cases.

Case 1. There exists $\widetilde{t} \in [a, +\infty)$ such that

$$\|x\|_{C_b([a,+\infty);\mathbb{R})} = x(\widetilde{t}). \qquad (4.2.15)$$

Setting $t = \widetilde{t}$ in (4.2.12), we get

$$\|x\|_{C_b([a,+\infty);\mathbb{R})} \leq e^{-\omega \widetilde{t}} \|x\|_{C_b([a,+\infty);\mathbb{R})} + \frac{\ell}{\omega}(1 - e^{-\omega \widetilde{t}})\left(\|y\|_{C_b([-\tau,+\infty);\mathbb{R})} + \frac{m}{\ell}\right)$$

and, since $\widetilde{t} \geq a > 0$, we deduce (4.2.14).

Case 2. If there is no $\widetilde{t} \in [a, +\infty)$ such that (4.2.15) holds true, then there does exist at least one sequence $(t_k)_k \subseteq [a, +\infty)$ with $\lim_k t_k = +\infty$ and

$$\|x\|_{C_b([a,+\infty);\mathbb{R})} = \lim_k x(t_k).$$

Setting $t = t_k$ in (4.2.12) and letting $k \to +\infty$, we again obtain (4.2.14).

Now, let us prove (4.2.13).

Here, we distinguish between three possible cases.

Case 1. There exists $\bar{t} \in [-\tau, 0]$ such that

$$\|x\|_{C_b([-\tau,+\infty);\mathbb{R})} = x(\bar{t}).$$

By (4.2.11), we get

$$x(\bar{t}) \leq \|x\|_{C_b([a,+\infty);\mathbb{R})}$$

and thus

$$\|x\|_{C_b([-\tau,+\infty);\mathbb{R})} = \|x\|_{C_b([a,+\infty);\mathbb{R})}$$

134   *Delay Differential Evolutions Subjected to Nonlocal Initial Conditions*

and by (4.2.14), we deduce (4.2.13).

Case 2. There exists $\bar{t} > 0$ such that

$$\|x\|_{C_b([-\tau,+\infty);\mathbb{R})} = x(\bar{t}).$$

Using (4.2.12), we conclude that

$$\|x\|_{C_b([-\tau,+\infty);\mathbb{R})} \le e^{-\omega\bar{t}}\|x\|_{C_b([-\tau,+\infty);\mathbb{R})} + \frac{\ell}{\omega}(1-e^{-\omega\bar{t}})\left(\|y\|_{C_b([-\tau,+\infty);\mathbb{R})} + \frac{m}{\ell}\right).$$

Since $\bar{t} > 0$, we easily get (4.2.13).

Case 3. If, for each $t \in [-\tau, +\infty)$,

$$x(t) < \|x\|_{C_b([-\tau,+\infty);\mathbb{R})},$$

then there exists $(t_k)_k$ with $\lim_k t_k = +\infty$ such that

$$\lim_k x(t_k) = \|x\|_{C_b([-\tau,+\infty);\mathbb{R})}.$$

Setting $t = t_k$ in (4.2.12) and letting $k \to +\infty$, we obtain (4.2.13).

So, in all possible cases, x and y satisfy (4.2.13) and this completes the proof. $\qquad\square$

4.2.3   Proof of Theorem 4.1.1

Proof. The proof is based on a fixed-point argument. Namely, let us define the operator $Q : C_b([-\tau, +\infty); X) \to C_b([-\tau, +\infty); X)$ by $Q(v) = u$, where $u \in C_b([-\tau, +\infty); X)$ is the unique C^0-solution of the problem

$$\begin{cases} u'(t) \in Au(t) + f(t, v_t), & t \in \mathbb{R}_+, \\ u(t) = g(u)(t), & t \in [-\tau, 0] \end{cases}$$

whose existence is ensured by Lemma 4.2.1. To complete the proof of the existence and uniqueness part, it would suffice to show that Q is a strict contraction. In order to prove this, let $v, w \in C_b([-\tau, +\infty); X)$ be arbitrary and let us observe that, for each $t \ge 0$, from (1.8.2) and (f_1), we get

$$\|Q(v)(t) - Q(w)(t)\| \le e^{-\omega t}\|Q(v)(0) - Q(w)(0)\|$$

$$+ \frac{\ell}{\omega}(1 - e^{-\omega t}) \sup_{s \in [-\tau,t]} \|v(s) - w(s)\|$$

$$\le e^{-\omega t}\|g(Q(v))(0) - g(Q(w))(0)\| + \frac{\ell}{\omega}(1 - e^{-\omega t})\|v - w\|_{C_b([-\tau,+\infty);X)}$$

$$\le e^{-\omega t}\|g(Q(v)) - g(Q(w))\|x + \frac{\ell}{\omega}(1 - e^{-\omega t})\|v - w\|_{C_b([-\tau,+\infty);X)}.$$

Nonlocal Initial Conditions: The Quasi-Autonomous Case 135

Thanks to (g_1), we conclude that, for each $t \geq 0$, we have

$$\|Q(v)(t) - Q(w)(t)\|$$

$$\leq e^{-\omega t}\|Q(v) - Q(w)\|_{C_b([a,+\infty);X)} + \frac{\ell}{\omega}(1 - e^{-\omega t})\|v - w\|_{C_b([-\tau,+\infty);X)}.$$

By (g_1), we have

$$\|Q(v)(t) - Q(w)(t)\| \leq \|Q(v) - Q(w)\|_{C_b([a,+\infty);X)}$$

for each $t \in [-\tau, 0]$. Denoting by

$$x(t) = \|Q(v)(t) - Q(w)(t)\|, \quad y(t) = \|v(t) - w(t)\|$$

for $t \in [-\tau, +\infty)$, with $\omega > 0$, $\ell > 0$ as above and $m = 0$, we have merely to observe that we are in the hypotheses of Lemma 4.2.2. Then, we conclude that

$$\|Q(v) - Q(w)\|_{C_b([-\tau,+\infty);X)} \leq \frac{\ell}{\omega}\|v - w\|_{C_b([-\tau,+\infty);X)},$$

which, in view of (H_c), i.e., $\ell < \omega$, shows that Q is a strict contraction.

In order to prove (4.1.2), let us observe that we have

$$\|u(t)\| \leq e^{-\omega t}\|u(0)\| + e^{-\omega t}\int_0^t e^{\omega s}\left(\|f(s, u_s) - f(s, 0)\| + \|f(s, 0)\|\right) ds$$

$$\leq e^{-\omega t}\|u(0)\| + \ell e^{-\omega t}\int_0^t e^{\omega s}\left(\|u_s\|_X + \frac{m}{\ell}\right) ds$$

$$\leq e^{-\omega t}\|u(0)\| + (1 - e^{-\omega t})\frac{\ell}{\omega}\left(\|u\|_{C_b([-\tau,+\infty);X)} + \frac{m}{\ell}\right),$$

for each $t \in \mathbb{R}_+$. This shows that we are in the hypotheses of Lemma 3.3.1, which implies that (4.1.2) holds true with k given by (4.1.3). Finally, if the stronger condition (f_5) is satisfied, then by Theorem 1.12.7 we deduce that the C^0-solution is globally asymptotically stable and this completes the proof of Theorem 4.1.1. $\qquad\square$

4.2.4   Proof of Theorem 4.1.2

Proof. We have only to observe that (4.1.4) can be equivalently rewritten as (4.1.1), where $g : C_b(\mathbb{R}_+; \overline{D(A)}) \to \mathcal{D}$ is defined by

$$g(u)(t) = u(T + t)$$

for each $u \in C_b(\mathbb{R}_+; X)$ and all $t \in [-\tau, 0]$.

In order to verify that g satisfies (g_1) with $a = T - \tau > 0$, let u, v be

136 *Delay Differential Evolutions Subjected to Nonlocal Initial Conditions*

arbitrary but fixed in $C_b(\mathbb{R}_+; \overline{D(A)})$. Clearly, there exists $t_{u,v} \in [-\tau, 0]$ such that

$$\|g(u) - g(v)\|_X = \|g(u)(t_{u,v}) - g(v)(t_{u,v})\| = \|u(T + t_{u,v}) - v(T + t_{u,v})\|.$$

But $T + t_{u,v} \geq T - \tau = a > 0$ and therefore

$$\|g(u) - g(v)\|_X \leq \|u - v\|_{C_b([a,+\infty);X)}.$$

So, g satisfies (g_1) in (H_g). Thus we are in the hypotheses of Theorem 4.1.1. Since (4.1.5) follows from (4.1.2) in conjunction with the periodicity condition and with $m_0 = \|g(0)\|_X = 0$, this completes the proof. $\square$

REMARK 4.2.1 With essentially the same proof, Theorem 4.1.1 can be easily extended to the more general case when f satisfies the hypothesis $(H_{\widetilde{f}})$ below.

$(H_{\widetilde{f}})$ The function $f : \mathbb{R}_+ \times X \to X$ is strongly measurable with respect to $t \in \mathbb{R}_+$ and satisfies

$(\widetilde{f}_1)$ there exists $\ell > 0$ such that

$$\|f(t, v) - f(t, \widetilde{v})\| \leq \ell \|v - \widetilde{v}\|_X$$

a.e. for $t \in \mathbb{R}_+$ and for each $v, \widetilde{v} \in X$

$(\widetilde{f}_2)$ there exists $m > 0$ such that

$$\|f(t, 0)\| \leq m$$

a.e. for $t \in \mathbb{R}_+$.

4.3 Nonlinear diffusion with nonlocal initial conditions

Let Ω be a nonempty, bounded domain in $\mathbb{R}^d$, $d \geq 2$, with C^1 boundary Σ, let $\tau \geq 0$, $\omega > 0$, and let Δ be the Laplace operator in the sense of distributions over the domain Ω. Let $\varphi : D(\varphi) \subseteq \mathbb{R} \leadsto \mathbb{R}$ be a maximal-monotone operator, let $X_1 = C([-\tau, 0,]; L^1(\Omega))$, let $h : \mathbb{R}_+ \times \Omega \times X_1 \to \mathbb{R}$ be continuous, and let $\psi \in X_1$. We will use the standard notations: $Q_+ = \mathbb{R}_+ \times \Omega$, $\Sigma_+ = \mathbb{R}_+ \times \Sigma$, $Q_\tau = [-\tau, 0] \times \Omega$. Let us consider the following problem with nonlocal initial conditions:

$$\begin{cases} \dfrac{\partial u}{\partial t}(t, x) = \Delta\varphi(u(t, x)) - \omega u(t, x) + h(t, x, u_t), & \text{in } Q_+, \\[2ex] \varphi(u(t, x)) = 0, & \text{on } \Sigma_+, \\[2ex] u(t, x) = \displaystyle\sum_{i=1}^{\infty} \alpha_i u(t_i + t, x) + \psi(t)(x), & \text{in } Q_\tau, \end{cases} \qquad (4.3.1)$$

Nonlocal Initial Conditions: The Quasi-Autonomous Case 137

where $\alpha_i \in [-1, 1]$, for $i = 1, 2, \ldots$, and $0 < t_1 < t_2 < \ldots$ are given.

The equation in (4.3.1) is simply a closed-loop system referring to some well-known mathematical models. As mentioned in Section 2.9, one of the most important is *the controlled porous media diffusion equation with delay*. See for instance Barbu [22, Section 5.3, p. 222] or Showalter [238, A2, pp. 253–254]. Here, the synthesis operator $(t, u) \mapsto -\omega u + h(t, \cdot, u_t)$, i.e., the reaction term, depends on both the cumulative past values of the state, u_t, as well as on the state, $-\omega u$, the latter term ensuring the global asymptotic stability of solutions whenever h is globally Lipschitz with respect to its first and third argument and the Lipschitz constant ℓ satisfies $\ell < \omega$. As in Remark 3.4.3, one can be easily see that the nonlocal initial condition imposed includes as specific cases: the T-periodicity condition as well as the T-anti-periodicity condition.

4.3.1   The main result

THEOREM 4.3.1 *Let Ω be a nonempty and bounded domain in $\mathbb{R}^d$, $d \geq 2$, with C^1 boundary Σ, let $\tau \geq 0$, $\omega > 0$ and let $\varphi : D(\varphi) \subseteq \mathbb{R} \rightsquigarrow \mathbb{R}$ be maximal-monotone with $0 \in \varphi(0)$. Let $h : \mathbb{R}_+ \times \Omega \times X_1 \to \mathbb{R}$, let $\psi \in X_1$ and let $\alpha_i \in [-1, 1]$, for $i = 1, 2, \ldots$ and $0 < t_1 < t_2 < \ldots$. If*

(h_1) *there exists $\ell_1 > 0$ such that*

$$|h(t, x, v) - h(t, x, \widetilde{v})| \leq \ell_1 \|v - \widetilde{v}\|_{X_1}$$

a.e. for $x \in \Omega$, for each $t \in \mathbb{R}_+$ and each $v, \widetilde{v} \in X_1$

(h_2) *there exists a nonnegative function $m_1 \in L^1(\Omega)$ such that*

$$|h(t, x, 0)| \leq m_1(x)$$

a.e. for $x \in \Omega$ and for each $t \in \mathbb{R}_+$

(α) $\displaystyle\sum_{i=1}^{\infty} |\alpha_i| \leq 1$ *and $\tau < t_1 < t_2 < \ldots$*

(c) $\ell_1 |\Omega| = \ell < \omega$

then the problem (4.3.1) has a unique C^0-solution, $u \in C_b([-\tau, +\infty); L^1(\Omega))$, that satisfies

$$\|u\|_{C_b([-\tau,+\infty);L^1(\Omega))} \leq \frac{m}{\omega - \ell} + k \cdot \|\psi\|_{X_1},$$

where $m = \|m_1\|_{L^1(\Omega)}$ and k is given by (3.2.3). If, instead of (h_1), the stronger condition,

(h_3) *there exists $\ell_1 > 0$ such that*

$$|h(t, x, v) - h(s, x, \widetilde{v})| \leq \ell_1 (|t - s| + \|v - \widetilde{v}\|_{X_1})$$

a.e. for $x \in \Omega$, for each $t, s \in \mathbb{R}_+$ and each $v, \widetilde{v} \in X_1$,

138   *Delay Differential Evolutions Subjected to Nonlocal Initial Conditions*

is satisfied, then the C^0-solution of (4.3.1) is globally asymptotically stable.

REMARK 4.3.1 As (4.3.1) rewrites as an abstract evolution equation in the function space $L^1(\Omega)$, the notion of C^0-solution for (4.3.1) should be understood by identification. Namely, a C^0-solution of (4.3.1) is simply a C^0-solution of the corresponding evolution problem in $L^1(\Omega)$. For details, see the proof of Theorem 4.3.1 below.

Proof. Let us define the operator $A : D(A) \subseteq L^1(\Omega) \rightsquigarrow L^1(\Omega)$ be given by $D(A) = D(\Delta\varphi)$, $Au = \Delta\varphi(u) - \omega u$, for each $u \in D(A)$, where $\Delta\varphi$ is defined as in Theorem 1.9.6. Let $f : \mathbb{R}_+ \times \mathfrak{X}_1 \to L^1(\Omega)$ and $g : C_b(\mathbb{R}_+; L^1(\Omega)) \to \mathfrak{X}_1$ be defined as

$$f(t, w)(x) = h(t, x, w)$$

for each $t \in \mathbb{R}_+$, each $w \in \mathfrak{X}_1$ and a.e. for $x \in \Omega$ and respectively by

$$[g(u)(t)](x) = \sum_{i=1}^{\infty} \alpha_i u(t_i + t, x) + \psi(t)(x)$$

for each $u \in C_b([-\tau, +\infty); L^1(\Omega))$, each $t \in [-\tau, 0]$ and a.e. for $x \in \Omega$. See (iv) in Remark 3.2.4.

We apply Theorem 4.1.1 in $X = L^1(\Omega)$. Namely, in view of Theorem 1.9.6, the operator A satisfies the hypothesis (H_A) in Theorem 4.1.1. From (h_1) and (h_2), we deduce that f satisfies (H_f). Furthermore, from (α), it follows that g satisfies (H_g), where a is given by $a = t_1 - \tau$. So, the problem (4.3.1) can be rewritten as an abstract problem of the form (4.1.1) satisfying the hypotheses of Theorem 4.1.1, from which comes the conclusion of Theorem 4.3.1. $\quad\square$

4.3.2   The periodic and the anti-periodic problem

Let us now consider the T-periodic problem for the nonlinear delay diffusion equation:

$$\begin{cases} \dfrac{\partial u}{\partial t}(t, x) = \Delta\varphi(u(t, x)) - \omega u(t, x) + h(t, x, u_t), & \text{in } Q_+, \\[2mm] \varphi(u(t, x)) = 0, & \text{on } \Sigma_+, \\[2mm] u(t, x) = u(t + T, x), & \text{in } Q_\tau. \end{cases} \qquad (4.3.2)$$

As an application of Theorem 4.3.1, we have:

THEOREM 4.3.2 *Let Ω be a nonempty, bounded domain in $\mathbb{R}^d$, $d \geq 1$, with C^1 boundary Σ, let $\tau \geq 0$, $\omega > 0$ and let $\varphi : D(\varphi) \subseteq \mathbb{R} \rightsquigarrow \mathbb{R}$ be a maximal-monotone operator with $0 \in \varphi(0)$. Let $h : \mathbb{R}_+ \times \Omega \times \mathfrak{X}_1 \to \mathbb{R}$ be continuous and T-periodic with respect to its first argument. If $T > \tau$ and the function h satisfies (h_1), (h_2) and (c_1) in Theorem 4.3.1, then the problem (4.3.2) has a unique C^0-solution $u \in C_b(\mathbb{R}_+; L^1(\Omega))$. If, instead of (h_1), the stronger condition (h_3) in Theorem 4.3.1 is satisfied, then the unique C^0-solution of (4.3.2) is globally asymptotically stable.*

Nonlocal Initial Conditions: The Quasi-Autonomous Case 139

Finally, let us consider the T-anti-periodic problem for the nonlinear delay diffusion equation subjected to anti-periodic conditions:

$$\begin{cases} \dfrac{\partial u}{\partial t}(t,x) = \Delta\varphi(u(t,x)) - \omega u(t,x) + h(t,x,u_t), & \text{in } Q_+, \\[2mm] \varphi(u(t,x)) = 0, & \text{on } \Sigma_+, \\[2mm] u(t,x) = -u(t+T,x), & \text{in } Q_\tau. \end{cases} \qquad (4.3.3)$$

Also from Theorem 4.3.1 combined with Lemma 4.1.2, we deduce:

THEOREM 4.3.3 *Let Ω be a nonempty, bounded domain in $\mathbb{R}^d$, $d \geq 1$, with C^1 boundary Σ, let $\tau \geq 0$, $\omega > 0$ and let $\varphi : D(\varphi) \subseteq \mathbb{R} \rightsquigarrow \mathbb{R}$ be a maximal-monotone operator with $0 \in \varphi(0)$. Let $h : \mathbb{R}_+ \times \Omega \times X_1 \to \mathbb{R}$ be continuous and satisfy*

$$-h(t+T,x,-u) = h(t,x,u)$$

a.e. for $x \in \Omega$, for each $t \in \mathbb{R}_+$ and each $u \in X_1$. If φ is odd, $T > \tau$ and the function h satisfies (h_1), (h_2) and (c) in Theorem 4.3.1, then the problem (4.3.3) has a unique C^0-solution $u \in C_b(\mathbb{R}_+; L^1(\Omega))$, which is $2T$-periodic. If, instead of (h_1), the stronger condition (h_3) in Theorem 4.3.1 is satisfied, then the unique C^0-solution of (4.3.3) is globally asymptotically stable.

4.4 Continuity with respect to the data

Our next goal is to prove the continuity of the C^0-solutions with respect to the right-hand side and the initial nonlocal condition, for a class of nonlinear delay differential evolution equations of the form

$$\begin{cases} u'(t) \in Au(t) + f(t,u_t), & t \in \mathbb{R}_+, \\[2mm] u(t) = g(u)(t), & t \in [-\tau,0]. \end{cases} \qquad (4.4.1)$$

Here, $A : D(A) \subseteq X \rightsquigarrow X$ is an m-dissipative operator in the Banach space X, $\tau \geq 0$, the function $f : \mathbb{R}_+ \times X \to X$ is continuous, $g : C_b(\mathbb{R}_+; \overline{D(A)}) \to \mathcal{D}$ is nonexpansive. We recall that, whenever $\tau = 0$, $X = C([-\tau,0];X)$ reduces to X and $\mathcal{D}$ to $\overline{D(A)}$. Continuity properties of the C^0-solution with respect to the data f and g are of great practical interest. As we already have mentioned, problems such as (4.4.1) represent mathematical models describing the time×spatial evolution of various phenomena. As one can easily understand, both f and g are partly obtained by direct measurements and partly by statistical considerations. So, in the process of these measurements, some errors can – and in fact do – occur. Therefore it is important to know that the models thus obtained are stable, i.e., are not too much affected by small perturbations, and thus reliable.

4.4.1 Statement of the main result

THEOREM 4.4.1 *Let $A: D(A) \subseteq X \rightsquigarrow X$ be an operator satisfying (H_A). Let $\{f_n : \mathbb{R}_+ \times \mathcal{X} \to X; \ n \in \mathbb{N}\}$ be a family of continuous functions satisfying the following:*

(h_1) *there exists $\ell > 0$ such that $\|f_n(t, x) - f_n(t, y)\| \le \ell \|x - y\|_{\mathcal{X}}$ for each $n \in \mathbb{N}$, each $t \in \mathbb{R}_+$ and $x, y \in \mathcal{X}$*

(h_2) *there exists $m > 0$ such that $\|f_n(t, 0)\| \le m$ for each $n \in \mathbb{N}$ and each $t \in \mathbb{R}_+$*

(h_3) $\lim\limits_{n} f_n(t, x) = f(t, x)$ *uniformly for $t \in \mathbb{R}_+$ (uniformly for t in bounded intervals in $\mathbb{R}_+$) and x in bounded subsets in $\mathcal{X}$.*

Let $\{g_n : C_b(\mathbb{R}_+; \overline{D(A)}) \to \mathcal{D}; \ n \in \mathbb{N}\}$ be a family of functions satisfying the following:

(h_4) *there exists $m_0 > 0$ such that $\|g_n(0)\|_{\mathcal{X}} \le m_0$ for each $n \in \mathbb{N}$*

(h_5) *there exists $a > 0$ such that for each $n \in \mathbb{N}$ and $u, \widetilde{u} \in C_b(\mathbb{R}_+; \overline{D(A)})$, we have*

$$\|g_n(u) - g_n(\widetilde{u})\|_{\mathcal{X}} \le \|u - \widetilde{u}\|_{C_b([a, +\infty); X)}$$

(there exist $a, d \in \mathbb{R}_+$ with $0 < a < d$ such that for each $n \in \mathbb{N}$ and $u, \widetilde{u} \in C_b(\mathbb{R}_+; \overline{D(A)})$, we have

$$\|g_n(u) - g_n(\widetilde{u})\|_{\mathcal{X}} \le \|u - \widetilde{u}\|_{C([a, d]; X)})$$

(h_6) $\lim\limits_{n} g_n(u) = g(u)$ *uniformly for u in bounded sets in $C_b(\mathbb{R}_+; \overline{D(A)})$.*

Let us assume further that (H_c) holds true. Let $(u_n)_n$ be the sequence of C^0-solutions of the problems

$$\begin{cases} u_n'(t) \in Au_n(t) + f_n(t, u_{n_t}), & t \in \mathbb{R}_+, \\ u_n(t) = g_n(u_n)(t), & t \in [-\tau, 0], \end{cases} \tag{4.4.2}$$

whose existence and uniqueness is ensured by Theorem 4.1.1. Then

$$\lim\limits_{n} u_n = u$$

in $C_b([-\tau, +\infty); X)$ (in $\widetilde{C}_b([-\tau, +\infty); X)^2$), where u is the unique C^0-solution of the problem (4.4.1).

REMARK 4.4.1 From (h_4) and (h_5) it follows that $(g_n)_n$ has uniform affine growth. More precisely, there exists $m_0 \ge 0$ such that for each $n \in \mathbb{N}$ and $u \in C_b([-\tau, +\infty); X)$, we have

$$\|g_n(u)\|_{\mathcal{X}} \le \|u\|_{C_b(\mathbb{R}_+; X)} + m_0.$$

---

[2]We recall that $\widetilde{C}_b([-\tau, +\infty); X)$ denotes the space $C_b([-\tau, +\infty); X)$ endowed with the uniform convergence on compacta topology.

4.4.2 Proof of the main result

Proof. We will focus our attention on the case of the norm-convergence of $(u_n)_n$ in $C_b([-\tau, +\infty); X)$, i.e., when, in (h_3), $\lim_n f_n(t, x) = f(t, x)$ uniformly for $t \in \mathbb{R}_+$ and x in bounded subsets in $\mathcal{X}$ and the first inequality in (h_5) is satisfied. We notice that the proof of the $\widetilde{C}_b([-\tau, +\infty); X)$ — convergence part is very similar and so we do not give details.

Let $n \in \mathbb{N}$ be arbitrary and let us observe that, by (H_A), (4.4.2), (4.4.1) and (1.8.2) in Theorem 1.8.1, we have

$$\|u_n(t) - u(t)\| \le e^{-\omega t}\|u_n(0) - u(0)\| + \int_0^t e^{-\omega(t-s)}\|f_n(s, u_{n_s}) - f(s, u_s)\|\, ds$$

$$(4.4.3)$$

for each $t \in \mathbb{R}_+$. Set

$$\mathcal{M} = \left\{ x \in C_b([-\tau, +\infty); \overline{D(A)});\ \|x\|_{C_b([-\tau, +\infty); X)} \le \frac{m}{\omega - \ell} + k \cdot m_0 \right\},$$

where k is given by (4.1.3) and m_0 by (h_4). From (h_1), (h_2), (h_4), (h_5), (H_A) and (H_c), we deduce that we are in the hypotheses of Theorem 4.1.1 and so, by (4.1.2), it follows that $u \in \mathcal{M}$ and $u_n \in \mathcal{M}$ for each $n \in \mathbb{N}$. Let $n \in \mathbb{N}$ and let us define

$$\alpha_n = \sup\{\|f_n(t, x) - f(t, x)\|;\ (t, x) \in \mathbb{R}_+ \times \mathcal{M}\}$$

and

$$\beta_n = \sup\{\|g_n(v) - g(v)\|_{\mathcal{X}};\ v \in \mathcal{M}\}.$$

By (h_3) and (h_6), we have

$$\lim_n \alpha_n = \lim_n \beta_n = 0.$$

From (4.4.3), (4.1.2), (h_1) and (h_5), after some obvious calculations, we get

$$\|u_n(t) - u(t)\| \le e^{-\omega t}\left[\beta_n + \|u_n - u\|_{C_b([a, +\infty); X)}\right]$$

$$+\alpha_n\left(1 - e^{-\omega t}\right)\frac{1}{\omega} + \left(1 - e^{-\omega t}\right)\frac{\ell}{\omega}\|u_n - u\|_{C_b([-\tau, +\infty); X)} \qquad (4.4.4)$$

for each $t \in \mathbb{R}_+$. Let us observe now that

$$\|u_n - u\|_{C_b([-\tau, +\infty); X)} \le \max\{\|u_n - u\|_{\mathcal{X}}, \|u_n - u\|_{C_b(\mathbb{R}_+; X)}\}$$

$$\le \max\{\|u_n - u\|_{C_b([a, +\infty); X)} + \beta_n, \|u_n - u\|_{C_b(\mathbb{R}_+; X)}\} \le \|u_n - u\|_{C_b(\mathbb{R}_+; X)} + \beta_n$$

for each $n \in \mathbb{N}$. Hence

$$\|u_n - u\|_{C_b([-\tau, +\infty); X)} \le \|u_n - u\|_{C_b(\mathbb{R}_+; X)} + \beta_n \qquad (4.4.5)$$

142   *Delay Differential Evolutions Subjected to Nonlocal Initial Conditions*

for each $n \in \mathbb{N}$. Denoting by

$$\varepsilon_n = \left(1 + \frac{\ell}{\omega}\right)\beta_n + \frac{\alpha_n}{\omega},$$

from (4.4.4) and (4.4.5), we deduce

$$\|u_n(t) - u(t)\| \le \varepsilon_n + e^{-\omega t}\|u_n - u\|_{C_b([a,+\infty);X)}$$

$$+ \left(1 - e^{-\omega t}\right)\frac{\ell}{\omega}\|u_n - u\|_{C_b(\mathbb{R}_+;X)} \qquad (4.4.6)$$

for each $n \in \mathbb{N}$ and $t \in \mathbb{R}_+$.

We prove first that

$$\lim_n \|u_n - u\|_{C_b(\mathbb{R}_+;X)} = 0.$$

To this end, we proceed by contradiction. So, let us assume that

$$\limsup_n \|u_n - u\|_{C_b(\mathbb{R}_+;X)} > 0.$$

Taking a subsequence and relabeling if necessary, we may assume with no loss of generality that

$$\lim_n \|u_n - u\|_{C_b(\mathbb{R}_+;X)} > 0. \qquad (4.4.7)$$

To get a contradiction, let us observe that $(u_n)_n$ is in at least one of the three cases described below.

Case 1. There exists an infinite set $\mathbb{N}_0 \subseteq \mathbb{N}$ such that, for each $n \in \mathbb{N}_0$, we have $\|u_n - u\|_{C_b(\mathbb{R}_+;X)} = \|u_n(0) - u(0)\|$. Then, by (h_5), we deduce that

$$\|u_n - u\|_{C_b(\mathbb{R}_+;X)} = \|g_n(u_n)(0) - g(u)(0)\|$$

$$\le \|g_n(u_n)(0) - g_n(u)(0)\| + \|g_n(u)(0) - g(u)(0)\| \le \|u_n - u\|_{C_b([a,+\infty);X)} + \beta_n.$$

So, for each $\varepsilon > 0$ there exists $t_{n,\varepsilon} \in [a, +\infty)$ such that

$$\|u_n - u\|_{C_b(\mathbb{R}_+;X)} \le \|u_n - u\|_{C_b([a,+\infty);X)} + \beta_n$$

$$\le \|u_n(t_{n,\varepsilon}) - u(t_{n,\varepsilon})\| + \beta_n + \varepsilon.$$

Again by (4.4.6), we have

$$\|u_n - u\|_{C_b(\mathbb{R}_+;X)} \le \varepsilon + \beta_n + \varepsilon_n$$

$$+ e^{-\omega t_{n,\varepsilon}}\|u_n - u\|_{C_b([a,+\infty);X)} + \left(1 - e^{-\omega t_{n,\varepsilon}}\right)\frac{\ell}{\omega}\|u_n - u\|_{C_b(\mathbb{R}_+;X)},$$

and thus

$$\left(1 - e^{-\omega t_{n,\varepsilon}}\right)\|u_n - u\|_{C_b(\mathbb{R}_+;X)} \le \varepsilon + \beta_n + \varepsilon_n + \left(1 - e^{-\omega t_{n,\varepsilon}}\right)\frac{\ell}{\omega}\|u_n - u\|_{C_b(\mathbb{R}_+;X)}.$$

Nonlocal Initial Conditions: The Quasi-Autonomous Case

Dividing on both sides of this inequality by $1 - e^{-\omega t_{n,\varepsilon}} \geq 1 - e^{-\omega a} > 0$ and passing to the $\limsup$ for $n \in \mathbb{N}_0$, we get

$$\limsup_{n,n \in \mathbb{N}_0} \|u_n - u\|_{C_b(\mathbb{R}_+;X)} \leq \frac{\ell}{\omega} \limsup_{n,n \in \mathbb{N}_0} \|u_n - u\|_{C_b(\mathbb{R}_+;X)} + \frac{\varepsilon}{1 - e^{-\omega a}}.$$

As $\varepsilon > 0$ is arbitrary and $\ell < \omega$, this shows that

$$\limsup_{n,n \in \mathbb{N}_0} \|u_n - u\|_{C_b(\mathbb{R}_+;X)} = 0 = \lim_{n,n \in \mathbb{N}_0} \|u_n - u\|_{C_b(\mathbb{R}_+;X)},$$

an equality that contradicts (4.4.7).

Case 2. There exists an infinite set $\mathbb{N}_0 \subseteq \mathbb{N}$ such that, for each $n \in \mathbb{N}_0$, there exists $t_n > 0$ such that

$$\|u_n(t_n) - u(t_n)\| = \|u_n - u\|_{C_b(\mathbb{R}_+;X)}.$$

Then, we distinguish between two subcases.

Subcase 2.1. $\liminf\limits_{n,n \in \mathbb{N}_0} t_n = 0$. From (4.4.3), we get

$$\|u_n - u\|_{C_b(\mathbb{R}_+;X)} \leq e^{-\omega t_n} \|u_n - u\|_{C_b(\mathbb{R}_+;X)}$$

$$+ \frac{\ell}{\omega} \left(1 - e^{-\omega t_n}\right) \|u_n - u\|_{C_b([-\tau,+\infty);X)} + \frac{1}{\omega} \left(1 - e^{-\omega t_n}\right) \alpha_n$$

for each $n \in \mathbb{N}$. Next, from (4.4.5), we deduce

$$\|u_n - u\|_{C_b(\mathbb{R}_+;X)} \leq e^{-\omega t_n} \|u_n - u\|_{C_b(\mathbb{R}_+;X)}$$

$$+ \frac{\ell}{\omega} \left(1 - e^{-\omega t_n}\right) \|u_n - u\|_{C_b(\mathbb{R}_+;X)} + \frac{\ell}{\omega} \left(1 - e^{-\omega t_n}\right) \beta_n + \frac{1}{\omega} \left(1 - e^{-\omega t_n}\right) \alpha_n$$

for each $n \in \mathbb{N}$. Since $1 - e^{-\omega t_n} > 0$ for each $n \in \mathbb{N}_0$, from this inequality, we get

$$\|u_n - u\|_{C_b(\mathbb{R}_+;X)} \leq \frac{\ell}{\omega} \|u_n - u\|_{C_b(\mathbb{R}_+;X)} + \frac{\ell}{\omega} \beta_n + \frac{1}{\omega} \alpha_n$$

for each $n \in \mathbb{N}_0$. So,

$$\limsup_n \|u_n - u\|_{C_b(\mathbb{R}_+;X)} = 0,$$

which obviously contradicts (4.4.7).

Subcase 2.2. $\liminf\limits_{n \to +\infty, n \in \mathbb{N}_0} t_n = \alpha > 0$, the case $\alpha = +\infty$ is not excluded. Setting $t = t_n$ in (4.4.6) and passing to the $\limsup$ on the subsequence associated with the $\liminf$ above, after some rearrangements, we get

$$\limsup_n \|u_n - u\|_{C_b(\mathbb{R}_+;X)} \leq \frac{\ell}{\omega} \limsup_n \|u_n - u\|_{C_b(\mathbb{R}_+;X)}.$$

144 *Delay Differential Evolutions Subjected to Nonlocal Initial Conditions*

But $\ell < \omega$, and thus $\limsup_n \|u_n - u\|_{C_b(\mathbb{R}_+;X)} = \lim_n \|u_n - u\|_{C_b(\mathbb{R}_+;X)} = 0$, which is in contradiction with (4.4.7).

Case 3. There exists $k_0 \in \mathbb{N}$ such that, for each $n \in \mathbb{N}$ with $n \geq k_0$ and each $t \in \mathbb{R}_+$, we have

$$\|u_n(t) - u(t)\| < \|u_n - u\|_{C_b(\mathbb{R}_+;X)}.$$

Then, there exists $(t_{n_k})_k$ with $k_0 \leq k \leq n_k$, $\lim_k t_{n_k} = +\infty$ such that

$$\lim_k \|u_n(t_{n,k}) - u(t_{n,k})\| = \|u_n - u\|_{C_b(\mathbb{R}_+;X)}.$$

Setting $t = t_{n_k}$ in (4.4.6) and passing to the limit for $k \to +\infty$, we get

$$\left(1 - \frac{\ell}{\omega}\right) \|u_n - u\|_{C_b(\mathbb{R}_+;X)} \leq \varepsilon_n$$

for each $n \in \mathbb{N}$. Since $\ell < \omega$, passing to $\limsup$ for $n \to +\infty$ in the inequality above, we get

$$\limsup_n \|u_n - u\|_{C_b(\mathbb{R}_+;X)} = 0.$$

Consequently $\|u_n - u\|_{C_b(\mathbb{R}_+;X)} = 0$, which again contradicts (4.4.7).

So, in all possible cases, the supposition (4.4.7) leads to a contradiction. This contradiction can be eliminated only if (4.4.7) is false. Thus

$$\lim_n \|u_n - u\|_{C_b(\mathbb{R}_+;X)} = 0.$$

Finally, from (4.4.5), we have $\lim_n \|u_n - u\|_{C_b(\mathbb{R}_+;X)} = \lim_n \|u_n - u\|_{C_b([-\tau,+\infty);X)}$ and the proof is complete. $\qquad\square$

4.4.3 The nondelayed case

As far as the nondelayed case, i.e., $\tau = 0$, is concerned, we have:

THEOREM 4.4.2 *Let* $A : D(A) \subseteq X \rightsquigarrow X$ *be an operator satisfying* (H_A). *Let* $\{f_n : \mathbb{R}_+ \times X \to X; \, n \in \mathbb{N}\}$ *be a family of continuous functions satisfying the following:*

$(h_1^{[\tau=0]})$ *there exists* $\ell > 0$ *such that* $\|f_n(t,x) - f_n(t,y)\| \leq \ell\|x - y\|$ *for each* $n \in \mathbb{N}$, *each* $t \in \mathbb{R}_+$ *and* $x, y \in X$

$(h_2^{[\tau=0]})$ *there exists* $m > 0$ *such that* $\|f_n(t,0)\| \leq m$ *for each* $n \in \mathbb{N}$ *and each* $t \in \mathbb{R}_+$

$(h_3^{[\tau=0]})$ $\lim_n f_n(t,x) = f(t,x)$ *uniformly for* $t \in \mathbb{R}_+$ *(uniformly for* t *in bounded intervals in* $\mathbb{R}_+$*) and* x *in bounded subsets in* X.

Nonlocal Initial Conditions: The Quasi-Autonomous Case

Let $\{g_n : C_b(\mathbb{R}_+; \overline{D(A)}) \to \overline{D(A)}; \ n \in \mathbb{N}\}$ be a family of functions satisfying the following:

$(h_4^{[\tau=0]})$ there exists $m_0 > 0$ such that $\|g_n(0)\| \leq m_0$ for each $n \in \mathbb{N}$;

$(h_5^{[\tau=0]})$ there exists $a > 0$ such that for each $n \in \mathbb{N}$ and $u, \widetilde{u} \in C_b(\mathbb{R}_+; \overline{D(A)})$, we have

$$\|g_n(u) - g_n(\widetilde{u})\| \leq \|u - \widetilde{u}\|_{C_b([\,a,+\infty);X)}$$

(there exist $a, d \in \mathbb{R}_+$ with $0 < a < d$ such that for each $n \in \mathbb{N}$ and $u, \widetilde{u} \in C_b(\mathbb{R}_+; \overline{D(A)})$, we have

$$\|g_n(u) - g_n(\widetilde{u})\| \leq \|u - \widetilde{u}\|_{C_b([\,a,d\,];X)})$$

$(h_6^{[\tau=0]})$ $\displaystyle\lim_n g_n(u) = g(u)$ uniformly for u in bounded subsets in $C_b(\mathbb{R}_+; \overline{D(A)})$.

Let us assume further that (H_c) holds true. Let $(u_n)_n$ be the sequence of C^0-solutions of the problems

$$\begin{cases} u_n'(t) \in Au_n(t) + f_n(t, u_n(t)), & t \in \mathbb{R}_+, \\ u_n(0) = g_n(u_n), \end{cases}$$

whose existence and uniqueness is ensured by Theorem 4.1.4.

Then

$$\lim_n u_n = u$$

in $C_b([-\tau, +\infty); X)$ (in $\widetilde{C}_b([-\tau, +\infty); X)$), where u is the unique C^0-solution of the problem

$$\begin{cases} u'(t) \in Au(t) + f(t, u(t)), & t \in \mathbb{R}_+, \\ u(0) = g(u). \end{cases}$$

4.4.4 A glance at periodic problems

An interesting variant of Theorem 4.4.2, which cannot be obtained as a simple consequence of the latter, refers to nondelayed periodic problems. More precisely, let us consider the following family of problems:

$$\begin{cases} u_n'(t) \in Au_n(t) + f_n(t, u_n(t)), & t \in \mathbb{R}_+, \\ u_n(t) = u_n(t + T_n), & t \in \mathbb{R}_+, \end{cases} \tag{4.4.8}$$

where, for each $n \in \mathbb{N}$, $f_n : \mathbb{R}_+ \times \overline{D(A)} \to X$ is a continuous function that is T_n-periodic with respect to its first argument, i.e., $f_n(t + T_n, u) = f_n(t, u)$ for each $t \in \mathbb{R}_+$, and $u \in \overline{D(A)}$. From Theorem 4.4.1, we deduce:

146   *Delay Differential Evolutions Subjected to Nonlocal Initial Conditions*

THEOREM 4.4.3 *Let* $A : D(A) \subseteq X \rightsquigarrow X$ *be an operator satisfying* (H_A). *Let* $\{f_n : \mathbb{R}_+ \times \overline{D(A)} \to X;\ n \in \mathbb{N}\}$ *be a family of continuous functions, and let* $(T_n)_n$ *be a sequence of strictly positive numbers satisfying*

$(\widetilde{h}_1^{[\tau=0]})$ *there exists* $\ell > 0$ *such that* $\|f_n(t+h, x) - f_n(t, y)\| \leq \ell[h + \|x - y\|]$
  for each $n \in \mathbb{N}$, *each* $t, h \in \mathbb{R}_+$ *and* $x, y \in \overline{D(A)}$

$(\widetilde{h}_2^{[\tau=0]})$ *there exists* $m > 0$ *such that* $\|f_n(t, 0)\| \leq m$ *for each* $n \in \mathbb{N}$ *and each*
  $t \in \mathbb{R}_+$

$(\widetilde{h}_3^{[\tau=0]})$ $\lim_n f_n(t, x) = f(t, x)$ *uniformly for* $t \in \mathbb{R}_+$ *and* x *in compact subsets*
  in $\overline{D(A)}$

$(\widetilde{h}_7^{[\tau=0]})$ $\lim_n T_n = T > 0$

$(\widetilde{h}_8^{[\tau=0]})$ $f_n(t, x) = f_n(t + T_n, x)$ *for each* $n \in \mathbb{N}$, *each* $t \in \mathbb{R}_+$ *and each*
  $x \in \overline{D(A)}$.

Let us assume further that $(I - A)^{-1}$ *is compact and* (H_c) *holds true. Let* $(u_n)_n$ *be the sequence of* T_n-*periodic* C^0-*solutions of the problems* (4.4.8) *whose existence and uniqueness is ensured by Theorem 4.1.5. Then*

$$\lim_n u_n = u$$

in $C_b(\mathbb{R}_+; X)$, *where* u *is the unique* T-*periodic* C^0-*solution of the problem*

$$\begin{cases} u'(t) \in Au(t) + f(t, u(t)), & t \in \mathbb{R}_+, \\ u(t) = u(t + T), & t \in \mathbb{R}_+. \end{cases}$$

To prove Theorem 4.4.3, we need the following lemma.

LEMMA 4.4.1 *Let* $A : D(A) \subseteq X \rightsquigarrow X$ *be an operator satisfying* (H_A). *Let* $\{f_n : \mathbb{R}_+ \times \overline{D(A)} \to X;\ n \in \mathbb{N}\}$ *be a family of functions satisfying* $(\widetilde{h}_1^{[\tau=0]})$ *and* $(\widetilde{h}_8^{[\tau=0]})$ *in Theorem 4.4.3. If* (H_c) *is satisfied, then the family* $\{u_n;\ n \in \mathbb{N}\}$, *where* u_n *is the unique* T_n-*periodic solution of the problem* (4.4.8), *satisfies*

$$\|u_n(t+h) - u_n(t)\| \leq \frac{\ell}{\omega - \ell} h$$

for each $n \in \mathbb{N}$, *each* $t \in \mathbb{R}_+$ *and* $h > 0$.

Proof. We will use a fixed-point argument as follows. Let

$$L = \frac{\ell}{\omega - \ell},$$

Nonlocal Initial Conditions: The Quasi-Autonomous Case

$$\mathfrak{C}_L = \{u \in C_b(\mathbb{R}_+; \overline{D(A)}); \ \|u(t+h) - u(t)\| \le Lh, \text{ for } t \in \mathbb{R}_+ \text{ and } h > 0\}$$

and

$$\mathcal{K}_n = \{u \in \mathfrak{C}_L; \ u(t + T_n) = u(t) \text{ for each } t \in \mathbb{R}_+\}.$$

Let $\widetilde{u} \in \mathcal{K}_n$ and let us consider the problem

$$\begin{cases} u_n'(t) \in Au_n(t) + f_n(t, \widetilde{u}(t)), & t \in \mathbb{R}_+, \\ u_n(t) = u_n(t + T_n), & t \in \mathbb{R}_+. \end{cases} \tag{4.4.9}$$

The problem above has a unique C^0-solution $u_n \in \mathcal{K}_n$. Indeed, let $n \in \mathbb{N}$ and let us define the Poincaré operator $P_n : \overline{D(A)} \to \overline{D(A)}$ by $P_n(\xi) = u_n(T_n)$, where $u_n : [0, T_n] \to \overline{D(A)}$ is the unique C^0-solution of the Cauchy problem

$$\begin{cases} u_n'(t) \in Au_n(t) + f_n(t, \widetilde{u}(t)), & t \in \mathbb{R}_+, \\ u_n(0) = \xi. \end{cases}$$

By virtue of (1.8.2) in Theorem 1.8.1, we have

$$\|P_n(\xi) - P_n(\eta)\| \le e^{-\omega T_n} \|\xi - \eta\|$$

for each $\xi, \eta \in \overline{D(A)}$ and thus P_n is a strict contraction. From the Banach Fixed-Point Theorem, it follows that P_n has a unique fixed point ξ_n which has the property that the corresponding C^0-solution of the Cauchy problem above with ξ replaced by ξ_n, satisfies $u_n(0) = u_n(T_n)$. Since $f_n(\cdot, \widetilde{u}(\cdot))$ is T_n-periodic, we easily conclude that u_n is a T_n-periodic C^0-solution of (4.4.9).

We will prove next that $u_n \in \mathcal{K}_n$. Thanks to the T_n-periodicity of u_n, it suffices to show that u_n is Lipschitz on $\mathbb{R}_+$ with Lipschitz constant $L = \dfrac{\ell}{\omega - \ell}$. So, let $t \in \mathbb{R}_+$, $h > 0$ and let us observe that, again from (1.8.2) in Theorem 1.8.1, recalling that $\widetilde{u} \in \mathcal{K}_n$, that u_n is T_n-periodic and f_n satisfies $(\widetilde{h}_1^{[\tau=0]})$, we get

$$\|u_n(t+h) - u_n(t)\| = \|u_n(T_n + t + h) - u_n(T_n + t)\|$$

$$\le e^{-\omega T_n} \|u_n(t+h) - u_n(t)\|$$

$$+ e^{-\omega T_n} \int_0^{T_n} e^{\omega s} \|f_n(T_n + s + h, \widetilde{u}(T_n + s + h)) - f_n(T_n + s, \widetilde{u}(T_n + s))\| \, ds$$

$$\le e^{-\omega T_n} \|u_n(t+h) - u_n(t)\|$$

$$+ e^{-\omega T_n} \int_0^{T_n} e^{\omega s} \ell \left(h + \|\widetilde{u}(T_n + s + h) - \widetilde{u}(T_n + s)\| \right) ds$$

$$\le e^{-\omega T_n} \|u_n(t+h) - u_n(t)\| + \frac{1 - e^{-\omega T_n}}{\omega} \ell \left(1 + \frac{\ell}{\omega - \ell} \right) h.$$

148   *Delay Differential Evolutions Subjected to Nonlocal Initial Conditions*

Since $1 - e^{-\omega T_n} > 0$ for each $n \in \mathbb{N}$, from the preceding inequality, we deduce that

$$\|u_n(t+h) - u_n(t)\| \le \frac{\ell}{\omega}\left(1 + \frac{\ell}{\omega - \ell}\right)h = \frac{\ell}{\omega - \ell}h.$$

Thus $u_n \in \mathcal{K}_n$. This shows that the operator $Q_n : \mathcal{K}_n \to C_b(\mathbb{R}_+; \overline{D(A)})$, defined by $Q_n(\widetilde{u}) = u_n$ – the unique C^0-solution of (4.4.9) – maps $\mathcal{K}_n$ into itself. To complete the proof, it suffices to show that Q_n is a strict contraction. So, its unique fixed point u_n, which is the unique C^0-solution of the problem (4.4.8), belongs to $\mathcal{K}_n$ and is therefore Lipschitz of constant $L = \dfrac{\ell}{\omega - \ell}$. To this end, let us observe that, for each $\widetilde{u}_1, \widetilde{u}_2 \in \mathcal{K}_n$ and each $t \in \mathbb{R}_+$, we have

$$\|Q_n(\widetilde{u}_1)(t + T_n) - Q_n(\widetilde{u}_2)(t + T_n)\| \le e^{-\omega T_n}\|Q_n(\widetilde{u}_1)(t) - Q_n(\widetilde{u}_2)(t)\|$$

$$+ \frac{\ell}{\omega}\left(1 - e^{-\omega T_n}\right)\|\widetilde{u}_1 - \widetilde{u}_2\|_{C_b(\mathbb{R}_+;X)}.$$

Since

$$\|Q_n(\widetilde{u}_1)(t + T_n) - Q_n(\widetilde{u}_2)(t + T_n)\| = \|Q_n(\widetilde{u}_1)(t) - Q_n(\widetilde{u}_2)(t)\|,$$

we get

$$\|Q_n(\widetilde{u}_1)(t) - Q_n(\widetilde{u}_2)(t)\| \le \frac{\ell}{\omega}\|\widetilde{u}_1 - \widetilde{u}_2\|_{C_b(\mathbb{R}_+;X)},$$

for each $t \in \mathbb{R}_+$. So, thanks to the periodicity condition, we conclude that

$$\|Q_n(\widetilde{u}_1) - Q_n(\widetilde{u}_2)\|_{C_b(\mathbb{R}_+;X)} \le \frac{\ell}{\omega}\|\widetilde{u}_1 - \widetilde{u}_2\|_{C_b(\mathbb{R}_+;X)},$$

for each $\widetilde{u}_1, \widetilde{u}_2 \in \mathcal{K}_n$. From (H_c), it follows that Q_n is a strict contraction and this completes the proof. $\qquad\square$

We can now prove Theorem 4.4.3.

Proof. We show that, from $(\widetilde{h}_7^{[\tau=0]})$, it follows that the family of functions $\{g_n : C_b(\mathbb{R}_+; \overline{D(A)}) \to \overline{D(A)}; \ n \in \mathbb{N}\}$, defined by $g_n(u) = u(T_n)$ for each $n \in \mathbb{N}$ and $u \in C_b(\mathbb{R}_+; \overline{D(A)})$ and the limit function $g : C_b(\mathbb{R}_+; \overline{D(A)}) \to \overline{D(A)}$, defined by $g(u) = u(T)$ for each $u \in C(\mathbb{R}_+; \overline{D(A)})$ satisfy $(h_4^{[\tau=0]})$, $(h_5^{[\tau=0]})$ in Theorem 4.4.2 and the "compact" variant of $(h_6^{[\tau=0]})$, i.e.,

$(\widetilde{h}_6^{[\tau=0]})$ $\lim\limits_{n} g_n(u) = g(u)$ uniformly with respect to u in compact subsets in $\widetilde{C}_b([\,a, +\infty); \overline{D(A)})$.

Since $(T_n)_n$ is a sequence of strictly positive numbers that converges to $T > 0$, we may assume with no loss of generality that there exists $a > 0$ such that

$$a \le T_n$$

Nonlocal Initial Conditions: The Quasi-Autonomous Case

for each $n \in \mathbb{N}$. The function g_n, $n \in \mathbb{N}$ satisfies $(h_4^{[\tau=0]})$ in Theorem 4.4.2, while $(h_5^{[\tau=0]})$ is satisfied simply because we have

$$\|g_n(u) - g_n(\widetilde{u})\| = \|u(T_n) - \widetilde{u}(T_n)\| \le \|u - \widetilde{u}\|_{C_b([a,+\infty);X)}$$

for each $u, \widetilde{u} \in C_b(\mathbb{R}_+; \overline{D(A)})$. Let now $\mathcal{K}$ be a subset in $C_b([a,+\infty); \overline{D(A)})$ which is compact in $\widetilde{C}_b([a,+\infty); X)$. By Arzelà–Ascoli Theorem 1.4.10, it follows that $\mathcal{K}$ is equicontinuous at each $t \in [a,+\infty)$. Since $\lim_n T_n = T > 0$, this implies that

$$\lim_n g_n(u) = \lim_n u(T_n) = u(T)$$

uniformly for $u \in \mathcal{K}$ and so $(\widetilde{h}_6^{[\tau=0]})$ also holds true.

Now, let us redefine β_n appearing in the proof of Theorem 4.4.1, as

$$\beta_n = \sup\{\|g_n(v) - g(v)\|; \ v \in \mathcal{K}\},$$

where

$$\mathcal{K} = \overline{\{u_n; \ n \in \mathbb{N}\}}.$$

Since $\mathcal{K}$ is compact in $\widetilde{C}_b([a,+\infty); X)$, we have $\lim_n \beta_n = 0$. So, to complete the proof, from now on, we can follow the very same arguments as those used in the proof of Theorem 4.4.1 by making use of $(\widetilde{h}_6^{[\tau=0]})$ instead of $(h_6^{[\tau=0]})$ which, in turn, reduces to (h_6) in the case, $\tau = 0$, here considered. $\qquad \square$

4.5 The case f continuous and g Lipschitz

In order to enlarge the class of problems that could be analyzed within the frame of the general theory of nonlocal initial-value problems, here we reconsider the case studied in Section 4.1 under different assumptions: some of them more general but compensated with some more restrictive. More precisely, we next allow f to be only continuous but we impose the extra assumption that the semigroup generated by A is compact. As we will see in the examples we will give later on, this general frame is suitable to handle various nonlinear parabolic problems on bounded domains.

4.5.1 Statement of the main result

(H_A) The operator $A : D(A) \subseteq X \rightsquigarrow X$ satisfies

(A_1) $0 \in D(A)$, $0 \in A0$ and A is ω-m-dissipative for some $\omega > 0$

(A_2) the semigroup generated by A is compact.

150 *Delay Differential Evolutions Subjected to Nonlocal Initial Conditions*

(H_f) The function $f: \mathbb{R}_+ \times X \to X$ is continuous and satisfies

(f_3) there exist $\ell > 0$ and $m > 0$ such that

$$\|f(t,v)\| \le \ell\|v\|_X + m$$

for each $t \in \mathbb{R}_+$ and $v \in X$

(f_4) the family of functions $\{f(s,\cdot);\ s \in \mathbb{R}_+\}$ is equi-uniformly continuous on X, i.e., for each $\varepsilon > 0$, there exists $\delta(\varepsilon) > 0$ such that, for each $s \in \mathbb{R}_+$ and $v, w \in X$ with $\|v - w\|_X \le \delta(\varepsilon)$, we have

$$\|f(s,v) - f(s,w)\| \le \varepsilon.$$

(H_c) The constants ℓ and ω satisfy

$$\ell < \omega.$$

(H_g) The function $g: C_b(\mathbb{R}_+; \overline{D(A)}) \to \mathcal{D}$ satisfies the following:

(g_1) there exists $a > 0$ such that for each $u, v \in C_b(\mathbb{R}_+; \overline{D(A)})$, we have

$$\|g(u) - g(v)\|_X \le \|u - v\|_{C_b([a,+\infty);X)}$$

(g_4) g is continuous from $C_b([a,+\infty); \overline{D(A)})$ endowed with the $\widetilde{C}_b([a,+\infty); X)$ topology to X, where a is given by (g_1).

REMARK 4.5.1 If g is defined as in Remark 3.2.4, then it satisfies the conditions (g_1) and (g_4) in (H_g).

THEOREM 4.5.1 *If (H_A), (H_f), (H_c) and (H_g) are satisfied, then the problem (4.1.1) has at least one C^0-solution, $u \in C_b([-\tau, +\infty); X)$. Moreover, each C^0-solution of (4.1.1) satisfies*

$$\|u\|_{C_b([-\tau,+\infty);X)} \le \frac{m}{\omega - \ell} + k \cdot m_0, \qquad (4.5.1)$$

where $m_0 = \|g(0)\|_X$ and k is given by (4.1.3).

4.5.2 Excursion to the nondelayed case

Taking $\tau = 0$ in Theorem 4.5.1, we get some existence results for nondelayed evolution equations of the form

$$\begin{cases} u'(t) \in Au(t) + f(t, u(t)), & t \in \mathbb{R}_+, \\ u(0) = g(u), \end{cases} \qquad (4.5.2)$$

where A is as above and f and g satisfy $(H_f^{[\tau=0]})$ and $(H_g^{[\tau=0]})$ below.

Nonlocal Initial Conditions: The Quasi-Autonomous Case 151

$(H_f^{[\tau=0]})$ The function $f : \mathbb{R}_+ \times X \to X$ is continuous and satisfies

$(f_3^{[\tau=0]})$ there exists $\ell > 0$ and $m > 0$ such that

$$\|f(t,v)\| \le \ell\|v\| + m$$

for each $t \in \mathbb{R}_+$ and $v \in X$

$(f_4^{[\tau=0]})$ the family of functions $\{f(s,\cdot);\ s \in \mathbb{R}_+\}$ is equi-uniformly continuous on X, i.e., for each $\varepsilon > 0$, there exists $\delta(\varepsilon) > 0$ such that, for each $s \in \mathbb{R}_+$ and $v, w \in X$ with $\|v - w\| \le \delta(\varepsilon)$, we have

$$\|f(s,v) - f(s,w)\| \le \varepsilon.$$

$(H_g^{[\tau=0]})$ The function $g : C_b(\mathbb{R}_+; \overline{D(A)}) \to \overline{D(A)}$ satisfies

$(g_1^{[\tau=0]})$ there exists $a > 0$ such that for each $u, v \in C_b(\mathbb{R}_+; \overline{D(A)})$, we have

$$\|g(u) - g(v)\| \le \|u - v\|_{C_b([a,+\infty);X)}$$

$(g_4^{[\tau=0]})$ g is continuous from $C_b([a,+\infty); \overline{D(A)})$ endowed with the $\widetilde{C}_b([a,+\infty); X)$ topology to X, where a is given by $(g_1^{[\tau=0]})$.

More precisely, we have:

THEOREM 4.5.2 *If* (H_A), $(H_f^{[\tau=0]})$, (H_c) *and* $(H_g^{[\tau=0]})$ *are satisfied, then the problem (4.5.2) has at least one C^0-solution, $u \in C_b(\mathbb{R}_+; \overline{D(A)})$, satisfying*

$$\|u\|_{C_b(\mathbb{R}_+;X)} \le \frac{m}{\omega - \ell} + k \cdot m_0,$$

where $m_0 = \|g(0)\|$ *and* k *is given by* (4.1.3).

Let us consider now the periodic problem

$$\begin{cases} u'(t) \in Au(t) + f(t, u(t)), & t \in \mathbb{R}_+, \\ u(t) = u(t+T), & t \in \mathbb{R}_+. \end{cases} \tag{4.5.3}$$

From Theorem 4.5.2, we deduce:

THEOREM 4.5.3 *Let us assume that* (H_A), $(H_f^{[\tau=0]})$ *and* (H_c) *are satisfied and* $f : \mathbb{R}_+ \times X \to X$ *is T-periodic with respect to its first argument. Then the periodic problem* (4.5.3) *has at least one C^0-solution, $u : \mathbb{R}_+ \to \overline{D(A)}$ satisfying*

$$\|u\|_{C_b(\mathbb{R}_+;X)} \le \frac{m}{\omega - \ell}. \tag{4.5.4}$$

152 Delay Differential Evolutions Subjected to Nonlocal Initial Conditions

More generally, from Theorem 4.5.2, we deduce an existence result for the two-point boundary problems of the form

$$\begin{cases} u'(t) \in Au(t) + f(t, u(t)), & t \in \mathbb{R}_+, \\ u(t) = h(u(t+T)), & t \in \mathbb{R}_+, \end{cases} \tag{4.5.5}$$

where $h : \overline{D(A)} \to \overline{D(A)}$ is nonexpansive. More precisely, we have

THEOREM 4.5.4 *Let us assume that* (H_A), $(H_f^{[\tau=0]})$ *and* (H_c) *are satisfied,* $h : \overline{D(A)} \to \overline{D(A)}$ *is nonexpansive and* $h(0) = 0$. *Then* (4.5.5) *has at least one* C^0-*solution,* $u \in C_b(\mathbb{R}_+; \overline{D(A)})$ *satisfying* (4.5.4).

Next, let us consider the T-anti-periodic problem

$$\begin{cases} u'(t) \in Au(t) + f(t, u(t)), & t \in \mathbb{R}_+, \\ u(t) = -u(t+T), & t \in \mathbb{R}_+. \end{cases} \tag{4.5.6}$$

THEOREM 4.5.5 *Let us assume that* (H_A), $(H_f^{[\tau=0]})$ *and* (H_c) *are satisfied,* $D(A)$ *is symmetric with respect to the origin, i.e.,* $D(A) = -D(A)$, $A(-\xi) = -A\xi$ *for each* $\xi \in D(A)$, f *satisfies*

$$-f(T + t, -v) = f(t, v)$$

for each $(t, v) \in \mathbb{R}_+ \times X$. *Then* (4.5.6) *has at least one* C^0-*solution,* $u \in C_b(\mathbb{R}_+; \overline{D(A)})$, *which is* $2T$-*periodic and satisfies* (4.5.4).

4.5.3 Proof of the main result

We can now pass to the proof of Theorem 4.5.1.

Proof. From Lemma 4.2.1, we know that for each $v \in C_b([-\tau, +\infty); X)$, the problem

$$\begin{cases} u'(t) \in Au(t) + f(t, v_t), & t \in \mathbb{R}_+, \\ u(t) = g(u)(t), & t \in [-\tau, 0] \end{cases} \tag{4.5.7}$$

has a unique C^0-solution $u \in C_b([-\tau, +\infty); X)$. We will complete the proof by using the Schaefer Fixed-Point Theorem 1.4.6 and an approximation procedure. More precisely, let $Q : C_b([-\tau, +\infty); X) \to C_b([-\tau, +\infty); X)$, be defined by

$$Q(v) = u$$

for each $v \in C_b([-\tau, +\infty); X)$, where u is the unique C^0-solution of (4.5.7) corresponding to v, let $k \in \mathbb{N}$ and let $\eta_k : \mathbb{R} \to \mathbb{R}$ be given by

$$\eta_k(t) = \begin{cases} 1, & t \in [-k, k], \\ 0, & t \in (-\infty, -k-1] \cup [k+1, +\infty), \\ t+k+1, & t \in (-k-1, -k), \\ -t+(k+1), & t \in (k, k+1). \end{cases}$$

Nonlocal Initial Conditions: The Quasi-Autonomous Case

Next, we define $Q_k = \eta_k Q$ and we show that it satisfies the hypotheses of the Schaefer Fixed-Point Theorem 1.4.6. Then, we will prove that there exists a sequence $(u_k)_k$, where, for each $k \in \mathbb{N}$, u_k is a fixed point of Q_k, converging in $\widetilde{C}_b([-\tau, +\infty); X)$ to some function u which turns out to be a C^0-solution of the problem (4.1.1).

To prove that, for each $k \in \mathbb{N}$, Q_k is continuous from $C_b([-\tau, +\infty); X)$ to $C_b([-\tau, +\infty); X)$, it suffices to show that Q is continuous from $C_b([-\tau, +\infty); X)$ to $\widetilde{C}_b([-\tau, +\infty); X)$.

Let $v \in C_b([-\tau, +\infty); X)$ and let $(v_n)_n$ be a sequence in $C_b([-\tau, +\infty); X)$ with

$$\lim_n v_n(t) = v(t)$$

uniformly for $t \in [-\tau, +\infty)$. Since $(v_n)_n$ is bounded, using Lemma 4.2.1, we conclude that $\{Q(v_n); \ n \in \mathbb{N}\} = \{u_n; \ n \in \mathbb{N}\}$ is bounded too. By virtue of (f_3), $\{f(\cdot, v_n.); \ n \in \mathbb{N}\}$ is bounded and therefore uniformly integrable in $L^1(0, T; X)$ for each $T > 0$. By Theorem 1.8.6, we conclude that, for each $T > 0$ and $\delta \in (0, T)$, $\{u_n; \ n \in \mathbb{N}\}$ is relatively compact in $C([\delta, T]; X)$. In particular, for each $T > a$, where a is given by (g_1), $\{u_n; \ n \in \mathbb{N}\}$ is relatively compact in $C([a, T]; X)$ and thus in $\widetilde{C}_b([a, +\infty); X)$. Next, by (H_g), we deduce that there exists $u \in C_b([-\tau, +\infty); X)$ such that

$$\lim_n g(u_n) = g(u)$$

in X. Again by Theorem 1.8.6 and the relation above, we conclude that

$$\lim_n u_n = u$$

in $\widetilde{C}_b([-\tau, +\infty); X)$. So, Q is continuous from the Banach space $C_b([-\tau, +\infty); X)$ to the locally convex space $\widetilde{C}_b([-\tau, +\infty); X)$. Now, it is a simple exercise to prove that, for each $k \in \mathbb{N}$, Q_k is continuous from $C_b([-\tau, +\infty); X)$ to $C_b([-\tau, +\infty); X)$.

Similarly, we deduce that the operator Q is compact from $C_b([-\tau, +\infty); X)$ to $\widetilde{C}_b([-\tau, +\infty); X)$. Consequently, for each $k \in \mathbb{N}$, Q_k is compact from $C_b([-\tau, +\infty); X)$ to $C_b([-\tau, +\infty); X)$.

In order to use the Schaefer Fixed-Point Theorem 1.4.6 it remains to verify that, for each $k \in \mathbb{N}$, the set

$$\mathcal{E}(Q_k) = \{v \in C_b([-\tau, +\infty); X); \ \exists \lambda \in [0, 1] \text{ such that } \lambda Q_k(v) = v\}$$

is bounded in $C_b([-\tau, +\infty); X)$. To this end, let us observe first that, for each $u \in C_b([-\tau, +\infty); X)$ with $\lambda Q_k(u) = u$ for some $\lambda \in [0, 1]$, we have

$$\|u(t)\| = \|\lambda Q_k(u)(t)\| \le \|Q(u)(t)\|$$

for each $t \in [-\tau, +\infty)$. From (A_1) in (H_A), (f_3) in (H_f), taking $x = y = 0$ in (1.8.3), we get

$$\|Q(u)(t)\| \le e^{-\omega t}\|Q(u)(0)\| + (1 - e^{-\omega t})\frac{\ell}{\omega}\left(\|u\|_{C_b([-\tau, +\infty); X)} + \frac{m}{\ell}\right)$$

154 *Delay Differential Evolutions Subjected to Nonlocal Initial Conditions*

for each $t \in \mathbb{R}_+$. Hence

$$\|Q(u)(t)\| \le e^{-\omega t}\|Q(u)(0)\| + (1 - e^{-\omega t})\frac{\ell}{\omega}\left(\|Q(u)\|_{C_b([-\tau,+\infty);X)} + \frac{m}{\ell}\right)$$

for each $t \in \mathbb{R}_+$. Since for $t \in [-\tau, 0]$, we have

$$Q(u)(t) = g(Q(u))(t),$$

we are in the hypotheses of Lemma 3.3.1. Hence $\{Q(u); \ u \in \mathcal{E}(Q_k)\}$ is bounded and thus $\mathcal{E}(Q_k)$ is bounded too.

Now, the Schaefer Fixed-Point Theorem 1.4.6 shows that, for each $k \in \mathbb{N}$, Q_k has at least one fixed point $u_k \in C_b([-\tau, +\infty); X)$, i.e., $u_k = \eta_k Q(u_k)$. Recalling the definition of Q, we conclude that, for each $k > \tau$, we have

$$u_k(t) = Q(u_k)(t)$$

for each $t \in [-\tau, k]$. So, for each $k > \tau$, u_k is a C^0-solution of the problem

$$\begin{cases} u'_k(t) \in Au_k(t) + f(t, u_{kt}), & t \in [0, k], \\ u_k(t) = g(Q(u_k))(t), & t \in [-\tau, 0]. \end{cases}$$

Finally, reasoning as before, we conclude that, for each $k \in \mathbb{N}$, $k > \tau$, we can choose a fixed point u_k such that, on a subsequence at least, $(u_k)_k$ converges to some function u in $\widetilde{C}_b([-\tau, +\infty); X)$. Using (g_4) and passing to the limit for $k \to +\infty$ in the problem above, we deduce that u is a C^0-solution of (4.1.1) satisfying (4.5.1). This completes the proof. $\qquad\square$

4.6 An example involving the p-Laplacian

Under the general setting in Section 3.7. let us consider the nonlinear delay parabolic problem subjected to a nonlocal initial condition:

$$\begin{cases} \dfrac{\partial u}{\partial t}(t, x) = \Delta_p^\lambda u(t, x) - \omega u(t, x) + f(t, u_t)(x), & \text{in } Q_+, \\[2mm] -\dfrac{\partial u}{\partial \nu_p}(t, x) \in \beta(u(t, x)), & \text{on } \Sigma_+, \qquad (4.6.1) \\[2mm] u(t, x) = \displaystyle\int_\tau^{+\infty} \mathcal{N}(u(t + s, x))\, d\mu(s) + \psi(t)(x), & \text{in } Q_\tau. \end{cases}$$

THEOREM 4.6.1 *Let Ω be a nonempty, bounded and open subset in $\mathbb{R}^d$, $d \ge 1$, with C^2 boundary Σ, let $\omega > 0$, $p \in [2, +\infty)$ and $\lambda > 0$. Let $\beta : D(\beta) \subseteq \mathbb{R} \rightsquigarrow \mathbb{R}$ be m-accretive with $0 \in D(\beta)$ and $0 \in \beta(0)$, let $\mathcal{N} : \mathbb{R} \to \mathbb{R}$, let $\tau \ge 0$, let $f : \mathbb{R}_+ \times \mathcal{X}_2 \to L^2(\Omega)$, and let $\psi \in \mathcal{X}_2$. Let us assume that*

Nonlocal Initial Conditions: The Quasi-Autonomous Case

(h_1) f is continuous on $\mathbb{R}_+ \times \mathfrak{X}_2$

(h_2) there exist $\ell > 0$ and $m > 0$ such that

$$\|f(t,v)\|_{L^2(\Omega)} \le \ell \|v\|_{\mathfrak{X}_2} + m$$

for each $(t,v) \in \mathbb{R}_+ \times \mathfrak{X}_2$

(h_3) for each $\varepsilon > 0$ there exists $\delta(\varepsilon) > 0$ such that, for each $t \in \mathbb{R}_+$ and each $v, w \in \mathfrak{X}_2$ with $\|v - w\|_{\mathfrak{X}_2} \le \delta(\varepsilon)$, we have

$$\|f(t,v) - f(t,w)\|_{L^2(\Omega)} \le \varepsilon$$

(c_1) $\ell < \omega$

(n_1) $|\mathcal{N}(u) - \mathcal{N}(\widetilde{u})| \le |u - \widetilde{u}|$ for each $u, \widetilde{u} \in \mathbb{R}$

(n_2) $\mathcal{N}(0) = 0$.

Let μ be a σ-finite and complete measure on $[\tau, +\infty)$ that is continuous with respect to the Lebesgue measure at $t = \tau$, i.e., $\lim_{\delta \downarrow \tau} \mu([\tau, \tau + \delta]) = 0$. Then, the problem (4.6.1) has at least one C^0-solution $u \in C_b([-\tau, \infty); L^2(\Omega))$ that satisfies

$$\|u\|_{C_b([-\tau, +\infty); L^2(\Omega))} \le \frac{m}{\omega - \ell} + k \cdot m_0,$$

where m is given by (h_2), $m_0 = \|\psi\|_{\mathfrak{X}_2}$ and k is given by (4.1.3),

$$\Delta_p^\lambda u \in L^2_{\mathrm{loc}}(\mathbb{R}_+; L^2(\Omega)),$$

$$-\frac{\partial u}{\partial \nu_p}(t,x) \in \beta(u(t,x)) \text{ for each } t \in \mathbb{R}_+ \text{ and a.e. for } x \in \Sigma,$$

$$u \in W^{1,2}_{\mathrm{loc}}(\mathbb{R}_+; L^2(\Omega)) \cap AC(\mathbb{R}_+; W^{1,p}(\Omega)).$$

Proof. Just repeat the arguments in the proof of Theorem 3.7.1 by using Theorem 4.5.1 instead of Theorem 3.6.1. $\qquad\square$

REMARK 4.6.1 We notice that the main differences between this example and its counterpart in Section 3.7 is that here, f is allowed to depend on t as well and it is not Lipschitz with respect to its last argument. Of course, as f is less regular than required by Theorem 1.12.7, under the hypotheses of Theorem 4.6.1, we cannot obtain the global asymptotic stability of the C^0-solution via the abstract result just mentioned.

156 *Delay Differential Evolutions Subjected to Nonlocal Initial Conditions*

4.7 The case f Lipschitz and g continuous

4.7.1 The general assumptions and the main theorem

(H_A) The operator $A:D(A) \subseteq X \rightsquigarrow X$ satisfies

(A_1) $0 \in D(A)$, $0 \in A0$ and A is ω-m-dissipative for some $\omega > 0$

(A_2) the semigroup generated by A is compact

(A_5) A is densely defined, i.e., $\overline{D(A)} = X$.

(H_f) The function $f : \mathbb{R}_+ \times X \to X$ is continuous and satisfies

(f_1) there exists $\ell > 0$ such that
$$\|f(t,v) - f(t,\widetilde{v})\| \le \ell\|v - \widetilde{v}\|_X$$
for each $t \in \mathbb{R}_+$ and $v, \widetilde{v} \in X$

(f_2) there exists $m > 0$ such that
$$\|f(t,0)\| \le m$$
for each $t \in \mathbb{R}_+$.

(H_c) The constants ℓ and ω satisfy the nonresonance condition
$$\ell < \omega.$$

(H_g) The function $g : C_b(\mathbb{R}_+; X) \to X$ is continuous and satisfies[3]

(g_2) g has affine growth, i.e., there exist $a > 0$ and $m_0 \ge 0$ such that for each $u \in C_b(\mathbb{R}_+; X)$, we have
$$\|g(u)\|_X \le \|u\|_{C_b([a,+\infty);X)} + m_0$$

(g_3) with a given by (g_2), for each $u, v \in C_b(\mathbb{R}_+; X)$ satisfying $u(t) = v(t)$ for each $t \in [a, +\infty)$, we have $g(u) = g(v)$

(g_4) g is continuous from $\widetilde{C}_b([a, +\infty); X)$ to X, where a is given by (g_2).

REMARK 4.7.1 If g is defined as in Remark 3.2.4, then it satisfies the conditions (g_2), (g_3) and (g_4) in (H_g).

Theorem 4.7.1 below extends Theorem 3.6.1 to the quasi-autonomous case.

THEOREM 4.7.1 *If (H_A), (H_f), (H_c) and (H_g) are satisfied, then the problem (4.1.1) has at least one C^0-solution, $u \in C_b([-\tau, +\infty); X)$. Moreover, each C^0-solution of (4.1.1) satisfies*
$$\|u\|_{C_b([-\tau,+\infty);X)} \le \frac{m}{\omega - \ell} + k \cdot m_0, \tag{4.7.1}$$
where k is given by (4.1.3).

---

[3] We notice that from the hypothesis (A_5) it follows that $\mathcal{D} = X$.

Nonlocal Initial Conditions: The Quasi-Autonomous Case 157

4.7.2 An auxiliary lemma

For the proof of Theorem 4.7.1 we need the following two lemmas.

LEMMA 4.7.1 *Let us assume that* (A_1) *in* (H_A), (H_f) *and* (H_c) *are satisfied. Then, for each* $\varphi \in \mathcal{D}$, *the problem*

$$\begin{cases} u'(t) \in Au(t) + f(t, u_t), & t \in \mathbb{R}_+, \\ u(t) = \varphi(t), & t \in [-\tau, 0], \end{cases} \tag{4.7.2}$$

has a unique C^0-*solution* $u \in C_b([-\tau, +\infty); X)$.

Proof. We shall use a fixed-point argument as follows. Let $\varphi \in \mathcal{D}$ be arbitrary but fixed, let

$$\mathcal{K} = \{v \in C_b([-\tau, +\infty); X); \ v(t) = \varphi(t), \ t \in [-\tau, 0]\},$$

let $v \in \mathcal{K}$ and let us consider the problem

$$\begin{cases} u'(t) \in Au(t) + f(t, v_t), & t \in \mathbb{R}_+, \\ u(t) = \varphi(t), & t \in [-\tau, 0]. \end{cases} \tag{4.7.3}$$

It is easy to see that, by Lemma 4.2.1, the problem (4.7.3) has a unique C^0-solution $u \in C_b([-\tau, +\infty); X)$. Indeed, u is the unique C^0-solution of the problem

$$\begin{cases} u'(t) \in Au(t) + h(t), & t \in \mathbb{R}_+, \\ u(t) = \varphi(t), & t \in [-\tau, 0], \end{cases}$$

where $h(t) = f(t, v_t)$ for each $t \in \mathbb{R}_+$.

Next, let $Q : \mathcal{K} \to C_b([-\tau, +\infty); X)$ be the operator defined by $Q(v) = u$, where u is the unique C^0-solution of (4.7.3) corresponding to v. Clearly Q maps $\mathcal{K}$ into itself and the latter is closed and thus a complete metric space. The idea of the proof is to show that the operator Q is a strict contraction on $\mathcal{K}$ and so, by the Banach Fixed-Point Theorem, it has a unique fixed point u, which is obviously a C^0-solution of (4.7.2).

To this end, let $v, w \in \mathcal{K}$ and let us observe that

$$\|Q(v)(t) - Q(w)(t)\| \leq \int_0^t e^{-\omega(t-s)} \ell \|v_s - w_s\|_{\mathcal{X}} ds$$

for each $t \in \mathbb{R}_+$. It follows that

$$\|Q(v)(t) - Q(w)(t)\| \leq \frac{\ell}{\omega}(1 - e^{-\omega t})\|v - w\|_{C_b([-\tau, +\infty); X)}$$

$$\leq \frac{\ell}{\omega}\|v - w\|_{C_b([-\tau, +\infty); X)}$$

158   *Delay Differential Evolutions Subjected to Nonlocal Initial Conditions*

for each $t \in \mathbb{R}_+$. Since for each $v, w \in \mathcal{K}$ we have $Q(v)(t) = Q(w)(t) = \varphi(t)$ for $t \in [-\tau, 0]$, the last inequality implies that

$$\|Q(v) - Q(w)\|_{C_b([-\tau, +\infty); X)} \leq \frac{\ell}{\omega} \|v - w\|_{C_b([-\tau, +\infty); X)}$$

for each $v, w \in \mathcal{K}$. Thanks to (H_c), this shows that Q is a contraction, as claimed. $\qquad\square$

4.7.3   The fixed-point argument

Next, let $\{S(t) : X \to X; \ t \in \mathbb{R}_+\}$ be the semigroup of nonexpansive operators generated by A, let $\varepsilon > 0$, and let us define the function $g_\varepsilon : C_b(\mathbb{R}_+; X) \to \mathcal{X}$ by

$$g_\varepsilon(u)(t) = S(\varepsilon)[g(u)(t)]$$

for each $u \in C_b(\mathbb{R}_+; X)$ and each $t \in [-\tau, 0]$. Let $v \in C_b(\mathbb{R}_+; X)$ and let us consider the problem

$$\begin{cases} u'(t) \in Au(t) + f(t, u_t), & t \in \mathbb{R}_+, \\ u(t) = g_\varepsilon(v)(t), & t \in [-\tau, 0]. \end{cases} \qquad (4.7.4)$$

Since by (H_A), $0 \in D(A)$ and $0 \in A0$, we have $S(\varepsilon)0 = 0$ and thus, by (g_2) in (H_g), it follows that

$$\|g_\varepsilon(u)\|_{\mathcal{X}} \leq \|g(u)\|_{\mathcal{X}} \leq \|u\|_{C_b([a, +\infty); X)} + m_0$$

for each $u \in C_b(\mathbb{R}_+; X)$. Therefore, g_ε satisfies all the properties that g does and, in addition, for each bounded set B in $C_b(\mathbb{R}_+; X)$, $g_\varepsilon(B)(0)$ is relatively compact because, by (H_A), $S(\varepsilon)$ is a compact operator. By virtue of Lemma 4.7.1, we know that, for each $v \in C_b(\mathbb{R}_+; X)$, (4.7.4) has a unique C^0-solution $u \in C_b([-\tau, +\infty); X)$. So, we can define the operator

$$\mathcal{S} : C_b(\mathbb{R}_+; X) \to C_b(\mathbb{R}_+; X)$$

by

$$\mathcal{S}(v) = u_{|\mathbb{R}_+},$$

where u is the unique C^0-solution of (4.7.4) corresponding to v.

LEMMA 4.7.2 *Let $\varepsilon > 0$ be arbitrary fixed. If (H_A), (H_f), (H_g) and (H_c) are satisfied, then the operator $\mathcal{S}$ defined as above satisfies the hypotheses of the Schaefer Fixed-Point Theorem 1.4.6 and thus the problem*

$$\begin{cases} u'(t) \in Au(t) + f(t, u_t), & t \in \mathbb{R}_+, \\ u(t) = g_\varepsilon(u)(t), & t \in [-\tau, 0] \end{cases} \qquad (4.7.5)$$

has at least one C^0-solution.

Nonlocal Initial Conditions: The Quasi-Autonomous Case

Proof. To complete the proof, we have to verify that $\mathcal{S}$ is continuous with respect to the norm topology of $C_b(\mathbb{R}_+; X)$, is compact, and

$$\mathcal{E}(\mathcal{S}) = \{u \in C_b(\mathbb{R}_+; X); \exists \lambda \in [0,1] \text{ such that } u = \lambda \mathcal{S}(u)\}$$

is bounded.

To prove the continuity of $\mathcal{S}$, let $v, \widetilde{v} \in C_b(\mathbb{R}_+; X)$ be arbitrary and let $u, \widetilde{u} \in C_b([-\tau, +\infty); X)$ be the C^0-solutions of (4.7.4) corresponding to v and $\widetilde{v}$, respectively. Set $u(t) = \mathcal{S}(v)(t)$ and $\widetilde{u}(t) = \mathcal{S}(\widetilde{v})(t)$ for $t \in \mathbb{R}_+$ and let us observe that

$$\|u(t) - \widetilde{u}(t)\| \leq e^{-\omega t}\|S(\varepsilon)g(v)(0) - S(\varepsilon)g(\widetilde{v})(0)\|$$

$$+ \int_0^t e^{-\omega(t-s)}\|f(s, u_s) - f(s, \widetilde{u}_s)\|\, ds$$

$$\leq e^{-\omega t}\|g(v)(0) - g(\widetilde{v})(0)\| + \ell \int_0^t e^{-\omega(t-s)}\|u_s - \widetilde{u}_s\|_X\, ds$$

for each $t \in \mathbb{R}_+$. So, we have

$$\|u(t) - \widetilde{u}(t)\| \leq e^{-\omega t}\|g(v)(0) - g(\widetilde{v})(0)\| + (1 - e^{-\omega t})\frac{\ell}{\omega}\|u - \widetilde{u}\|_{C_b([-\tau,+\infty);X)}$$

for each $t \in \mathbb{R}_+$. Hence

$$\|u - \widetilde{u}\|_{C_b(\mathbb{R}_+;X)} \leq \|g(v) - g(\widetilde{v})\|_X + \frac{\ell}{\omega}\|u - \widetilde{u}\|_{C_b([-\tau,+\infty);X)}.$$

On the other hand, by the nonlocal initial condition, we have

$$\|u - \widetilde{u}\|_X = \|g_\varepsilon(v) - g_\varepsilon(\widetilde{v})\|_X \leq \|g(v) - g(\widetilde{v})\|_X \qquad (4.7.6)$$

for each $v, \widetilde{v} \in C_b(\mathbb{R}_+; X)$. From the last two inequalities, we deduce

$$\|u - \widetilde{u}\|_{C_b(\mathbb{R}_+;X)} \leq \|g(v) - g(\widetilde{v})\|_X + \frac{\ell}{\omega}\max\{\|g(v) - g(\widetilde{v})\|_X, \|u - \widetilde{u}\|_{C_b(\mathbb{R}_+;X)}\}$$

for each $v, \widetilde{v} \in C_b(\mathbb{R}_+; X)$. Next, analyzing the two cases when either

$$\max\{\|g(v) - g(\widetilde{v})\|_X, \|u - \widetilde{u}\|_{C_b(\mathbb{R}_+;X)}\} = \|g(v) - g(\widetilde{v})\|_X,$$

or

$$\max\{\|g(v) - g(\widetilde{v})\|_X, \|u - \widetilde{u}\|_{C_b(\mathbb{R}_+;X)}\} = \|u - \widetilde{u}\|_{C_b(\mathbb{R}_+;X)},$$

we get

$$\|u - \widetilde{u}\|_{C_b(\mathbb{R}_+;X)} \leq \max\left\{\frac{\omega}{\omega - \ell}, \frac{\omega + \ell}{\omega}\right\}\|g(v) - g(\widetilde{v})\|_X$$

for each $v, \widetilde{v} \in C_b(\mathbb{R}_+; X)$. From this inequality, (4.7.6) and (H_g), it follows that the solution operator $\mathcal{S}$ is continuous on $C_b(\mathbb{R}_+; X)$.

160 *Delay Differential Evolutions Subjected to Nonlocal Initial Conditions*

The next step is to show that, for each bounded set $\mathcal{K}$ in $C_b(\mathbb{R}_+; X)$, $\mathcal{S}(\mathcal{K})$ is relatively compact in $C_b(\mathbb{R}_+; X)$. To this end, let $\mathcal{K}$ be a bounded set in $C_b(\mathbb{R}_+; X)$, let $(v_k)_k$ be an arbitrary sequence in $\mathcal{K}$, and let $(u_k)_k$ be the sequence of the C^0-solutions of (4.7.4) corresponding to $(v_k)_k$. We show first that $(u_k)_k$ is bounded in $C_b([-\tau, +\infty); X)$. Indeed

$$\|u_k(t)\| \le e^{-\omega t}\|u_k(0)\| + (1 - e^{-\omega t})\frac{\ell}{\omega}\left(\|u_k\|_{C_b([-\tau, +\infty); X)} + \frac{m}{\ell}\right)$$

for each $t \in (0, +\infty)$. By (g_2) in (H_g), it follows that for each $k \in \mathbb{N}$ and each $t \in [-\tau, 0]$, we have

$$\|u_k(t)\| = \|g_\varepsilon(v_k)(t)\| \le \|g(v_k)\|_X \le \|v_k\|_{C_b(\mathbb{R}_+; X)} + m_0.$$

Since $(v_k)_k$ is bounded in $C_b(\mathbb{R}_+; X)$, there exists $m_1 > 0$ such that

$$\|u_k\|_X \le m_1$$

for each $k \in \mathbb{N}$. On the other hand, we have

$$\|u_k\|_{C_b([-\tau, +\infty); X)} = \max\{\|u_k\|_X, \|u_k\|_{C_b(\mathbb{R}_+; X)}\}$$

$$\le m_1 + \|u_k\|_{C_b(\mathbb{R}_+; X)} \tag{4.7.7}$$

for each $k \in \mathbb{N}$. Accordingly,

$$\|u_k(t)\| \le e^{-\omega t}m_1 + (1 - e^{-\omega t})\frac{\ell}{\omega}\left(\|u_k\|_{C_b(\mathbb{R}_+; X)} + m_1 + \frac{m}{\ell}\right)$$

$$\le m_1 + \frac{\ell}{\omega}\left(\|u_k\|_{C_b(\mathbb{R}_+; X)} + m_1 + \frac{m}{\ell}\right)$$

for each $k \in \mathbb{N}$ and $t \in (0, +\infty)$. Hence

$$\left(1 - \frac{\ell}{\omega}\right)\|u_k\|_{C_b(\mathbb{R}_+; X)} \le m_1\left(1 + \frac{\ell}{\omega}\right) + \frac{m}{\omega}$$

for each $k \in \mathbb{N}$. Finally, we deduce

$$\|u_k\|_{C_b([-\tau, +\infty); X)} \le m_1 + \frac{\omega + \ell}{\omega - \ell} \cdot m_1 + \frac{m}{\omega - \ell}$$

for each $k \in \mathbb{N}$. From the last inequality and (4.7.7), we conclude that $(u_k)_k$ is bounded and consequently $\{s \mapsto f(\cdot, u_{ks}); \ k \in \mathbb{N}\}$ is uniformly bounded on $\mathbb{R}_+$ and thus uniformly integrable in $L^1(0, T; X)$ for each $T > 0$.

To complete the proof of the compactness of $\mathcal{S}(\mathcal{K})$ in $C_b(\mathbb{R}_+; X)$ it suffices to show that the sequence $(\mathcal{S}(v_k))_k$ has at least one convergent subsequence in the norm topology of $C_b(\mathbb{R}_+; X)$.

First, by Theorem 1.8.6, we conclude that $\{\mathcal{S}(v_k); \ k \in \mathbb{N}\}$ is relatively compact in $C([\delta, T]; X)$ for each $T > 0$ and each $\delta \in (0, T)$. Therefore, $\mathcal{S}(\mathcal{K})$

Nonlocal Initial Conditions: The Quasi-Autonomous Case 161

is relatively compact in $C([\,\delta,\,T\,];X)$. Since $\{S(\varepsilon)g(v_k)(0);\ k\in\mathbb{N}\}$ is relatively compact, again by Theorem 1.8.6, we deduce that $\{S(v_k);\ k\in\mathbb{N}\}$ is relatively compact even in $C([\,0,T\,];X)$ for each $T>0$. In view of the compactness arguments above, we conclude that $\{S(v_k);\ k\in\mathbb{N}\}$ is relatively compact in $\widetilde{C}_b(\mathbb{R}_+;X)$. At this point, let us observe that, in view of (g_4) in (H_g), it follows that $\{g_\varepsilon(v_k);\ k\in\mathbb{N}\}$ is relatively compact in $\mathcal{X}$. Accordingly, on a subsequence at least, $(g_\varepsilon(v_k))_k$ is convergent in $\mathcal{X}$. Moreover, we may assume with no loss of generality that the very same subsequence – denoted for simplicity again by $(u_{k|\mathbb{R}_+})_k = (S(v_k))_k$ – is convergent in $\widetilde{C}_b(\mathbb{R}_+;X)$ to some function u. Clearly

$$\|u_k(t) - u_p(t)\| \le e^{-\omega t}\|g_\varepsilon(v_k)(0) - g_\varepsilon(v_p)(0)\|$$

$$+ (1 - e^{-\omega t})\frac{\ell}{\omega}\|u_k - u_p\|_{C_b(([-\tau,+\infty);X)}$$

for each $k, p \in \mathbb{N}$ and each $t \in \mathbb{R}_+$. Since

$$\|u_k - u_p\|_{C_b(([-\tau,+\infty);X)} \le \max\{\|u_k - u_p\|_{C_b(\mathbb{R}_+;X)}, \|u_k - u_p\|_{\mathcal{X}}\}$$

$$\le \|u_k - u_p\|_{C_b(\mathbb{R}_+;X)} + \|g_\varepsilon(v_k) - g_\varepsilon(v_p)\|_{\mathcal{X}}$$

for each $k, p \in \mathbb{N}$, we deduce

$$\|u_k - u_p\|_{C_b(([-\tau,+\infty);X)} \le \frac{\omega + \ell}{\omega - \ell}\|g_\varepsilon(v_k) - g_\varepsilon(v_p)\|_{\mathcal{X}}$$

for each $k, p \in \mathbb{N}$. Since $(g_\varepsilon(v_k))_k$ is fundamental in $\mathcal{X}$ being convergent, from the last inequality, it follows that $(u_k)_k$ is fundamental too. Hence $S(\mathcal{K})$ is relatively compact in $C_b(\mathbb{R}_+;X)$, as claimed.

To complete the proof, we have merely to prove that $\mathcal{E}(S)$ is bounded. Let $u \in \mathcal{E}(S)$, i.e.,

$$u = \lambda S(u)$$

for some $\lambda \in [\,0,1\,]$. Hence

$$\|u(t)\| \le \|S(u)(t)\|$$

for each $t \in \mathbb{R}_+$. But

$$\|S(u)(t)\| \le e^{-\omega t}\|S(u)(0)\| + \int_0^t e^{-\omega(t-s)}\|f(s, w_s)\|\, ds$$

for each $t \in \mathbb{R}_+$, where $w \in C_b([-\tau,+\infty);X)$ is defined by

$$w(t) = \begin{cases} g_\varepsilon(u)(t), & t \in [\,-\tau, 0\,] \\ S(u)(t), & t \in \mathbb{R}_+. \end{cases}$$

By (H_f) and (g_2) in (H_g), we get

$$\|S(u)(t)\| \le e^{-\omega t}\|S(u)(0)\| + (1 - e^{-\omega t})\frac{\ell}{\omega}\left(\|w\|_{C_b([-\tau,+\infty);X)} + \frac{m}{\ell}\right)$$

162   *Delay Differential Evolutions Subjected to Nonlocal Initial Conditions*

for each $t \in \mathbb{R}_+$. Now, from Lemma 3.3.1, we deduce that $\mathcal{E}(\mathcal{S})$ is bounded. Thus the Schaefer Fixed-Point Theorem 1.4.6 applies, implying that $\mathcal{S}$ has at least one fixed point that is a C^0-solution of the problem (4.7.5). This completes the proof. $\qquad\square$

4.7.4 Proof of the main theorem

We can now proceed with the proof of Theorem 4.7.1.

Proof. For each $\varepsilon \in (0,1)$ let us fix a C^0-solution u_ε of the problem (4.7.5). We first prove that the set $\{u_\varepsilon;\ \varepsilon \in (0,1)\}$ is relatively compact in $\widetilde{C}_b([-\tau,+\infty);X)$. As a consequence, there exists a sequence $\varepsilon_n \downarrow 0$ such that the corresponding sequence $(u_{\varepsilon_n})_n$ converges in $\widetilde{C}_b([-\tau,+\infty);X)$ to a function $u \in C_b([-\tau,+\infty);X)$ which turns out to be a C^0-solution of (4.1.1).

From Lemma 3.3.1, we easily deduce that $\{u_\varepsilon;\ \varepsilon \in (0,1)\}$ is bounded in $C_b([-\tau,+\infty);X)$. From Theorem 1.8.6, we conclude that $\{u_\varepsilon;\ \varepsilon \in (0,1)\}$ is relatively compact in $C([\delta,T];X)$ for each $T > 0$ and each $\delta \in (0,T)$. Consequently, it is relatively compact in $\widetilde{C}_b([a,+\infty);X)$. In view of (g_3) in (H_g), Remark 3.2.3 applies and so, by (g_4) in (H_g), it follows that the set $\{g_\varepsilon(u_\varepsilon);\ \varepsilon \in (0,1)\}$ is relatively compact in X. Accordingly, there exists a sequence $\varepsilon_n \downarrow 0$ such that $(u_{\varepsilon_n})_n$ – denoted for simplicity by $(u_n)_n$ – converges in $\widetilde{C}_b([a,+\infty);X)$ to some function $u \in C_b([a,+\infty);X)$ and the restriction of $(u_n)_n$ to $[-\tau,0]$, i.e., $(g_{\varepsilon_n}(u_n))_n$, converges in X to some element $v \in X$, i.e.,

$$\lim_n g_{\varepsilon_n}(u_n) = v.$$

We have

$$\|u_n(t) - u_p(t)\| \le e^{-\omega t}\|u_n(0) - u_p(0)\| + (1 - e^{-\omega t})\frac{\ell}{\omega}\|u_n - u_p\|_{C_b([-\tau,+\infty);X)}$$

for each $n, p \in \mathbb{N}$ and each $t \in \mathbb{R}_+$. Since

$$\|u_n - u_p\|_{C_b([-\tau,+\infty);X)} = \max\{\|u_n - u_p\|_X, \|u_n - u_p\|_{C_b(\mathbb{R}_+;X)}\} \qquad (4.7.8)$$

$$\le \|u_n - u_p\|_X + \|u_n - u_p\|_{C_b(\mathbb{R}_+;X)}$$

$$= \|g_{\varepsilon_n}(u_n) - g_{\varepsilon_p}(u_p)\|_X + \|u_n - u_p\|_{C_b(\mathbb{R}_+;X)},$$

we deduce

$$\|u_n(t) - u_p(t)\| \le e^{-\omega t}\|g_{\varepsilon_n}(u_n) - g_{\varepsilon_p}(u_p)\|_X$$

$$+(1 - e^{-\omega t})\frac{\ell}{\omega}\left[\|g_{\varepsilon_n}(u_n) - g_{\varepsilon_p}(u_p)\|_X + \|u_n - u_p\|_{C_b(\mathbb{R}_+;X)}\right]$$

for each $n, p \in \mathbb{N}$ and $t \in \mathbb{R}_+$. Hence

$$\|u_n - u_p\|_{C_b(\mathbb{R}_+;X)} \le \frac{\omega + \ell}{\omega - \ell}\|g_{\varepsilon_n}(u_n) - g_{\varepsilon_p}(u_p)\|_X$$

Nonlocal Initial Conditions: The Quasi-Autonomous Case 163

for each $n, p \in \mathbb{N}$. Observing that $(g_{\varepsilon_n}(u_n))_n$ is fundamental in $\mathfrak{X}$ being convergent, from the last inequality, it follows that $(u_n)_n$ is fundamental in $C_b(\mathbb{R}_+; X)$. By virtue of (4.7.8), it follows that $(u_n)_n$ is fundamental even in $C_b([-\tau, +\infty); X)$ and so convergent in this space.

Finally, let us observe that (4.7.1) follows from Remark 3.2.2 and Lemma 3.2.1 and this completes the proof. $\qquad\square$

4.8 The case A linear, f compact, and g nonexpansive

In this section we reconsider the problem (4.1.1) under the specific assumption that A is linear. As we will see, in this case, we can substitute the compactness condition on the semigroup generated by A imposed in the preceding section, by the compactness of f. Furthermore, as we have already seen in Section 3.5, this general frame perfectly fits the second-order PDEs of hyperbolic type.

Let $\tau \geq 0$. Clearly, since A is linear, we have $\overline{D(A)} = X$ and consequently $\mathfrak{X} = \mathcal{D}$. We recall that if $\tau = 0$, $\mathfrak{X}$, and thus $\mathcal{D}$, identifies with X.

DEFINITION 4.8.1 The function $f : \mathbb{R}_+ \times \mathfrak{X} \to X$ is called *compact* if it is continuous and maps bounded subsets in $\mathbb{R}_+ \times \mathfrak{X}$ into relatively compact subsets in X.

4.8.1 Statement of the main result

(H_A) The operator $A : D(A) \subseteq X \rightsquigarrow X$ satisfies

(A_1) $0 \in D(A)$, $0 \in A0$ and A is ω-m-dissipative for some $\omega > 0$

(A_4) A is single-valued and linear.

(H_f) The function $f : \mathbb{R}_+ \times \mathfrak{X} \to X$ is compact and satisfies

(f_3) there exist $\ell > 0$ and $m > 0$ such that

$$\|f(t, v)\| \leq \ell \|v\|_{\mathfrak{X}} + m$$

for each $t \in \mathbb{R}_+$ and $v \in \mathfrak{X}$.

(H_c) The constants ℓ and ω satisfy the nonresonance condition

$$\ell < \omega.$$

(H_g) The function $g : C_b(\mathbb{R}_+; X) \to \mathfrak{X}$ satisfies

164   *Delay Differential Evolutions Subjected to Nonlocal Initial Conditions*

(g_1) there exists $a > 0$ such that for each $u, v \in C_b(\mathbb{R}_+; X)$, we have

$$\|g(u) - g(v)\|_X \le \|u - v\|_{C_b([a, +\infty); X)}$$

(g_4) g is continuous from $\widetilde{C}_b([a, +\infty); X)$ to X, where a is given by (g_1).

THEOREM 4.8.1 *If* (H_A), (H_f), (H_g) *and* (H_c) *are satisfied, then* (4.1.1) *has at least one mild solution,* $u \in C_b([-\tau, +\infty); X)$, *i.e.,*

$$\begin{cases} u(t) = S(t)[g(u)(0)] + \displaystyle\int_0^t S(t - s)f(s, u_s)\, ds, & t \in \mathbb{R}_+, \\ u(t) = g(u)(t), & t \in [-\tau, 0]. \end{cases}$$

Moreover, each mild solution of (4.1.1) *satisfies*

$$\|u\|_{C_b([-\tau, +\infty); X)} \le \frac{m}{\omega - \ell} + k \cdot m_0,$$

where k *is given by* (4.1.3) *and* $m_0 = \|g(0)\|_X$.

4.8.2   The nondelayed case

Taking $\tau = 0$ in Theorem 4.8.1, we get some existence results for nondelayed evolution equations of the form

$$\begin{cases} u'(t) = Au(t) + f(t, u(t)), & t \in \mathbb{R}_+, \\ u(0) = g(u), \end{cases} \tag{4.8.1}$$

where A is as above and f and g satisfy (H_A) and $(H_g^{[\tau=0]})$ below.

$(H_f^{[\tau=0]})$ The function $f : \mathbb{R}_+ \times X \to X$ is compact and satisfies

$(f_3^{[\tau=0]})$ there exist $\ell > 0$ and $m > 0$ such that

$$\|f(t, v)\| \le \ell\|v\| + m$$

for each $t \in \mathbb{R}_+$ and $v \in X$.

$(H_g^{[\tau=0]})$ The function $g : C_b(\mathbb{R}_+; X) \to X$ satisfies

$(g_1^{[\tau=0]})$ there exists $a > 0$ such that for each $u, v \in C_b(\mathbb{R}_+; X)$, we have

$$\|g(u) - g(v)\| \le \|u - v\|_{C_b([a, +\infty); X)}$$

$(g_4^{[\tau=0]})$ g is continuous from $\widetilde{C}_b([a, +\infty); X)$ to X, where a is given by $(g_1^{[\tau=0]})$.

Nonlocal Initial Conditions: The Quasi-Autonomous Case 165

More precisely, we have:

THEOREM 4.8.2 *If* (H_A), $(H_f^{[\tau=0]})$, $(H_g^{[\tau=0]})$ *and* (H_c) *are satisfied, then the problem* (4.8.1) *has at least one mild solution,* $u \in C_b(\mathbb{R}_+; X)$, *i.e.,*

$$u(t) = S(t)g(u) + \int_0^t S(t-s)f(s, u(s))\, ds, \quad t \in \mathbb{R}_+.$$

Moreover, each mild solution of (4.8.1) *satisfies*

$$\|u\|_{C_b(\mathbb{R}_+;X)} \leq \frac{m}{\omega - \ell} + k \cdot m_0,$$

where k *is given by* (4.1.3) *and* $m_0 = \|g(0)\|$.

Let us consider now the periodic problem

$$\begin{cases} u'(t) = Au(t) + f(t, u(t)), & t \in \mathbb{R}_+, \\ u(t) = u(t+T), & t \in \mathbb{R}_+. \end{cases} \tag{4.8.2}$$

From Theorem 4.8.2, we deduce:

THEOREM 4.8.3 *If* (H_A), $(H_f^{[\tau=0]})$, (H_c) *are satisfied and* $f : \mathbb{R}_+ \times X \to X$ *is* T-*periodic with respect to its first argument, then the periodic problem* (4.8.2) *has at least one mild solution,* $u : \mathbb{R}_+ \to X$, *i.e.,*

$$u(t) = S(t)u(T) + \int_0^t S(t-s)f(s, u(s))\, ds, \quad t \in \mathbb{R}_+,$$

satisfying

$$\|u\|_{C_b(\mathbb{R}_+;X)} \leq \frac{m}{\omega - \ell}. \tag{4.8.3}$$

More generally, from Theorem 4.8.2, we deduce an existence result for the two-point boundary problems of the form

$$\begin{cases} u'(t) = Au(t) + f(t, u(t)), & t \in \mathbb{R}_+, \\ u(0) = h(u(T)), \end{cases} \tag{4.8.4}$$

where $h : X \to X$ is nonexpansive. More precisely, we have

THEOREM 4.8.4 *If* (H_A), $(H_f^{[\tau=0]})$ *and* (H_c) *are satisfied,* $h : X \to X$ *is nonexpansive and* $h(0) = 0$, *then the problem* (4.8.4) *has at least one mild solution,* $u \in C_b(\mathbb{R}_+; X)$, *i.e.,*

$$u(t) = S(t)h(u(T)) + \int_0^t S(t-s)f(s, u(s))\, ds, \quad t \in \mathbb{R}_+.$$

Moreover, each mild solution of (4.8.4) *satisfies* (4.8.3).

166   *Delay Differential Evolutions Subjected to Nonlocal Initial Conditions*

Next, let us consider the T-anti-periodic problem

$$\begin{cases} u'(t) = Au(t) + f(t, u(t)), & t \in \mathbb{R}_+, \\ u(t) = -u(t + T), & t \in \mathbb{R}_+. \end{cases} \qquad (4.8.5)$$

Since A is linear, $D(A)$ is a linear subspace of X and thus it is always symmetric with respect to the origin and $A(-\xi) = -A\xi$ for each $\xi \in D(A)$. Therefore, from Theorem 4.8.2, we deduce:

THEOREM 4.8.5 *If* (H_A), $(H_f^{[\tau=0]})$ *and* (H_c) *are satisfied, f satisfies*

$$-f(T + t, -v) = f(t, v)$$

for each $(t, v) \in \mathbb{R}_+ \times X$, *then* (4.8.5) *has at least one mild solution,* $u \in C_b(\mathbb{R}_+; X)$, *i.e.,*

$$u(t) = -S(t)u(T) + \int_0^t S(t - s)f(s, u(s)) \, ds, \quad t \in \mathbb{R}_+,$$

which is $2T$-periodic and satisfies (4.8.3).

4.8.3   Proof of the main result

We can now pass to the proof of Theorem 4.8.1.

Proof. From Lemma 4.2.1, we know that for each $v \in C_b(\mathbb{R}_+; X)$, the problem

$$\begin{cases} u'(t) = Au(t) + f(t, v_t), & t \in \mathbb{R}_+, \\ u(t) = g(u)(t), & t \in [-\tau, 0] \end{cases}$$

has a unique mild solution $u \in C_b([-\tau, +\infty); X)$.

From now on, the proof follows the very same lines as those of the proof of Theorem 4.5.1 with the sole exception that here, the compactness arguments are based on both Lemma 1.7.1 and Theorem 1.7.3, avoiding the use of Theorem 1.8.6, which cannot be applied in this setting.   □

4.9   The case f Lipschitz and compact, g continuous

For the sake of completeness, we reconsider here the problem (4.1.1) under the hypothesis that g is merely continuous. As expected, the lack of nonexpansivity of g had to be counterbalanced by some stronger conditions on both A and f. More precisely, in this section we will assume that A is linear, while f is compact and Lipschitz with respect to its second argument.

Nonlocal Initial Conditions: The Quasi-Autonomous Case

4.9.1 The main assumptions and some preliminaries

(H_X) There exists a family of linear operators $\{I_\varepsilon;\ \varepsilon \in (0,1)\} \subseteq \mathcal{L}(X)$, with I_ε compact and $\|I_\varepsilon\|_{\mathcal{L}(X)} \leq 1$ for each $\varepsilon \in (0,1)$, and

$$\lim_{\varepsilon \downarrow 0} I_\varepsilon x = x$$

for each $x \in X$.

(H_A) The operator $A: D(A) \subseteq X \rightsquigarrow X$ satisfies

(A_1) $0 \in D(A)$, $0 \in A0$ and A is ω-m-dissipative for some $\omega > 0$

(A_4) A is single-valued and linear.

(H_f) The function $f: \mathbb{R}_+ \times X \to X$ is compact and satisfies

(f_1) there exists $\ell > 0$ such that

$$\|f(t,v) - f(t,\widetilde{v})\| \leq \ell \|v - \widetilde{v}\|_X$$

for each $t \in \mathbb{R}_+$ and $v, \widetilde{v} \in X$

(f_2) there exists $m > 0$ such that

$$\|f(t,0)\| \leq m$$

for each $t \in \mathbb{R}_+$.

(H_c) The constants ℓ and ω satisfy the nonresonance condition

$$\ell < \omega.$$

(Hg) The function $g: C_b(\mathbb{R}_+; X) \to X$ is continuous and satisfies

(g_2) g has affine growth, i.e., there exist $a > 0$ and $m_0 \geq 0$ such that, for each $u \in C_b(\mathbb{R}_+; X)$, we have

$$\|g(u)\|_X \leq \|u\|_{C_b([a,+\infty);X)} + m_0$$

(g_3) with a given by (g_2), for each $u, v \in C_b(\mathbb{R}_+; X)$ satisfying $u(t) = v(t)$ for each $t \in [a, +\infty)$, we have $g(u) = g(v)$

(g_4) g is continuous from $\widetilde{C}_b([a,+\infty); X)$ to X, where a is given by (g_2).

REMARK 4.9.1 If X is a separable Hilbert space with inner product $\langle \cdot, \cdot \rangle$, then it satisfies (H_X). Indeed, in this case it admits an orthogonal basis $\{e_i;\ i \in \mathbb{N}\}$, with $\langle e_i, e_j \rangle = \delta_{ij}$, for $i, j \in \mathbb{N}$, where δ_{ij} denotes the Kronecker delta. Let $\varepsilon > 0$ and let us define $I_\varepsilon : X \to X$ by

$$I_\varepsilon x = \sum_{m=0}^{\infty} e^{-\varepsilon m} \langle e_m, x \rangle x$$

168 *Delay Differential Evolutions Subjected to Nonlocal Initial Conditions*

for each $x \in X$. Clearly $I_\varepsilon \in \mathcal{L}(X)$, $\|I_\varepsilon\| \le 1$, I_ε is compact because it is the limit, in the uniform operator topology, of a sequence of linear bounded operators with finite dimensional range, i.e.,

$$I_\varepsilon = \lim_n I_{\varepsilon,n}, \qquad I_{\varepsilon,m}x = \sum_{p=0}^{m} e^{-\varepsilon p}\langle e_p, x\rangle x$$

for each $m \in \mathbb{N}$ and $x \in X$. Also, if there exists a linear m-dissipative operator A on X which generates a compact C_0-semigroup $\{S(t) : X \to X; \ t \in \mathbb{R}_+\}$, then (H_X) is satisfied with $I_\varepsilon = S(\varepsilon)$ for each $\varepsilon \in (0,1)$. More generally, if the operator $(I - A)^{-1}$ is compact, it is easy to see that (H_X) is also satisfied with $I_\varepsilon = J_\varepsilon = (I - \varepsilon A)^{-1}$ for each $\varepsilon \in (0,1)$. So, all the function spaces $L^p(\Omega)$, where Ω is a nonempty and bounded domain in $\mathbb{R}^d$ and $p \in [1, +\infty)$, and $C(\overline{\Omega})$ enjoy the property assumed in (H_X).

REMARK 4.9.2 If g is defined as in Remark 3.2.4, then it satisfies the conditions (g_2), (g_3) and (g_4) in (Hg).

4.9.2 The main theorem

THEOREM 4.9.1 *If (H_X), (H_A), (H_f), (Hg) and (H_c) hold true, then the problem (4.1.1) has at least one mild solution, $u \in C_b([-\tau, +\infty); X)$. Moreover, for each mild solution of (4.1.1), we have*

$$\|u\|_{C_b([-\tau,+\infty);X)} \le \frac{m}{\omega - \ell} + k \cdot m_0, \qquad (4.9.1)$$

where k is given by (4.1.3). If, in addition, instead of (f_1) in (H_f) the stronger condition

(f_5) *there exists $\ell > 0$ such that*

$$\|f(t,v) - f(s,\tilde{v})\| \le \ell\left(|t - s| + \|v - \tilde{v}\|_x\right)$$

for each $t, s \in \mathbb{R}_+$ and $v, \tilde{v} \in \mathfrak{X}$

is satisfied, then the mild solution of (4.1.1) is unique and globally asymptotically stable.

Proof. We shall use a combination of two fixed-point theorems along with an approximation procedure. Namely, let $\varepsilon > 0$, let $g_\varepsilon : C_b(\mathbb{R}_+; X) \to \mathfrak{X}$,

$$g_\varepsilon(u)(t) = I_\varepsilon[g(u)(t)]$$

for each $u \in C_b(\mathbb{R}_+; X)$ and each $t \in [-\tau, 0]$, where I_ε is given by (H_X) and let us consider the auxiliary ε-approximate problem

$$\begin{cases} u'(t) = Au(t) + f(t, u_t), & t \in \mathbb{R}_+, \\ u(t) = g_\varepsilon(u)(t), & t \in [-\tau, 0]. \end{cases} \qquad (4.9.2)$$

Nonlocal Initial Conditions: The Quasi-Autonomous Case 169

In order to prove that (4.9.2) has at least one mild solution, we consider first the problem

$$\begin{cases} u'(t) = Au(t) + f(t, u_t), & t \in \mathbb{R}_+, \\ u(t) = g_\varepsilon(v)(t), & t \in [-\tau, 0]. \end{cases} \tag{4.9.3}$$

By virtue of Theorem 1.12.5, it follows that, for each $v \in C_b(\mathbb{R}_+; X)$, (4.9.3) has a unique C^0-solution, $u \in C_b([-\tau, +\infty); X)$. So, we can define the operator

$$\mathcal{S}_\varepsilon : C_b(\mathbb{R}_+; X) \to C_b(\mathbb{R}_+; X)$$

by

$$\mathcal{S}_\varepsilon(v) = u_{|\mathbb{R}_+},$$

where u is the unique mild solution of (4.9.3) corresponding to v. To complete the proof, it would suffice to show that the operator $\mathcal{S}_\varepsilon$ defined as above satisfies the hypotheses of the Schaefer Fixed-Point Theorem 1.4.6 and thus (4.9.2) has at least one mild solution. So, we have to verify that $\mathcal{S}_\varepsilon$ is continuous with respect to norm topology of $C_b(\mathbb{R}_+; X)$, is compact, and

$$\mathcal{E}(\mathcal{S}_\varepsilon) = \{ u \in C_b(\mathbb{R}_+; X); \exists \lambda \in [0,1], \text{ such that } u = \lambda \mathcal{S}_\varepsilon(u) \}$$

is bounded.

To prove the continuity of the operator $\mathcal{S}_\varepsilon$ let $v, \widetilde{v} \in C_b(\mathbb{R}_+; X)$ and let $u, \widetilde{u} \in C_b([-\tau, +\infty); X)$ be the mild solutions of (4.9.3) corresponding to v and $\widetilde{v}$, respectively. Let us observe that

$$\|\mathcal{S}_\varepsilon(v)(t) - \mathcal{S}_\varepsilon(\widetilde{v})(t)\| \le e^{-\omega t} \|g_\varepsilon(v)(0) - g_\varepsilon(\widetilde{v})(0)\|$$

$$+ \int_0^t e^{-\omega(t-s)} \|f(s, u_s) - f(s, \widetilde{u}_s)\| \, ds$$

$$\le e^{-\omega t} \|g(v)(0) - g(\widetilde{v})(0)\| + \ell \int_0^t e^{-\omega(t-s)} \|u_s - \widetilde{u}_s\|_X \, ds$$

for each $t \in \mathbb{R}_+$. So, we have

$$\|\mathcal{S}_\varepsilon(v)(t) - \mathcal{S}_\varepsilon(\widetilde{v})(t)\| \le e^{-\omega t} \|g(v)(0) - g(\widetilde{v})(0)\|$$

$$+ (1 - e^{-\omega t}) \frac{\ell}{\omega} \|u - \widetilde{u}\|_{C_b([-\tau, +\infty); X)}$$

for each $t \in \mathbb{R}_+$.

But

$$\|u - \widetilde{u}\|_{C_b([-\tau, +\infty); X)} \le \|u - \widetilde{u}\|_X + \|u - \widetilde{u}\|_{C_b(\mathbb{R}_+; X)}.$$

On the other hand, by the nonlocal initial condition, we have

$$\|u - \widetilde{u}\|_X = \|g_\varepsilon(v) - g_\varepsilon(\widetilde{v})\|_X \le \|g(v) - g(\widetilde{v})\|_X$$

170 *Delay Differential Evolutions Subjected to Nonlocal Initial Conditions*

and thus

$$\|\mathcal{S}_\varepsilon(v) - \mathcal{S}_\varepsilon(\widetilde{v})\|_{C_b(\mathbb{R}_+;X)} \le \frac{\omega + \ell}{\omega - \ell} \|g(v) - g(\widetilde{v})\|_x$$

for each $v, \widetilde{v} \in C_b(\mathbb{R}_+; X)$. It follows that

$$\|\mathcal{S}_\varepsilon(v) - \mathcal{S}_\varepsilon(\widetilde{v})\|_{C_b(\mathbb{R}_+;X)} \le \|u - \widetilde{u}\|_{C_b([-\tau,+\infty);X)} \le \frac{2\omega}{\omega - \ell} \|g(v) - g(\widetilde{v})\|_x$$

for each $v, \widetilde{v} \in C_b(\mathbb{R}_+; X)$, which, thanks to (Hg), shows that the solution operator $\mathcal{S}_\varepsilon$ is continuous on $C_b(\mathbb{R}_+; X)$.

The next step is to show that, for each bounded set $\mathcal{K}$ in $C_b(\mathbb{R}_+; X)$, $\mathcal{S}_\varepsilon(\mathcal{K})$ is relatively compact in $C_b(\mathbb{R}_+; X)$. To this end, let $\mathcal{K}$ be a bounded set in $C_b(\mathbb{R}_+; X)$, let $(v_k)_k$ be an arbitrary sequence in $\mathcal{K}$, and let $(u_k)_k$ be the sequence of the mild solutions of (4.9.3) corresponding to $(v_k)_k$. For each $k \in \mathbb{N}$ we have

$$\|u_k(t)\| \le e^{-\omega t}\|u_k(0)\| + (1 - e^{-\omega t})\frac{\ell}{\omega}\left(\|u_k\|_{C_b([-\tau,+\infty);X)} + \frac{m}{\ell}\right)$$

for each $t \in (0, +\infty)$. For $t \in [-\tau, 0]$, we have

$$\|u_k(t)\| = \|g_\varepsilon(v_k)(t)\| \le \|v_k\|_{C_b(\mathbb{R}_+;X)} + m_0$$

for each $k \in \mathbb{N}$ and, since $(v_k)_k$ is bounded, there exists $m_1 > 0$ such that

$$\|u_k\|_x \le m_1$$

for each $k \in \mathbb{N}$. Accordingly,

$$\|\mathcal{S}_\varepsilon(v_k)(t)\| \le e^{-\omega t}m_1 + (1 - e^{-\omega t})\frac{\ell}{\omega}\left(\|u_k\|_{C_b(\mathbb{R}_+;X)} + m_1 + \frac{m}{\ell}\right)$$

$$\le m_1 + \frac{\ell}{\omega}\left[\|\mathcal{S}_\varepsilon(v_k)\|_{C_b(\mathbb{R}_+;X)} + m_1 + \frac{m}{\ell}\right]$$

for each $k \in \mathbb{N}$ and $t \in \mathbb{R}_+$. Hence

$$\left(1 - \frac{\ell}{\omega}\right)\|\mathcal{S}_\varepsilon(v_k)\|_{C_b(\mathbb{R}_+;X)} \le m_1\left(1 + \frac{\ell}{\omega}\right) + \frac{m}{\omega}$$

for each $k \in \mathbb{N}$. Finally, we deduce

$$\|\mathcal{S}_\varepsilon(v_k)\|_{C_b(\mathbb{R}_+;X)} \le \frac{\omega + \ell}{\omega - \ell} \cdot m_1 + \frac{m}{\omega - \ell}$$

for each $k \in \mathbb{N}$. So, $(\mathcal{S}_\varepsilon(v_k))_k$ is bounded in $C_b(\mathbb{R}_+; X)$ and thus $(u_k)_k$ is bounded in $C_b([-\tau, +\infty); X)$. Accordingly, from (H_f), we conclude that, for each $T > 0$, the set

$$\{f(t, \mathcal{S}_\varepsilon(v_k)_t); \ k \in \mathbb{N}, t \in [0, T]\}$$

Nonlocal Initial Conditions: The Quasi-Autonomous Case

is relatively compact. An appeal to Theorem 1.7.3, shows that $\{\mathcal{S}_\varepsilon(v_k);\ k \in \mathbb{N}\}$ is relatively compact in $C([\delta, T]; X)$ for each $T > 0$ and each $\delta \in (0, T)$.

By (H_X), it follows that $\{\mathcal{S}_\varepsilon(v_k)(0);\ k \in \mathbb{N}\} = \{g_\varepsilon(v_k)(0);\ k \in \mathbb{N}\}$ is relatively compact in X, simply because $(v_k)_k$ is bounded and I_ε is compact. Using once again Theorem 1.7.3, we conclude that $\{\mathcal{S}_\varepsilon(v_k);\ k \in \mathbb{N}\}$ is relatively compact in $C([0, T]; X)$ for each $T > 0$ and thus in $\widetilde{C}_b(\mathbb{R}_+; X)$.

To conclude the compactness of $\mathcal{S}_\varepsilon(\mathcal{K})$ in $C_b(\mathbb{R}_+; X)$ it suffices to show that the sequence $(\mathcal{S}_\varepsilon(v_k))_k$ has at least one convergent subsequence in the norm topology of $C_b(\mathbb{R}_+; X)$.

Since $\{\mathcal{S}_\varepsilon(v_k);\ k \in \mathbb{N}\}$ is relatively compact in $\widetilde{C}_b(\mathbb{R}_+; X)$, we may assume without loss of generality that $(\mathcal{S}_\varepsilon(v_k))_k$ is convergent in $\widetilde{C}_b(\mathbb{R}_+; X)$ to some function u. Denoting for simplicity $\mathcal{S}_\varepsilon(v_k) = u_k$ (in fact $\mathcal{S}_\varepsilon(v_k) = u_{k|\mathbb{R}_+}$), we get

$$\|u_k(t) - u(t)\| \le e^{-\omega(t-\tau)}\|u_k(\tau) - u(\tau)\| + \int_\tau^t \ell e^{-\omega(t-s)}\|u_{k\,s} - u_s\|_X\, ds$$

for each $t \in [\tau, +\infty)$. Taking

$$y(t) = e^{\omega(t-\tau)}\|u_k(t) - u(t)\|, \quad \alpha_0(t) = \|u_k(\tau) - u(\tau)\| \quad \text{and} \quad \beta = \ell$$

in Lemma 1.13.3 applied on the shifted intervals $\mathbb{R}_+$ and $[\tau, +\infty)$, after some simple calculations and recalling that $\ell < \omega$, we deduce

$$\|u_k(t) - u(t)\| \le e^{(\ell-\omega)(t-\tau)}(\|u_k(\tau) - u(\tau)\| + \|u_k - u\|_{C([0,\tau];X)})$$

for each $t \in [\tau, +\infty)$. Since $(u_k)_k$ is bounded, this shows that $\lim_k u_k = u$ in $C_b([\tau, +\infty); X)$. Furthermore, from $\lim_k u_k = u$ in $\widetilde{C}_b(\mathbb{R}_+; X)$, it follows that $(\mathcal{S}_\varepsilon(v_k))_k$ is convergent even in $C_b(\mathbb{R}_+; X)$. Thus $\mathcal{S}_\varepsilon(\mathcal{K})$ is relatively compact in $C_b(\mathbb{R}_+; X)$, as claimed.

To show that $\mathcal{E}(\mathcal{S}_\varepsilon)$ is bounded, let $u \in \mathcal{E}(\mathcal{S}_\varepsilon)$, i.e., there exists $\lambda \in [0, 1]$ such that

$$u = \lambda \mathcal{S}_\varepsilon(u).$$

Clearly

$$\|u(t)\| \le \|\mathcal{S}_\varepsilon(u)(t)\|$$

for each $t \in \mathbb{R}_+$. But

$$\|\mathcal{S}_\varepsilon(u)(t)\| \le e^{-\omega t}\|\mathcal{S}_\varepsilon(u)(0)\| + \int_0^t e^{-\omega(t-s)}\|f(s, w_s)\|\, ds$$

for each $t \in \mathbb{R}_+$, where $w \in C_b([-\tau, +\infty); X)$ is defined by

$$w(t) = \begin{cases} g_\varepsilon(u)(t), & t \in [-\tau, 0] \\ \mathcal{S}_\varepsilon(u)(t), & t \in \mathbb{R}_+ \end{cases}$$

172   *Delay Differential Evolutions Subjected to Nonlocal Initial Conditions*

and thus, by (H_f), we get

$$\|\mathcal{S}_\varepsilon(u)(t)\| \le e^{-\omega t}\|\mathcal{S}_\varepsilon(u)(0)\| + (1 - e^{-\omega t})\frac{\ell}{\omega}\left(\|w\|_{C_b([-\tau,+\infty);X)} + \frac{m}{\ell}\right)$$

for each $t \in \mathbb{R}_+$. Since $w(t) = g_\varepsilon(u)(t)$ for each $t \in [-\tau, 0]$, we are in the hypotheses of Lemma 3.3.1 which shows that $\mathcal{E}(\mathcal{S}_\varepsilon)$ is bounded. Thus the Schaefer Fixed-Point Theorem 1.4.6 applies, implying that $\mathcal{S}_\varepsilon$ has at least one fixed point, $u_{|\mathbb{R}_+}$, where u is a mild solution of the problem (4.9.2) as claimed.

We can now proceed with the final part of the proof of Theorem 4.9.1.

For each $\varepsilon \in (0, 1)$ let us fix a mild solution u_ε of the problem (4.9.2). We first prove that the set $\{u_\varepsilon;\ \varepsilon \in (0, 1)\}$ is relatively compact in $\widetilde{C}_b(\mathbb{R}_+; X)$. As a consequence, there exists a sequence $\varepsilon_n \downarrow 0$ such that the corresponding sequence $(u_{\varepsilon_n})_n$ converges in $\widetilde{C}_b(\mathbb{R}_+; X)$ to a function $u \in C_b(\mathbb{R}_+; X)$, which turns out to be a mild solution of (4.1.1).

From Lemma 3.3.1, we easily deduce that $\{u_\varepsilon;\ \varepsilon \in (0, 1)\}$ is bounded in $C_b(\mathbb{R}_+; X)$. From Theorem 1.7.3, we conclude that $\{u_\varepsilon;\ \varepsilon \in (0, 1)\}$ is relatively compact in $C([\delta, T]; X)$ for each $T > 0$ and each $\delta \in (0, T)$. So, it is relatively compact in $\widetilde{C}_b([a, +\infty); X)$. In view of (g_4) in (H_g), it follows that the set $\{g_\varepsilon(u_\varepsilon);\ \varepsilon \in (0, 1)\}$ is relatively compact in $\mathfrak{X}$. Hence, there exists a subsequence of $\varepsilon_n \downarrow 0$, denoted again by $\varepsilon_n \downarrow 0$, such that both $(u_{\varepsilon_n})_n$ – denoted for simplicity also by $(u_n)_n$ – converges in $\widetilde{C}_b([a, +\infty); X)$ to some function $u \in C_b([a, +\infty); X)$ and the restriction of $(u_n)_n$ to $[-\tau, 0]$, i.e., $(g_{\varepsilon_n}(u_n))_n$, converges in $\mathfrak{X}$ to some element $v \in \mathfrak{X}$. Again from Theorem 1.7.3, it follows that $\{u_\varepsilon;\ \varepsilon \in (0, 1)\}$ is relatively compact in $C([\delta, T]; X)$ for each $T > 0$. So

$$\begin{cases} \lim_n u_n = u & \text{in } \widetilde{C}_b(\mathbb{R}_+; X) \\ \lim_n u_n = \lim_n g_{\varepsilon_n}(u_n) = \lim_n I_{\varepsilon_n} g(u_n) = v & \text{in } \mathfrak{X}. \end{cases}$$

We have

$$\|u_n(t) - u_p(t)\| \le e^{-\omega t}\|u_n(0) - u_p(0)\| + (1 - e^{-\omega t})\frac{\ell}{\omega}\|u_n - u_p\|_{C_b([-\tau,+\infty);X)}$$

for each $n, p \in \mathbb{N}$ and each $t \in \mathbb{R}_+$. Since

$$\|u_n - u_p\|_{C_b([-\tau,+\infty);X)} \le \|u_n - u_p\|_{\mathfrak{X}} + \|u_n - u_p\|_{C_b(\mathbb{R}_+;X)}$$

$$= \|g_{\varepsilon_n}(u_n) - g_{\varepsilon_p}(u_p)\|_{\mathfrak{X}} + \|u_n - u_p\|_{C_b(\mathbb{R}_+;X)}, \tag{4.9.4}$$

we deduce

$$\|u_n(t) - u_p(t)\| \le e^{-\omega t}\|g_{\varepsilon_n}(u_n) - g_{\varepsilon_p}(u_p)\|_{\mathfrak{X}}$$

$$+(1 - e^{-\omega t})\frac{\ell}{\omega}\left(\|g_{\varepsilon_n}(u_n) - g_{\varepsilon_p}(u_p)\|_{\mathfrak{X}} + \|u_n - u_p\|_{C_b(\mathbb{R}_+;X)}\right)$$

for each $n, p \in \mathbb{N}$ and $t \in \mathbb{R}_+$. Hence

$$\|u_n - u_p\|_{C_b(\mathbb{R}_+;X)} \le \frac{\omega + \ell}{\omega - \ell}\|g_{\varepsilon_n}(u_n) - g_{\varepsilon_p}(u_p)\|_{\mathfrak{X}}$$

Nonlocal Initial Conditions: The Quasi-Autonomous Case 173

for each $n, p \in \mathbb{N}$. Observing that $(g_{\varepsilon_n}(u_n))_n$ is fundamental in $\mathfrak{X}$ being convergent, from the last inequality, if follows that $(u_n)_n$ is fundamental in $C_b(\mathbb{R}_+; X)$. By virtue of (4.9.4), it follows that $(u_n)_n$ is fundamental even in $C_b([-\tau, +\infty); X)$ and so it is convergent in this space.

Finally, the estimate (4.9.1) follows from Remark 3.2.2 and Lemma 3.3.1 and this completes the proof. $\qquad\square$

4.10 The damped wave equation revisited

Here, we reconsider the example in Section 3.5 in a different setting, partly more general, partly more particular. To be more specific, here we do not assume any Lipschitz condition of h but we impose that it depends merely on u but not on the partial time derivative of u. Furthermore, here we allow h to depend on t as well. Let Ω be a nonempty bounded and open subset in $\mathbb{R}^d$, $d \geq 1$, with C^1 boundary Σ, let $Q_+ = \mathbb{R}_+ \times \Omega$, $Q_\tau = [-\tau, 0] \times \Omega$, $\Sigma_+ = \mathbb{R}_+ \times \Sigma$, and let $\omega > 0$.

Let us denote both $\mathcal{H}_0^1 = C([-\tau, 0]; H_0^1(\Omega))$ and $\mathfrak{X}_2 = C([-\tau, 0]; L^2(\Omega))$, and let us consider the following damped wave equation with delay, subjected to nonlocal initial conditions:

$$\begin{cases} \dfrac{\partial^2 u}{\partial t^2} = \Delta u - 2\omega \dfrac{\partial u}{\partial t} - \omega^2 u + h\left(t, u_t\right), & \text{in } Q_+, \\[2mm] u(t, x) = 0, & \text{on } \Sigma_+, \\[2mm] u(t, x) = \displaystyle\int_\tau^{+\infty} \alpha(s) u(t+s, x)\, ds + \psi_1(t)(x), & \text{in } Q_\tau, \\[2mm] \dfrac{\partial u}{\partial t}(t, x) = \displaystyle\int_\tau^{+\infty} \mathcal{N}\left(s, u(t+s, x), \dfrac{\partial u}{\partial t}(t+s, x)\right) ds + \psi_2(t)(x), & \text{in } Q_\tau, \end{cases} \tag{4.10.1}$$

where $h : \mathbb{R}_+ \times \mathcal{H}_0^1 \to L^2(\Omega)$, $\alpha \in L^2(\mathbb{R}_+)$, $\mathcal{N} : \mathbb{R}_+ \times \mathbb{R} \times \mathbb{R} \to \mathbb{R}$, while $\psi_1 \in \mathcal{H}_0^1$ and $\psi_2 \in \mathfrak{X}_2$.

As we already noted, the main difference from the case considered in Section 3.5 is that here, h is supposed to be merely continuous. In order to compensate for the lack of the Lipschitz condition of h, we have to assume instead that it does not depend on the history of the partial derivative of u with respect to t. Then, the proof will essentially rest on the compactness of the embedding $H_0^1(\Omega) \subseteq L^2(\Omega)$ and on the fact that the reaction term h in (4.10.1) is shifted down in the system of first-order partial differential equations, which is equivalent to (4.10.1). This simple trick will allow us to take advantage of this compact embedding.

174 *Delay Differential Evolutions Subjected to Nonlocal Initial Conditions*

4.10.1 The first existence result

THEOREM 4.10.1 *Let Ω be a nonempty bounded and open subset in $\mathbb{R}^d$, $d \geq 1$, with C^1 boundary Σ, let $\tau \geq 0$ and let $\omega > 0$. Let further $\psi_1 \in \mathcal{H}_0^1$, $\psi_2 \in \mathcal{X}_2$ and let us assume that $h : \mathbb{R}_+ \times \mathcal{H}_0^1 \to L^2(\Omega)$, $\alpha \in L^2(\mathbb{R}_+)$ and $\mathcal{N} : \mathbb{R}_+ \times \mathbb{R} \times \mathbb{R} \to \mathbb{R}$ are continuous and satisfy*

(h_1) *there exist $\ell > 0$ and $m > 0$ such that*

$$\|h(t, w)\|_{L^2(\Omega)} \leq \ell \|w\|_{\mathcal{H}_0^1} + m$$

for each $t \in \mathbb{R}_+$ and each $w \in \mathcal{H}_0^1$

(n_1) *there exists a nonnegative continuous function $\eta \in L^2(\mathbb{R}_+)$ such that*

$$|\mathcal{N}(t, u, v)| \leq \eta(t)(|u| + |v|),$$

for each $t \in \mathbb{R}_+$ and $u, v \in \mathbb{R}$

(n_2) *we have*

$$|\mathcal{N}(t, u, v) - \mathcal{N}(t, \widetilde{u}, \widetilde{v})| \leq \eta(t)(|u - \widetilde{u}| + |v - \widetilde{v}|),$$

for each $t \in \mathbb{R}_+$ and $u, \widetilde{u}, v, \widetilde{v} \in \mathbb{R}$, where η is given by (n_1).

Let λ_1 be the first eigenvalue of $-\Delta$ and let us assume that

(n_3) $\begin{cases} \|\eta\|_{L^2(\mathbb{R}_+)} \leq 1 \\ (1 + \lambda_1^{-1}\omega)\|\alpha\|_{L^2(\mathbb{R}_+)} + \lambda_1^{-1}(1 + \omega)\|\eta\|_{L^2(\mathbb{R}_+)} \leq 1 \end{cases}$

(n_4) *there exists $b > \tau$ such that $\alpha(t) = \eta(t) = 0$ for each $t \in [0, b]$.*

Let us assume, in addition, that

(c_1) $\ell < \omega$.

Then, (4.10.1) has at least one mild solution[4] $u \in C_b([-\tau, +\infty); H_0^1(\Omega))$ with $\dfrac{\partial u}{\partial t} \in C_b([-\tau, +\infty); L^2(\Omega))$. In addition, u satisfies

$$\|u\|_{C_b([-\tau, +\infty); H_0^1(\Omega))} + \left\|\frac{\partial u}{\partial t}\right\|_{C_b([-\tau, +\infty); L^2(\Omega))} \leq \frac{m}{\omega - \ell} + k \cdot m_0,$$

where k is given by (4.1.3) and $m_0 = \|\psi_1\|_{\mathcal{H}_0^1} + \omega\|\psi_1\|_{\mathcal{X}_2} + \|\psi_2\|_{\mathcal{X}_2}$.

---

[4]For the precise meaning of mild solution in this specific case, see Remark 2.2.1.

Nonlocal Initial Conditions: The Quasi-Autonomous Case 175

Proof. As in the proof of Theorem 3.5.1, (4.10.1) can be equivalently rewritten in the form (4.1.1) in the Hilbert space $X = \begin{pmatrix} H_0^1(\Omega) \\ \times \\ L^2(\Omega) \end{pmatrix}$, endowed with the usual inner product

$$\left\langle \begin{pmatrix} u \\ v \end{pmatrix}, \begin{pmatrix} \tilde{u} \\ \tilde{v} \end{pmatrix} \right\rangle = \int_\Omega \nabla u(x) \cdot \nabla \tilde{u}(x)\, dx + \int_\Omega v(x)\tilde{v}(x)\, dx$$

for each $\begin{pmatrix} u \\ v \end{pmatrix}, \begin{pmatrix} \tilde{u} \\ \tilde{v} \end{pmatrix} \in X$, where A, f and g are defined as follows. First, let us define the linear operator $A : D(A) \subseteq X \to X$ by

$$D(A) = \begin{pmatrix} H_0^1(\Omega) \cap H^2(\Omega) \\ \times \\ H_0^1(\Omega) \end{pmatrix}, \qquad A \begin{pmatrix} u \\ v \end{pmatrix} = \begin{pmatrix} -\omega u + v \\ \Delta u - \omega v \end{pmatrix}$$

for each $\begin{pmatrix} u \\ v \end{pmatrix} \in D(A)$. Further, let us define $f : \mathbb{R}_+ \times X \to X$ by

$$f \left(t, \begin{pmatrix} u \\ v \end{pmatrix} \right) = \begin{pmatrix} 0 \\ h(t,u) \end{pmatrix}$$

for each $t \in \mathbb{R}_+$ and $\begin{pmatrix} u \\ v \end{pmatrix} \in X$.

Finally, in order to define the nonlocal function $g : C_b(\mathbb{R}_+; X) \to X$, let us denote by

$$\mathcal{M}(t,u,w) = \mathcal{N}(t,u,w) + \omega \alpha(t)u, \quad \psi_3 = \psi_2 + \omega \psi_1 \quad \text{and} \quad w = v - \omega u$$

for each $(t,u,v) \in \mathbb{R}_+ \times \mathbb{R} \times \mathbb{R}$. Then, g is given by

$$\left[g \begin{pmatrix} u \\ v \end{pmatrix} (t) \right](x) = \begin{pmatrix} \displaystyle\int_\tau^{+\infty} \alpha(s)u(t+s,x)\, ds + \psi_1(t)(x) \\ \displaystyle\int_\tau^{+\infty} \mathcal{M}(s,u(t+s,x),w(t+s,x))\, ds + \psi_3(t,x) \end{pmatrix}$$

for each $\begin{pmatrix} u \\ v \end{pmatrix} \in C_b(\mathbb{R}_+; X)$, each $t \in [-\tau, 0]$ and a.e. for $x \in \Omega$.

From now on, up to a certain point, the proof follows exactly the same lines as those in the proof of Theorem 3.5.1. The only additional fact we have to prove here is that f is compact. But this follows from Lemma 2.2.1. Hence, the conclusion follows from Theorem 4.8.1 and this completes the proof. □

4.10.2 The second existence result

From Theorem 4.9.1, using Remark 4.9.1 combined with similar arguments as in the proof of Theorem 4.10.1, we deduce:

176   *Delay Differential Evolutions Subjected to Nonlocal Initial Conditions*

THEOREM 4.10.2 *Let Ω be a nonempty bounded and open subset in $\mathbb{R}^d$, $d \geq 1$, with C^1 boundary Σ, let $\tau \geq 0$ and let $\omega > 0$. Let further $\psi_1 \in \mathcal{H}_0^1$, $\psi_2 \in \mathcal{X}_2$ and let us assume that $h : \mathbb{R}_+ \times \mathcal{H}_0^1 \to L^2(\Omega)$, $\alpha \in L^2(\mathbb{R}_+)$ and $\mathcal{N} : \mathbb{R}_+ \times \mathbb{R} \times \mathbb{R} \to \mathbb{R}$ are continuous and satisfy*

(h_1) *there exists $\ell > 0$ such that*

$$\|h(t, u) - h(t, v)\|_{L^2(\Omega)} \leq \ell \|u - v\|_{\mathcal{H}_0^1}$$

for each $t \in \mathbb{R}_+$ and $u, v \in \mathcal{H}_0^1$

(h_2) *there exists $m > 0$ such that*

$$\|h(t, 0)\|_{L^2(\Omega)} \leq m$$

for each $t \in \mathbb{R}_+$

(n_1) *there exists a nonnegative continuous function $\eta \in L^2(\mathbb{R}_+)$ such that*

$$|\mathcal{N}(t, u, v)| \leq \eta(t)(|u| + |v|),$$

for each $t \in \mathbb{R}_+$ and $u, v \in \mathbb{R}$.

Let λ_1 be the first eigenvalue of $-\Delta$ and let us assume that

(n_3) $\begin{cases} \|\eta\|_{L^2(\mathbb{R}_+)} \leq 1 \\ (1 + \lambda_1^{-1}\omega)\|\alpha\|_{L^2(\mathbb{R}_+)} + \lambda_1^{-1}(1 + \omega)\|\eta\|_{L^2(\mathbb{R}_+)} \leq 1 \end{cases}$

(n_4) *there exists $b > \tau$ such that $\alpha(t) = \eta(t) = 0$ for each $t \in [0, b]$*

(c_1) $\ell < \omega$.

Then the problem (4.10.1) has at least one mild solution[5] $u \in C_b([-\tau, +\infty); H_0^1(\Omega))$ with $\dfrac{\partial u}{\partial t} \in C_b([-\tau, +\infty); L^2(\Omega))$. In addition, u satisfies

$$\|u\|_{C_b([-\tau, +\infty); H_0^1(\Omega))} + \left\|\frac{\partial u}{\partial t}\right\|_{C_b([-\tau, +\infty); L^2(\Omega))} \leq \frac{m}{\omega - \ell} + k \cdot m_0,$$

where k is given by (4.1.3) and $m_0 = \|\psi_1\|_{\mathcal{H}_0^1} + \omega\|\psi_1\|_{\mathcal{X}_2} + \|\psi_2\|_{\mathcal{X}_2}$.

4.11   Further investigations in the case $\ell = \omega$

Here we analyze the case in which f and g are Lipschitz but the Lipschitz constant ℓ of f coincides with the decay exponent ω of the nonlinear semigroup generated by A, i.e., when, in a certain sense, a resonance situation may occur. In this limiting case, we assume the extra condition $g(0) = 0$, which means that the nonlocal initial condition is purely nonlocal. See Definition 3.1.2.

---

[5] For the meaning of the mild solution see Remark 2.2.1.

Nonlocal Initial Conditions: The Quasi-Autonomous Case

4.11.1 The assumptions

(H_A) The operator $A : D(A) \subseteq X \rightsquigarrow X$ satisfies

(A_1) $0 \in D(A)$, $0 \in A0$ and A is ω-m-dissipative for some $\omega > 0$

(A_2) A generates a compact semigroup

(A_3) $\overline{D(A)}$ is convex.

(H_f) The function $f : \mathbb{R}_+ \times X \to X$ is continuous and

(f_1) it is Lipschitz of constant ω with respect to its second argument, i.e.,
$$\|f(t, v) - f(t, \widetilde{v})\| \leq \omega \|v - \widetilde{v}\|_X$$
for each $t \in \mathbb{R}_+$ and $v, \widetilde{v} \in X$

(f_2) there exists $m > 0$ such that
$$\|f(t, 0)\| \leq m$$
for each $t \in \mathbb{R}_+$

(f_6) there exists $r > 0$ so that, for each $w \in X$ with $\|w(0)\| \geq r$, we have
$$[w(0), f(t, w)]_+ \leq 0.$$

(H_g) The function $g : C_b(\mathbb{R}_+; \overline{D(A)}) \to \mathcal{D}$ satisfies

(g_1) there exists $a > 0$ such that for each $u, v \in C_b(\mathbb{R}_+; \overline{D(A)})$, we have
$$\|g(u) - g(v)\|_X \leq \|u - v\|_{C_b([a, +\infty); X)}$$

(g_4) g is continuous from $C_b([a, +\infty); \overline{D(A)})$ endowed with the $\widetilde{C}_b([a, +\infty); X)$ topology to X, where a is given by (g_1)

(g_5) $g(0) = 0$.

REMARK 4.11.1 In the absence of the condition $\ell < \omega$, which was essential in our preceding analysis, here we imposed (f_6), which is simply a tangency-type condition. It ensures that each C^0-solution of the local initial-value problem, associated with (4.1.1), issuing from a point ψ in the closed ball of center 0 and radius $r > 0$, $D(0, r)$, in $\mathcal{D}$, does not escape $D(0, r)$ as long as it exists.

4.11.2 The main result

Concerning the purely nonlocal delay problem (4.1.1) – see Definition 3.1.2 – we have:

THEOREM 4.11.1 *If (H_A), (H_f) and (H_g) are satisfied, then the problem (4.1.1) has at least one mild solution, $u \in C_b([-\tau, +\infty); X)$. If, instead of (f_1) in (H_f), the stronger condition*

178   *Delay Differential Evolutions Subjected to Nonlocal Initial Conditions*

(f_7) *f is jointly Lipschitz of constant ω, i.e.,*

$$\|f(t,v) - f(s,\widetilde{v})\| \le \omega \left(|t - s| + \|v - \widetilde{v}\|_X\right)$$

for each $t, s \in \mathbb{R}_+$ and $v, \widetilde{v} \in X$

is satisfied, then each C^0-solution of (4.1.1) is stable.

As concerns the regularity of the C^0-solutions in the case in which X is a Hilbert space and $A = -\partial\varphi$, where $\partial\varphi$ is the subdifferential of an l.s.c., proper and convex function $\varphi : X \to \mathbb{R}_+ \cup \{+\infty\}$, see Theorem 1.10.3.

4.11.3   Periodic solutions

If we choose $\tau \in (0, T)$ and $g(u)(t) = u(t + T)$ for each $t \in [-\tau, 0]$, from Theorem 4.11.1, we obtain an existence result for the periodic problem

$$\begin{cases} u'(t) \in Au(t) + f(t, u_t), & t \in \mathbb{R}_+, \\ u(t) = u(t + T), & t \in [-\tau, 0]. \end{cases} \tag{4.11.1}$$

More precisely, we have:

THEOREM 4.11.2 *If the conditions (H_A) and (H_f) are satisfied and the function $f : \mathbb{R}_+ \times X \to X$ is T-periodic with respect to its first argument, then (4.11.1) has at least one T-periodic C^0-solution, $u \in C_b(\mathbb{R}_+; X)$. If, instead of (f_1) in (H_f), the stronger condition (f_7) in Theorem 4.11.1 is satisfied, then each T-periodic C^0-solution of (4.11.1) is stable.*

The conclusion of Theorem 4.11.2 follows from Theorem 4.11.1 and the following:

LEMMA 4.11.1 *If (H_A), (H_f) are satisfied and the function $f : \mathbb{R}_+ \times X \to X$ is T-periodic with respect to its first argument, then each C^0-solution of the problem (4.11.1) is T-periodic.*

Since the proof is almost the same as the proof of Lemma 4.1.1, we do not give details.

4.11.4   Anti-periodic solutions

If $\tau \in (0, T)$ and $g(u)(t) = -u(t + T)$ for each $t \in [-\tau, 0]$, from Theorem 4.11.1, we obtain an existence result for the T-anti-periodic C^0-solutions, which are at the same time $2T$-periodic. More precisely, let us consider the problem

$$\begin{cases} u'(t) \in Au(t) + f(t, u_t), & t \in \mathbb{R}_+, \\ u(t) = -u(t + T), & t \in [-\tau, 0]. \end{cases} \tag{4.11.2}$$

More precisely, we have:

Nonlocal Initial Conditions: The Quasi-Autonomous Case

THEOREM 4.11.3 *Let us assume that (H_A) and (H_f) are satisfied, $D(A)$ is symmetric with respect to the origin, i.e., $D(A) = -D(A)$, $A(-\xi) = -A\xi$ for each $\xi \in D(A)$, f satisfies*

$$-f(t+T, -v) = f(t, v), \quad t \in \mathbb{R}_+, \ v \in C([-\tau, 0]; \overline{D(A)}). \qquad (4.11.3)$$

Then $(4.11.2)$ has at least one C^0-solution, $u \in C_b([-\tau, +\infty); X)$, which is $2T$-periodic. If, instead of (f_1) in (H_f), the stronger condition (f_7) in Theorem 4.11.1 is satisfied, then each T-anti-periodic C^0-solution of $(4.11.2)$ is stable.

The conclusion of Theorem 4.11.3 follows from Theorem 4.11.1 along with:

LEMMA 4.11.2 *If (H_A), (H_f) are satisfied, $D(A) = -D(A)$, $A(-\xi) = -A\xi$ for each $\xi \in D(A)$ and $f : \mathbb{R}_+ \times X \to X$ satisfies $(4.11.3)$, then each C^0-solution of the problem $(4.11.1)$ is T-anti-periodic and $2T$-periodic.*

The proof of Lemma 4.11.2 is a simple copy of the proof of Lemma 4.1.2 and therefore we omit it.

4.11.5 The nondelayed case

Taking $\tau = 0$, we get some existence results for evolution equations subjected to nonlocal initial conditions. More precisely, let us consider

$$\begin{cases} u'(t) \in Au(t) + f(t, u(t)), & t \in \mathbb{R}_+, \\ u(0) = g(u), \end{cases} \qquad (4.11.4)$$

where A is as above and f, g satisfy the hypotheses below.

$(H_f^{[\tau=0]})$ The function $f : \mathbb{R}_+ \times \overline{D(A)} \to X$ is continuous and satisfies

$(f_1^{[\tau=0]})$ it is Lipschitz of constant $\omega > 0$ with respect to its second argument, i.e.,
$$\|f(t, v) - f(t, \tilde{v})\| \le \omega \|v - \tilde{v}\|$$
for each $t \in \mathbb{R}_+$ and $v, \tilde{v} \in \overline{D(A)}$

$(f_2^{[\tau=0]})$ there exists $m > 0$ such that
$$\|f(t, 0)\| \le m$$
for each $t \in \mathbb{R}_+$

$(f_4^{[\tau=0]})$ there exists $r > 0$ such that
$$[u, f(t, u)]_+ \le 0$$
for each $t \in \mathbb{R}_+$ and each $u \in \overline{D(A)}$ with $\|u\| \ge r$.

180   *Delay Differential Evolutions Subjected to Nonlocal Initial Conditions*

$(H_g^{[\tau=0]})$ The function $g : C_b(\mathbb{R}_+; \overline{D(A)}) \to \overline{D(A)}$ satisfies

$(g_1^{[\tau=0]})$ there exists $a > 0$ such that for each $u, v \in C_b(\mathbb{R}_+; \overline{D(A)})$, we have

$$\|g(u) - g(v)\| \leq \|u - v\|_{C_b([a,+\infty);X)}$$

$(g_4^{[\tau=0]})$ g is continuous from $C_b([a, +\infty); \overline{D(A)})$ endowed to the $\widetilde{C}_b([a, +\infty); X)$ topology to X, where a is given by $(g_1^{[\tau=0]})$

$(g_5^{[\tau=0]})$ $g(0) = 0$.

An immediate consequence of Theorem 4.11.1 is:

THEOREM 4.11.4 *Let us assume that* (H_A), $(H_f^{[\tau=0]})$ *and* $(H_g^{[\tau=0]})$ *are satisfied. Then the problem* (4.11.4) *has at least one* C^0-*solution* $u \in C_b(\mathbb{R}_+; \overline{D(A)})$. *If, instead of* $(f_1^{[\tau=0]})$ *in* (H_f), *the stronger condition*

$(f_7^{[\tau=0]})$ f *is jointly Lipschitz of constant* ω, *i.e.,*

$$\|f(t, v) - f(s, \widetilde{v})\| \leq \omega \left(|t - s| + \|v - \widetilde{v}\|\right)$$

for each $t, s \in \mathbb{R}_+$ *and* $v, \widetilde{v} \in \overline{D(A)}$,

is satisfied, then each C^0-*solution of* (4.1.1) *is stable.*

As far as the periodic problem

$$\begin{cases} u'(t) \in Au(t) + f(t, u(t)), & t \in \mathbb{R}_+, \\ u(0) = u(T) \end{cases} \tag{4.11.5}$$

is concerned, from Theorem 4.11.4, we deduce:

THEOREM 4.11.5 *Let us assume that* (H_A) *and* $(H_f^{[\tau=0]})$ *are satisfied and* $f : \mathbb{R}_+ \times \overline{D(A)} \to X$ *is* T-*periodic with respect to its first argument. Then the periodic problem* (4.11.5) *has at least one* C^0-*solution* $u \in C_b(\mathbb{R}_+; \overline{D(A)})$. *If, instead of* $(f_1^{[\tau=0]})$ *in* (H_f), *the stronger condition* $(f_7^{[\tau=0]})$ *in Theorem* 4.11.4 *is satisfied, then each* T-*periodic* C^0-*solution of the problem* (4.11.5) *is stable.*

More generally, from Theorem 4.11.4, we deduce an existence result for two-point boundary value problems of the form

$$\begin{cases} u'(t) \in Au(t) + f(t, u(t)), & t \in \mathbb{R}_+, \\ u(0) = h(u(T)), \end{cases} \tag{4.11.6}$$

where $h : \overline{D(A)} \to \overline{D(A)}$ is nonexpansive. Namely, we have

Nonlocal Initial Conditions: The Quasi-Autonomous Case 181

THEOREM 4.11.6 *Let us assume that (H_A) and $(H_f^{[\tau=0]})$ are satisfied and $h : \overline{D(A)} \to \overline{D(A)}$ is nonexpansive and satisfies $h(0) = 0$. Then the two-point boundary value problem $(4.11.6)$ has at least one C^0-solution $u \in C_b(\mathbb{R}_+; \overline{D(A)})$. If, instead of $(f_1^{[\tau=0]})$ in (H_f), the stronger condition $(f_7^{[\tau=0]})$ in Theorem 4.11.4 is satisfied, then each C^0-solution of the problem $(4.11.6)$ is stable.*

Next, let us consider the T-anti-periodic problem

$$\begin{cases} u'(t) \in Au(t) + f(t, u(t)), & t \in \mathbb{R}_+, \\ u(0) = -u(T). \end{cases} \tag{4.11.7}$$

THEOREM 4.11.7 *Let us assume that (H_A) and $(H_f^{[\tau=0]})$ are satisfied, $D(A)$ is symmetric with respect to the origin, i.e., $D(A) = -D(A)$, $A(-\xi) = -A\xi$ for each $\xi \in D(A)$ and the function f satisfies $-f(t + T, -v) = f(t, v)$ for each $(t, v) \in [-T, 0] \times \overline{D(A)}$. Then $(4.11.7)$ has at least one C^0-solution, $u \in C_b(\mathbb{R}_+; \overline{D(A)})$ which is $2T$-periodic. If, instead of $(f_1^{[\tau=0]})$ in (H_f), the stronger condition $(f_7^{[\tau=0]})$ in Theorem 4.11.4 is satisfied, then each T-anti-periodic C^0-solution of the problem $(4.11.7)$ is stable.*

4.11.6 Proof of the main result

We can now proceed with the proof of Theorem 4.11.1.

Proof. The idea is to consider a sequence of approximate problems satisfying the hypotheses of Theorem 4.1.1, to get a sequence of C^0-approximate solutions and then to pass to the limit on a suitably chosen subsequence.

Let $\varepsilon \in (0, 1)$ and let the approximate functions $f_\varepsilon : \mathbb{R}_+ \times \mathcal{X} \to X$ and $g_\varepsilon : C_b(\mathbb{R}_+; \overline{D(A)}) \to \mathcal{D}$ be defined by

$$\begin{cases} f_\varepsilon(t, v) = (1 - \varepsilon)f(t, v), & (t, v) \in \mathbb{R}_+ \times \mathcal{X} \\ g_\varepsilon(u) = (1 - \varepsilon)g(u), & u \in C_b(\mathbb{R}_+; \overline{D(A)}). \end{cases}$$

We emphasize first that, due to (A_1) and (A_3), it readily follows that for each $u \in C_b(\mathbb{R}_+; \overline{D(A)})$, $(1 - \varepsilon)g(u) \in \mathcal{D}$ and thus g_ε is well-defined. So, f_ε and g_ε satisfy the hypotheses of Theorem 4.1.1, f_ε having the Lipschitz constant $\ell_\varepsilon = (1 - \varepsilon)\omega < \omega$, while g_ε being a strict contraction of constant $(1 - \varepsilon)$. Now, let us consider the auxiliary approximate problem

$$\begin{cases} u'(t) \in Au(t) + f_\varepsilon(t, u_t), & t \in \mathbb{R}_+, \\ u(t) = g_\varepsilon(u)(t), & t \in [-\tau, 0]. \end{cases} \tag{4.11.8}$$

From Theorem 4.1.1, it follows that the problem $(4.11.8)$ has a unique C^0-solution $u_\varepsilon \in C_b([-\tau, +\infty); X)$. More than this, in our case we have

$$\|u_\varepsilon(t)\| \leq r \tag{4.11.9}$$

182   Delay Differential Evolutions Subjected to Nonlocal Initial Conditions

for each $t \in \mathbb{R}_+$, where r is given by (f_6) in (H_f).

To prove (4.11.9), let us observe that u_ε is the unique fixed point of the operator $Q_\varepsilon : C_b(\mathbb{R}_+; \overline{D(A)}) \to C_b(\mathbb{R}_+; \overline{D(A)})$, defined by $Q_\varepsilon(v) = u_{|\mathbb{R}_+}$, where u is the unique C^0-solution of the problem

$$\begin{cases} u'(t) \in Au(t) + f_\varepsilon(t, u_t), & t \in \mathbb{R}_+, \\ u(t) = g_\varepsilon(v)(t), & t \in [-\tau, 0] \end{cases} \tag{4.11.10}$$

whose existence and uniqueness is ensured by Theorem 2.6.3.

First, let us remark that Q_ε maps $D(0, r)$, i.e., the closed ball with center 0 and radius r in $C_b(\mathbb{R}_+; \overline{D(A)})$, into itself. Let $v \in D(0, r)$ and let u be the C^0-solution of (4.11.10) corresponding to v. Then

$$\|u(t)\| = (1 - \varepsilon)\|g(v)(t)\| < r,$$

for each $t \in [-\tau, 0]$. If $\|u(t)\| \leq r$ for each $t \in \mathbb{R}_+$, we have nothing to prove. So, let us assume by contradiction that there exists at least one $t_1 > 0$ such that $\|u(t_1)\| > r$. Then, there exists $\overline{t} \in (0, t_1)$ such that

$$\|u(\overline{t})\| = r$$

and

$$\|u(s)\| > r$$

for each $s \in (\overline{t}, t_1)$.

In fact, $\overline{t}$ is the last point in $(0, t_1)$ at which the function $s \mapsto \|u(s)\|$ attains the value r.

Now, let us consider the problem

$$\begin{cases} u'(t) \in (A + \omega I)u(t) + f_\varepsilon(t, u_t) - \omega u(t), & t \in \mathbb{R}_+, \\ u(t) = g_\varepsilon(v)(t), & t \in [-\tau, 0] \end{cases}$$

which, obviously, is equivalent to (4.11.10). Taking into account that, by (H_A), $A + \omega I$ is dissipative, $x = 0 \in D(A + \omega I)$ and $y = 0 \in (A + \omega I)0$, from (1.8.4) in Theorem 1.8.2 and (f_6) in (H_f), we deduce

$$r < \|u(t_1)\| \leq \|u(\overline{t})\| + \int_{\overline{t}}^{t_1} [u(s), f(s, u_s) - \omega u(s)]_+ \, ds$$

$$= r + \int_{\overline{t}}^{t_1} [u(s), f(s, u_s)]_+ \, ds - \omega \int_{\overline{t}}^{t_1} \|u(s)\| \, ds < r,$$

which is a contradiction. This contradiction can be eliminated only if u satisfies $\|u(t)\| \leq r$ for each $t \in \mathbb{R}_+$, which proves that $Q_\varepsilon(D(0, r)) \subseteq D(0, r)$.

We prove next that Q_ε is a strict contraction on $D(0, r)$. To this end, let $v, \widetilde{v} \in D(0, r)$ and let u and $\widetilde{u}$ be the C^0-solutions of (4.11.10) corresponding to v and $\widetilde{v}$, respectively. We have

$$\|u(t) - \widetilde{u}(t)\| \leq (1 - \varepsilon)e^{-\omega t}\|g(v)(0) - g(\widetilde{v})(0)\|$$

Nonlocal Initial Conditions: The Quasi-Autonomous Case

$$+(1-\varepsilon)\int_0^t e^{-\omega(t-s)}\|f(s,u_s)-f(s,\widetilde{u}_s)\|\,ds$$

for each $t \in \mathbb{R}_+$. Then

$$\|u(t)-\widetilde{u}(t)\| \le (1-\varepsilon)e^{-\omega t}\|v-\widetilde{v}\|_{C_b(\mathbb{R}_+;X)} \tag{4.11.11}$$

$$+(1-\varepsilon)(1-e^{-\omega t})\max\left\{\|u-\widetilde{u}\|_{C_b(\mathbb{R}_+;X)}, \|v-\widetilde{v}\|_{C_b(\mathbb{R}_+;X)}\right\}$$

for each $t \in \mathbb{R}_+$. We distinguish between two cases.

Case 1. If

$$\max\left\{\|u-\widetilde{u}\|_{C_b(\mathbb{R}_+;X)}, \|v-\widetilde{v}\|_{C_b(\mathbb{R}_+;X)}\right\} = \|u-\widetilde{u}\|_{C_b(\mathbb{R}_+;X)},$$

then there exists $\bar{t} \in \mathbb{R}_+$ such that

$$\|u-\widetilde{u}\|_{C_b(\mathbb{R}_+;X)} = \|u(\bar{t})-\widetilde{u}(\bar{t})\|.$$

Indeed, if this is not true, it follows that

$$\lim_{t\to+\infty}\|u(t)-\widetilde{u}(t)\| = \|u-\widetilde{u}\|_{C_b(\mathbb{R}_+;X)},$$

from which, passing to the limit for $t \to +\infty$ in (4.11.11), we get

$$\|u-\widetilde{u}\|_{C_b(\mathbb{R}_+;X)} \le (1-\varepsilon)\|u-\widetilde{u}\|_{C_b(\mathbb{R}_+;X)},$$

which shows that $\|u-\widetilde{u}\|_{C_b(\mathbb{R}_+;X)} = 0$.

So, for $\bar{t} \in \mathbb{R}_+$ for which $\|u-\widetilde{u}\|_{C_b(\mathbb{R}_+;X)} = \|u(\bar{t})-\widetilde{u}(\bar{t})\|$, from (4.11.11) we deduce

$$\|u(\bar{t})-\widetilde{u}(\bar{t})\| \le \frac{(1-\varepsilon)e^{-\omega\bar{t}}}{(1-\varepsilon)e^{-\omega\bar{t}}+\varepsilon}\|v-\widetilde{v}\|_{C_b(\mathbb{R}_+;X)}.$$

Taking into account that the function $x \mapsto \dfrac{(1-\varepsilon)x}{(1-\varepsilon)x+\varepsilon}$ is strictly increasing on $[0,1]$, from the last inequality, we obtain

$$\|u-\widetilde{u}\|_{C_b(\mathbb{R}_+;X)} \le (1-\varepsilon)\|v-\widetilde{v}\|_{C_b(\mathbb{R}_+;X)}. \tag{4.11.12}$$

Case 2. If

$$\max\left\{\|u-\widetilde{u}\|_{C_b(\mathbb{R}_+;X)}, \|v-\widetilde{v}\|_{C_b(\mathbb{R}_+;X)}\right\} = \|v-\widetilde{v}\|_{C_b(\mathbb{R}_+;X)},$$

from (4.11.11), we get

$$\|u(t)-\widetilde{u}(t)\| \le (1-\varepsilon)\|v-\widetilde{v}\|_{C_b(\mathbb{R}_+;X)}$$

for each $t \in \mathbb{R}_+$ and thus (4.11.12) holds true.

Consequently, in both cases, Q_ε is a contraction of constant $(1-\varepsilon) \in (0,1)$ from $D(0,r)$ into itself. By the Banach Fixed-Point Theorem, it follows that

184   *Delay Differential Evolutions Subjected to Nonlocal Initial Conditions*

Q_ε has a unique fixed point $u \in D(0,r)$ and $\widetilde{u} : [-\tau, +\infty) \to \overline{D(A)}$, defined by

$$\widetilde{u}(t) = \begin{cases} g(u)(t), & t \in [-\tau, 0), \\ u(t), & t \in \mathbb{R}_+, \end{cases}$$

is the unique C^0-solution of (4.11.8) and satisfies (4.11.9).

Let $(\varepsilon_n)_n$ in $(0,1)$ with $\lim_n \varepsilon_n = 0$ and let us consider the corresponding sequence of solutions for the approximate problems (4.11.8) for $\varepsilon = \varepsilon_n$. For the sake of simplicity, we denote the sequences by u_{ε_n}, g_{ε_n} and f_{ε_n} by $(u_n)_n$, f_n and g_n, respectively. From the above proof, we know that

$$\|u_n\|_{C_b(\mathbb{R}_+;X)} \leq r$$

for each $n \in \mathbb{N}$, where r is given by (f_6) in (H_f). So, the family of functions $\{s \mapsto f_n(s, u_{n_s}); \ n \in \mathbb{N}\}$ is bounded on $[0, T]$ for each $T > 0$ and $\{u_n(0); \ n \in \mathbb{N}\}$ is bounded in $\overline{D(A)}$. By Theorem 1.8.6 it follows that the set $\{u_n; \ n \in \mathbb{N}\}$ is relatively compact in $C([\delta, T]; X)$ for each $T > 0$ and $\delta \in (0, T)$. In particular, $\{u_n; \ n \in \mathbb{N}\}$ is relatively compact in $C([a, T]; X)$ for each $T > a$, where a is given by (g_1). By (H_g) it follows that $\{u_n; \ n \in \mathbb{N}\} = \{g_n(u_n); \ n \in \mathbb{N}\}$ is relatively compact in $\mathcal{X}$. Thus $\{u_n(0); \ n \in \mathbb{N}\} = \{g_n(u_n)(0); \ n \in \mathbb{N}\}$ is relatively compact in X. Using once again Theorem 1.8.6, we conclude that $\{u_n; \ n \in \mathbb{N}\}$ is relatively compact even in $C([0, T]; X)$ for each $T > 0$ and thus in $\widetilde{C}_b(\mathbb{R}_+; X)$. By (g_4), we conclude that the set of histories, $\{g_n(u_n); \ n \in \mathbb{N}\}$, is relatively compact in $\mathcal{X}$. So, we have

$$\lim_n u_n = u$$

in $\widetilde{C}_b([-\tau, +\infty); X)$,

$$\lim_n g_n(u_n) = g(u)$$

in $\mathcal{X}$ and

$$\lim_n f_n(t, u_{n_t}) = f(t, u_t)$$

uniformly for t in compact intervals in $\mathbb{R}_+$. Passing to the limit for $n \to +\infty$ in the approximate equation (4.11.8) with $\varepsilon = \varepsilon_n$, we conclude that u is a C^0-solution of (4.1.1). As the stability part follows from (c_1) in Theorem 1.12.5, this completes the proof. $\qquad\square$

4.12   The nonlinear diffusion equation revisited

4.12.1   Statement of the main result

As before, Ω is a nonempty, bounded domain in $\mathbb{R}^d$, $d \geq 1$, with C^1 boundary Σ, $\tau \geq 0$ and $\omega > 0$. Let Δ be the Laplace operator in the sense of distributions

Nonlocal Initial Conditions: The Quasi-Autonomous Case

over the domain Ω, let $\varphi : D(\varphi) \subseteq \mathbb{R} \rightsquigarrow \mathbb{R}$ be a maximal-monotone operator, let $h : \mathbb{R}_+ \times \Omega \times \mathbb{R} \times \mathbb{R} \to \mathbb{R}$, let $h_0 \in L^1(\Omega)$, $\lambda_i \in [-1, 1]$, for $i = 1, 2, \ldots$, and let $0 < t_1 < t_2 < \ldots$. We denote by $Q_+ = \mathbb{R}_+ \times \Omega$, $\Sigma_+ = \mathbb{R}_+ \times \Sigma$ and $Q_\tau = [-\tau, 0] \times \Omega$.

We recall that $\mathfrak{X}_1 = C([-\tau, 0]; L^1(\Omega))$ and we reconsider the following nonlinear diffusion equation subjected to nonlocal initial conditions, already discussed in Section 4.3, in a different general frame.

$$
\begin{cases}
\dfrac{\partial u}{\partial t}(t, x) = \Delta\varphi(u(t, x)) - \omega u(t, x) + H(u_t)(t, x), & \text{in } Q_+, \\[2mm]
\varphi(u(t, x)) = 0, & \text{on } \Sigma_+, \\[2mm]
u(t, x) = \displaystyle\sum_{i=1}^{\infty} \lambda_i u(t_i + t, x), & \text{in } Q_\tau,
\end{cases}
\tag{4.12.1}
$$

where $H(v)(t, x)$ is defined by

$$
H(v)(t, x) = h\left(t, x, v(0)(x), \int_{-\tau}^{0} v(s)(x)\, ds\right) + h_0(x)
\tag{4.12.2}
$$

for each $v \in \mathfrak{X}_1$ and $(t, x) \in Q_+$.

THEOREM 4.12.1 *Let Ω be a nonempty and bounded domain in $\mathbb{R}^d$, $d \geq 1$, with C^1 boundary Σ, let $\tau \geq 0$, $\omega > 0$ and let $\varphi : D(\varphi) \subseteq \mathbb{R} \rightsquigarrow \mathbb{R}$ be a maximal-monotone with $0 \in \varphi(0)$. Let $h : \mathbb{R}_+ \times \Omega \times \mathbb{R} \times \mathbb{R} \to \mathbb{R}$ be continuous, let $h_0 \in L^1(\Omega)$ with $\|h_0\|_{L^1(\Omega)} > 0$, let $\lambda_i \in [-1, 1]$, for $i = 1, 2, \ldots$, and $0 < t_1 < t_2 < \ldots$. If the following hypotheses are satisfied*

(φ_1) *$\varphi : \mathbb{R} \to \mathbb{R}$ is continuous on $\mathbb{R}$ and C^1 on $\mathbb{R} \setminus \{0\}$ and there exist two constants $C > 0$ and $\alpha > 0$ if $d \leq 2$ and $\alpha > (d-2)/d$ if $d \geq 3$ such that*

$$
\varphi'(r) \geq C|r|^{\alpha-1}
$$

for each $r \in \mathbb{R} \setminus \{0\}$

(h_1) *there exists $\ell > 0$ such that*

$$
|h(t, x, u, v) - h(t, x, \widetilde{u}, \widetilde{v})| \leq \ell\left(|u - \widetilde{u}| + |v - \widetilde{v}|\right)
$$

for each $t \in \mathbb{R}_+$, a.e. for $x \in \Omega$ and each $u, \widetilde{u}, v, \widetilde{v} \in \mathbb{R}$

(h_2) *there exists $m \geq 0$ such that for each $t \in \mathbb{R}_+$, each $u \in \mathbb{R}$ and a.e. for $x \in \Omega$, we have*

$$
|h(t, x, 0, u)| \leq m
$$

(h_3) *there exists $c > 0$ such that, for each $t \in \mathbb{R}_+$, each $u, v \in \mathbb{R}$ and a.e. for $x \in \Omega$, we have*

$$
\operatorname{sign} v \cdot h(t, x, v, u) \leq -c|v|
$$

186   *Delay Differential Evolutions Subjected to Nonlocal Initial Conditions*

(λ_1) $\lambda_i \in [-1,1]$, $i = 1, 2, \ldots$, *satisfy* $\sum_{i=1}^{\infty} |\lambda_i| \leq 1$ *and* $\tau < t_1 < t_2 < \ldots$

(c_1) $\ell = \omega$,

then (4.12.1) *has at least one C^0-solution that satisfies*

$$\|u\|_{C_b([-\tau,+\infty);L^1(\Omega))} \leq r,$$

where r is given by $r = c^{-1}\|h_0\|_{L^1(\Omega)}$.
   If, instead of (h_1), the stronger condition,

(h_4) *there exists a nonnegative function $\ell_1 \in L^1(\Omega)$ such that*

$$|h(t,x,u,v) - h(t,x,\widetilde{u},\widetilde{v})| \leq \ell_1(x)\left(|t-s| + |u-\widetilde{u}| + |v-\widetilde{v}|\right)$$

for each $t, s \in \mathbb{R}_+$, a.e. for $x \in \Omega$ and each $u, \widetilde{u}, v, \widetilde{v} \in \mathbb{R}$

is satisfied, then each C^0-solution of (4.12.1) is stable.

REMARK 4.12.1 Comparing Theorem 4.12.1 with Theorem 4.3.1, we observe that hypothesis (c), i.e., $\ell < \omega$, in the latter was relaxed to the weaker condition (c_1) in the former. The price paid for this change was to assume a stronger condition on φ, i.e., (φ_1), which, in view of Theorem 1.9.6, implies that $\Delta\varphi$ generates a compact semigroup. Also, under the hypothesis (h_4), the C^0-solutions, even stable, are not necessarily globally asymptotically stable.

REMARK 4.12.2 The key invariance condition (h_3) has the following simple geometrical interpretation. It amounts to saying that for the continuous function $h : \mathbb{R}_+ \times \Omega \times \mathbb{R} \times \mathbb{R} \to \mathbb{R}$, the family of the partial functions

$$\{v \mapsto h(t,x,v,u); \ (t,x,u) \in \mathbb{R}_+ \times \Omega \times \mathbb{R}\}$$

enjoys the following property: there exists $c > 0$ such that the graph of every function in the family is situated in $R_1 \cup R_2$, where

$$R_1 = \{(v,h) \in \mathbb{R}^2; \ v \leq 0 \ \text{ and } \ -h \leq cv\},$$

$$R_2 = \{(v,h) \in \mathbb{R}^2; \ v \geq 0 \ \text{ and } \ h \leq -cv\}.$$

4.12.2   Proof of the main result

Proof. Take $X = L^1(\Omega)$ and let the operator $A : D(A) \subseteq L^1(\Omega) \rightsquigarrow L^1(\Omega)$ be given by $D(A) = D(\Delta\varphi)$, $Au = \Delta\varphi(u) - \omega u$, for each $u \in D(A)$, where $\Delta\varphi$ is defined as in Theorem 1.9.6. Let $f : \mathbb{R}_+ \times \mathcal{X}_1 \to L^1(\Omega)$ and $g : C_b(\mathbb{R}_+; L^1(\Omega)) \to \mathcal{X}_1$ be defined as

$$f(t,v)(x) = h\left(t,x,v(0)(x), \int_{-\tau}^0 v(s)(x)\,ds\right) + h_0(x)$$

Nonlocal Initial Conditions: The Quasi-Autonomous Case

for each $t \in \mathbb{R}_+$, each $x \in \Omega$ and $v \in \mathfrak{X}_1$ and respectively by

$$[g(u)(t)](x) = \sum_{i=1}^{\infty} \lambda_i u(t_i + t, x)$$

for each $u \in C_b([-\tau, +\infty); L^1(\Omega))$, each $t \in [-\tau, 0]$ and a.e. for $x \in \Omega$.

Clearly g satisfies (H_g) in Theorem 4.11.1.

To verify that f satisfies (H_f) in Theorem 4.11.1, let us observe first that, for the simplicity of writing, it is more convenient to decompose f as

$$f(t, v)(x) = F(v)(t, x) + h_0(x),$$

where

$$F(v)(t, x) = h\left(t, x, v(0)(x), \int_{-\tau}^{0} v(s)(x)\right) ds$$

for each $v \in \mathfrak{X}_1$, each $t \in \mathbb{R}_+$ and a.e. for $x \in \Omega$.

We apply Theorem 4.11.1. Namely, in view of Theorem 1.9.6, the operator A satisfies hypothesis (H_A) in Theorem 4.11.1. From (h_1), (h_2) and (h_3), we deduce that f satisfies (H_f).

Since (f_1) and (f_2) follow from (h_1), (h_2) and (ii) in Lemma 1.3.1, we confine ourselves only to the proof of (f_4). So, let

$$r = c^{-1}\|h_0\|_{L^1(\Omega)}.$$

We will show next that, for each $(t, v) \in \mathbb{R}_+ \times \mathfrak{X}_1$, with

$$\|v(0)(\cdot)\|_{L^1(\Omega)} \geq r,$$

we have

$$[v(0)(\cdot), f(t, v)(\cdot)]_+ \leq 0.$$

In view of Example 1.8.1, we have

$$[v(0)(\cdot), f(t, v)(\cdot)]_+ = \int_{\{y \in \Omega; v(0)(y) > 0\}} f(t, v)(x) \, dx - \int_{\{y \in \Omega; v(0)(y) < 0\}} f(t, v)(x) \, dx$$

$$+ \int_{\{y \in \Omega; v(0)(y) = 0\}} |f(t, v)(x)| \, dx.$$

Since $f(t, v)(x) = F(v)(t, x) + h_0(x)$, we deduce

$$[v(0)(\cdot), f(t, v)(\cdot)]_+ \qquad (4.12.3)$$

$$\leq \int_{\{y \in \Omega; v(0)(y) > 0\}} F(v)(t, x) \, dx - \int_{\{y \in \Omega; v(0)(y) < 0\}} F(v)(t, x) \, dx$$

$$+ \int_{\{y \in \Omega; v(0)(y) = 0\}} |F(v)(t, x)| \, dx + \int_{\{y \in \Omega; v(0)(y) > 0\}} h_0(x) \, dx$$

188   *Delay Differential Evolutions Subjected to Nonlocal Initial Conditions*

$$- \int_{\{y \in \Omega; v(0)(y) < 0\}} h_0(x)\, dx + \int_{\{y \in \Omega; v(0)(y) = 0\}} |h_0(x)|\, dx.$$

Let us observe that, from (h_3) and the continuity of h, we have

$$h(t, x, 0, u) = 0$$

for each $t \in \mathbb{R}_+$, each $u \in \mathbb{R}$ and a.e. for $x \in \Omega$. Therefore,

$$F(v)(t, x) = 0$$

a.e. for those $x \in \Omega$ for which $v(0)(x) = 0$. Hence, (4.12.3) in conjunction with (h_3), yields

$$[\, v(0)(\cdot), f(t, v)(\cdot)\,]_+ \leq \int_\Omega \operatorname{sign}[\, v(0)(x)\,] F(v)(t, x)\, dx + \int_\Omega |h_0(x)|\, dx$$

$$\leq -c \int_\Omega |v(0)(x)|\, dx + \int_\Omega |h_0(x)|\, dx \leq 0$$

and thus f satisfies (f_4).

So, the problem (4.12.1) can be rewritten as an abstract problem of the form (4.1.1) satisfying the hypotheses of Theorem 4.11.1, from which the conclusion of Theorem 4.12.1. Since the stability of C^0-solutions follows from Theorem 1.12.6, the proof is complete.   □

We consider now the nonlinear delay diffusion equation subjected to T-periodic conditions:

$$\begin{cases} \dfrac{\partial u}{\partial t}(t, x) = \Delta \varphi(u(t, x)) - \omega u(t, x) + H(u_t)(t, x), & \text{in } Q_+, \\[2mm] \varphi(u(t, x)) = 0, & \text{on } \Sigma_+, \\[2mm] u(t, x) = u(t + T, x), & \text{in } Q_\tau, \end{cases} \qquad (4.12.4)$$

where H is defined as in (4.12.2).

From Theorem 4.11.2, we deduce the following variant of Theorem 4.3.2:

THEOREM 4.12.2 *Let Ω be a nonempty, bounded domain in $\mathbb{R}^d$, $d \geq 1$, with C^1 boundary Σ, let $\tau \geq 0$, $\omega > 0$ and let $\varphi : D(\varphi) \subseteq \mathbb{R} \rightsquigarrow \mathbb{R}$ be a maximal-monotone operator with $0 \in \varphi(0)$. Let $h : \mathbb{R}_+ \times \Omega \times \mathbb{R} \times \mathbb{R} \to \mathbb{R}$ be continuous and T-periodic with respect to its first argument. If (φ_1) is satisfied, $T > \tau$ and the function h satisfies (h_1), (h_2) and (h_3) in Theorem 4.12.1, then the problem (4.12.4) has at least one C^0-solution $u \in C_b(\mathbb{R}_+; L^1(\Omega))$ that satisfies*

$$\|u\|_{C_b(\mathbb{R}_+; L^1(\Omega))} \leq r,$$

where r is given by $r = c^{-1}\|h_0\|_{L^1(\Omega)}$.

If, instead of (h_1), the stronger condition (h_4) in Theorem 4.12.1 is satisfied, then each solution of the problem (4.12.4) is stable.

Nonlocal Initial Conditions: The Quasi-Autonomous Case   189

We conclude this section with an existence result concerning T-anti-periodic C^0-solutions. Namely, let us consider the nonlinear delay diffusion equation subjected to anti-periodic conditions:

$$\begin{cases} \dfrac{\partial u}{\partial t}(t,x) = \Delta\varphi(u(t,x)) - \omega u(t,x) + H(u_t)(t,x), & \text{in } Q_+, \\[2mm] \varphi(u(t,x)) = 0, & \text{on } \Sigma_+, \\[2mm] u(t,x) = -u(t+T,x), & \text{in } Q_\tau, \end{cases} \qquad (4.12.5)$$

where, as before, H is defined by (4.12.2).

From Theorem 4.11.3 and Lemma 4.11.2, we get the following variant of Theorem 4.3.3:

THEOREM 4.12.3 *Let Ω be a nonempty, bounded domain in $\mathbb{R}^d$, $d \geq 1$, with C^1 boundary Σ, let $\omega > 0$ and let $\varphi : D(\varphi) \subseteq \mathbb{R} \rightsquigarrow \mathbb{R}$ be a maximal-monotone operator with $0 \in \varphi(0)$. Let $h : \mathbb{R}_+ \times \Omega \times \mathbb{R} \times \mathbb{R} \to \mathbb{R}$ be continuous and let us assume that it satisfies*

$$-h(t+T,x,-u,-v) - h_0(x) = h(t,x,u,v) + h_0(x)$$

a.e. for $x \in \Omega$, for each $t \in \mathbb{R}_+$ and each $u,v \in \mathbb{R}$. If (φ_1) is satisfied, φ is odd, $T > \tau$ and the function h satisfies (h_1), (h_2) and (h_3) in Theorem 4.12.1, then the problem (4.12.5) has at least one C^0-solution $u \in C_b(\mathbb{R}_+; L^1(\Omega))$ which is $2T$-periodic and satisfies

$$\|u\|_{C_b(\mathbb{R}_+; L^1(\Omega))} \leq r,$$

where r is given by $r = c^{-1}\|h_0\|_{L^1(\Omega)}$.

If, instead of (h_1), the stronger condition (h_4) in Theorem 4.12.1 is satisfied then each solution of the problem (4.12.5) is stable.

4.13 Bibliographical notes and comments

Section 4.1 A class of semilinear delay evolution equation subjected to non-local initial conditions was considered by Byszewski and Acka [67] who have proved some existence and uniqueness results under natural Lipschitz assumptions on both the forcing term f and the nonlocal function g. They also obtained a continuous dependence theorem with respect to the local part ψ of the nonlocal initial data $g(u) + \psi$. See also Byszewski [65], Boucherif and Precup [44], Ntouyas and Tsamatos [199]. For related global existence results on fully nonautonomous semilinear evolutions subjected to nonlocal initial conditions, we refer to Wang and Zhu [263]. Theorem 4.1.1 extends the main result

190   *Delay Differential Evolutions Subjected to Nonlocal Initial Conditions*

of Vrabie [261] which, in turn, is a generalization of the main result of Burlică and Roşu [58], covering the case in which the history function, instead of linear growth has affine growth. This theorem and its proof are new and appear for the first time here. In turn, the main result of Burlică and Roşu [58] generalizes an existence result in Vrabie [257] referring to nonlinear purely nonlocal initial-value problems – see Definition 3.1.2 – of the form

$$
\begin{cases}
u'(t) = Au(t) + f(t, u(t), u(t - \tau_1), \ldots, u(t - \tau_n)), & t \in \mathbb{R}_+, \\
u(t) = g(u)(t), & t \in [-\tau, 0]
\end{cases}
$$

allowing f to depend on t and on u_t. See Remark 3.1.3. The main result of Vrabie [257] extends the main results of Y. Li [171] to the fully nonlinear case and to purely general nonlocal initial conditions in an abstract Banach space frame. Under some additional continuity assumptions on f, a specific form of Theorem 4.1.1 can be derived from some general abstract results of Dyson and Villela Bressan [106], Ghavidel [125], [126] and Ruess [229]. Theorem 4.1.2 referring to periodic conditions is due to Burlică and Roşu [58] and extends the existence periodic counterpart for $(N.2.2)$ in Vrabie [257]. This can be easily adapted to handle infinite delays as considered by Henríquez and Lizama [139]. In fact, in the latter paper it was shown that, under Lipschitz conditions on f, there is no need to assume the compactness of the semigroup in either the linear or the nonlinear case. Compare our Theorem 4.1.2 with Theorem 1.2 in Y. Li [171]. It should be mentioned that if, in addition, we assume that X is Hilbert and A is a subdifferential, we can recover most of the regularity properties of the C^0-solution proved by Y. Li [171] via the analyticity of the generated semigroup. Theorem 4.1.3 is new, Lemma 4.1.1 is from Burlică and Roşu [58], while Lemma 4.1.2 and Theorems 4.1.4~4.1.7 are new.

Section 4.2 Lemma 4.2.1 is a slight extension of a result in Vrabie [261], but the proof given here for the existence and uniqueness part differs from its counterpart in Vrabie [261] and rests heavily on Remark 3.2.3. It should be noted that the proofs of Theorems 4.1.1 and 4.1.2 are based on an interplay between compactness arguments and metric fixed-point techniques developed in both Paicu and Vrabie [204] and Y. Li [171]. Lemma 4.2.2 is also from Vrabie [261].

Section 4.3 The first illustrative application referring to the porous medium equation, i.e., Theorem 4.3.1, is inspired by both Burlică and Roşu [58] and García–Falset and Reich [123]. We emphasize that in Theorem 4.3.1, we assume a specific nonlocal initial condition that reduces, in the nondelayed case, to the one introduced by Byszewski [64]. Theorem 4.3.2 is also inspired by Burlică and Roşu [58] as well as by Vrabie [251]. For a similar result referring to a reaction term f of a specific form, see Vrabie [257]. For a multi-valued nondelayed related result, see Paicu and Vrabie [204].

Section 4.4 Theorem 4.4.1 is new, but the $C_b([-\tau, +\infty); X)$-convergence part, under the additional hypothesis $g_n(0) = 0$ for each $n \in \mathbb{N}$, was proved in

Nonlocal Initial Conditions: The Quasi-Autonomous Case 191

Burlică, Roșu and Vrabie [61]. As far as we know, the latter is the first general continuity result with respect to the data for problems of the form (4.1.1). As we already have mentioned, previous continuity results with respect to the local initial part of $g(u) + \psi$, in the semilinear case, are due to Byszewski and Acka [67] and Balachandran and Park [17]. The $\widetilde{C}_b([-\tau, +\infty); X)$-convergence part is also new and appears for the first time here. Theorem 4.4.2 is a simple consequence of Theorem 4.4.1 covering the nondelayed case. Theorem 4.4.3 as well as Lemma 4.4.1 are new.

Section 4.5 Theorems 4.5.1 $\sim$ 4.5.5 extend some results in Vrabie [261]. The proof of Theorem 4.5.1, based on the Schaefer Fixed-Point Theorem 1.4.6, is inspired by Aizicovici and McKibben [3], McKibben [182]. However, it should be noted that this idea goes back, among others, to Prüss [220] who, in the semilinear case, uses the Leray-Schauder Fixed-Point Theorem in order to produce a mild T-periodic solution, by showing that each one of the members of a suitably defined family of nonlinear operators has at least one fixed point. It should be noted also that in Aizicovici, Papageorgiou and Staicu [4], the authors make use of the Schaefer Fixed-Point Theorem 1.4.6 to prove an existence result for periodic solutions for fully nonlinear evolution equations. An averaging principle for the translation operator associated with a nonlinear hyperbolic evolution system was proved by Ćwiszewski and Kokocki [93]. They applied this principle to the study of a damped hyperbolic partial differential equation.

Section 4.6 In its present form, Theorem 4.6.1 is new but somehow related to some earlier results obtained in the nondelayed case by García–Falset and Reich [123] and by G. Avalishvili and M. Avalishvili [14], the latter result referring to a problem whose forcing term is independent of the state variable.

Section 4.7 As far as we know, Theorem 4.7.1 and Lemmas 4.7.1$\sim$4.7.2 are new and appear for the first time here.

Section 4.8 Theorems 4.8.1 and 4.8.2 are new. Theorem 4.8.3 is an extension of a result of Y. Li [171] to general Banach spaces and to infinitesimal generators of possibly non-analytic semigroups. We notice that it encompasses several earlier existence results for periodic solutions referring to semilinear evolution equations. Here, one should mention the papers of Prüss [220], already discussed in connection with Section 4.5 and that of Becker [25], who, in a Hilbert space frame, exploiting the compactness and the exponential decay of the C_0-semigroup involved, uses the Schauder Fixed-Point Theorem 1.4.4 to prove the existence of at least one periodic solution to a semilinear evolution equation in a real Hilbert space. Theorems 4.8.4 and 4.8.5 are new.

Section 4.9 Theorem 4.9.1, in which the compactness condition on the semigroup is moved on the forcing term f but g is only continuous and A is linear, was proved by Vrabie [260]. The case when f is a multifunction that satisfies a condition involving a certain measure of non-compactness was considered by Basova and Obukhovskii [23] and by Benedetti and Rubbioni [31].

192 *Delay Differential Evolutions Subjected to Nonlocal Initial Conditions*

A variant without delay, but with impulses, was analyzed by Cardinali and Rubbioni [70]. An existence result referring to the semilinear evolution equation with measures subjected to nonlocal initial conditions

$$\begin{cases} du(t) = \{Au(t) + f(t, u_t)\}dt + dh(t), & t \in \mathbb{R}_+, \\ u(t) = g(u)(t), & t \in [-\tau, 0] \end{cases}$$

was obtained by Benedetti, Malaguti, Taddei and Vrabie [30]. Here $\tau \geq 0$, $A : D(A) \subseteq X \to X$ is the infinitesimal generator of a C_0-semigroup, $\mathcal{R}([-\tau, 0]; X)$ denotes the space of all functions from $[-\tau, 0]$ to X having only discontinuities of the first kind, $\mathcal{R}_b(\mathbb{R}_+; X)$ is the space of all bounded functions belonging to $\mathcal{R}(\mathbb{R}_+; X)$, and $BV_{\text{loc}}(\mathbb{R}_+; X)$ the space of all functions from $\mathbb{R}_+$ to X which are of bounded variation on compact intervals. Furthermore, $f : \mathbb{R}_+ \times \mathcal{R}([-\tau, 0]; X) \to X$ is continuous, $g : \mathcal{R}_b(\mathbb{R}_+; X) \to \mathcal{R}([-\tau, 0]; X)$ is nonexpansive, and $h \in BV_{\text{loc}}(\mathbb{R}_+; X)$.

Section 4.10 Theorem 4.10.1 referring to the damped wave equation and encompassing a large variety of problems is also from Vrabie [260]. For related results referring to nondelayed higher-order hyperbolic problems subjected to nonlocal initial conditions, see G. Avalishvili and M. Avalishvili [15].

Section 4.11 Theorem 4.11.1, considering the critical case $\omega = \ell$, is new and inspired by Paicu and Vrabie [204]. The hypothesis (f_6) is in fact a tangency condition ensuring that the C^0-solutions of the problem

$$u'(t) \in Au(t) + f(t, u_t), \quad t \in \mathbb{R}_+$$

cannot escape from the "moving set" $t \mapsto \mathcal{K}(t)$, where

$$\mathcal{K}(t) = \{(t, \varphi) \in \mathbb{R}_+ \times X; \ \|\varphi(0)\| \leq r\}.$$

For related results, see Ghavidel [125], Ruess [229] and Necula, Popescu and Vrabie [194], [195]. For another result of this type, see Chapter 9. The existence results in the case of periodic problems, i.e., Theorem 4.11.2, Lemma 4.11.1, as well as Theorem 4.11.3 and Lemma 4.11.2 are also new. The nondelayed case, i.e., Theorem 4.11.4, is also inspired by Paicu and Vrabie [204]. Theorem 4.11.5, referring to the T-periodic case, Theorem 4.11.6, concerning the two point boundary value problem (4.11.6), and Theorem 4.11.7, referring to the T-anti-periodic case, are new.

Section 4.12 As far as we know, Theorem 4.12.1, as well as its T-periodic variant, Theorem 4.12.2, and its T-anti-periodic counterpart, Theorem 4.12.3, all are new and appear for the first time here. For other related results referring to the multi-valued nondelayed case, see Paicu and Vrabie [204].

Chapter 5

Almost Periodic Solutions

Overview

We consider the nonlinear delay differential evolution equation

$$\begin{cases} u'(t) \in Au(t) + f(t, u_t), & t \in \mathbb{R}_+, \\ u(t) = g(u)(t), & t \in [-\tau, 0], \end{cases}$$

where $\tau \geq 0$, X is a real Banach space, A is an ω-m-dissipative operator for some $\omega > 0$, $\mathcal{X} = C([-\tau, 0]; X)$, $\mathcal{D} = \{\varphi \in \mathcal{X}; \ \varphi(0) \in \overline{D(A)}\}$, and the function $f : \mathbb{R}_+ \times \mathcal{X} \to X$ is jointly continuous. We prove that if f is Lipschitz with respect to its second argument and its Lipschitz constant ℓ satisfies the condition $\ell e^{\omega \tau} < \omega$, $g : C_b(\mathbb{R}_+; \overline{D(A)}) \to \mathcal{D}$ is nonexpansive and $(I - A)^{-1}$ is compact, then the unique C^0-solution of the problem above is almost periodic.

5.1 Almost periodic functions

Let us consider the following nonlinear delay differential evolution equation subjected to a nonlocal initial condition:

$$\begin{cases} u'(t) \in Au(t) + f(t, u_t), & t \in \mathbb{R}_+, \\ u(t) = g(u)(t), & t \in [-\tau, 0]. \end{cases} \tag{5.1.1}$$

Here, $A : D(A) \subseteq X \rightsquigarrow X$ is an m-dissipative operator in the real Banach space X which generates a nonlinear semigroup of contractions on $\overline{D(A)}$, $\{S(t) : \overline{D(A)} \to \overline{D(A)}; \ t \in \mathbb{R}_+\}$, $\tau \geq 0$ is arbitrary but fixed, $f : \mathbb{R}_+ \times \mathcal{X} \to X$ is jointly continuous and Lipschitz with respect to its second argument with constant $\ell > 0$, while the nonlocal initial constraint $g : C_b(\mathbb{R}_+; \overline{D(A)}) \to \mathcal{D}$ is nonexpansive. In this chapter, we prove that, whenever $(I - A)^{-1}$ is compact and $\ell e^{\omega \tau} < \omega$, the unique C^0-solution of (5.1.1) is almost periodic. If $\tau = 0$, i.e., if the delay is absent, the condition $\ell e^{\omega \tau} < \omega$ reduces to $\ell < \omega$, which is the

193

194   *Delay Differential Evolutions Subjected to Nonlocal Initial Conditions*

key hypothesis in the Poincaré–Lyapunov Theorem on asymptotic stability. See Vrabie [254, Theorem 5.3.1. (Poincaré–Lyapunov), p. 170].

Since there are several definitions of almost periodicity, we introduce first a definition clarifying the precise meaning of this concept exactly in the sense we will use. As we are working with functions defined on intervals of the form $[c, +\infty)$, with $c \in \mathbb{R}$, we will rephrase all the definitions (introduced for functions defined on $\mathbb{R}$) in this specific case.

Hereafter, denote by

$$\begin{cases} s = \{(r_n)_n;\ r_n \in \mathbb{R}_+,\ n \in \mathbb{N}\} \\ s_\infty = \{(r_n)_n;\ r_n \in \mathbb{R}_+,\ n \in \mathbb{N},\ \lim_n r_n = +\infty\}. \end{cases}$$

DEFINITION 5.1.1 Let $c \in \mathbb{R}$. A function $u : [c, +\infty) \to X$ is called *almost periodic* if for each $(r_n)_n \in s$, the sequence $(t \mapsto u(t + r_n))_n$ has at least one convergent subsequence in $C_b([c, +\infty); X)$, i.e., there exist $(t \mapsto u(t + r_{n_k}))_k$ and $\widetilde{u} \in C_b([c, +\infty); X)$ such that

$$\lim_k u(t + r_{n_k}) = \widetilde{u}(t)$$

uniformly for $t \in [c, +\infty)$.

For $c = 0$, some authors call this property *asymptotic almost periodicity* and keep the name of "almost periodic" only for those functions $u : \mathbb{R} \to X$ satisfying the following property: for each sequence $(r_n)_n$, the sequence $(t \mapsto u(t + r_n))_n$ has at least one convergent subsequence in $C_b(\mathbb{R}; X)$. See Fréchet [118] and [119].

REMARK 5.1.1 Clearly, if u is continuous and almost periodic, it is uniformly continuous. Furthermore, if $u : [c, +\infty) \to X$ is uniformly continuous, then it is almost periodic if and only if for each $(r_n)_n \in s_\infty$, the sequence of translates $(t \mapsto u(t + r_n))_n$ has at least one convergent subsequence in $C_b([c, +\infty); X)$.

DEFINITION 5.1.2 Let Y be a Banach space, $Z \subseteq Y$ and $f : [c, +\infty) \times Y \to X$ a continuous function. The family $\{t \mapsto f(t, v);\ v \in Z\}$ is called *uniformly almost periodic on* $[c, +\infty)$ if for each $(r_n)_n \in s_\infty$, there exists a function

$$\widetilde{f} : [c, +\infty) \times Z \to X$$

and a subsequence $(r_{n_k})_k$ of $(r_n)_n$ such that

$$\lim_k f(t + r_{n_k}, v) = \widetilde{f}(t, v)$$

uniformly for $(t, v) \in [c, +\infty) \times Z$.

REMARK 5.1.2 If $f(t, v) = g(t)h(v)$, where $g : [c, +\infty) \to \mathbb{R}$ and $h : Y \to X$ are continuous, g is almost periodic and h is bounded on Z, then the family of functions $\{t \mapsto f(t, v);\ v \in Z\}$ is uniformly almost periodic.

Similarly, if $f(t, v) = \widetilde{g}(t) + h(v)$, $\widetilde{g} : [c, +\infty) \to X$ and $h : Y \to X$ are continuous and $\widetilde{g}$ is almost periodic, then the family $\{t \mapsto f(t, v);\ v \in Y\}$ is uniformly almost periodic.

5.2 The main results

Our main assumptions on A, f and g are listed below.

(H_A) The operator $A \colon D(A) \subseteq X \rightsquigarrow X$ satisfies

(A_1) $0 \in D(A)$, $0 \in A0$ and A is $\omega\text{-}m$-dissipative for some $\omega > 0$.

(H_f) The function $f \colon \mathbb{R}_+ \times X \to X$ is continuous and satisfies

(f_1) there exists $\ell > 0$ such that

$$\|f(t,v) - f(t,\widetilde{v})\| \le \ell \|v - \widetilde{v}\|_X$$

for each $t \in \mathbb{R}_+$ and each $v, \widetilde{v} \in X$

(f_2) there exists $m > 0$ such that

$$\|f(t,0)\| \le m$$

for each $t \in \mathbb{R}_+$.

(H_c) The constants ℓ, τ and ω satisfy one of these two conditions:

(c_1) $\ell < \omega$

(c_2) $\ell e^{\omega \tau} < \omega$.

(H_g) The function $g \colon C_b(\mathbb{R}_+; \overline{D(A)}) \to \mathcal{D}$ satisfies

(g_1) there exists $a > 0$ such that for each $u, v \in C_b(\mathbb{R}_+; \overline{D(A)})$, we have

$$\|g(u) - g(v)\|_X \le \|u - v\|_{C_b([a,+\infty);X)}.$$

$(H_{\overline{f}})$ The function f satisfies

$(\overline{f}_1)$ for each bounded set $\mathcal{C} \subseteq X$, the family

$$\{t \mapsto f(t,v); \ v \in \mathcal{C}\}$$

is uniformly equicontinuous on $\mathbb{R}_+$

$(\overline{f}_2)$ for each bounded set $\mathcal{C} \subseteq X$, the family

$$\{t \mapsto f(t,v); \ v \in \mathcal{C}\}$$

is uniformly almost periodic on $\mathbb{R}_+$. See Definition 5.1.2.

The main result in this chapter is:

196   *Delay Differential Evolutions Subjected to Nonlocal Initial Conditions*

THEOREM 5.2.1 *If (H_A), (H_f), (H_g), (c_2) in (H_c), $(\overline{f}_1)$ in $(H_{\overline{f}})$ are satisfied and $(I - A)^{-1}$ is compact, then the C^0-solution u of the problem (5.1.1), whose existence and uniqueness is ensured by Theorem 4.1.1, has a relatively compact trajectory. If, in addition $(\overline{f}_2)$ in $(H_{\overline{f}})$ is satisfied, then the C^0-solution, u, is almost periodic.*

REMARK 5.2.1 If A generates a compact semigroup and f is Lipschitz with respect to both arguments, i.e., there exists $\ell > 0$ such that

$$\|f(t,v) - f(\widetilde{t},\widetilde{v})\| \le \ell \left[|t - \widetilde{t}| + \|v - \widetilde{v}\|_X \right]$$

for each $t, \widetilde{t} \in \mathbb{R}_+$ and each $v, \widetilde{v} \in X$, then $(I - A)^{-1}$ is compact and f satisfies $(\overline{f}_1)$ in $(H_{\overline{f}})$. It should be noted that there are m-dissipative operators, A, with $(I - A)^{-1}$ compact but, nevertheless, the corresponding generated semigroup is not compact. See Vrabie [252, Example 2.3.2, p. 50] and the examples in Sections 5.6 and 5.7.

REMARK 5.2.2 We notice that, by the resolvent identity, i.e.,

$$(I - \lambda A)^{-1} = (I - \mu A)^{-1} \left[\frac{\mu}{\lambda} I + \frac{\lambda - \mu}{\lambda} (I - \lambda A)^{-1} \right],$$

for each $\lambda > 0$ and $\mu > 0$, it follows that $(I - A)^{-1}$ is compact if and only if, for each $\lambda > 0$, $(I - \lambda A)^{-1}$ is compact.

5.3   Auxiliary lemmas

A direct consequence of Lemma 1.13.3 that we need in what follows is:

LEMMA 5.3.1 *Let $y : [-\tau, +\infty) \to \mathbb{R}_+$ be continuous, let $\lambda \in \mathbb{R}_+$ and let $\alpha_0, \ell, \omega, \tau \in (0, +\infty)$. If $\ell e^{\omega\tau} < \omega$ and*

$$y(t) \le e^{-\omega t}\alpha_0 + \lambda + \int_0^t \ell e^{-\omega(t-s)} \|y_s\|_{\mathcal{R}} \, ds \tag{5.3.1}$$

for each $t \in \mathbb{R}_+$, then

$$y(t) \le (\alpha_0 + \|y_0\|_{\mathcal{R}}) \, e^{-(\omega - b)t} + \frac{\lambda\omega}{\omega - b}, \tag{5.3.2}$$

for each $t \in \mathbb{R}_+$, where $b = \ell e^{\omega\tau}$ and $\mathcal{R} = C([-\tau, 0]; \mathbb{R})$.

Proof. From (5.3.1), we get

$$e^{\omega t} y(t) \le \alpha_0 + \lambda e^{\omega t} + \ell e^{\omega\tau} \int_0^t \sup_{\theta \in [-\tau, 0]} \{ e^{\omega(s+\theta)} y(s+\theta) \} \, ds$$

Almost Periodic Solutions 197

for each $t \in \mathbb{R}_+$. Setting $b = \ell e^{\omega \tau}$ in the last inequality and using Lemma 1.13.3, we get

$$e^{\omega t} y(t) \le \alpha_0 + \|y_0\|_{\mathcal{R}} + \lambda e^{\omega t} + b \int_0^t \left(\alpha_0 + \|y_0\|_{\mathcal{R}} + \lambda e^{\omega s} \right) e^{b(t-s)} \, ds$$

for each $t \in \mathbb{R}_+$. Integrating the last term on the right-hand side, after some obvious computations, we get (5.3.2). $\qquad\square$

Let us consider the nonlinear delay evolution equation

$$\begin{cases} u'(t) \in Au(t) + f(t, u_t), & t \in \mathbb{R}_+, \\ u(t) = \varphi(t), & t \in [-\tau, 0]. \end{cases} \tag{5.3.3}$$

From Theorems 2.6.3 and 2.8.1, we easily get:

PROPOSITION 5.3.1 *If (H_A), (H_f) and (c_1) in (H_c) are satisfied, then, for each initial history $\varphi \in \mathfrak{X}$, there exists a unique C^0-solution of the problem (5.3.3), satisfying*

$$\|u\|_{C_b([-\tau, +\infty); X)} \le \max \left\{ \|\varphi\|_{\mathfrak{X}}, \frac{m}{\omega - \ell} \right\}. \tag{5.3.4}$$

If $\varphi \in \mathfrak{X}$, we denote by $u(\cdot, \varphi) \in C_b([-\tau, +\infty); \overline{D(A)})$ the unique C^0-solution of the problem (5.3.3), whose existence is ensured by Proposition 5.3.1.

LEMMA 5.3.2 *Let $\{S(t) : \overline{D(A)} \to \overline{D(A)}; \ t \in \mathbb{R}_+\}$ be the semigroup of non-expansive mappings generated by A and $C \subset \overline{D(A)}$ be a compact set. Then, for each $t \in \mathbb{R}_+$, we have*

$$\lim_{h \downarrow 0} \|S(t+h)\xi - S(t)\xi\| = 0$$

uniformly for $\xi \in C$.

Proof. Let $t \in \mathbb{R}_+$ and let $\varepsilon > 0$. Since C is compact, there exists a finite family $\{\xi_1, \xi_2, \ldots \xi_{n(\varepsilon)}\}$ in C such that, for each $\xi \in C$ there exists $i \in \{1, 2, \ldots n(\varepsilon)\}$ such that

$$\|\xi - \xi_i\| \le \varepsilon.$$

Since the family $\{S(\cdot)\xi_i; \ i = 1, 2, \ldots n(\varepsilon)\}$ is finite and continuous on $\mathbb{R}_+$, it is equicontinuous at t. So, there exists $\delta(\varepsilon) > 0$ such that

$$\|S(t+h)\xi_i - S(t)\xi_i\| \le \varepsilon$$

for each $i \in \{1, 2, \ldots n(\varepsilon)\}$ and each $h \in (0, \delta(\varepsilon)]$. Let $\xi \in C$ be arbitrary and let ξ_i be as above. We have

$$\|S(t+h)\xi - S(t)\xi\| \le \|S(t+h)\xi - S(t+h)\xi_i\| + \|S(t+h)\xi_i - S(t)\xi_i\| + \|S(t)\xi_i - S(t)\xi\|$$

$$\le 2\|\xi - \xi_i\| + \|S(t+h)\xi_i - S(t)\xi_i\| \le 3\varepsilon$$

for each $h \in (0, \delta(\varepsilon)]$, which completes the proof. $\qquad\square$

198   *Delay Differential Evolutions Subjected to Nonlocal Initial Conditions*

LEMMA 5.3.3 *Let us assume that* (H_A), (H_f) *and* (c_1) *in* (H_c) *are satisfied and the operator* $(I - A)^{-1}$ *is compact. Let* $\varphi \in \mathfrak{X}$. *Then the trajectory of the unique* C^0-*solution* u *of the problem* (5.3.3), *i.e.,* $\{u(t); \ t \in [-\tau, +\infty)\}$, *is relatively compact if and only if* u *is uniformly continuous from the right*[1] *on* $\mathbb{R}_+$.

Proof. We begin with the necessity. So, let $\varphi \in \mathfrak{X}$, u be the unique C^0-solution of the problem (5.3.3), and let us assume that the trajectory of u, i.e., $\{u(t); \ t \in [-\tau, +\infty)\}$, is relatively compact. Then, by Lemma 1.8.1 for each $t \in \mathbb{R}_+$ and $h > 0$, we have

$$\|u(t+h) - u(t)\| \le \|u(t+h) - S(h)u(t)\| + \|S(h)u(t) - u(t)\|$$

$$\le \int_t^{t+h} \|f(s, u_s)\| \, ds + \|S(h)u(t) - u(t)\|.$$

By (5.3.4) in Proposition 5.3.1, u is bounded by

$$k(\varphi) = \max \left\{ \|\varphi\|_{\mathfrak{X}}, \frac{m}{\omega - \ell} \right\} \tag{5.3.5}$$

and therefore

$$\|f(t, u_t)\| \le \ell k(\varphi) + m \tag{5.3.6}$$

for each $t \in \mathbb{R}_+$. Hence

$$\|u(t+h) - u(t)\| \le [\ell k(\varphi) + m] h + \|S(h)u(t) - u(t)\|,$$

for each $t \in \mathbb{R}_+$ and $h > 0$, where $k(\varphi)$ is given by (5.3.5). Now, Lemma 5.3.2 comes into play and shows that u is uniformly continuous from the right on $\mathbb{R}_+$ or equivalently, is uniformly continuous on $\mathbb{R}_+$. This completes the proof of the necessity.

In order to prove the sufficiency, we observe first that, due to the continuity of u, it is enough to show that if u is uniformly continuous from the right on $\mathbb{R}_+$, then its positive trajectory, i.e., $\{u(t); \ t \in \mathbb{R}_+\}$, is relatively compact. So, let us assume that u is uniformly continuous from the right on $\mathbb{R}_+$. Let $\{S(t) : \overline{D(A)} \to \overline{D(A)}; \ t \in \mathbb{R}_+\}$ be the semigroup of contractions generated by A, let $\lambda > 0$ and let $J_\lambda = (I - \lambda A)^{-1}$. From Lemma 1.8.2, we know that

$$\|J_\lambda u(t) - u(t)\| \le \frac{4}{\lambda} \int_0^\lambda \left[\|S(s)u(t) - u(t+s)\| + \|u(t+s) - u(t)\| \right] ds. \tag{5.3.7}$$

Next, let us observe that, by Lemma 1.8.1, we have

$$\|S(s)u(t) - u(t+s)\| \le \int_t^{t+s} \|f(\theta, u_\theta)\| \, d\theta \tag{5.3.8}$$

---

[1]In fact, in our case, u is uniformly continuous from the right on $\mathbb{R}_+$ if and only if it is uniformly continuous on $\mathbb{R}_+$.

Almost Periodic Solutions

for each $t, s \in \mathbb{R}_+$.

Let $\delta : \mathbb{R}_+ \to \mathbb{R}_+$ be the modulus of uniform continuity from the right of u on $[-\tau, +\infty)$, i.e.,

$$\delta(\lambda) = \sup\{\|u(t+s) - u(t)\|; \ t \in [-\tau, +\infty), \ s \in (0, \lambda]\}$$

for each $\lambda \geq 0$.

From (5.3.7), (5.3.8) and (5.3.6), we deduce

$$\|J_\lambda u(t) - u(t)\| \leq \frac{4}{\lambda} \int_0^\lambda \{[\ell k(\varphi) + m]s + \delta(\lambda)\} \ ds \leq 4\left\{[\ell k(\varphi) + m]\frac{\lambda}{2} + \delta(\lambda)\right\}$$

for each $t \in \mathbb{R}_+$ and $\lambda > 0$. But u is uniformly continuous from the right on $\mathbb{R}_+$ and so $\lim_{\lambda \downarrow 0} \delta(\lambda) = 0$. Thus

$$\lim_{\lambda \downarrow 0} \|J_\lambda u(t) - u(t)\| = 0$$

uniformly for $t \in \mathbb{R}_+$. As $(I - A)^{-1}$ is compact, by virtue of Remark 5.2.2, it follows that, for each $\lambda > 0$, J_λ is compact. Since u is bounded, we conclude that the positive trajectory of u is relatively compact and this completes the proof of the sufficiency. $\qquad\square$

We also need the following lemma:

LEMMA 5.3.4 *If* (H_A), (H_f), (c_2) *in* (H_c) *and* $(\overline{f}_1)$ *in* $(H_{\overline{f}})$ *are satisfied, then, for each* $\varphi \in X$, *the unique* C^0-*solution* u *of* (5.3.3) *is uniformly continuous on* $[-\tau, +\infty)$.

Proof. As u is continuous, it suffices to show that it is uniformly continuous from the right on $\mathbb{R}_+$. To this end, let us observe that, in view of (1.8.2), for each $t \in \mathbb{R}_+$ and each $h > 0$, we have

$$\|u(t+h) - u(t)\| \leq e^{-\omega t}\|u(h) - u(0)\| + \int_0^t e^{-\omega(t-s)}\|f(s+h, u_{s+h}) - f(s, u_s)\| \ ds.$$

Let

$$\mathcal{B}(\omega, \ell) = \left\{v \in X; \ \|v\|_X \leq \frac{m}{\omega - \ell}\right\}$$

and let $\gamma : \mathbb{R}_+ \to \mathbb{R}_+$ be the modulus of equi-uniform continuity of the family

$$\mathcal{F}(\omega, \ell) = \{t \mapsto f(t, v); \ v \in \mathcal{B}(\omega, \ell)\},$$

i.e.,

$$\gamma(h) = \sup\{\|f(t+\theta, v) - f(t, v)\|; \ v \in \mathcal{B}(\omega, \ell), \ t \in \mathbb{R}_+, \ \theta \in (0, h]\},$$

for each $h > 0$. Since, by the hypothesis $(\overline{f}_1)$ in $(H_{\overline{f}})$, $\mathcal{F}(\omega, \ell)$ is uniformly equicontinuous on $\mathbb{R}_+$, it follows that

$$\lim_{h \downarrow 0} \gamma(h) = 0.$$

200   *Delay Differential Evolutions Subjected to Nonlocal Initial Conditions*

Then

$$\|u(t+h)-u(t)\| \le e^{-\omega t}\|u(h)-u(0)\| + \int_0^t e^{-\omega(t-s)} \left[\gamma(h) + \ell\|u_{s+h} - u_s\|_x\right] ds$$

$$\le e^{-\omega t}\|u(h)-u(0)\| + \frac{1}{\omega}\gamma(h) + \int_0^t \ell e^{-\omega(t-s)}\|u_{s+h}-u_s\|_x \, ds$$

for each $t \in \mathbb{R}_+$ and $h > 0$. So, for each fixed $h > 0$, we are in the hypotheses of Lemma 5.3.1 with

$$\begin{cases} y(t) = \|u(t+h) - u(t)\|, \quad t \in [-\tau, +\infty), \\ \alpha_0 = \|u(h) - u(0)\|, \\ \lambda = \dfrac{\gamma(h)}{\omega}. \end{cases}$$

Accordingly, we get

$$\|u(t+h) - u(t)\| \le \left[\|u(h) - u(0)\| + \|u_h - u_0\|_x\right] e^{-(\omega-b)t} + \frac{\gamma(h)}{\omega - b}$$

$$\le 2\|u_h - u_0\|_x + \frac{\gamma(h)}{\omega - b}$$

for each $t \in \mathbb{R}_+$ and $h > 0$, where $b = \ell e^{\omega \tau}$. As

$$\lim_{h \downarrow 0} \gamma(h) = \lim_{h \downarrow 0} \|u_h - u_0\|_x = 0,$$

from the last inequality, we conclude that u is uniformly continuous from the right on $\mathbb{R}_+$ and this completes the proof. $\qquad \square$

5.4   Proof of Theorem 5.2.1

Proof. Let u be the C^0-solution of (5.1.1) and let $\varphi = g(u)$ be the initial history appearing in (5.3.3). From Lemmas 5.3.3 and 5.3.4, it follows that the trajectory $\{u(t);\ t \in [-\tau, +\infty)\}$ is relatively compact, which proves the first assertion in Theorem 5.2.1.

To complete the proof, it remains only to show that, under the additional hypothesis $(\overline{f}_2)$ in $(H_{\overline{f}})$, u is almost periodic. In view of Remark 5.1.1, it suffices to prove that for each sequence $(r_n)_n \in s_\infty$, there exists $\widetilde{u} \in C_b([-\tau, +\infty); X)$ such that we have

$$\lim_n u(t + r_n) = \widetilde{u}(t)$$

Almost Periodic Solutions 201

uniformly for $t \in \mathbb{R}_+$. To this end, let $(r_n)_n \in s_\infty$ be arbitrary. Since, by (4.1.2) in Theorem 4.1.1, u is bounded on $[-\tau, +\infty)$, it follows that the set $\mathcal{C} = \{u_t;\ t \in \mathbb{R}_+\}$ is bounded in $\mathcal{X}$. From $(\overline{f}_2)$ in $(H_{\overline{f}})$, it follows that there exists a function $\widetilde{f} : \mathbb{R}_+ \times \mathcal{C} \to X$ such that, on a subsequence at least,

$$\lim_n f(t + r_n, v) = \widetilde{f}(t, v)$$

uniformly for $t \in \mathbb{R}_+$ and $v \in \mathcal{C}$. In particular,

$$\lim_n f(s + r_n, u_s) = \widetilde{f}(s, u_s) \tag{5.4.1}$$

uniformly for $s \in \mathbb{R}_+$. Moreover, let us observe that, from the uniform continuity of u, it readily follows that the family of functions $\{s \mapsto u(s + r_n);\ n \in \mathbb{N}\}$ is equicontinuous from $[-\tau, +\infty)$ to X. In addition, in as much as u has relatively compact trajectory, we conclude that the family above has relatively compact cross sections in X. From Arzelà–Ascoli's Theorem 1.4.10, we deduce that the family of functions $\{s \mapsto u(s + r_n);\ n \in \mathbb{N}\}$ is relatively compact in $\widetilde{C}_b([-\tau, +\infty); X)$, i.e., in $C_b([-\tau, +\infty); X)$ endowed with the uniform convergence on compacta topology. So, we may assume with no loss of generality – by extracting another subsequence if necessary – that there exists $\widetilde{u} \in C_b([-\tau, +\infty); X)$ such that

$$\lim_n u(t + r_n) = \widetilde{u}(t) \tag{5.4.2}$$

uniformly for $t \in [-\tau, k]$, for $k = 0, 1, \ldots$. We will complete the proof by showing that, on that sub-subsequence, $(u(\cdot + r_n))_n$ converges in fact in the norm of $C_b(\mathbb{R}_+; X)$. To simplify the notation, relabeling if necessary, we can assume with no loss of generality that the sequence $(u(\cdot + r_n))_n$ has all the above mentioned properties.

By (5.4.2) and the continuity of f, it follows that

$$\lim_n f(s, u_{r_n+s}) = f(s, \widetilde{u}_s)$$

uniformly for $s \in [0, k]$, $k = 1, 2, \ldots$. Since $(s \mapsto f(s, u_{r_n+s}))_n$ is bounded (we recall that $(u_{r_n})_n$ does) from (5.4.1) and (5.4.2), we conclude that $\widetilde{u}$ is the unique C^0-solution of the problem

$$\begin{cases} \widetilde{u}'(t) \in A\widetilde{u}(t) + f(t, \widetilde{u}_t), & t \in \mathbb{R}_+, \\ \widetilde{u}(t) = \lim_n u(t + r_n), & t \in [-\tau, 0]. \end{cases}$$

At this point, let us observe that, by (1.8.2), we have

$$\|u(t+r_n) - \widetilde{u}(t)\| \leq e^{-\omega t}\|u(r_n) - \widetilde{u}(0)\| + \int_0^t e^{-\omega(t-s)}\|f(s + r_n, u_{s+r_n}) - f(s, \widetilde{u}_s)\|\, ds$$

$$\leq e^{-\omega t}\|u(r_n) - \widetilde{u}(0)\| + \int_0^t e^{-\omega(t-s)}\|f(s + r_n, u_{s+r_n}) - f(s + r_n, \widetilde{u}_s)\|\, ds$$

202 *Delay Differential Evolutions Subjected to Nonlocal Initial Conditions*

$$+e^{-\omega t} \int_0^t e^{\omega s} \|f(s + r_n, \widetilde{u}_s) - f(s, \widetilde{u}_s)\| \, ds$$

for each $n \in \mathbb{N}$ and $t \in \mathbb{R}_+$. From the compactness arguments mentioned above, we conclude that the sequence $(a_n)_n$ defined by

$$a_n = \|u(r_n) - \widetilde{u}(0)\|,$$

for $n \in \mathbb{N}$, satisfies

$$\lim_n a_n = 0.$$

Set $\mathcal{C} = \{\widetilde{u}_s; \ s \in \mathbb{R}_+\}$. Since $\widetilde{u} \in C_b([-\tau, +\infty); X)$, it follows that $\mathcal{C}$ is bounded in $\mathcal{X}$. Next, recalling that f satisfies $(\overline{f}_2)$ in $(H_{\overline{f}})$, it follows that the sequence $(b_n)_n$, defined by

$$b_n = \sup\{\|f(s + r_n, v) - f(s, v)\|; \ v \in \mathcal{C}, \ s \in \mathbb{R}_+\}$$

for $n \in \mathbb{N}$, satisfies

$$\lim_n b_n = 0.$$

Now, let us observe that the last inequality yields

$$\|u(t + r_n) - \widetilde{u}(t)\| \leq a_n e^{-\omega t} + \frac{1}{\omega}(1 - e^{-\omega t})b_n + \int_0^t \ell e^{-\omega(t-s)} \|u_{s+r_n} - \widetilde{u}_s\|_{\mathcal{X}} \, ds$$

$$\leq a_n e^{-\omega t} + \frac{1}{\omega} b_n + \int_0^t \ell e^{-\omega(t-s)} \|u_{s+r_n} - \widetilde{u}_s\|_{\mathcal{X}} \, ds$$

for each $n \in \mathbb{N}$ and $t \in \mathbb{R}_+$.

Let $n \in \mathbb{N}$ be arbitrary but fixed and let us define

$$\begin{cases} y(t) = \|u(t + r_n) - \widetilde{u}(t)\|, \quad t \in [-\tau, +\infty), \\[2mm] \alpha_0(t) = a_n, \\[2mm] \beta = \ell, \\[2mm] \lambda = \frac{1}{\omega} b_n. \end{cases}$$

We observe that we are in the hypotheses of Lemma 5.3.1, from which we deduce

$$\|u(t + r_n) - \widetilde{u}(t)\| \leq (a_n + \|u_{r_n} - u_0\|_{\mathcal{X}}) \, e^{-(\omega - b)t} + \frac{1}{\omega - b} b_n$$

$$\leq a_n + \|u_{r_n} - u_0\|_{\mathcal{X}} + \frac{1}{\omega - b} b_n,$$

for each $n \in \mathbb{N}$ and $t \in \mathbb{R}_+$, where $b = \ell e^{\omega \tau}$. Since

$$\lim_n a_n = \lim_n b_n = \lim_n \|u_{r_n} - u_0\|_{\mathcal{X}} = 0,$$

we conclude that

$$\lim_n \|u(t + r_n) - \widetilde{u}(t)\| = 0$$

uniformly for $t \in \mathbb{R}_+$ and this completes the proof. $\qquad\square$

Almost Periodic Solutions 203

REMARK 5.4.1 Under the hypotheses of Theorem 5.2.1, we have a stronger conclusion, i.e., for each sequence $(r_n)_n \in s_\infty$, there exists at least one function $\widetilde{u} \in C_b([-\tau, +\infty); X)$ and a subsequence of $(r_n)_n$, denoted for simplicity, again, by $(r_n)_n$, such that

$$\lim_n \|u(t + r_n) - \widetilde{u}(t)\| = 0 \qquad (5.4.3)$$

uniformly for $t \in [-\tau, +\infty)$. Indeed, from Theorem 5.2.1, we know that there exists $u^* \in C_b([-\tau, +\infty); X)$ such that

$$\lim_n \|u(t + r_n) - u^*(t)\| = 0$$

uniformly for $t \in \mathbb{R}_+$. Let us fix an arbitrary $k \in \mathbb{N}$ satisfying $r_k > \tau$. Then, for each $t \in [-\tau, +\infty)$, we have $r_k + t \in \mathbb{R}_+$. Therefore,

$$\lim_n \|u(t + r_k + r_n) - u^*(t + r_k)\| = 0$$

uniformly for $t \in [-\tau, +\infty)$. So, the function $\widetilde{u}$, defined by $\widetilde{u}(t) = u^*(t + r_k)$ for each $t \in [-\tau, +\infty)$, where $k \in \mathbb{N}$ is fixed as above, satisfies (5.4.3).

The above proof clearly suggests that $\widetilde{u}$ could not be unique.

5.5   The ω-limit set

DEFINITION 5.5.1 Let $\varphi \in X$ and let us assume that (H_A), (H_f), and (c_1) in (H_c) are satisfied. Let $u(\cdot, \varphi)$ be the unique C^0-solution of the problem (5.3.3) whose existence and uniqueness is ensured by Proposition 5.3.1. The *omega-limit set* of φ is the set of all elements $\varphi^* \in X$ for which there exists $(r_n)_n \in s_\infty$ such that

$$\lim_n u(r_n + t, \varphi) = \varphi^*(t),$$

uniformly for $t \in [-\tau, 0]$. We denote the omega-limit set of φ by $\overline{\omega}(\varphi)$.

For $t \in [-\tau, 0]$, we denote by

$$\overline{\omega}(\varphi)(t) = \{\varphi^*(t); \ \varphi^* \in \overline{\omega}(\varphi)\}.$$

DEFINITION 5.5.2 We say that $f : \mathbb{R}_+ \times X \to X$ is *asymptotically autonomous on bounded sets* if there exists $\widetilde{f} : X \to X$ such that, for each bounded set $V \subseteq X$, we have

$$\lim_{t \to +\infty} f(t, \varphi) = \widetilde{f}(\varphi)$$

uniformly for $\varphi \in V$.

REMARK 5.5.1 If f satisfies (H_f) and is asymptotically autonomous on bounded sets, then it satisfies $(H_{\widetilde{f}})$.

204   *Delay Differential Evolutions Subjected to Nonlocal Initial Conditions*

Let $f : \mathbb{R}_+ \times \mathcal{X} \to X$ be asymptotically autonomous on bounded sets, let $\widetilde{f} : \mathcal{X} \to X$ be the function whose existence is ensured by Definition 5.5.2 and let us assume that f satisfies (H_f). Then $\widetilde{f}$ is globally Lipschitz on $\mathcal{X}$ and therefore, for each $s \in \mathbb{R}_+$ and each $\varphi \in \mathcal{D}$, the problem

$$\begin{cases} \widetilde{u}'(t) \in A\widetilde{u}(t) + \widetilde{f}(\widetilde{u}_t), & t \in [s, +\infty), \\ \widetilde{u}(t) = \varphi(t - s), & t \in [-\tau + s, s] \end{cases} \qquad (5.5.1)$$

has a unique C^0-solution defined on $[-\tau + s, +\infty)$, denoted by $\widetilde{u}(\cdot, s, \varphi(\cdot - s))$. Since $\widetilde{f}$ does not depend on t and $u_t(\theta) = u(t + \theta)$ for each $\theta \in [-\tau, 0]$, the problem above is translation invariant, i.e.,

$$\widetilde{u}(t, s, \varphi(\cdot - s)) = \widetilde{u}(t - s, 0, \varphi).$$

Let $V \subseteq \mathcal{X}$ and let $\mathcal{S} : \{(t, s); 0 \le s \le t < +\infty\} \times V \to \overline{D(A)}$ be the operator defined by

$$\mathcal{S}(t, s, \varphi) = \widetilde{u}(t, s, \varphi(\cdot - s))$$

for each $(t, s, \varphi) \in \{(t, s); 0 \le s \le t < +\infty\} \times V$.

DEFINITION 5.5.3 The set $V \subseteq \mathcal{X}$ is called *invariant under the autonomous delay equation* $u' \in Au + \widetilde{f}(u_t)$ if for each $\psi \in V$, we have $\mathcal{S}(\cdot, s, \psi)_t \in V$ for each $t \ge s$.

THEOREM 5.5.1 *If (H_A), (H_f) and (c_2) in (H_c) are satisfied, $(I - A)^{-1}$ is compact, and f is asymptotically autonomous on bounded sets, then for each $\varphi \in \mathcal{X}$, the set $\overline{\omega}(\varphi)$ is nonempty and compact in $\mathcal{X}$. Moreover, $\overline{\omega}(\varphi)$ is independent of $\varphi \in \mathcal{X}$ and is invariant under the autonomous delay equation $u' \in Au + \widetilde{f}(u_t)$, where $\widetilde{f}$ is the unique function given by Definition 5.5.2.*

Proof. Let $\varphi \in \mathcal{D}$ be arbitrary but fixed. The fact that the set $\overline{\omega}(\varphi)$ is nonempty follows from Theorem 4.1.1, Remark 5.5.1 and Theorem 5.2.1. Indeed, let $(r_n)_n \in s_\infty$. Let $g : C_b(\mathbb{R}_+; \overline{D(A)}) \to \mathcal{D}$ be any fixed function satisfying (H_g). An example of such a function is

$$g(u)(t) = u(t + T)$$

for each $t \in [-\tau, 0]$, where $T > \tau$ is a given fixed number. See (i) in Remark 3.2.4. By virtue of Remark 5.4.1, we may assume with no loss of generality that there exists $\widetilde{u} \in C([-\tau, +\infty); \overline{D(A)})$ such that

$$\lim_n u(r_n + t, g(u)) = \widetilde{u}(t)$$

uniformly for $t \in [-\tau, +\infty)$. Let us observe that $\widetilde{\varphi} = \widetilde{u}_{|[-\tau,0]}$ belongs to $\overline{\omega}(\varphi)$ and thus the latter is nonempty. Indeed, if $\varphi = g(u)$, then the statement above holds true. If $\varphi \ne g(u)$, the assertion follows from the fact that u is globally asymptotically stable and the preceding remark.

Almost Periodic Solutions 205

To show that, for each $\varphi, \psi \in \mathcal{X}$, $\overline{\omega}(\varphi) = \overline{\omega}(\psi)$ it suffices to check out that for each $\varphi, \psi \in \mathcal{X}$, $\overline{\omega}(\varphi) \subseteq \overline{\omega}(\psi)$. So, let $\varphi^* \in \overline{\omega}(\varphi)$ be arbitrary. Then, there exists $(r_n)_n \in s_\infty$ such that

$$\lim_n u(r_n + t, \varphi) = \varphi^*(t)$$

uniformly for $t \in [-\tau, 0]$. By Theorem 5.2.1 and Remark 5.4.1, we know that the unique C^0-solution of (5.1.1), u, which coincides with the unique C^0-solution of (5.3.3) for $\varphi = g(u)$, i.e., $u = u(\cdot, g(u))$, with g any function satisfying (H_g), is almost periodic. This means that there exists $\widetilde{u} \in C([-\tau, +\infty); \overline{D(A)})$ such that, at least for a subsequence, we have

$$\lim_n u(r_n + t, g(u)) = \widetilde{u}(t)$$

uniformly for $t \in [-\tau, +\infty)$. On the other hand, by Theorem 1.12.3, it follows that the unique solution v of the problem

$$\begin{cases} v'(t) \in Av(t) + \widetilde{f}(v_t), & t \in \mathbb{R}_+, \\ v(t) = g(u)(t), & t \in [-\tau, 0] \end{cases}$$

is globally asymptotically stable. As $\lim_{t \to +\infty} f(t, w) = \widetilde{f}(w)$ uniformly for w in bounded subsets in $\mathcal{X}$ and $u(\cdot, g(u))$ is bounded, it readily follows that $u(\cdot, g(u))$ is uniformly asymptotically stable, which implies that

$$\lim_n \|u(r_n + t, g(u)) - u(r_n + t, \varphi)\| = 0$$

uniformly for $t \in [-\tau, +\infty)$. This shows that $\widetilde{u}_{|[-\tau, 0]} = \varphi^*$ and consequently $\varphi^* \in \overline{\omega}(g(u))$. Again from the uniform asymptotic stability of $u(\cdot, g(u))$, for the very same $(r_n)_n$ and every $\eta \in \mathcal{X}$, we get

$$\lim_n \|u(r_n + t, g(u)) - u(r_n + t, \eta)\| = 0$$

uniformly for $t \in [-\tau, +\infty)$. So, there exists $\lim_n u(r_n + t, \eta) = \eta^*(t)$ uniformly for $t \in [-\tau, 0]$ and consequently $\varphi^* = \eta^* \in \overline{\omega}(\psi)$.

By Theorem 5.2.1, $u(\cdot, g(u))$ is uniformly equicontinuous on $[-\tau, +\infty)$ and its trajectory is relatively compact. From Arzelà–Ascoli's Theorem 1.4.10, it follows that $\overline{\omega}(g(u))$ is relatively compact. Since the closedness of $\overline{\omega}(g(u))$ is obvious, this completes the proof. $\qquad\square$

If $\xi \in X$ and $C, D \subseteq X$, we denote by

$$\begin{cases} \text{dist}\,(\xi; C) = \inf_{\eta \in C} \text{dist}\,(\xi; \eta) = \inf_{\eta \in C} \|\xi - \eta\|, \\ \text{dist}\,(C; D) = \inf_{(\xi, \eta) \in C \times D} \text{dist}\,(\xi; \eta). \end{cases} \qquad (5.5.2)$$

THEOREM 5.5.2 *Let us assume that (H_A), (H_f), (c_2) in (H_c), $(H_{\widetilde{f}})$ are satisfied and $(I - A)^{-1}$ is compact. Let $C = \overline{\omega}(\eta)$, where $\eta \in \mathcal{X}$ is arbitrary but fixed and let V be any bounded subset in $\mathcal{X}$. Then*

$$\lim_{\substack{t \to +\infty \\ \varphi \in V}} \sup \text{dist}\,(u_t(\cdot, \varphi); C) = 0. \qquad (5.5.3)$$

206   Delay Differential Evolutions Subjected to Nonlocal Initial Conditions

Proof. We proceed by contradiction. So, if we assume that (5.5.3) does not hold, there exists $\varepsilon > 0$ such that, for each $n \in \mathbb{N}$, there exist $r_n \in [n, +\infty)$ and $\varphi_n \in V$ such that

$$\varepsilon \le \text{dist}\,(u_{r_n}(\cdot, \varphi_n); C). \tag{5.5.4}$$

Let $\varphi \in V$ be arbitrary and let us observe that

$$\text{dist}\,(u_{r_n}(\cdot, \varphi_n); C) \le \text{dist}\,(u_{r_n}(\cdot, \varphi_n); u_{r_n}(\cdot, \varphi)) + \text{dist}\,(u_{r_n}(\cdot, \varphi); C)$$

$$= \|u_{r_n}(\cdot, \varphi_n) - u_{r_n}(\cdot, \varphi)\| + \text{dist}\,(u_{r_n}(\cdot, \varphi); C). \tag{5.5.5}$$

At this point, as in the proof of Theorem 5.5.1, we consider an auxiliary function $g\colon C_b(\mathbb{R}_+; \overline{D(A)}) \to X$ satisfying (H_g). Now, by virtue of Theorems 5.2.1 and 5.5.1, on a subsequence at least, we successively have

$$\lim_n \text{dist}\,(u_{r_n}(\cdot, g(u)); C) = 0, \quad \lim_n \|u_{r_n}(\cdot, g(u)) - u_{r_n}(\cdot, \varphi)\|_X = 0$$

and

$$\lim_n \text{dist}\,(u_{r_n}(\cdot, \varphi); C) = 0. \tag{5.5.6}$$

But, on the other hand, from (1.8.2), we get

$$\|u(t, \varphi_n) - u(t, \varphi)\| \le e^{-\omega t}\|\varphi_n(0) - \varphi(0)\|$$

$$+ \int_0^t e^{-\omega(t-s)}\ell\|u(\cdot, \varphi_n)_s - u(\cdot, \varphi)_s\|_X\,ds$$

for each $n \in \mathbb{N}$ and $t \in \mathbb{R}_+$. Clearly, we are in the hypotheses of Lemma 5.3.1, with $\alpha_0 = \|\varphi_n(0) - \varphi(0)\|$, $\lambda = 0$ and $\ell > 0$, $\omega > 0$ and $\tau > 0$, as specified in (H_f), (H_g) and (c_2) in (H_c). Accordingly, we have (we recall that $b = \ell e^{\omega\tau}$)

$$\|u(t, \varphi_n) - u(t, \varphi)\| \le 2e^{-(\omega-b)t}\|u(\cdot, \varphi_n)_0 - u(\cdot, \varphi)_0\|_X \tag{5.5.7}$$

for each $n \in \mathbb{N}$ and $t \in \mathbb{R}_+$.

Next, let us observe that, by virtue of (H_A), (1.8.2) and (H_f), we have

$$\|u(t, \varphi_n)\| \le e^{-\omega t}\|\varphi_n(0)\| + \frac{m}{\omega} + \int_0^t \ell e^{-\omega(t-s)}\|u(\cdot, \varphi_n)_s\|_X\,ds$$

for each $n \in \mathbb{N}$ and $t \in \mathbb{R}_+$. Using once again Lemma 5.3.1 with $\alpha_0 = \|\varphi_n(0)\|$, $\lambda = \frac{m}{\omega}$ and $\ell > 0$, $\omega > 0$, $\tau \ge 0$ as before and $b = \ell e^{\omega\tau}$, we conclude that

$$\|u(t, \varphi_n)\| \le 2e^{-(\omega-b)t}\|u(\cdot, \varphi_n)_0\|_X + \frac{m}{\omega - b}$$

for each $n \in \mathbb{N}$ and $t \in \mathbb{R}_+$. Since V is bounded in X, from Proposition 5.3.1, if follows that $\{u(\cdot, \varphi_n); \ n \in \mathbb{N}\}$ is bounded in $C_b([-\tau, +\infty); X)$. Setting $t = r_n + s$ in (5.5.7) with $s \in [-\tau, 0]$ and $n \in \mathbb{N}$, $r_n > \tau$, taking into account

Almost Periodic Solutions 207

the last remark, the condition $b < \omega$ and passing to the limit for $n \to +\infty$, we conclude that

$$\lim_n \|u(r_n + s, \varphi_n) - u(r_n + s, \varphi)\| = 0,$$

uniformly for $s \in [-\tau, 0]$. But this relation, along with (5.5.5) and (5.5.6), contradicts (5.5.4). This contradiction can be eliminated only if (5.5.3) holds true and this completes the proof. $\square$

DEFINITION 5.5.4 Let $V \subseteq X$ be a nonempty set and let $S : \mathbb{R}_+ \times V \to \overline{D(A)}$. The mapping S is called *asymptotically compact* if for each sequence $((r_n, \varphi_n))_n$ in $\mathbb{R}_+ \times V$, with $(r_n)_n \in s_\infty$, there exists a subsequence $((r_{n_k}, \varphi_{n_k}))_k$ of $((r_n, \varphi_n))_n$ such that $(S(r_{n_k}, \varphi_{n_k}))_k$ is convergent.

THEOREM 5.5.3 *Let us assume that* (H_A), (H_f), (c_2) *in* (H_c), $(H_{\overline{f}})$ *are satisfied and* $(I - A)^{-1}$ *is compact. Let* $V \subseteq X$ *be a nonempty and bounded set. Then, the mapping* $S : \mathbb{R}_+ \times V \to \overline{D(A)}$, *defined by*

$$S(t, \varphi) = u(t, \varphi)$$

for each $(t, \varphi) \in \mathbb{R}_+ \times V$, *where* $u(\cdot, \varphi)$ *is the* C^0-*solution of the problem* (5.3.3), *is asymptotically compact.*

Proof. Let $V \subseteq X$ be bounded and let $((r_n, \varphi_n))_n$ be an arbitrary sequence in $\mathbb{R}_+ \times V$, with $\lim_n r_n = +\infty$. Let $\varphi \in V$ be arbitrary but fixed and let us observe that, by Theorem 5.5.1, there exist $u^* \in \overline{D(A)}$ and at least one subsequence of $(r_n)_n$, denoted for simplicity again by $(r_n)_n$, such that

$$\lim_n S(r_n, \varphi) = u^*.$$

Now, repeating the very same arguments as in the proof of Theorem 5.5.2, we conclude that $\lim_n \|S(r_n, \varphi_n) - S(r_n, \varphi)\| = 0$, which completes the proof. $\square$

5.6 The transport equation in one dimension

Let C_π be the space of all continuous and π-periodic functions from $\mathbb{R}$ to $\mathbb{R}$, endowed with the norm $\| \cdot \|_\pi$, defined by

$$\|u\|_\pi = \|u\|_{C([0,\pi];\mathbb{R})}$$

for each $u \in C_\pi$. Let $c > 0$, $\omega > 0$ and let $\mathcal{C}_\pi = C([-\tau, 0]; C_\pi)$. Furthermore, let $f : \mathbb{R}_+ \times \mathcal{C}_\pi \to C_\pi$ be a continuous function, $g : C_b(\mathbb{R}_+; C_\pi) \to \mathcal{C}_\pi$ a

208 *Delay Differential Evolutions Subjected to Nonlocal Initial Conditions*

nonexpansive mapping, and let us consider the following transport equation with delay:

$$\begin{cases} \dfrac{\partial u}{\partial t}(t,x) = -c\dfrac{\partial u}{\partial x}(t,x) - \omega u(t,x) + f(t,u_t)(x), & \text{in } \mathbb{R}_+ \times \mathbb{R}, \\[2mm] u(t,x) = u(t,x+\pi), & \text{in } \mathbb{R}_+ \times \mathbb{R}, \\[2mm] u(t,x) = g(u)(t)(x), & \text{in } [-\tau,0] \times \mathbb{R}. \end{cases} \quad (5.6.1)$$

THEOREM 5.6.1 *Let $c > 0$, $\tau \geq 0$, $\omega > 0$ and let $f : \mathbb{R}_+ \times \mathcal{C}_\pi \to C_\pi$ and $g : C_b(\mathbb{R}_+; C_\pi) \to \mathcal{C}_\pi$. We assume that*

(h_1) *there exist $\ell > 0$ and $m > 0$ such that*

$$\|f(t,v) - f(s,w)\|_\pi \leq \ell\, [|t-s| + \|v-w\|_{\mathcal{C}_\pi}],$$

$$\|f(t,v)\|_\pi \leq \ell\|v\|_{\mathcal{C}_\pi} + m$$

for each $t, s \in \mathbb{R}_+$ and each $v, w \in \mathcal{C}_\pi$

(h_2) *$\ell e^{\omega\tau} < \omega$*

(h_3) *for each bounded set $\mathcal{C} \subseteq \mathcal{C}_\pi$, the family $\{t \mapsto f(t,v);\ v \in \mathcal{C}\}$ is uniformly equicontinuous on $\mathbb{R}_+$*

(h_4) *for each bounded set $\mathcal{C} \subseteq \mathcal{C}_\pi$, the family $\{t \mapsto f(t,v);\ v \in \mathcal{C}\}$ is uniformly almost periodic in the sense of Definition 5.1.2*

(h_5) *there exists $m_0 \geq 0$ such that*

$$\|g(u)\|_{\mathcal{C}_\pi} \leq \|u\|_{C_b(\mathbb{R}_+; C_\pi)} + m_0$$

and there exists $a > 0$ such that

$$\|g(u) - g(v)\|_{\mathcal{C}_\pi} \leq \|u-v\|_{C_b([a,+\infty); C_\pi)}$$

for each $u, v \in C_b([-\tau,+\infty); C_\pi)$.

Then, the problem (5.6.1) has a unique almost periodic solution u. Moreover, u is globally asymptotically stable and its orbit is compact.

Proof. Let $B : D(B) \subseteq C_\pi \to C_\pi$ be defined by

$$\begin{cases} D(B) = \{u \in C_\pi;\ u' \in C_\pi\}, \\[2mm] Bu = -cu', \quad \text{for each} \quad u \in D(B). \end{cases}$$

Clearly, (5.6.1) can be written in the abstract form (5.1.1) with f and g as above and $A = B - \omega I$. By Vrabie [252, Problem 3.4, p. 74], we know that B generates a C_0-group of isometries $\{T(t);\ t \in \mathbb{R}\}$, given by

$$[T(t)\xi](x) = \xi(x - ct)$$

Almost Periodic Solutions 209

for each $\xi \in C_\pi$, each $x \in \mathbb{R}$ and each $t \in \mathbb{R}$. Thus A generates a C_0-semigroup of contractions $\{S(t);\ t \in \mathbb{R}_+\}$, with

$$S(t) = e^{-\omega t}T(t)$$

for each $t \in \mathbb{R}_+$. An appeal to the infinite dimensional version of the Arzelà–Ascoli Theorem 1.4.10 shows that, for each $k > 0$, the set

$$\{u \in D(A);\ \|u\|_\pi + \|Au\|_\pi \le k\}$$

is relatively compact in C_π. This clearly implies that $(I - A)^{-1}$ is compact. So, we are in the hypotheses of Theorem 5.2.1, from which the conclusion. $\square$

REMARK 5.6.1 We emphasize that although $(I - A)^{-1}$ is compact, A does not generate a compact semigroup. Indeed, as we have already seen, A generates a C_0-group on C_π, i.e., $\{e^{-\omega t}T(t);\ t \in \mathbb{R}\}$, where $\{T(t);\ t \in \mathbb{R}\}$ is defined as above. So, if we assume that the corresponding C_0-semigroup, i.e., $\{e^{-\omega t}T(t);\ t \in \mathbb{R}_+\}$ is compact, then it follows that $I = [e^{\omega t}T(-t)]\circ[e^{-\omega t}T(t)]$ is compact, which is impossible as long as C_π is infinite dimensional.

Finally, let us consider the problem

$$\begin{cases} \dfrac{\partial u}{\partial t}(t,x) = -c\dfrac{\partial u}{\partial x}(t,x) - \omega u(t,x) + f(t,u_t)(x), & \text{in } \mathbb{R}_+ \times \mathbb{R}, \\[2mm] u(t,x) = u(t,x+\pi), & \text{in } \mathbb{R}_+ \times \mathbb{R}, \\[2mm] u(t,x) = \varphi(t)(x), & \text{in } [-\tau,0] \times \mathbb{R}, \end{cases} \qquad (5.6.2)$$

where $\varphi \in \mathcal{C}_\pi$. Obviously, (5.6.2) can be written in the abstract form (5.4.1) with A and f as above. Theorems 5.5.1, 5.5.2 and 5.5.3 yield:

THEOREM 5.6.2 *Let $c > 0$, $\tau \ge 0$, $\omega > 0$, $f : \mathbb{R}_+ \times \mathcal{C}_\pi \to C_\pi$ and let us assume that $(h_1) \sim (h_4)$ in Theorem 5.6.1 are satisfied. Then, $\overline{\omega}(\eta)$ is nonempty, compact, and independent of $\eta \in \mathcal{C}_\pi$. Let V be a nonempty and bounded subset in $\mathcal{C}_\pi$ and let $C = \overline{\omega}(\eta)$ for some $\eta \in \mathcal{C}_\pi$. Then*

$$\lim_{t \to +\infty} \sup_{\varphi \in V} \operatorname{dist}(u(t,\varphi);C) = 0.$$

In addition, the mapping $\mathcal{S} : \mathbb{R}_+ \times V \to C_\pi$, defined by

$$\mathcal{S}(t,\varphi) = u(t,\varphi)$$

for each $(t,\varphi) \in \mathbb{R}_+ \times V$, where $u(\cdot,\varphi)$ is the C^0-solution of the problem (5.6.2), is asymptotically compact.

5.7 An application to the damped wave equation

Let Ω be a nonempty bounded and open subset in $\mathbb{R}^d$, $d \geq 1$, with C^1 boundary Σ, let $Q_+ = \mathbb{R}_+ \times \Omega$, $Q_\tau = [-\tau, 0] \times \Omega$, $\Sigma_+ = \mathbb{R}_+ \times \Sigma$, and let $\omega > 0$. Let $\mathcal{H}_0^1 = C([-\tau, 0]; H_0^1(\Omega))$, $\mathcal{X}_2 = C([-\tau, 0]; L^2(\Omega))$ and let us consider the following damped wave equation with delay, subjected to nonlocal initial conditions:

$$
\begin{cases}
\dfrac{\partial^2 u}{\partial t^2} = \Delta u - 2\omega \dfrac{\partial u}{\partial t} - \omega^2 u + h\left(t, u_t, \left(\dfrac{\partial u}{\partial t}\right)_t\right), & \text{in } Q_+, \\[2ex]
u(t, x) = 0, & \text{on } \Sigma_+, \\[2ex]
u(t, x) = \displaystyle\int_\tau^{+\infty} \alpha(s) u(t+s, x)\, ds + \psi_1(t)(x), & \text{in } Q_\tau, \\[2ex]
\dfrac{\partial u}{\partial t}(t, x) = \displaystyle\int_\tau^{+\infty} \mathcal{N}\left(s, u(t+s, x), \dfrac{\partial u}{\partial t}(t+s, x)\right) ds + \psi_2(t)(x), & \text{in } Q_\tau,
\end{cases}
\tag{5.7.1}
$$

where $h : \mathbb{R}_+ \times \mathcal{H}_0^1 \times \mathcal{X}_2 \to L^2(\Omega)$, $\alpha \in L^2(\mathbb{R}_+)$, $\mathcal{N} : \mathbb{R}_+ \times \mathbb{R} \times \mathbb{R} \to \mathbb{R}$ and the local parts of the initial data $\psi_1 \in \mathcal{H}_0^1$ and $\psi_2 \in \mathcal{X}_2$.

THEOREM 5.7.1 *Let Ω be a nonempty bounded and open subset in $\mathbb{R}^d$, $d \geq 1$, with C^1 boundary Σ, let $\tau \geq 0$, $\omega > 0$, let $\psi_1 \in \mathcal{H}_0^1$ and $\psi_2 \in \mathcal{X}_2$. Finally, let $h : \mathbb{R}_+ \times \mathcal{H}_0^1 \times \mathcal{X}_2 \to L^2(\Omega)$, $\alpha \in L^2(\mathbb{R}_+)$ and $\mathcal{N} : \mathbb{R}_+ \times \mathbb{R} \times \mathbb{R} \to \mathbb{R}$ be continuous functions satisfying*

(h_1) *there exists $\tilde{\ell} > 0$ such that*

$$\|h(t, w, y) - h(s, \widetilde{w}, \widetilde{y})\|_{L^2(\Omega)}$$

$$\leq \tilde{\ell} \left(|t - s| + \|w - \widetilde{w}\|_{\mathcal{H}_0^1} + \|y - \widetilde{y}\|_{\mathcal{X}_2}\right)$$

for each $t, s \in \mathbb{R}_+$, each $w, \widetilde{w}$ in $\mathcal{H}_0^1$ and each $y, \widetilde{y} \in \mathcal{X}_2$

(h_2) *there exists $m \geq 0$ such that*

$$\|h(t, 0, 0)\|_{L^2(\Omega)} \leq m$$

for each $t \in \mathbb{R}_+$

$(\overline{h}_1)$ *for each bounded set $\mathcal{C}$ in $\mathcal{H}_0^1 \times \mathcal{X}_2$, the family $\{t \mapsto h(t, w, y); \ (w, y) \in \mathcal{C}\}$ is uniformly equicontinuous on $\mathbb{R}_+$*

$(\overline{h}_2)$ *for each bounded set $\mathcal{C}$ in $\mathcal{H}_0^1 \times \mathcal{X}_2$, the family $\{t \mapsto h(t, w, y); \ (w, y) \in \mathcal{C}\}$ is uniformly almost periodic on $\mathbb{R}_+$ in the sense of Definition 5.1.2*

Almost Periodic Solutions

(n_1) *there exists a continuous and nonnegative function $\eta \in L^2(\mathbb{R}_+)$ such that*

$$|\mathcal{N}(t, u, v)| \leq \eta(t)(|u| + |v|),$$

for each $t \in \mathbb{R}_+$ and $u, v \in \mathbb{R}$

(n_2) *we have*

$$|\mathcal{N}(t, u, v) - \mathcal{N}(t, \widetilde{u}, \widetilde{v})| \leq \eta(t)(|u - \widetilde{u}| + |v - \widetilde{v}|),$$

for each $t \in \mathbb{R}_+$ and $u, \widetilde{u}, v, \widetilde{v} \in \mathbb{R}$, where η is given by (n_1).

Let λ_1 be the first eigenvalue of $-\Delta$ and let us assume that

(n_3)
$$\begin{cases} \|\eta\|_{L^2(\mathbb{R}_+)} \leq 1 \\ (1 + \lambda_1^{-1}\omega)\|\alpha\|_{L^2(\mathbb{R}_+)} + \lambda_1^{-1}(1 + \omega)\|\eta\|_{L^2(\mathbb{R}_+)} \leq 1 \end{cases}$$

(n_4) *there exists $c > 0$ such that $\alpha(t) = \eta(t) = 0$ for each $t \in [0, c]$.*

Let us assume, in addition, that

(c_1) $\ell = \widetilde{\ell}(1 + \omega\lambda_1^{-1})e^{\omega\tau} < \omega.$

Then, the problem (5.7.1) has a unique almost periodic solution. Moreover, u is globally asymptotically stable and its orbit is compact.

Proof. As we already have noted in Section 3.5, (5.7.1) can be equivalently rewritten as a first-order system of partial differential equations of the form

$$\begin{cases} \dfrac{\partial u}{\partial t}(t, x) = v(t, x) - \omega u(t, x), & \text{in } Q_+ \\[2mm] \dfrac{\partial v}{\partial t}(t, x) = \Delta u(t, x) - \omega v(t, x) + h\,(t, u_t, w_t)\,, & \text{in } Q_+ \\[2mm] u(t, x) = 0, & \text{on } \Sigma_+ \\[2mm] u(t, x) = \displaystyle\int_\tau^{+\infty} \alpha(s)u(t + s, x)\,ds + \psi_1(t)(x), & \text{in } Q_\tau, \\[2mm] v(t, x) = \displaystyle\int_\tau^{+\infty} \mathcal{M}\,(s, u(t + s, x), w(t + s, x))\,ds + \psi_3(t)(x), & \text{in } Q_\tau, \end{cases} \quad (5.7.2)$$

where $w = v - \omega u$, $\mathcal{M}(t, u, w) = \omega\alpha(t)u + \mathcal{N}(t, u, w)$ and $\psi_3 = \omega\psi_1 + \psi_2$, in the

product space $X = \begin{pmatrix} H_0^1(\Omega) \\ \times \\ L^2(\Omega) \end{pmatrix}$ endowed with the usual inner product

$$\left\langle \begin{pmatrix} u \\ v \end{pmatrix}, \begin{pmatrix} \widetilde{u} \\ \widetilde{v} \end{pmatrix} \right\rangle = \int_\Omega \nabla u(x) \cdot \nabla \widetilde{u}(x)\,dx + \int_\Omega v(x)\widetilde{v}(x)\,dx$$

212   *Delay Differential Evolutions Subjected to Nonlocal Initial Conditions*

for each $\left(\begin{array}{c} u \\ v \end{array}\right), \left(\begin{array}{c} \tilde{u} \\ \tilde{v} \end{array}\right) \in X$. Further, (5.7.2) can be rewritten as an abstract evolution equation subjected to nonlocal initial conditions of the form (4.1.1) in X, where the linear operator A is defined as in Section 3.5, while the functions $f : \mathbb{R}_+ \times X \to X$ and $g : C_b(\mathbb{R}_+; X) \to X$ are given by

$$ f\left(t, \left(\begin{array}{c} z \\ y \end{array}\right)\right) = \left(\begin{array}{c} 0 \\ h(t, z, y - \omega z) \end{array}\right) $$

for each $t \in \mathbb{R}_+$ and each $\left(\begin{array}{c} z \\ y \end{array}\right) \in X$ and, respectively, by

$$ \left[g\left(\begin{array}{c} u \\ v \end{array}\right)(t) \right](x) = \left(\begin{array}{c} \displaystyle\int_{\tau}^{+\infty} \alpha(s) u(t+s, x)\, ds + \psi_1(t)(x) \\[2ex] \displaystyle\int_{\tau}^{+\infty} \mathcal{M}\left(s, u(t+s, x), w(t+s, x)\right) ds + \psi_3(t)(x) \end{array}\right) $$

for each $\left(\begin{array}{c} u \\ v \end{array}\right) \in C_b(\mathbb{R}_+; X)$, each $t \in [-\tau, 0]$, and a.e. for $x \in \Omega$, with $w, \mathcal{M}$ and ψ_3 defined as above.

By Theorem 4.1.1, we deduce that the problem (5.7.1) has a unique bounded and globally asymptotically stable solution. To conclude the proof with the help of Theorem 5.2.1, we have merely to show that the operator $(I - A)^{-1}$ is compact, i.e., it maps the closed unit ball, $D_X(0,1)$, in X into a relatively compact subset in X. So, let $\left(\begin{array}{c} f_1 \\ f_2 \end{array}\right) \in X$ and let us consider the equation

$$ (I - A)\left(\begin{array}{c} u \\ v \end{array}\right) = \left(\begin{array}{c} f_1 \\ f_2 \end{array}\right) $$

which reduces to the following system

$$ \begin{cases} u + \omega u - v = f_1 \\ v + \omega v - \Delta u = f_2. \end{cases} \tag{5.7.3} $$

Multiplying both sides in the first equation by $1 + \omega$ and adding, side by side, the equation thus obtained to the second one, we get

$$ (1 + \omega)^2 u - \Delta u = (1 + \omega) f_1 + f_2. $$

So, by Theorem 1.9.1, the set

$$ \{ u \in H_0^1(\Omega); \ (1 + \omega)^2 u - \Delta u = (1 + \omega) f_1 + f_2, \ \|f_1\|_{H_0^1(\Omega)} + \|f_2\|_{L^2(\Omega)} \leq 1 \} $$

which is bounded in $H^2(\Omega)$, is relatively compact in $H_0^1(\Omega)$. Now, coming back to the first equation in (5.7.3), again by Theorem 1.9.1, we deduce that

$$ \{ v \in L^2(\Omega); \ v = (1 + \omega) u - f_1, \ \|f_1\|_{H_0^1(\Omega)} + \|f_2\|_{L^2(\Omega)} \leq 1 \} $$

Almost Periodic Solutions 213

being bounded in $H_0^1(\Omega)$ is relatively compact in $L^2(\Omega)$. So, the set

$$\left\{ \begin{pmatrix} u \\ v \end{pmatrix} \in X; \ \begin{pmatrix} u \\ v \end{pmatrix} \ \text{satisfies (5.7.3)}, \ \begin{pmatrix} f_1 \\ f_2 \end{pmatrix} \in D_X(0,1) \right\},$$

which coincides with $(I - A)^{-1} D_X(0,1)$, is relatively compact in X.

Hence, the conclusion follows from Theorem 5.2.1. $\qquad\square$

Next, let us consider the problem

$$\begin{cases} \dfrac{\partial^2 u}{\partial t^2} = \Delta u - 2\omega \dfrac{\partial u}{\partial t} - \omega^2 u + h\left(t, u_t, \left(\dfrac{\partial u}{\partial t}\right)_t\right), & \text{in } Q_+, \\[2mm] u(t,x) = 0, & \text{on } \Sigma_+, \\[2mm] u(t,x) = \varphi_1(t)(x), & \text{in } Q_\tau, \\[2mm] \dfrac{\partial u}{\partial t}(t,x) = \varphi_2(t)(x), & \text{in } Q_\tau, \end{cases} \qquad (5.7.4)$$

where $\varphi_1 \in \mathcal{H}_0^1$ and $\varphi_2 \in X_2$. Let us define X as before, and let us denote by
$\varphi = \begin{pmatrix} \varphi_1 \\ \varphi_2 \end{pmatrix} \in X$ and $\psi = \begin{pmatrix} \psi_1 \\ \psi_2 \end{pmatrix} \in X$.

From Theorems 5.5.1, 5.5.2 and 5.5.3, we deduce:

THEOREM 5.7.2 *Let Ω be a nonempty bounded and open subset in $\mathbb{R}^d$, $d \geq 1$, with C^1 boundary Σ, let $\tau \geq 0$, $\omega > 0$, let $\varphi_1 \in \mathcal{H}_0^1$, $\varphi_2 \in X_2$, and let us assume that $h : \mathbb{R}_+ \times \mathcal{H}_0^1 \times X_2 \to L^2(\Omega)$ satisfies all the hypotheses of Theorem 5.7.1. Then, the ω-limit set of ν, $\overline{\omega}(\nu)$, defined in Section 5.5, is nonempty, compact and independent of $\nu \in X$.*

Furthermore, let V be any bounded subset in X and let $C = \overline{\omega}(\psi)$ for some $\psi \in X$. Then

$$\lim_{t \to +\infty} \sup_{\varphi \in V} \text{dist}\,(u(t,\varphi); C) = 0.$$

Finally, the mapping $S : \mathbb{R}_+ \times V \to X$, defined by

$$S(t,\varphi) = u(t,\varphi)$$

for each $(t,\varphi) \in \mathbb{R}_+ \times V$, where $u(\cdot,\varphi)$ is the C^0-solution of the problem (5.7.4), is asymptotically compact.

5.8 Bibliographical notes and comments

Section 5.1 The class of functions in Definition 5.1.1 was introduced by Bochner [36] under the name *normal functions*. He also showed that it co-incides with the class of *uniformly almost periodic* functions introduced by

214   *Delay Differential Evolutions Subjected to Nonlocal Initial Conditions*

Bohr [39]. See Corduneanu [86, V, p. 31, Proposition 36, p. 56, Proposition 37, p. 58]. See also Zaidman [269, 2. Normal Functions, pp. 25–34] or Levitan and Zhikov [169, 1. Bochner's theorem, p. 4]. For the basic theory of almost periodic functions see the classical monograph of Corduneanu [84]. For a systematic study of almost periodic functions in Banach spaces the reader is referred to Zaidman [269], while for the classical theory of almost periodic solutions to functional equations, see Amerio and Prouse [8].

The existence problem of almost periodic solutions for semilinear functional differential equations with infinite delay was studied by Henríquez and Vásquez [140], Hernández and Pelicer [141], and Hino, Murakami and Yoshizawa [145]. The case of neutral equations was considered by Maqbul [179]. The relationship between the almost periodicity of the function u and that of $t \mapsto u_t$ was analyzed by Ghavidel [125, Sections 4.3, 4.4, pp. 76–87]. For results concerning weighted pseudo-almost periodic solutions to neutral functional differential equations, see Damak, Ezzimbi and Souden [94].

The existence part in Proposition 5.3.1 could be known, but we did not find any explicit reference to it in its present formulation. Lemma 5.3.3 is inspired by Thieme and Vrabie [243, Theorem 1.1] and is based on some compactness arguments developed in Vrabie [249] in the case of continuous forcing terms and by Mitidieri and Vrabie [185] in the general case.

Section 5.2. The main result concerning the existence of almost periodic solutions, i.e., Theorem 5.2.1 which is simply a slight generalization of a previous result of Vrabie [259], appears for the first time here.

Section 5.3 Lemma 5.3.1 and Proposition 5.3.1 are from Vrabie [259], Lemma 5.3.2 is a simple variant of a result of Brezis [50], while Lemmas 5.3.3 and 5.3.4 are also from Vrabie [259].

Section 5.4 The proof of Theorem 5.2.1 is very closely related to the proof of the main result in Vrabie [259].

Section 5.5 The main results referring to ω-limit sets and global attractors, i.e., Theorems 5.5.1, 5.5.2 and 5.5.3, are also immediate extensions of some similar results established in Vrabie [259].

Section 5.6 The example here included referring to the transport equation, as well as Theorems 5.6.1 and 5.6.2, are slightly more general than the corresponding ones in Vrabie [259].

Section 5.7 The example referring to the semilinear damped wave equation and Theorems 5.6.1 and 5.6.2 are new and appear for the first time here.

Chapter 6

Evolution Systems with Nonlocal Initial Conditions

Overview

In this chapter, we consider a class of nonlinear delay reaction–diffusion systems subjected to nonlocal initial conditions and, using the results in Chapter 4, we prove some sufficient conditions for the existence and global uniform asymptotic stability of C^0-solutions. Some applications to specific reaction–diffusion systems are included.

6.1 Single-valued perturbed systems

In this chapter, we prove two existence and uniform asymptotic stability results for C^0-solutions to the abstract nonlinear delay reaction–diffusion system with nonlocal initial data

$$
\begin{cases}
u'(t) \in Au(t) + F(t, u_t, v_t), & t \in \mathbb{R}_+, \\[4pt]
v'(t) \in Bv(t) + G(t, u_t, v_t), & t \in \mathbb{R}_+, \\[4pt]
u(t) = \displaystyle\sum_{i=1}^{n} \alpha_i u(t_i + t) + \psi_1(t), & t \in [-\tau, 0], \\[4pt]
v(t) = \displaystyle\sum_{i=1}^{p} \beta_i v(s_i + t) + \psi_2(t), & t \in [-\tau, 0].
\end{cases}
\tag{6.1.1}
$$

Here, X, Y are Banach spaces, $A : D(A) \subseteq X \rightsquigarrow X$, $B : D(B) \subseteq Y \rightsquigarrow Y$ are m-dissipative operators, and $\tau \geq 0$. For the simplicity of writing, we denote by $\| \cdot \|$ the norms on both X and Y, by

$$
\mathfrak{X} = C([-\tau, 0]; X), \quad \mathfrak{Y} = C([-\tau, 0]; Y)
\tag{6.1.2}
$$

and also by

$$
\mathfrak{D} = \{\varphi \in \mathfrak{X}; \ \varphi(0) \in \overline{D(A)}\}, \quad \mathcal{E} = \{\psi \in \mathfrak{Y}; \ \psi(0) \in \overline{D(B)}\}.
\tag{6.1.3}
$$

215

216   *Delay Differential Evolutions Subjected to Nonlocal Initial Conditions*

Furthermore, $F : \mathbb{R}_+ \times \mathcal{X} \times \mathcal{Y} \to X$ and $G : \mathbb{R}_+ \times \mathcal{X} \times \mathcal{Y} \to Y$ are continuous, $\psi_1 \in \mathcal{D}$ and $\psi_2 \in \mathcal{E}$, while $(t_i)_{i=1}^n$, $(s_i)_{i=1}^p$, $(\alpha_i)_{i=1}^n$ and $(\beta_i)_{i=1}^p$ are four systems of points satisfying: $0 < t_1 < t_2 < \cdots < t_n$, $0 < s_1 < s_2 < \cdots < s_p$, $\alpha_i \in [-1, 1]$ for $i = 1, 2, \ldots, n$ and $\beta_i \in [-1, 1]$ for $i = 1, 2, \ldots, p$.

REMARK 6.1.1 We notice that the problem (6.1.1) contains, as particular cases, various reaction–diffusion systems. For instance, an important specific case refers to a reaction–diffusion system subjected to T-periodic conditions, i.e.,

$$\begin{cases} u'(t) \in Au(t) + F(t, u_t, v_t), & t \in \mathbb{R}_+, \\ v'(t) \in Bv(t) + G(t, u_t, v_t), & t \in \mathbb{R}_+, \\ u(t) = u(t+T), & t \in [-\tau, 0], \\ v(t) = v(t+T), & t \in [-\tau, 0]. \end{cases} \tag{6.1.4}$$

It is easy to see that this case corresponds to the particular choice: $n = 1$, $t_1 = T$, $\alpha_1 = 1$, $p = 1$, $s_1 = T$, $\beta_1 = 1$, $\psi_1 \equiv 0$ and $\psi_2 \equiv 0$. If, in addition, both F and G are T-periodic with respect to their first argument, then the solution of (6.1.4) – if any – is necessarily T-periodic.

We would like to add that, trying to keep the right balance between the clarity of presentation and the generality of the hypotheses, we have decided to confine ourselves only to these simple types of nonlocal initial conditions. First, they are general enough to handle many relevant specific cases such as T-periodic or T-anti-periodic, and sufficiently simple not to bring into discussion cumbersome technicalities that reveal almost nothing about the main idea of the proof. In fact, from a heuristic point of view, both this simple case and the general case – to be briefly discussed a little bit later – are completely similar, although at a first glance they look rather different.

DEFINITION 6.1.1 By a C^0-*solution* of (6.1.1) we mean a continuous function $(u, v) : [-\tau, +\infty) \to X \times Y$ satisfying both $u(t) = \sum_{i=1}^n \alpha_i u(t_i + t) + \psi_1(t)$ and

$v(t) = \sum_{i=1}^p \beta_i v(s_i + t) + \psi_2(t)$ for each $t \in [-\tau, 0]$ and, for each $T > 0$, (u, v) is a C^0-solution on $[0, T]$ in the sense of Definition 1.8.2 in the space $X \times Y$, for

$$\begin{cases} u'(t) \in Au(t) + f(t), & t \in \mathbb{R}_+, \\ v'(t) \in Bv(t) + g(t), & t \in \mathbb{R}_+, \end{cases}$$

where $f(t) = F(t, u_t, v_t)$ and $g(t) = G(t, u_t, v_t)$ for $t \in [0, T]$.

6.2 The main result

The assumptions we need in what follows are listed below.

(H_A) The operator $A : D(A) \subseteq X \rightsquigarrow X$ satisfies

(A_1) $0 \in D(A)$, $0 \in A0$ and A is ω-m-dissipative for some $\omega > 0$

(A_5) $\overline{D(A)}$ is a linear subspace in X.

(H_B) The operator $B : D(B) \subseteq Y \rightsquigarrow Y$ satisfies

(B_1) $0 \in D(B)$, $0 \in B0$ and B is γ-m-dissipative for some $\gamma > 0$

(B_2) B generates a compact semigroup

(B_5) $\overline{D(B)}$ is a linear subspace in Y.

(H_F) The function $F : \mathbb{R}_+ \times X \times Y \to X$ is continuous and satisfies

(F_1) there exists $\ell > 0$ such that

$$\|F(t, u, v) - F(t, \widetilde{u}, \widetilde{v})\| \le \ell \max\{\|u - \widetilde{u}\|_X, \|v - \widetilde{v}\|_Y\}$$

for all $(t, u, v), (t, \widetilde{u}, \widetilde{v}) \in \mathbb{R}_+ \times X \times Y$

(F_2) there exists $m > 0$ such that $\|F(t, u, v)\| \le \ell \|u\|_X + m$ for each $(t, u, v) \in \mathbb{R}_+ \times X \times Y$, where ℓ is given by (F_1).

(H_G) The function $G : \mathbb{R}_+ \times X \times Y \to Y$ is continuous and satisfies

(G_1) with ℓ and m given by (F_1) and (F_2), we have

$$\|G(t, u, v)\| \le \ell \max\{\|u\|_X, \|v\|_Y\} + m$$

for each $(t, u, v) \in \mathbb{R}_+ \times X \times Y$

(G_2) for each $T > 0$, the family of functions $\{G(t, \cdot, \cdot); \ t \in [0, T]\}$ is uniformly equicontinuous on $X \times Y$, i.e., for each $\varepsilon > 0$ there exists $\eta(\varepsilon) > 0$ such that

$$\|G(t, u, v) - G(t, \widetilde{u}, \widetilde{v})\| \le \varepsilon$$

for all $(t, u, v), (t, \widetilde{u}, \widetilde{v}) \in \mathbb{R}_+ \times X \times Y$ satisfying $\|u - \widetilde{u}\|_X \le \eta(\varepsilon)$ and $\|v - \widetilde{v}\|_Y \le \eta(\varepsilon)$.

(H_c) The constants ℓ and $\delta = \min\{\omega, \gamma\}$ satisfy $\ell < \delta$.

(H_p) The families $(t_i)_{i=1}^n$ and $(\alpha_i)_{i=1}^n$ are such that $\tau < t_1 < t_2 < \cdots < t_n$ and $\sum_{i=1}^n |\alpha_i| \le 1$.

218   Delay Differential Evolutions Subjected to Nonlocal Initial Conditions

(H_q) The families $(s_i)_{i=1}^{p}$ and $(\beta_i)_{i=1}^{p}$ are such that $\tau < s_1 < s_2 < \cdots < s_p$ and $\sum_{i=1}^{p} |\beta_i| \le 1$.

(H_ψ) $\psi_1 \in \mathcal{D}$ and $\psi_2 \in \mathcal{E}$.

REMARK 6.2.1 Roughly speaking, the meaning of the conditions $\tau < t_1$ and $\tau < s_1$ in (H_p), and respectively in (H_q), is that we do operate the necessary initial nonlocal measurements starting at the moments t_1 for u and s_1 for v, only after a waiting period strictly greater than the delay τ. This condition is pretty natural as the simplest case when $\tau = 0$, $t_1 = s_1 = T$ and $\alpha_1 = \beta_1 = 1$ shows, when we obtain the classical T-periodic condition.

The first main result in this chapter is:

THEOREM 6.2.1 *If (H_A), (H_B), (H_F), (H_G), (H_c), (H_p), (H_q) and (H_ψ) are satisfied, then (6.1.1) has at least one C^0-solution,*

$$(u, v) \in C_b([-\tau, +\infty); X) \times C_b([-\tau, +\infty); Y)$$

with $(u(t), v(t)) \in \overline{D(A)} \times \overline{D(B)}$ for each $t \in \mathbb{R}_+$ and satisfying

$$\begin{cases} \|u\|_{C_b([-\tau,+\infty);X)} \le \dfrac{m}{\omega - \ell} + \left[\dfrac{\omega}{\omega - \ell} \left(\dfrac{1}{e^{\omega a} - 1} + \dfrac{\ell}{\omega} \right) + 1 \right] \cdot m, \\[4mm] \|v\|_{C_b([-\tau,+\infty);Y)} \le \dfrac{m}{\gamma - \ell} + \left[\dfrac{\gamma}{\gamma - \ell} \left(\dfrac{1}{e^{\gamma a} - 1} + \dfrac{\ell}{\gamma} \right) + 1 \right] \cdot m, \end{cases} \tag{6.2.1}$$

where

$$a = \min\{t_1 - \tau, s_1 - \tau\} > 0.$$

If, in addition, G satisfies

$$\|G(t, u, v) - G(t, \widetilde{u}, \widetilde{v})\| \le \ell \max\left\{ \|u - \widetilde{u}\|_X, \|v - \widetilde{v}\|_Y \right\} \tag{6.2.2}$$

for all $(t, u, v), (t, \widetilde{u}, \widetilde{v}) \in \mathbb{R}_+ \times C([-\tau, 0]; \overline{D(A)}) \times C([-\tau, 0]; \overline{D(B)})$, then the C^0-solution of (6.1.1) is unique. Finally, if both F and G are Lipschitz with respect to all three variables with Lipschitz constant ℓ, the unique C^0-solution of (6.1.1) is globally asymptotically stable.

As far as the nondelayed case is concerned, i.e., the case in which $\tau = 0$, we consider the system

$$\begin{cases} u'(t) \in Au(t) + F(t, u(t), v(t)), & t \in \mathbb{R}_+, \\[2mm] v'(t) \in Bv(t) + G(t, u(t), v(t)), & t \in \mathbb{R}_+, \\[2mm] u(0) = \displaystyle\sum_{i=1}^{n} \alpha_i u(t_i) + \psi_1, \\[2mm] v(0) = \displaystyle\sum_{i=1}^{p} \beta_i v(s_i) + \psi_2. \end{cases} \tag{6.2.3}$$

Evolution Systems with Nonlocal Initial Conditions 219

In this specific case, $\mathcal{X}$ and $\mathcal{Y}$ reduce to X and Y respectively, while $\mathcal{D}$ and $\mathcal{E}$ reduce to $\overline{D(A)}$ and $\overline{D(B)}$ respectively. So, we need to reformulate the hypotheses (H_F), (H_G) and (H_ψ) as follows:

$(H_F^{[\tau=0]})$ The function $F : \mathbb{R}_+ \times X \times Y \to X$ is continuous and satisfies

$(F_1^{[\tau=0]})$ there exists $\ell > 0$ such that

$$\|F(t, u, v) - F(t, \widetilde{u}, \widetilde{v})\| \leq \ell \max\{\|u - \widetilde{u}\|, \|v - \widetilde{v}\|\}$$

for each $(t, u, v), (t, \widetilde{u}, \widetilde{v}) \in \mathbb{R}_+ \times X \times Y$

$(F_2^{[\tau=0]})$ there exists $m > 0$ such that

$$\|F(t, u, v)\| \leq \ell \|u\| + m$$

for each $(t, u, v) \in \mathbb{R}_+ \times X \times Y$, where ℓ is given by $(F_1^{[\tau=0]})$.

$(H_G^{[\tau=0]})$ The function $G : \mathbb{R}_+ \times X \times Y \to Y$ is continuous and satisfies

$(G_1^{[\tau=0]})$ with ℓ and m given by (F_1) and (F_2), we have

$$\|G(t, u, v)\| \leq \ell \max\{\|u\|, \|v\|\} + m$$

for each $(t, u, v) \in \mathbb{R}_+ \times X \times Y$

$(G_2^{[\tau=0]})$ for each $T > 0$, the family of functions $\{G(t, \cdot, \cdot); \ t \in [0, T]\}$ is uniformly equicontinuous on $X \times Y$, i.e., for each $\varepsilon > 0$ there exists $\eta(\varepsilon) > 0$ such that

$$\|G(t, u, v) - G(t, \widetilde{u}, \widetilde{v})\| \leq \varepsilon$$

for each $(t, u, v), (t, \widetilde{u}, \widetilde{v}) \in [0, T] \times X \times Y$ satisfying $\|u - \widetilde{u}\| \leq \eta(\varepsilon)$ and $\|v - \widetilde{v}\| \leq \eta(\varepsilon)$.

$(H_\psi^{[\tau=0]})$ $\psi_1 \in \overline{D(A)}$ and $\psi_2 \in \overline{D(B)}$.

THEOREM 6.2.2 *If (H_A), (H_B), $(H_F^{[\tau=0]})$, $(H_G^{[\tau=0]})$, (H_c), (H_p), (H_q) and $(H_\psi^{[\tau=0]})$ are satisfied, then (6.2.3) has at least one C^0-solution,*

$$(u, v) \in C_b(\mathbb{R}_+; \overline{D(A)}) \times C_b(\mathbb{R}_+; \overline{D(B)}),$$

satisfying

$$\begin{cases} \|u\|_{C_b(\mathbb{R}_+;X)} \leq \dfrac{m}{\omega - \ell} + \left[\dfrac{\omega}{\omega - \ell} \left(\dfrac{1}{e^{\omega a} - 1} + \dfrac{\ell}{\omega} \right) + 1 \right] \cdot m \\[4mm] \|v\|_{C_b(\mathbb{R}_+;Y)} \leq \dfrac{m}{\gamma - \ell} + \left[\dfrac{\gamma}{\gamma - \ell} \left(\dfrac{1}{e^{\gamma a} - 1} + \dfrac{\ell}{\gamma} \right) + 1 \right] \cdot m, \end{cases} \qquad (6.2.4)$$

220 *Delay Differential Evolutions Subjected to Nonlocal Initial Conditions*

where
$$a = \min\{t_1 - \tau, s_1 - \tau\} > 0.$$

If, in addition, G satisfies

$$\|G(t, u, v) - G(t, \widetilde{u}, \widetilde{v})\| \le \ell \max\{\|u - \widetilde{u}\|, \|v - \widetilde{v}\|\}$$

for all $(t, u, v), (t, \widetilde{u}, \widetilde{v}) \in \mathbb{R}_+ \times X \times Y$, *then the C^0-solution of (6.2.4) is unique. Finally, if both F and G are Lipschitz with respect to all three variables with Lipschitz constant ℓ, the unique C^0-solution of (6.2.4) is globally asymptotically stable.*

REMARK 6.2.2 One may easily see that, from Theorem 6.2.1, we can obtain as simple consequences, existence and uniqueness results referring to various systems whose forcing terms are either of the form $F(t, u_t, v(t))$ and $G(t, u_t, v(t))$ or of the form $F(t, u(t), v_t)$ and $G(t, u(t), v_t)$ with the nonlocal initial conditions accordingly defined.

6.3 The idea of the proof

We will use a fixed-point argument. Namely, let (u, v) be arbitrary but fixed in $C_b([-\tau, +\infty); X) \times C_b([-\tau, +\infty); Y)$ and let us consider the auxiliary problem

$$\begin{cases} \widetilde{u}'(t) \in A\widetilde{u}(t) + F(t, \widetilde{u}_t, \widetilde{v}_t), & t \in \mathbb{R}_+, \\ \widetilde{v}'(t) \in B\widetilde{v}(t) + G(t, u_t, v_t), & t \in \mathbb{R}_+, \\ \widetilde{u}(t) = \sum_{i=1}^{n} \alpha_i \widetilde{u}(t_i + t) + \psi_1(t), & t \in [-\tau, 0], \\ \widetilde{v}(t) = \sum_{i=1}^{p} \beta_i \widetilde{v}(s_i + t) + \psi_2(t), & t \in [-\tau, 0]. \end{cases} \qquad (6.3.1)$$

By Lemma 4.2.1, it follows that the problem

$$\begin{cases} \widetilde{v}'(t) \in B\widetilde{v}(t) + G(t, u_t, v_t), & t \in \mathbb{R}_+, \\ \widetilde{v}(t) = \sum_{i=1}^{p} \beta_i \widetilde{v}(s_i + t) + \psi_2(t), & t \in [-\tau, 0] \end{cases}$$

has a unique C^0-solution, $\widetilde{v} \in C_b([-\tau, +\infty); Y) \cap C_b(\mathbb{R}_+; \overline{D(B)})$, satisfying

$$\|\widetilde{v}\|_{C_b([-\tau, +\infty); Y)} \le \frac{e^{\delta a}}{e^{\delta a} - 1} m_2 + \frac{1}{\delta} \|G(\cdot, u_{(\cdot)}, v_{(\cdot)})\|_{C_b(\mathbb{R}_+; Y)}, \qquad (6.3.2)$$

Evolution Systems with Nonlocal Initial Conditions 221

where $\delta = \min\{\omega, \gamma\}$ and $m_2 = \|\psi_2\|_Y$. Next we consider the problem

$$
\begin{cases}
\widetilde{u}'(t) \in A\widetilde{u}(t) + F(t, \widetilde{u}_t, \widetilde{v}_t), & t \in \mathbb{R}_+, \\
\widetilde{u}(t) = \sum_{i=1}^{n} \alpha_i \widetilde{u}(t_i + t) + \psi_1(t), & t \in [-\tau, 0].
\end{cases}
\tag{6.3.3}
$$

From Theorem 4.1.1, we deduce that the problem (6.3.3) has a unique C^0-solution, $\widetilde{u} \in C_b([-\tau, +\infty); X) \cap C_b(\mathbb{R}_+; \overline{D(A)})$, satisfying

$$
\|\widetilde{u}\|_{C_b([-\tau, +\infty); X)} \leq \frac{m}{\delta - \ell} + \left[\frac{\delta}{\delta - \ell} \left(\frac{1}{e^{\delta a} - 1} + \frac{\ell}{\delta} \right) + 1 \right] \cdot m_1,
\tag{6.3.4}
$$

where $m_1 = \|\psi_1\|_X$. From (6.3.2) and (6.3.4), it readily follows that, for each pair (u, v) belonging to the set $C_b([-\tau, +\infty); X) \times C_b([-\tau, +\infty); Y)$ the pair $(\widetilde{u}, \widetilde{v})$, defined as above, belongs to $C_b([-\tau, +\infty); X) \times C_b([-\tau, +\infty); Y)$. So, we can define the operator Γ from $\widetilde{C}_b([-\tau, +\infty); X) \times \widetilde{C}_b([-\tau, +\infty); Y)$ into itself by

$$
\Gamma(u, v) = (\widetilde{u}, \widetilde{v}),
\tag{6.3.5}
$$

where $(\widetilde{u}, \widetilde{v})$ is the unique C^0-solution of the problem (6.3.1). We will show that Γ maps a suitably defined nonempty, closed, bounded and convex subset $\mathcal{C}$ in $\widetilde{C}_b([-\tau, +\infty); X) \times \widetilde{C}_b([-\tau, +\infty); Y)$ into itself, is continuous and compact. Thus, by the Tychonoff Fixed-Point Theorem 1.4.5, it will follow that Γ has a fixed point which is a C^0-solution of the problem (6.3.1).

6.4 An auxiliary lemma

We will carry out the proposed plan with the help of the next lemmas. From Lemma 4.2.1, it follows that, for each $(f, h) \in C_b(\mathbb{R}_+; X) \times C_b(\mathbb{R}_+; Y)$, the decoupled system

$$
\begin{cases}
u'(t) \in Au(t) + f(t), & t \in \mathbb{R}_+, \\
v'(t) \in Bv(t) + h(t), & t \in \mathbb{R}_+, \\
u(t) = \sum_{i=1}^{n} \alpha_i u(t_i + t) + \psi_1(t), & t \in [-\tau, 0], \\
v(t) = \sum_{i=1}^{p} \beta_i v(s_i + t) + \psi_2(t), & t \in [-\tau, 0],
\end{cases}
\tag{6.4.1}
$$

has a unique C^0-solution, $(u^f, v^h) \in C_b([-\tau, +\infty); X) \times C_b([-\tau, +\infty); Y)$. The main point in the lemma below is a precise evaluation of the Lipschitz constant of the operator $(f, h) \mapsto (u^f, v^h)$ with respect to the max-norm on both domain and range.

222 *Delay Differential Evolutions Subjected to Nonlocal Initial Conditions*

LEMMA 6.4.1 *If (H_A), (B_1) and (B_5) in (H_B), (H_p), (H_q) and (H_ψ) are satisfied then, for each $(f, h) \in L^\infty(\mathbb{R}_+; X) \times L^\infty(\mathbb{R}_+; Y)$, the problem (6.4.1) has a unique C^0-solution $(u^f, v^h) \in C_b([-\tau, +\infty); X) \times C_b([-\tau, +\infty); Y)$. In addition, the mapping $(f, h) \mapsto (u^f, v^h)$ is Lipschitz continuous from $L^\infty(\mathbb{R}_+; X) \times L^\infty(\mathbb{R}_+; Y)$ to $C_b([-\tau, +\infty); X) \times C_b([-\tau, +\infty); Y)$, with Lipschitz constant $L = \delta^{-1}$, where $\delta = \min\{\omega, \gamma\} > 0$, both domain and range being endowed with the max-norm of the corresponding factors.*

The proof of Lemma 6.4.1 is essentially based on the following remark.

REMARK 6.4.1 Let X, Y be two Banach spaces and let $Z = X \times Y$. Endowed with the max-norm,

$$\|z\| = \|(u, v)\| = \max\{\|u\|, \|v\|\},$$

for each $z = (u, v) \in Z$, Z is a Banach space. Then, for each $z = (u, v)$ and $w = (f, h)$ in Z, the semi-inner product $[z, w]_+$, i.e., the right directional derivative of the max-norm satisfies

$$[(u, v), (f, h)]_+ \leq \max\{[u, f]_+, [v, h]_+\}. \tag{6.4.2}$$

Indeed, let $\lambda > 0$, $u, f \in X$ and $v, h \in Y$ be arbitrary. We distinguish between three cases.

Case 1. If $\|u\| > \|v\|$, then

$$[(u, v), (f, h)]_+ = \lim_{\lambda \downarrow 0} \frac{\|(u, v) + \lambda(f, h)\| - \|(u, v)\|}{\lambda}$$

$$= \lim_{\lambda \downarrow 0} \frac{\max\{\|u + \lambda f\|, \|v + \lambda h\|\} - \max\{\|u\|, \|v\|\}}{\lambda}$$

$$= \lim_{\lambda \downarrow 0} \frac{\|u + \lambda f\| - \|u\|}{\lambda} = [u, f]_+ \leq \max\{[u, f]_+, [v, h]_+\},$$

simply because for $\lambda > 0$, small enough, we have $\|u + \lambda f\| > \|v + \lambda h\|$.

Case 2. If $\|u\| < \|v\|$, using the very same arguments, we get

$$[(u, v), (f, h)]_+ = [v, h]_+ \leq \max\{[u, f]_+, [v, h]_+\}.$$

Case 3. If $\|u\| = \|v\|$, let us denote by $\alpha = \|u\| = \|v\|$.

$$[(u, v), (f, h)]_+ = \lim_{\lambda \downarrow 0} \frac{\max\{\|u + \lambda f\|, \|v + \lambda h\|\} - \alpha}{\lambda}$$

$$= \lim_{\lambda \downarrow 0} \frac{\max\{\|u + \lambda f\| - \alpha, \|v + \lambda h\| - \alpha\}}{\lambda}$$

$$= \lim_{\lambda \downarrow 0} \max\left\{\frac{\|u + \lambda f\| - \|u\|}{\lambda}, \frac{\|v + \lambda h\| - \|v\|}{\lambda}\right\}$$

Evolution Systems with Nonlocal Initial Conditions

$$= \max \left\{ \lim_{\lambda \downarrow 0} \frac{\|u + \lambda f\| - \|u\|}{\lambda}, \lim_{\lambda \downarrow 0} \frac{\|v + \lambda h\| - \|v\|}{\lambda} \right\}$$

$$= \max\{[\, u, f\,]_+, [\, v, h\,]_+\}.$$

Here lim and max commute because all the three limits involved exist and are finite.

Indeed, if $[\, u, f\,]_+ > [\, v, h\,]_+$, then for $\lambda > 0$ small enough, we have

$$\max \left\{ \frac{\|u + \lambda f\| - \|u\|}{\lambda}, \frac{\|v + \lambda h\| - \|v\|}{\lambda} \right\} = \frac{\|u + \lambda f\| - \|u\|}{\lambda}.$$

Analogously, if $[\, u, f\,]_+ < [\, v, h\,]_+$, for $\lambda > 0$ small enough, we have

$$\max \left\{ \frac{\|u + \lambda f\| - \|u\|}{\lambda}, \frac{\|v + \lambda h\| - \|v\|}{\lambda} \right\} = \frac{\|v + \lambda h\| - \|v\|}{\lambda}.$$

If $[\, u, f\,]_+ = [\, v, h\,]_+$ we proceed by contradiction.
Namely, let us assume that

$$\alpha = [\, (u, v), (f, h)\,]_+ \neq [\, u, f\,]_+ = [\, v, h\,]_+.$$

Then there exists a neighborhood V of α such that for $\lambda > 0$ small enough none of the two ratios

$$\frac{\|u + \lambda f\| - \|u\|}{\lambda}, \frac{\|v + \lambda h\| - \|v\|}{\lambda}$$

belong to V and thus

$$\max \left\{ \frac{\|u + \lambda f\| - \|u\|}{\lambda}, \frac{\|v + \lambda h\| - \|v\|}{\lambda} \right\}$$

does not belong to V, thereby contradicting the definition of α.

So, in each one of the three possible cases, (6.4.2) holds true.

We can now pass to the proof of Lemma 6.4.1.

Proof. We will use Remark 6.4.1 and Lemma 4.2.1. First, we write (6.4.1) as an evolution equation subjected to nonlocal initial conditions in the product space $Z = X \times Y$, endowed with the max-norm, i.e., $\|(u, v)\| = \max\{\|u\|, \|v\|\}$. Namely, let us define $\mathcal{A} : D(\mathcal{A}) \subseteq Z \rightsquigarrow Z$ by

$$\begin{cases} D(\mathcal{A}) = D(A) \times D(B); \\ \mathcal{A}(u, v) = (Au, Bv), \quad (u, v) \in D(\mathcal{A}), \end{cases}$$

$\mathcal{F} : \mathbb{R}_+ \to Z$, by $\mathcal{F}(t) = (f(t), h(t))$, for each $t \in \mathbb{R}_+$ and the history function $\mathcal{G} : C_b(\mathbb{R}_+; \overline{D(\mathcal{A})}) \to \mathcal{M}$, by

$$\mathcal{G}(u, v) = \left(\sum_{i=1}^{n} \alpha_i u(t_i + \cdot) + \psi_1(\cdot), \sum_{i=1}^{p} \beta_i v(s_i + \cdot) + \psi_2(\cdot) \right)$$

224   *Delay Differential Evolutions Subjected to Nonlocal Initial Conditions*

for each $(u, v) \in C_b(\mathbb{R}_+; \overline{D(\mathcal{A})})$, where

$$\mathcal{M} = \{\psi \in C([-\tau, 0]; Z); \ \psi(0) \in \overline{D(\mathcal{A})}\}.$$

Then, with $z = (u, v)$, (6.4.1) rewrites as

$$\begin{cases} z'(t) \in \mathcal{A}z(t) + \mathcal{F}(t), & t \in \mathbb{R}_+, \\ z(t) = \mathcal{G}(z)(t), & t \in [-\tau, 0]. \end{cases} \tag{6.4.3}$$

Since the normalized semi-inner product $[\cdot, \cdot]_+$ on $Z = X \times Y$ satisfies (6.4.2) and A, B are m-dissipative, from (iv) in Proposition 1.8.1, it follows that $\mathcal{A}$ is m-dissipative, too. In addition, since A and B satisfy (A_1) in (H_A) and (B_1) in (H_B), it follows that $\mathcal{A}$ is δ-m-dissipative and satisfies (A_1) in (H_A) in Section 4.1. Moreover, from (H_p) and (H_q), we deduce that $g = \mathcal{G}$ satisfies the hypothesis (g_1) in (H_g) in Section 4.1. Now, let us observe that, thanks to Lemma 4.2.1, (6.4.3) has a unique C^0-solution $z \in C_b([-\tau, +\infty); Z)$ and the mapping $\mathcal{F} \mapsto z$ is Lipschitz from $L^\infty(\mathbb{R}_+; Z)$ to $C_b([-\tau, +\infty); Z)$ with Lipschitz constant $1/\delta$. This completes the proof. $\qquad \square$

Let us also consider the corresponding unperturbed system

$$\begin{cases} z'(t) \in Az(t), & t \in \mathbb{R}_+, \\ w'(t) \in Bw(t), & t \in \mathbb{R}_+, \\ z(t) = \sum_{i=1}^{n} \alpha_i z(t_i + t) + \psi_1(t), & t \in [-\tau, 0], \\ w(t) = \sum_{i=1}^{p} \beta_i w(s_i + t) + \psi_2(t), & t \in [-\tau, 0]. \end{cases} \tag{6.4.4}$$

An immediate consequence of Lemma 6.4.1 is:

COROLLARY 6.4.1 *Let* $(f, g) \in C_b(\mathbb{R}_+; X) \times C_b(\mathbb{R}_+; Y)$. *If* (H_A), (B_1) *and* (B_5) *in* (H_B), (H_p), (H_q) *and* (H_ψ) *are satisfied, then, with* $\delta = \min\{\omega, \gamma\}$, *the unique* C^0-*solution* (u, v) *of the problem* (6.4.1) *satisfies*

$$\begin{cases} \|u - z\|_{C_b([-\tau, +\infty); X)} \leq \dfrac{1}{\delta} \max\big\{\|f\|_{C_b(\mathbb{R}_+; X)}, \|g\|_{C_b(\mathbb{R}_+; Y)}\big\}, \\ \|v - w\|_{C_b([-\tau, +\infty); Y)} \leq \dfrac{1}{\delta} \max\big\{\|f\|_{C_b(\mathbb{R}_+; X)}, \|g\|_{C_b(\mathbb{R}_+; Y)}\big\}, \end{cases}$$

where (z, w) *is the unique* C^0-*solution of* (6.4.4) *whose existence and uniqueness is ensured by Lemma 6.4.1.*

6.5   Proof of Theorem 6.2.1

We begin with some basic lemmas.

Evolution Systems with Nonlocal Initial Conditions

LEMMA 6.5.1 *Let us assume that* (H_A), (H_B), (H_F), (H_G), (H_c), (H_p), (H_q) *and* (H_ψ) *are satisfied. Let* $a = \min\{t_1 - \tau, s_1 - \tau\} > 0$ *and let* $\widetilde{m} > 0$ *be such that*

$$\begin{cases} \ell\left(\dfrac{\widetilde{m}}{\gamma} + m\right) + m \leq \widetilde{m}, \\ r_1 \leq r_2, \end{cases} \tag{6.5.1}$$

where

$$\begin{cases} r_1 = \dfrac{m}{\omega - \ell} + \left[\dfrac{\omega}{\omega - \ell}\left(\dfrac{1}{e^{\omega a} - 1} + \dfrac{\ell}{\omega}\right) + 1\right] m, \\ r_2 = \dfrac{\widetilde{m}}{\gamma} + m. \end{cases} \tag{6.5.2}$$

Let $\widetilde{\mathcal{Z}} = \widetilde{C}_b([-\tau, +\infty); X) \times \widetilde{C}_b([-\tau, +\infty); Y)$ *and let*

$$\mathcal{C} = \left\{(u, v) \in \widetilde{\mathcal{Z}}; \; \|u\|_{C_b([-\tau, +\infty); X)} \leq r_1, \; \|v\|_{C_b([-\tau, +\infty); Y)} \leq r_2\right\}. \tag{6.5.3}$$

Then $\mathcal{C}$ *is nonempty, closed and convex in* $\widetilde{\mathcal{Z}}$ *and the operator* Γ, *defined by* (6.3.5), *maps* $\mathcal{C}$ *into itself, is continuous, and* $\Gamma(\mathcal{C})$ *is relatively compact.*

REMARK 6.5.1 We notice that, since $\ell < \gamma$, we can always choose a constant $\widetilde{m} > 0$, large enough to satisfy (6.5.1).

Proof. Clearly, the set $\mathcal{C}$, defined by (6.5.3), is nonempty, closed and convex in $C_b([-\tau, +\infty); X) \times C_b([-\tau, +\infty); Y)$. Let $(u, v) \in \mathcal{C}$ and let $(\widetilde{u}, \widetilde{v}) = \Gamma(u, v)$. From (F_2) in (H_F), (4.1.2) and (4.1.3) with $m_0 = m$ (we may always assume that $m_0 = \max\{\|\psi_1\|_X, \|\psi_2\|_Y\} = m$, by increasing m if necessary), it follows that

$$\|\widetilde{u}\|_{C_b([-\tau, +\infty); X)} \leq \dfrac{m}{\omega - \ell} + \left[\dfrac{\omega}{\omega - \ell}\left(\dfrac{1}{e^{\omega a} - 1} + \dfrac{\ell}{\omega}\right) + 1\right] m = r_1.$$

From (G_1) in (H_G), the definition of r_2 – see (6.5.2) – and (6.5.1), we deduce

$$\|G(s, u_s, v_s)\| \leq \ell \max\{r_1, r_2\} + m \leq \ell\left(\dfrac{\widetilde{m}}{\gamma} + m\right) + m \leq \widetilde{m}$$

for each $s \in \mathbb{R}_+$. Next, for $t \in (0, +\infty)$, we obtain

$$\|\widetilde{v}(t)\| \leq e^{-\gamma t}\|\widetilde{v}(0)\| + \dfrac{1 - e^{-\gamma t}}{\gamma}\widetilde{m} \leq e^{-\gamma t}\|\widetilde{v}\|_{C_b(\mathbb{R}_+; Y)} + \dfrac{1 - e^{-\gamma t}}{\gamma}\widetilde{m}.$$

Reasoning as in **Case 2** and **Case 3** in Lemma 4.2.1, we get

$$\|\widetilde{v}\|_{C_b(\mathbb{R}_+; Y)} \leq \dfrac{\widetilde{m}}{\gamma} \leq r_2.$$

226   *Delay Differential Evolutions Subjected to Nonlocal Initial Conditions*

On the other hand, for each $t \in [-\tau, 0]$, we have

$$\|\widetilde{v}(t)\| \le \left\| \sum_{i=1}^{p} \beta_i \widetilde{v}(s_i + t) + \psi_2(t) \right\| \le \|\widetilde{v}\|_{C_b([a,+\infty);Y)} + m \le \frac{\widetilde{m}}{\gamma} + m = r_2,$$

where $a = \min\{t_1 - \tau, s_1 - \tau\} > 0$. From these inequalities, we conclude that

$$\|\widetilde{v}\|_{C_b([-\tau,+\infty);Y)} \le r_2$$

and thus, from the definition of $\mathcal{C}$ – see (6.5.3) – we conclude that $\Gamma(\mathcal{C}) \subseteq \mathcal{C}$.

We prove next that Γ is continuous. To this end, let $((u_k, v_k))_k$ be a sequence in $\mathcal{C}$ with

$$\begin{cases} \lim_k u_k = u & \text{in } \widetilde{C}_b([-\tau, +\infty); X), \\ \lim_k v_k = v & \text{in } \widetilde{C}_b([-\tau, +\infty); Y). \end{cases} \tag{6.5.4}$$

Then $\Gamma(u_k, v_k) = (\widetilde{u}_k, \widetilde{v}_k)$ satisfies

$$\begin{cases} \widetilde{u}_k'(t) \in A\widetilde{u}_k(t) + F(t, \widetilde{u}_{k_t}, \widetilde{v}_{k_t}), & t \in \mathbb{R}_+, \\ \widetilde{v}_k'(t) \in B\widetilde{v}_k(t) + G(t, u_{k_t}, v_{k_t}), & t \in \mathbb{R}_+, \\ \widetilde{u}_k(t) = \sum_{i=1}^{n} \alpha_i \widetilde{u}_k(t_i + t) + \psi_1(t), & t \in [-\tau, 0], \\ \widetilde{v}_k(t) = \sum_{i=1}^{p} \beta_i \widetilde{v}_k(s_i + t) + \psi_2(t), & t \in [-\tau, 0] \end{cases} \tag{6.5.5}$$

for each $k \in \mathbb{N}$. Clearly, for each $T > 0$,

$$\begin{cases} \lim_k u_{k_t} = u_t & \text{in } \mathcal{X}, \\ \lim_k v_{k_t} = v_t & \text{in } \mathcal{Y}, \end{cases}$$

uniformly for $t \in [0, T]$ and therefore, thanks to (G_2) in (H_G), we conclude that, for each $T > 0$,

$$\lim_k G(t, u_{k_t}, v_{k_t}) = G(t, u_t, v_t) \tag{6.5.6}$$

uniformly for $t \in [0, T]$.

Since $((u_k, v_k))_k$ is bounded in $C_b([-\tau, +\infty); X) \times C_b([-\tau, +\infty); Y)$, by (G_1) in (H_G), it follows that the family of mappings

$$\{G(\cdot, u_{k_{(\cdot)}}, v_{k_{(\cdot)}}); \ k \in \mathbb{N}\}$$

is uniformly bounded in $C_b(\mathbb{R}_+; Y)$. Moreover, since the sequence $((u_k, \widetilde{v}_k))_k$ is bounded in $C_b([-\tau, +\infty); X) \times C_b([-\tau, +\infty); Y)$ because its terms belong to $\mathcal{C}$, which is bounded, by (H_q), it follows that the set

$$\left\{ \sum_{i=1}^{p} \beta_i \widetilde{v}_k(t_i + \cdot); \ k \in \mathbb{N} \right\}$$

Evolution Systems with Nonlocal Initial Conditions

is bounded in $\mathcal{Y}$. At this point, let us observe that the system (6.5.5) can be decoupled as

$$
\begin{cases}
\widetilde{v}'_k(t) \in B\widetilde{v}_k(t) + G(t, u_{k_t}, v_{k_t}), & t \in \mathbb{R}_+, \\
\widetilde{v}_k(t) = \displaystyle\sum_{i=1}^{p} \beta_i \widetilde{v}_k(t_i + t) + \psi_2(t), & t \in [-\tau, 0]
\end{cases}
\tag{6.5.7}
$$

and

$$
\begin{cases}
\widetilde{u}'_k(t) \in A\widetilde{u}_k(t) + F(t, \widetilde{u}_{k_t}, \widetilde{v}_{k_t}), & t \in \mathbb{R}_+, \\
\widetilde{u}_k(t) = \displaystyle\sum_{i=1}^{n} \alpha_i \widetilde{u}_k(t_i + t) + \psi_1(t), & t \in [-\tau, 0].
\end{cases}
\tag{6.5.8}
$$

Taking into account that by (B_2) in (H_B), B generates a compact semigroup, the family $\{G(\cdot, u_{k_{(\cdot)}}, v_{k_{(\cdot)}}); \ k \in \mathbb{N}\}$ is uniformly bounded in $C_b(\mathbb{R}_+; Y)$ and

$$
\left\{ \sum_{i=1}^{p} \beta_i \widetilde{v}_k(t_i) + \psi_2(0); \ k \in \mathbb{N} \right\}
$$

is bounded in Y, in view of Theorem 1.8.6, it follows that $\{\widetilde{v}_k; \ k \in \mathbb{N}\}$ is relatively compact in $C([\sigma, k]; Y)$, for each $k = 1, 2, \ldots$ and each $\sigma \in (0, k]$. Therefore, it is relatively compact in $\widetilde{C}_b([\sigma, +\infty); Y)$, for each $\sigma > 0$ and thus in $\widetilde{C}_b([a, +\infty); Y)$. Now, from (H_q), it follows that

$$
\left\{ \sum_{i=1}^{p} \beta_i \widetilde{v}_k(t_i + \cdot) + \psi_2(\cdot); \ k \in \mathbb{N} \right\}
$$

is relatively compact in $C([-\tau, 0,]; Y)$. Hence

$$
\left\{ \sum_{i=1}^{p} \beta_i \widetilde{v}_k(t_i); \ k \in \mathbb{N} \right\}
$$

is relatively compact in Y. Using once again Theorem 1.8.6, we conclude that $\{\widetilde{v}_k; \ k \in \mathbb{N}\}$ is relatively compact $\widetilde{C}_b(\mathbb{R}_+; Y)$ and so in $\widetilde{C}_b([-\tau, +\infty); Y)$. Therefore, there exists $\widetilde{v} \in C_b([-\tau, +\infty); Y)$ such that on a subsequence, denoted for simplicity again by $(\widetilde{v}_k)_k$, we have

$$
\lim_k \widetilde{v}_k = \widetilde{v}
$$

in $\widetilde{C}_b([-\tau, +\infty); Y)$.

Using (1.8.2) and (6.5.6), reasoning as before and passing to the limit for $k \to +\infty$ in (6.5.7), we conclude that $\widetilde{v}$ is the unique C^0-solution of the problem

$$
\begin{cases}
\widetilde{v}'(t) \in B\widetilde{v}(t) + G(t, u_t, v_t), & t \in \mathbb{R}_+, \\
\widetilde{v}(t) = \displaystyle\sum_{i=1}^{p} \beta_i \widetilde{v}(t_i + t) + \psi_2(t), & t \in [-\tau, 0].
\end{cases}
\tag{6.5.9}
$$

228   Delay Differential Evolutions Subjected to Nonlocal Initial Conditions

At this point, using the last remark, by (6.5.4), (H_F) and (H_p), it follows that the functions $f_k : \mathbb{R}_+ \times \mathcal{X} \to X$ and $g_k : C_b(\mathbb{R}_+; \overline{D(A)}) \to \mathcal{D}$ given by

$$
\begin{cases}
f_k(t,u) = F(t,u,\widetilde{v}_{k_t}), & (t,u) \in \mathbb{R}_+ \times \mathcal{X} \\[2mm]
g_k(u) = \displaystyle\sum_{i=1}^{n} \alpha_i u(t_i + \cdot) + \psi_1(\cdot), & u \in C_b(\mathbb{R}_+; \overline{D(A)}),
\end{cases}
$$

for $k \in \mathbb{N}$, satisfy the hypotheses of the $\widetilde{C}_b([-\tau,+\infty);X)$ convergence part in Theorem 4.4.1 for the problem (6.5.8). So, we deduce that there exists

$$
\lim_k \widetilde{u}_k = \widetilde{u}
$$

in $\widetilde{C}_b([-\tau,+\infty);X)$, where $\widetilde{u} \in \widetilde{C}_b([-\tau,+\infty);X)$ is the unique C^0-solution of the problem

$$
\begin{cases}
\widetilde{u}'(t) \in A\widetilde{u}(t) + F(t,\widetilde{u}_t,\widetilde{v}_t), & t \in \mathbb{R}_+, \\[2mm]
\widetilde{u}(t) = \displaystyle\sum_{i=1}^{n} \alpha_i \widetilde{u}(t_i + t) + \psi_1(t), & t \in [-\tau,0].
\end{cases}
\tag{6.5.10}
$$

Since the C^0-solution of the coupled system problem (6.5.9) and (6.5.10) is unique, from the above compactness argument, we deduce that $((\widetilde{u}_k,\widetilde{v}_k))_k$ itself is convergent in $\widetilde{C}_b([-\tau,+\infty);X) \times \widetilde{C}_b([-\tau,+\infty);Y)$ to $(\widetilde{u},\widetilde{v})$. So, Γ is continuous.

We can pass next to the proof of the relative compactness of $\Gamma(\mathcal{C})$. To this end, let $((u_k,v_k))_k$ be an arbitrary sequence in $\mathcal{C}$, let $k \in \mathbb{N}$, and let us denote by

$$
(\widetilde{u}_k,\widetilde{v}_k) = \Gamma(u_k,v_k).
$$

In view of (6.3.5), $\widetilde{v}_k$ is the unique C^0-solution of the problem (6.5.7), while $\widetilde{u}_k$ is the unique C^0-solution of the problem (6.5.8). Reasoning as before, we deduce that $\{\widetilde{v}_k;\ k \in \mathbb{N}\}$ is relatively compact in $\widetilde{C}_b([a,+\infty);Y)$. Furthermore, from (H_q), we conclude that

$$
\{\widetilde{v}_k(0);\ k \in \mathbb{N}\} = \left\{ \sum_{i=1}^{p} \beta_i \widetilde{v}_k(t_i);\ k \in \mathbb{N} \right\}
$$

is relatively compact in Y. Therefore, by Theorem 1.8.6, it follows that $\{\widetilde{v}_k;\ k \in \mathbb{N}\}$ is relatively compact in $\widetilde{C}_b(\mathbb{R}_+;Y)$. By (H_q), we deduce that $\{\widetilde{v}_k;\ k \in \mathbb{N}\}$ is relatively compact in $\mathcal{Y}$, too. This implies that $\{\widetilde{v}_k;\ k \in \mathbb{N}\}$ is relatively compact in $\widetilde{C}_b([-\tau,+\infty);Y)$. As a consequence, there exists $\widetilde{v} \in C_b([-\tau,+\infty);Y)$ such that on a subsequence, denoted for simplicity again by $(\widetilde{v}_k)_k$, we have

$$
\lim_k \widetilde{v}_k = \widetilde{v}
$$

Evolution Systems with Nonlocal Initial Conditions   229

in $\widetilde{C}_b([-\tau, +\infty); Y)$.

As before, from Theorem 4.4.1, we deduce that

$$\lim_k \widetilde{u}_k = \widetilde{u},$$

in $\widetilde{C}_b([-\tau, +\infty); X)$ and, in addition, $\widetilde{u}$ is the unique C^0-solution of the problem (6.5.10).

Now, coming back to (6.5.7), we conclude that

$$\lim_k \widetilde{v}_k = \widetilde{v}$$

in $\mathcal{Y}$. So

$$\lim_k \widetilde{u}_k = \widetilde{u}, \quad \lim_k \widetilde{v}_k = \widetilde{v}$$

in $\widetilde{C}_b([-\tau, +\infty); X)$ and respectively in $\widetilde{C}_b([-\tau, +\infty); Y)$. Consequently, $\Gamma(\mathcal{C})$ is relatively compact in $\widetilde{\mathcal{Z}}$ and this completes the proof.   □

We can now proceed to the proof of Theorem 6.2.1.

Proof. By Lemma 6.5.1, it follows that $\mathcal{C}$ is nonempty, closed and convex in $\widetilde{C}_b([-\tau, +\infty); X) \times \widetilde{C}_b([-\tau, +\infty); Y)$ and the operator $\Gamma : \mathcal{C} \to \mathcal{C}$ is continuous and compact. So, by the Tychonoff Fixed-Point Theorem 1.4.5, it follows that Γ has at least one fixed point $(u, v) \in \mathcal{C}$, which obviously is a C^0-solution of the problem (6.1.1). Both inequalities in (6.2.1) follow from Theorem 4.1.1.

In order to show that, under the additional hypothesis (6.2.2), the C^0-solution of (6.1.1) is unique, we have to write the system as an evolution equation in the product space, $Z = X \times Y$, endowed with the maximum norm and to apply Theorem 4.1.1. Similarly, we deduce the global asymptotic stability. This completes the proof of Theorem 6.2.1.   □

6.6   Application to a reaction–diffusion system in $L^2(\Omega)$

Let Ω be a nonempty, bounded domain in $\mathbb{R}^d$, $d \geq 1$, with C^1 boundary Σ, let $\tau \geq 0$, $\omega > 0$, $\gamma > 0$, let $Q_+ = \mathbb{R}_+ \times \Omega$, $\Sigma_+ = \mathbb{R}_+ \times \Sigma$, $Q_\tau = [-\tau, 0] \times \Omega$, let $\alpha : D(\alpha) \subseteq \mathbb{R} \rightsquigarrow \mathbb{R}$ and $\beta : D(\beta) \subseteq \mathbb{R} \rightsquigarrow \mathbb{R}$ be maximal-monotone operators and let us denote by $\mathcal{X}_2 = C([-\tau, 0]; L^2(\Omega))$. Let $F, G : \mathbb{R}_+ \times \mathcal{X}_2 \times \mathcal{X}_2 \to L^2(\Omega)$ be continuous. Let $0 < t_1 < t_2 < \cdots < t_n$, $(\alpha_i)_{i=1}^n$, $0 < s_1 < s_2 < \cdots < s_p$ and $(\beta_i)_{i=1}^p$.

We consider the following system subjected to mixed nonlocal initial con-

230 *Delay Differential Evolutions Subjected to Nonlocal Initial Conditions*

ditions:

$$
\begin{cases}
\dfrac{\partial u}{\partial t}(t,x) = \Delta u(t,x) - \omega u(t,x) + F(t, u_t, v_t)(x), & \text{in } Q_+, \\[2mm]
\dfrac{\partial v}{\partial t}(t,x) = \Delta v(t,x) - \gamma v(t,x) + G(t, u_t, v_t)(x), & \text{in } Q_+, \\[2mm]
-\dfrac{\partial u}{\partial \nu}(t,x) \in \alpha(u(t,x)), \quad -\dfrac{\partial v}{\partial \nu}(t,x) \in \beta(u(t,x)), & \text{on } \Sigma_+, \\[2mm]
u(t,x) = \displaystyle\sum_{i=1}^{n} \alpha_i u(t_i + t, x) + \psi_1(t)(x), & \text{in } Q_\tau, \\[2mm]
v(t,x) = \displaystyle\sum_{i=1}^{p} \beta_i v(s_i + t, x) + \psi_2(t)(x), & \text{in } Q_\tau,
\end{cases}
\tag{6.6.1}
$$

where, as usual, Δ is the Laplace operator in the sense of distributions over Ω and $\dfrac{\partial u}{\partial \nu}$ denotes the outer normal derivative of u on Σ.

Our main result concerning the system (6.6.1) is:

THEOREM 6.6.1 *Let Ω be a nonempty, bounded, open subset in $\mathbb{R}^d$, $d \geq 1$, with C^1 boundary Σ, let $\tau \geq 0$, $\omega > 0$, $\gamma > 0$, let $\alpha : D(\alpha) \subseteq \mathbb{R} \rightsquigarrow \mathbb{R}$ and $\beta : D(\beta) \subseteq \mathbb{R} \rightsquigarrow \mathbb{R}$ be maximal-monotone operators with $0 \in D(\alpha)$, $0 \in \alpha(0)$, $0 \in D(\beta)$, and $0 \in \beta(0)$. Let $F, G : \mathbb{R}_+ \times \mathfrak{X}_2 \times \mathfrak{X}_2 \to L^2(\Omega)$ be continuous and let $\psi_i \in \mathfrak{X}_2$, $i = 1, 2$. Let us assume that*

(h_1) *there exist ℓ and m such that*

$$\|F(t,u,v) - F(t,\widetilde{u},\widetilde{v})\|_{L^2(\Omega)} \leq \ell \max\{\|u - \widetilde{u}\|_{\mathfrak{X}_2}, \|v - \widetilde{v}\|_{\mathfrak{X}_2}\},$$

$$\|F(t,u,v)\|_{L^2(\Omega)} \leq \ell\|u\|_{\mathfrak{X}_2} + m,$$

$$\|G(t,u,v)\|_{L^2(\Omega)} \leq \ell \max\{\|u\|_{\mathfrak{X}_2}, \|v\|_{\mathfrak{X}_2}\} + m,$$

for each $(t,u,v), (t,\widetilde{u},\widetilde{v}) \in \mathbb{R}_+ \times \mathfrak{X}_2 \times \mathfrak{X}_2$

(h_2) *the family of functions $\{G(t,\cdot,\cdot)(x); \ (t,x) \in \mathbb{R}_+ \times \Omega\}$ is uniformly equicontinuous on $\mathfrak{X}_2 \times \mathfrak{X}_2$*

(h_3) *the constants ℓ and $\delta = \min\{\omega, \gamma\}$ satisfy $\ell < \delta$*

(h_4) *$\tau < t_1 < t_2 < \cdots < t_n$ and $\sum_{i=1}^{n} |\alpha_i| \leq 1$*

(h_5) *$\tau < s_1 < s_2 < \cdots < s_p$ and $\sum_{i=1}^{p} |\beta_i| \leq 1$.*

Then, (6.6.1) has at least one C^0-solution. If, in addition, G satisfies

$$\|G(t,u,v) - G(t,\widetilde{u},\widetilde{v})\|_{L^2(\Omega)} \leq \ell \max\{\|u - \widetilde{u}\|_{\mathfrak{X}_2}, \|v - \widetilde{v}\|_{\mathfrak{X}_2}\}$$

for each $(t,u,v), (t,\widetilde{u},\widetilde{v}) \in \mathbb{R}_+ \times \mathfrak{X}_2 \times \mathfrak{X}_2$, then the C^0-solution of (6.6.1) is unique. If both F and G are Lipschitz with respect to all three variables with Lipschitz constant ℓ, the unique C^0-solution of (6.6.1) is globally asymptotically stable.

Evolution Systems with Nonlocal Initial Conditions 231

Proof. We will show that (6.6.1) can be written in the abstract form (6.1.1) in $X = Y = L^2(\Omega)$ and all the hypotheses of Theorem 6.2.1 are satisfied. To this end, let us define $A : D(A) \subseteq L^2(\Omega) \to L^2(\Omega)$ by

$$
\begin{cases}
D(A) = \left\{ u \in H^2(\Omega), \; -\dfrac{\partial u}{\partial \nu}(x) \in \alpha(u(x)) \text{ a.e. for } x \in \Sigma \right\}, \\
Au = \Delta u - \omega u, \quad \text{for } u \in D(A).
\end{cases}
$$

Similarly, we define $B : D(B) \subseteq L^2(\Omega) \to L^2(\Omega)$ by

$$
\begin{cases}
D(B) = \left\{ v \in H^2(\Omega), \; -\dfrac{\partial v}{\partial \nu}(x) \in \beta(v(x)) \text{ a.e. for } x \in \Sigma \right\}, \\
Bv = \Delta v - \gamma v, \quad \text{for } v \in D(B).
\end{cases}
$$

Since $0 \in \alpha(0)$ and $0 \in \beta(0)$, it follows that $C_0^\infty(\Omega) \subseteq D(A) \cap D(B)$. So, $\overline{D(A)} = \overline{D(B)} = L^2(\Omega)$. Next, from Theorem 1.9.8 with $p = 2$ and $\lambda = 0$, we deduce that A, $A + \omega I$, B and $B + \gamma I$ are m-dissipative on $L^2(\Omega)$, $0 \in D(A)$, $0 \in A0$, $0 \in D(B)$, $0 \in B(0)$ and both generate compact semigroups. So, the hypotheses (H_A) and (H_B) in Theorem 6.2.1 are satisfied.

Further, from (h_1) and (h_2), it follows that F and G satisfy (H_F) and, respectively, (H_G). Since (h_3), (h_4) and (h_5) are exactly (H_c), (H_p) and (H_q), we are in the hypothesis of Theorem 6.2.1. This completes the proof. $\qquad \square$

6.7 Nonlocal initial conditions with linear growth

In this section we reconsider the problem of the existence and uniform asymptotic stability for C^0-solutions for another class of abstract nonlinear delay reaction–diffusion systems with nonlocal initial data

$$
\begin{cases}
u'(t) \in Au(t) + F(t, u_t, v_t), & t \in \mathbb{R}_+, \\
v'(t) \in Bv(t) + G(t, u_t, v_t), & t \in \mathbb{R}_+, \\
u(t) = p(u, v)(t), & t \in [-\tau, 0], \\
v(t) = q(u, v)(t), & t \in [-\tau, 0].
\end{cases}
\tag{6.7.1}
$$

Here, X, Y are Banach spaces, $A : D(A) \subseteq X \rightsquigarrow X$, $B : D(B) \subseteq Y \rightsquigarrow Y$ are m-dissipative operators, $\tau \geq 0$, $\mathcal{X}$, $\mathcal{Y}$ are given by (6.1.2) while $\mathcal{D}$ and $\mathcal{E}$ are given by (6.1.3). The functions $F : \mathbb{R}_+ \times \mathcal{X} \times \mathcal{Y} \to X$ and $G : \mathbb{R}_+ \times \mathcal{X} \times \mathcal{Y} \to Y$ are continuous while the functions $p : C_b(\mathbb{R}_+; \overline{D(A)}) \times C_b(\mathbb{R}_+; \overline{D(B)}) \to \mathcal{D}$ and $q : C_b(\mathbb{R}_+; \overline{D(A)}) \times C_b(\mathbb{R}_+; \overline{D(B)}) \to \mathcal{E}$ are nonexpansive and have linear growth. We notice that, in the case here considered, we can relax the growth condition on F and G. Moreover, this setting allows us to give a different proof

232 *Delay Differential Evolutions Subjected to Nonlocal Initial Conditions*

of the existence result, whose main idea is very important because it could be adapted to handle the multivalued case as well. In addition, here, we can get rid of the assumption (B_5) on $\overline{D(B)}$ as well as the condition (G_2) in (H_G) both assumed to hold true in Section 6.2.

As usual, if $u \in C([-\tau, +\infty); X)$, $v \in C([-\tau, +\infty); Y)$ and $t \in \mathbb{R}_+$, $u_t \in \mathcal{X}$ and $v_t \in \mathcal{Y}$ are defined by

$$\begin{cases} u_t(s) = u(t+s), & s \in [-\tau, 0], \\ v_t(s) = v(t+s), & s \in [-\tau, 0]. \end{cases}$$

Moreover, if $u \in C_b([-\tau, +\infty); X)$ and $v \in C_b([-\tau, +\infty); Y)$ are such that $u(t) \in \overline{D(A)}$ and $v(t) \in \overline{D(A)}$ for each $t \in \mathbb{R}_+$, we denote by

$$\begin{cases} p(u, v) = p(u_{|\mathbb{R}_+}, v_{|\mathbb{R}_+}), \\ q(u, v) = q(u_{|\mathbb{R}_+}, v_{|\mathbb{R}_+}). \end{cases}$$

DEFINITION 6.7.1 By a C^0-*solution* of (6.7.1) we mean a continuous function $(u, v) : [-\tau, +\infty) \to X \times Y$ satisfying $u(t) = p(u, v)(t)$ and $v(t) = q(u, v)(t)$ for each $t \in [-\tau, 0]$ and, for each $T > 0$, (u, v) is a C^0-solution on $[0, T]$ in the sense of Definition 1.8.2 in the space $X \times Y$, for

$$\begin{cases} u'(t) \in Au(t) + f(t), & t \in \mathbb{R}_+, \\ v'(t) \in Bv(t) + g(t), & t \in \mathbb{R}_+, \end{cases}$$

where $f(t) = F(t, u_t, v_t)$ and $g(t) = G(t, u_t, v_t)$ for $t \in [0, T]$.

In this section, we will use the hypotheses below, classified, for simplicity, as follows. It should be noted that, although some of the hypotheses on the history functions p and q seem to be rather technical, they are all very natural and they are all satisfied in several important specific cases.

Hypotheses on the operators A and B

(H_A) The operator $A : D(A) \subseteq X \rightsquigarrow X$ satisfies

$\quad (A_1)$ $0 \in D(A)$, $0 \in A0$ and A is ω-m-dissipative for some $\omega > 0$

$\quad (A_3)$ $\overline{D(A)}$ is convex.

(H_B) The operator $B : D(B) \subseteq Y \rightsquigarrow Y$ satisfies

$\quad (B_1)$ $0 \in D(B)$, $0 \in B0$ and B is γ-m-dissipative for some $\gamma > 0$

$\quad (B_2)$ B generates a compact semigroup on $\overline{D(B)}$.

Hypotheses on the functions F and G

Evolution Systems with Nonlocal Initial Conditions

(H_F) The function $F : \mathbb{R}_+ \times X \times Y \to X$ is continuous and satisfies

(F_1) there exists $\ell > 0$ such that

$$\|F(t, u, v) - F(t, \widetilde{u}, \widetilde{v})\| \leq \ell \max\{\|u - \widetilde{u}\|_X, \|v - \widetilde{v}\|_Y\}$$

for all $(t, u, v), (t, \widetilde{u}, \widetilde{v}) \in \mathbb{R}_+ \times X \times Y$

(F_2) there exists $m > 0$ such that $\|F(t, u, v)\| \leq \ell\|u\|_X + m$ for each $(t, u, v) \in \mathbb{R}_+ \times X \times Y$, where ℓ is given by (F_1).

(H_G) The function $G : \mathbb{R}_+ \times X \times Y \to Y$ is continuous and satisfies

(G_1) with ℓ and m given by (F_1) and (F_2), we have

$$\|G(t, u, v)\| \leq \ell \max\{\|u\|_X, \|v\|_Y\} + m$$

for each $(t, u, v) \in \mathbb{R}_+ \times X \times Y$, where ℓ is given by (F_1).

Hypotheses on the constants ℓ, ω and γ

$(\widetilde{H}_c)$ The constants ℓ, ω and γ satisfy

$$\ell < \frac{\omega\gamma}{\omega + \gamma}.$$

REMARK 6.7.1 One may easily verify that $(\widetilde{H}_c)$ implies (H_c) in Section 6.1.

Hypotheses on the functions p and q

(H_p) The function $p : C_b(\mathbb{R}_+; \overline{D(A)}) \times C_b(\mathbb{R}_+; \overline{D(B)}) \to \mathcal{D}$ is continuous from its domain endowed with the $\widetilde{C}_b(\mathbb{R}_+; X) \times \widetilde{C}_b(\mathbb{R}_+; Y)$ topology to X and there exists $a > 0$ such that the conditions below are satisfied:

(p_1) for each $u, \widetilde{u} \in C_b(\mathbb{R}_+; \overline{D(A)})$ and $v, \widetilde{v} \in C_b(\mathbb{R}_+; \overline{D(B)})$, we have

$$\|p(u, v) - p(\widetilde{u}, \widetilde{v})\|_X$$

$$\leq \max\{\|u - \widetilde{u}\|_{C_b([a, +\infty); X)}, \|v - \widetilde{v}\|_{C_b([a, +\infty); Y)}\}$$

(p_2) for each $u \in C_b(\mathbb{R}_+; \overline{D(A)})$ and each $v \in C_b(\mathbb{R}_+; \overline{D(B)})$, we have

$$\|p(u, v)\|_X \leq \|u\|_{C_b([a, +\infty); X)}$$

(p_3) for each bounded subset $\mathcal{U}$ in $C_b(\mathbb{R}_+; \overline{D(A)})$, the family of functions $\{p(u, \cdot); \ u \in \mathcal{U}\}$ is equicontinuous from $C_b(\mathbb{R}_+; \overline{D(B)})$ endowed with $\widetilde{C}_b(\mathbb{R}_+; Y)$ topology to X.

234   *Delay Differential Evolutions Subjected to Nonlocal Initial Conditions*

(H_q) The function $q : C_b(\mathbb{R}_+; \overline{D(A)}) \times C_b(\mathbb{R}_+; \overline{D(B)}) \to \mathcal{E}$ is continuous from its domain endowed with the $\widetilde{C}_b(\mathbb{R}_+; X) \times \widetilde{C}_b(\mathbb{R}_+; Y)$ topology to $\mathcal{Y}$ and, with a given by (H_p), the conditions below are satisfied:

(q_1) for each $u, \widetilde{u} \in C_b(\mathbb{R}_+; \overline{D(A)})$ and $v, \widetilde{v} \in C_b(\mathbb{R}_+; \overline{D(B)})$, we have

$$\|q(u,v) - q(\widetilde{u}, \widetilde{v})\|_{\mathcal{Y}}$$

$$\leq \max\{\|u - \widetilde{u}\|_{C_b([a,+\infty);X)}, \|v - \widetilde{v}\|_{C_b([a,+\infty);Y)}\}$$

(q_2) for each $u \in C_b(\mathbb{R}_+; \overline{D(A)})$ and each $v \in C_b(\mathbb{R}_+; \overline{D(B)})$, we have

$$\|q(u,v)\|_{\mathcal{Y}} \leq \|v\|_{C_b([a,+\infty);Y)}$$

(q_3) for each bounded set $\mathcal{U}$ in $C_b(\mathbb{R}_+; \overline{D(A)})$ and each set $\mathcal{V}$ in $C_b([a,+\infty); \overline{D(B)})$ that is relatively compact in $\widetilde{C}_b([a,+\infty); Y)$, the set $q(\mathcal{U}, \mathcal{V})$ is relatively compact in $\mathcal{Y}$.

The main result in this section is:

THEOREM 6.7.1 *If (H_A), (H_B), (H_F), (H_G), (H_p), (H_q) and $(\widetilde{H}_c)$ are satisfied, then (6.7.1) has at least one C^0-solution,*

$$(u,v) \in C_b([-\tau, +\infty); X) \times C_b([-\tau, +\infty); Y),$$

with $(u(t), v(t)) \in \overline{D(A)} \times \overline{D(B)}$ for each $t \in \mathbb{R}_+$ and satisfying

$$\begin{cases} \|u\|_{C_b([-\tau,+\infty);X)} \leq \dfrac{m}{\omega - \ell}, \\[2mm] \|v\|_{C_b([-\tau,+\infty);Y)} \leq \dfrac{m}{\gamma - \ell}. \end{cases} \tag{6.7.2}$$

If, in addition, G satisfies

$$\|G(t,u,v) - G(t,\widetilde{u},\widetilde{v})\| \leq \ell \max\{\|u - \widetilde{u}\|_X, \|v - \widetilde{v}\|_Y\} \tag{6.7.3}$$

for all $(t,u,v), (t,\widetilde{u},\widetilde{v}) \in \mathbb{R}_+ \times X \times Y$, then the C^0-solution of (6.7.1) is unique. Moreover, if both F and G are Lipschitz with respect to all three variables with Lipschitz constant ℓ, then the unique C^0-solution of (6.7.1) is globally asymptotically stable.

REMARK 6.7.2 We can relax the continuity assumption on G by imposing only that G is strongly-weakly continuous. However, in this more general frame for G, in order to compensate for the lack of strong-strong continuity, we have to assume in addition that B is of complete continuous type. We refrain from giving details on this case here because it will be considered in a slightly different setting in Chapter 9.

Evolution Systems with Nonlocal Initial Conditions 235

As far as the nondelayed case is concerned, i.e., when $\tau = 0$, and $\mathcal{X}$, $\mathcal{Y}$ reduce to X and Y, respectively, while $\mathcal{D}$ and $\mathcal{E}$ reduce to $\overline{D(A)}$ and $\overline{D(B)}$, respectively, we consider the system

$$\begin{cases} u'(t) \in Au(t) + F(t, u(t), v(t)), & t \in \mathbb{R}_+, \\ v'(t) \in Bv(t) + G(t, u(t), v(t)), & t \in \mathbb{R}_+, \\ u(0) = p(u, v), \\ v(0) = q(u, v). \end{cases} \tag{6.7.4}$$

For this specific case, we need to reformulate the hypotheses (H_F), (H_G), (H_p) and (H_q) as follows.

Hypotheses on the functions F and G

$(H_F^{[\tau=0]})$ The function $F : \mathbb{R}_+ \times X \times Y \to X$ is continuous and satisfies

$(F_1^{[\tau=0]})$ there exists $\ell > 0$ such that

$$\|F(t, u, v) - F(t, \widetilde{u}, \widetilde{v})\| \le \ell \max\{\|u - \widetilde{u}\|, \|v - \widetilde{v}\|\}$$

for each $(t, u, v), (t, \widetilde{u}, \widetilde{v}) \in \mathbb{R}_+ \times X \times Y$

$(F_2^{[\tau=0]})$ there exists $m > 0$ such that

$$\|F(t, u, v)\| \le \ell \|u\| + m$$

for each $(t, u, v) \in \mathbb{R}_+ \times X \times Y$, where ℓ is given by $(F_1^{[\tau=0]})$.

$(H_G^{[\tau=0]})$ The function $G : \mathbb{R}_+ \times X \times Y \to Y$ is continuous and satisfies

$(G_1^{[\tau=0]})$ with ℓ and m given by $(F_1^{[\tau=0]})$ and $(F_2^{[\tau=0]})$, we have

$$\|G(t, u, v)\| \le \ell \max\{\|u\|, \|v\|\} + m$$

for each $(t, u, v) \in \mathbb{R}_+ \times X \times Y$.

Hypotheses on the functions p and q

$(H_p^{[\tau=0]})$ The function $p : C_b(\mathbb{R}_+; \overline{D(A)}) \times C_b(\mathbb{R}_+; \overline{D(B)}) \to \overline{D(A)}$ is continuous from its domain endowed with the $\widetilde{C}_b(\mathbb{R}_+; X) \times \widetilde{C}_b(\mathbb{R}_+; Y)$ topology to X and there exists $a > 0$ such that the next conditions are satisfied:

$(p_1^{[\tau=0]})$ for each $u, \widetilde{u} \in C_b(\mathbb{R}_+; \overline{D(A)})$ and each $v, \widetilde{v} \in C_b(\mathbb{R}_+; \overline{D(B)})$, we have

$$\|p(u, v) - p(\widetilde{u}, \widetilde{v})\| \le \max\{\|u - \widetilde{u}\|_{C_b([a, +\infty); X)}, \|v - \widetilde{v}\|_{C_b([a, +\infty); Y)}\}$$

236 *Delay Differential Evolutions Subjected to Nonlocal Initial Conditions*

$(p_2^{[\tau=0]})$ for each $u \in C_b(\mathbb{R}_+; \overline{D(A)})$ and each $v \in C_b(\mathbb{R}_+; \overline{D(B)})$, we have

$$\|p(u, v)\| \le \|u\|_{C_b([\,a, +\infty);X)}$$

$(p_3^{[\tau=0]})$ for each bounded subset $\mathcal{U}$ in $C_b(\mathbb{R}_+; \overline{D(A)})$, the family of functions $\{p(u, \cdot); \ u \in \mathcal{U}\}$ is equicontinuous from $\widetilde{C}_b(\mathbb{R}_+; \overline{D(B)})$ to X.

$(H_q^{[\tau=0]})$ The function $q : C_b(\mathbb{R}_+; \overline{D(A)}) \times C_b(\mathbb{R}_+; \overline{D(B)}) \to \overline{D(B)}$ is continuous from its domain endowed with the $\widetilde{C}_b(\mathbb{R}_+; X) \times \widetilde{C}_b(\mathbb{R}_+; Y)$ topology to Y and, with a given by $(H_p^{[\tau=0]})$, the conditions below are satisfied:

$(q_1^{[\tau=0]})$ for each $u, \widetilde{u} \in C_b(\mathbb{R}_+; \overline{D(A)})$ and $v, \widetilde{v} \in C_b(\mathbb{R}_+; \overline{D(B)})$, we have

$$\|q(u, v) - q(\widetilde{u}, \widetilde{v})\| \le \max\{\|u - \widetilde{u}\|_{C_b([\,a, +\infty);X)}, \|v - \widetilde{v}\|_{C_b([\,a, +\infty);Y)}\}$$

$(q_2^{[\tau=0]})$ for each $(u, v) \in C_b(\mathbb{R}_+; \overline{D(A)}) \times C_b(\mathbb{R}_+; \overline{D(B)})$, we have

$$\|q(u, v)\| \le \|v\|_{C_b([\,a, +\infty);Y)}$$

$(q_3^{[\tau=0]})$ for each bounded set $\mathcal{U}$ in $C_b(\mathbb{R}_+; \overline{D(A)})$ and each set $\mathcal{V}$ in $C_b([\,a, +\infty); \overline{D(B)})$ that is relatively compact in $\widetilde{C}_b([\,a, +\infty); Y)$, the set $q(\mathcal{U}, \mathcal{V})$ is relatively compact in Y.

From Theorem 6.7.1, we deduce:

THEOREM 6.7.2 *If* (H_A), (H_B), $(H_F^{[\tau=0]})$, $(H_G^{[\tau=0]})$, $(H_p^{[\tau=0]})$, $(H_q^{[\tau=0]})$ *and* $(\widetilde{H}_c)$ *are satisfied, then (6.7.4) has at least one* C^0-*solution,*

$$(u, v) \in C_b(\mathbb{R}_+; \overline{D(A)}) \times C_b(\mathbb{R}_+; \overline{D(B)}),$$

satisfying

$$\begin{cases} \|u\|_{C_b(\mathbb{R}_+;X)} \le \dfrac{m}{\omega - \ell}, \\[2ex] \|v\|_{C_b(\mathbb{R}_+;Y)} \le \dfrac{m}{\gamma - \ell}. \end{cases}$$

If, in addition, G satisfies

$$\|G(t, u, v) - G(t, \widetilde{u}, \widetilde{v})\| \le \ell \max\{\|u - \widetilde{u}\|, \|v - \widetilde{v}\|\}$$

for each $(t, u, v), (t, \widetilde{u}, \widetilde{v}) \in \mathbb{R}_+ \times \overline{D(A)} \times \overline{D(B)}$, *then the* C^0-*solution of (6.7.4) is unique. If both F and G are Lipschitz with respect to all three variables with Lipschitz constant ℓ, then the unique* C^0-*solution of (6.7.4) is globally asymptotically stable.*

REMARK 6.7.3 One may easily see that, from Theorem 6.7.1, we can obtain as simple consequences, existence, uniqueness, and global asymptotic results referring to various systems whose forcing terms are either of the form $F(t, u_t, v(t))$ and $G(t, u_t, v(t))$ or of the form $F(t, u(t), v_t)$ and $G(t, u(t), v_t)$ and, in either case, p and q are defined accordingly.

6.8 The idea of the proof

Let $\varepsilon \in (0, 1)$ and let us consider the approximate problem

$$\begin{cases} u'(t) \in Au(t) + F(t, u_t, v_t), & t \in \mathbb{R}_+, \\ v'(t) \in Bv(t) + \eta_\varepsilon(t)G(t, u_t, v_t), & t \in \mathbb{R}_+, \\ u(t) = p(u, v)(t), & t \in [-\tau, 0], \\ v(t) = q(u, v)(t), & t \in [-\tau, 0], \end{cases} \tag{6.8.1}$$

where $\eta_\varepsilon : \mathbb{R}_+ \to [0, 1]$ is defined by

$$\eta_\varepsilon(t) = \begin{cases} 1, & t \in [0, 1/\varepsilon), \\ 1 + 1/\varepsilon - t, & t \in [1/\varepsilon, 1 + 1/\varepsilon), \\ 0, & t \in [1 + 1/\varepsilon, +\infty). \end{cases} \tag{6.8.2}$$

First, we will show that the problem (6.8.1) has at least one C^0-solution $(u_\varepsilon, v_\varepsilon)$. Then, for each $\varepsilon \in (0, 1)$, we fix such a C^0-solution $(u_\varepsilon, v_\varepsilon)$ and we will show that the family $\{(u_\varepsilon, v_\varepsilon); \varepsilon \in (0, 1)\}$ is relatively compact in the product space $\widetilde{C}_b([-\tau, +\infty); X) \times \widetilde{C}_b([-\tau, +\infty); Y)$.

So, there exist $(u, v) \in C_b([-\tau, +\infty); X) \times C_b([-\tau, +\infty); Y)$ and two sequences: $(\varepsilon_n)_n$ with $\lim_n \varepsilon_n = 0$ and $((u_{\varepsilon_n}, v_{\varepsilon_n}))_n$ with $\lim_n(u_{\varepsilon_n}, v_{\varepsilon_n}) = (u, v)$, in the space $\widetilde{C}_b([-\tau, +\infty); X) \times \widetilde{C}_b([-\tau, +\infty); Y)$. By observing that (u, v) is a C^0-solution of (6.7.1), we will complete the proof of Theorem 6.7.1.

In order to prove that (6.8.1) has at least one C^0-solution, we fix an arbitrary $(u, h) \in C_b([-\tau, +\infty); X) \times C_b(\mathbb{R}_+; Y)$ with $u(s) \in \overline{D(A)}$ for each $s \in \mathbb{R}_+$ and we consider the problem

$$\begin{cases} \widetilde{v}'(t) \in B\widetilde{v}(t) + h(t), & t \in \mathbb{R}_+, \\ \widetilde{v}(t) = q(u, \widetilde{v})(t), & t \in [-\tau, 0], \end{cases} \tag{6.8.3}$$

which, in view of Lemma 4.2.1, has a unique C^0-solution $\widetilde{v}$. Now, we consider the problem

$$\begin{cases} \widetilde{u}'(t) \in A\widetilde{u}(t) + F(t, \widetilde{u}_t, \widetilde{v}_t), & t \in \mathbb{R}_+, \\ \widetilde{u}(t) = p(\widetilde{u}, \widetilde{v})(t), & t \in [-\tau, 0]. \end{cases} \tag{6.8.4}$$

By virtue of Theorem 4.1.1, the problem thus obtained has a unique C^0-solution $\widetilde{u}$. Next, let us define

$$\Gamma_\varepsilon(u, h)(t) = (\widetilde{u}(t), G_\varepsilon(t, \widetilde{u}_t, \widetilde{v}_t)), \tag{6.8.5}$$

for each $(u, h) \in C_b([-\tau, +\infty); X) \times C_b(\mathbb{R}_+; Y)$, with $u(s) \in \overline{D(A)}$ for each $s \in \mathbb{R}_+$, and for each $t \in [-\tau, +\infty)$, where

$$G_\varepsilon(t, u, v) = \eta_\varepsilon(t)G(t, u, v)$$

238   *Delay Differential Evolutions Subjected to Nonlocal Initial Conditions*

for each $(t, u, v) \in \mathbb{R}_+ \times X \times Y$.

We will show that the restriction of the function Γ_ε to a suitably defined set $\mathcal{K} \subseteq C_b([-\tau, +\infty); X) \times C_b(\mathbb{R}_+; Y)$ satisfies the hypotheses of the Schauder Fixed-Point Theorem 1.4.4 and, accordingly, it has a fixed point $(\widetilde{u}, \widetilde{h}) \in \mathcal{K}$. Finally, we have merely to observe that $(\widetilde{u}, \widetilde{v})$, satisfying (6.8.3) and (6.8.4), is a C^0-solution for (6.8.1) if and only if $(\widetilde{u}, \widetilde{h})$ is a fixed point of Γ_ε.

6.9   Auxiliary results

LEMMA 6.9.1 *Let us assume that* (H_A), (H_B), (H_F), (H_G), (H_p), (H_q) *and* $(\widetilde{H}_c)$ *are satisfied and let*

$$r = \frac{m\omega\gamma}{\omega\gamma - \ell(\omega + \gamma)}. \tag{6.9.1}$$

Then, for each $(u, h) \in C_b([-\tau, +\infty); X) \times C_b(\mathbb{R}_+; Y)$, *with* $u(s) \in \overline{D(A)}$ *for each* $s \in \mathbb{R}_+$, *satisfying*

$$\begin{cases} \|u\|_{C_b([-\tau, +\infty); X)} \leq \dfrac{r}{\omega} \\[2mm] \|h\|_{C_b(\mathbb{R}_+; Y)} \leq r, \end{cases}$$

the pair $(\widetilde{u}, \widetilde{v})$, *where* $\widetilde{v}$ *is the unique* C^0-*solution of* (6.8.3) *and* $\widetilde{u}$ *is the unique* C^0-*solution of* (6.8.4), *satisfies*

$$\|\widetilde{v}\|_{C_b([-\tau, +\infty); Y)} \leq \frac{r}{\gamma} \tag{6.9.2}$$

$$\|\widetilde{u}\|_{C_b([-\tau, +\infty); X)} \leq \frac{r}{\omega} \tag{6.9.3}$$

and

$$\|G(t, \widetilde{u}_t, \widetilde{v}_t)\| \leq r \tag{6.9.4}$$

for each $t \in \mathbb{R}_+$.

Proof. We show first that $\widetilde{v}$ satisfies (6.9.2). Indeed, from (B_1) in (H_B) and (1.8.3), we get

$$\|\widetilde{v}(t)\| \leq e^{-\gamma t}\|\widetilde{v}(0)\| + \int_0^t e^{-\gamma(t-s)}\|h(s)\|\, ds \leq e^{-\gamma t}\|\widetilde{v}\|_{C_b([a, +\infty); Y)} + \frac{r}{\gamma}\left(1 - e^{-\gamma t}\right)$$

for each $t \in \mathbb{R}_+$. Now, arguing as in Lemma 4.2.1, we obtain

$$\|\widetilde{v}\|_{C_b([a, +\infty); Y)} \leq \frac{r}{\gamma}.$$

Evolution Systems with Nonlocal Initial Conditions

Recalling that, by (q_2)

$$\|\widetilde{v}(t)\| \leq \|q(u,\widetilde{v})\|_{\mathcal{Y}} \leq \|\widetilde{v}\|_{C_b([a,+\infty);Y)},$$

for each $t \in [-\tau, 0]$, from the preceding inequality, we deduce (6.9.2).

We will show next that (6.9.3) is also satisfied. To this end, let us observe that, by (A_1) in (H_A), (1.8.3) and (F_1) in (H_F), we have

$$\|\widetilde{u}(t)\| \leq e^{-\omega t}\|\widetilde{u}(0)\| + \int_0^t e^{-\omega(t-s)} \left(\ell \max\{\|\widetilde{u}_s\|_{\mathcal{X}}, \|\widetilde{v}_s\|_{\mathcal{Y}}\} + m\right) ds,$$

for each $t \in \mathbb{R}_+$, which implies

$$\|\widetilde{u}(t)\| \leq e^{-\omega t}\|\widetilde{u}(0)\| + \int_0^t e^{-\omega(t-s)} \left(\ell \max\left\{\|\widetilde{u}_s\|_{\mathcal{X}}, \frac{r}{\gamma}\right\} + m\right) ds. \qquad (6.9.5)$$

Since, for each $t \in [-\tau, 0]$, we have

$$\|\widetilde{u}(t)\| = \|p(\widetilde{u},\widetilde{v})(t)\| \leq \|\widetilde{u}\|_{C_b([a,+\infty);X)},$$

we deduce

$$\|\widetilde{u}_s\|_{\mathcal{X}} = \sup_{\theta \in [-\tau,0]} \|\widetilde{u}(s+\theta)\| \leq \|\widetilde{u}\|_{C_b([-\tau,+\infty);X)} \leq \|\widetilde{u}\|_{C_b(\mathbb{R}_+;X)},$$

for each $s \in \mathbb{R}_+$. Hence, from (6.9.5), we get

$$\|\widetilde{u}(t)\| \leq e^{-\omega t}\|\widetilde{u}\|_{C_b(\mathbb{R}_+;X)} + \frac{1-e^{-\omega t}}{\omega}\left(\ell \max\left\{\|\widetilde{u}\|_{C_b(\mathbb{R}_+;X)}, \frac{r}{\gamma}\right\} + m\right)$$

for each $t \in (0,+\infty)$. Next, since for each $x, y \in \mathbb{R}_+$, $\max\{x,y\} \leq x+y$, we get

$$\|\widetilde{u}(t)\| \leq e^{-\omega t}\|\widetilde{u}\|_{C_b(\mathbb{R}_+;X)} + \frac{1-e^{-\omega t}}{\omega}\left[\ell\left(\|\widetilde{u}\|_{C_b(\mathbb{R}_+;X)} + \frac{r}{\gamma}\right) + m\right] \qquad (6.9.6)$$

for each $t \in (0,+\infty)$.

If there exists $\bar{t} \in \mathbb{R}_+$ such that

$$\|\widetilde{u}(\bar{t})\| = \|\widetilde{u}\|_{C_b(\mathbb{R}_+;X)},$$

then, if $\bar{t} = 0$, we deduce that

$$\|\widetilde{u}\|_{C_b(\mathbb{R}_+;X)} = \|\widetilde{u}(0)\| \leq \|\widetilde{u}\|_{C_b([a,+\infty);X)} \leq \|\widetilde{u}\|_{C_b(\mathbb{R}_+;X)}$$

and thus

$$\|\widetilde{u}\|_{C_b(\mathbb{R}_+;X)} = \|\widetilde{u}\|_{C_b([a,+\infty);X)}.$$

If $\bar{t} > 0$, taking $t = \bar{t} > 0$ in (6.9.6), we get

$$\|\widetilde{u}\|_{C_b(\mathbb{R}_+;X)} \leq e^{-\omega\bar{t}}\|\widetilde{u}\|_{C_b(\mathbb{R}_+;X)}$$

$$+ \left(1 - e^{-\omega \bar{t}}\right) \frac{1}{\omega} \left[\ell \left(\|\widetilde{u}\|_{C_b(\mathbb{R}_+;X)} + \frac{r}{\gamma} \right) + m \right]. \tag{6.9.7}$$

From (6.9.7), we deduce

$$\|\widetilde{u}\|_{C_b(\mathbb{R}_+;X)} \le \frac{\ell r + m\gamma}{\gamma(\omega - \ell)}.$$

Recalling the definition of r – see (6.9.1) – we get

$$\frac{\ell r + m\gamma}{\gamma(\omega - \ell)} = \frac{1}{\gamma(\omega - \ell)} \cdot \left[\frac{\ell m \omega \gamma}{\omega \gamma - \ell(\omega + \gamma)} + m\gamma \right] = \frac{m\gamma}{\omega \gamma - \ell(\omega + \gamma)} = \frac{r}{\omega}$$

and so

$$\|\widetilde{u}\|_{C_b(\mathbb{R}_+;X)} \le \frac{r}{\omega}.$$

From (p_2), we have

$$\|\widetilde{u}(t)\| = \|p(\widetilde{u}, \widetilde{v})(t)\| \le \|\widetilde{u}\|_{C_b([a,+\infty);X)}$$

for each $t \in [-\tau, 0]$ inequality which, along with the preceding one, implies (6.9.3).

Now, from (6.9.2), (6.9.3) and (H_G), we get

$$\|G(t, \widetilde{u}_t, \widetilde{v}_t)\| \le \ell \max\left\{ \frac{r}{\omega}, \frac{r}{\gamma} \right\} + m \le \ell \left(\frac{r}{\omega} + \frac{r}{\gamma} \right) + m$$

$$\le \ell \cdot \frac{m\omega^2 \gamma + m\omega\gamma^2}{\omega\gamma \left[\omega\gamma - \ell(\omega + \gamma) \right]} + m = r,$$

a relation that leads to (6.9.4).

In the complementary case, i.e., if there is no $\bar{t} \in \mathbb{R}_+$ such that

$$\|\widetilde{u}(\bar{t})\| = \|\widetilde{u}\|_{C_b(\mathbb{R}_+;X)},$$

then there exists $(t_k)_k$ such that $\lim_k t_k = +\infty$ and

$$\lim_k \|\widetilde{u}(t_k)\| = \|\widetilde{u}\|_{C_b(\mathbb{R}_+;X)}.$$

Setting $t = t_k$ in (6.9.6) and passing to the limit for $k \to +\infty$, after some calculations like the preceding ones, we get (6.9.4) and this completes the proof. $\square$

LEMMA 6.9.2 *Let us assume that* (H_A), (H_B), (H_F), (H_G), (H_p), (H_q) *and* $(\widetilde{H}_c)$ *are satisfied, let* $r > 0$ *given by* (6.9.7) *and let* $\rho = r/\omega$. *Let*

$$\mathcal{K}_\varepsilon = K_\rho \times K_r,$$

where K_ρ *is the intersection of the closed ball with center* 0 *and radius* ρ

Evolution Systems with Nonlocal Initial Conditions 241

in the space $C_b([-\tau, +\infty); X)$ with $C_b(\mathbb{R}_+; \overline{D(A)})$ and K_r is the closed ball with center 0 and radius r in $C_b(\mathbb{R}_+; Y)$ multiplied by the function η_ε, defined by (6.8.2). Then $\mathcal{K}_\varepsilon$ is nonempty, closed and convex in $C_b([-\tau, +\infty); X) \times C_b(\mathbb{R}_+; Y)$ and Γ_ε, given by (6.8.5), maps $\mathcal{K}_\varepsilon$ into itself and is continuous from $\mathcal{K}_\varepsilon$ to $\mathcal{K}_\varepsilon$ when both the domain and the range of Γ_ε are endowed with the topology of the product space $C_b([-\tau, +\infty); X) \times C_b(\mathbb{R}_+; Y)$.

REMARK 6.9.1 If $\varepsilon \in (0, 1)$ and η_ε is defined by (6.8.2), then

$$\lim_{\varepsilon \downarrow 0} \eta_\varepsilon(t) = 1$$

in $\widetilde{C}_b(\mathbb{R}_+; \mathbb{R})$.

We can now proceed with the proof of Lemma 6.9.2.

Proof. By (A_3) in (H_A), it follows that $\mathcal{K}_\varepsilon$, which is obviously nonempty and closed, is convex too. The fact that Γ_ε maps $\mathcal{K}_\varepsilon$ into itself is an immediate consequence of Lemma 6.9.1.

To prove that Γ_ε is continuous from $\mathcal{K}_\varepsilon$ to $\mathcal{K}_\varepsilon$ if both domain and range are endowed with the topology of $C_b([-\tau, +\infty); X) \times C_b(\mathbb{R}_+; Y)$, let $((u_n, h_n))_n$ be an arbitrary sequence in $\mathcal{K}_\varepsilon$ and $((\widetilde{u}_n, \widetilde{h}_n))_n$ with $(\widetilde{u}_n, \widetilde{h}_n) = \Gamma_\varepsilon(u_n, h_n)$ for each $n \in \mathbb{N}$ and

$$\begin{cases} \lim_n u_n = u & \text{in } C_b([-\tau, +\infty); X), \\ \lim_n h_n = h & \text{in } C_b(\mathbb{R}_+; Y). \end{cases}$$

So, we have

$$\begin{cases} \widetilde{v}_n'(t) \in B\widetilde{v}_n(t) + h_n(t), & t \in \mathbb{R}_+, \\ \widetilde{v}_n(t) = q(u_n, \widetilde{v}_n)(t), & t \in [-\tau, 0] \end{cases} \tag{6.9.8}$$

and

$$\begin{cases} \widetilde{u}_n'(t) \in A\widetilde{u}_n(t) + F(t, \widetilde{u}_{n_t}, \widetilde{v}_{n_t}), & t \in \mathbb{R}_+, \\ \widetilde{u}_n(t) = p(\widetilde{u}_n, \widetilde{v}_n)(t), & t \in [-\tau, 0], \end{cases} \tag{6.9.9}$$

for $n \in \mathbb{N}$. Let us observe that, by the definition of $\mathcal{K}_\varepsilon$, the set $\{h_n; \ n \in \mathbb{N}\}$ is bounded in $C_b(\mathbb{R}_+; Y)$ by r. From Lemma 6.9.1 we get

$$\|\widetilde{v}_n\|_{C_b(\mathbb{R}_+; Y)} \le \frac{r}{\gamma}$$

for each $n \in \mathbb{N}$. In particular, $\{\widetilde{v}_n(0); \ n \in \mathbb{N}\}$ is bounded in Y. Since $\{h_n; \ n \in \mathbb{N}\}$ is bounded in $C_b(\mathbb{R}_+; Y)$, recalling that, by (B_2), B generates a compact semigroup, by virtue of Theorem 1.8.5, we know that $\{\widetilde{v}_n; \ n \in \mathbb{N}\}$ is relatively compact in $\widetilde{C}_b([\sigma, +\infty); Y)$ for each $\sigma > 0$. As $\{u_n; \ n \in \mathbb{N}\}$ is bounded in $C_b(\mathbb{R}_+; \overline{D(A)})$ and $\{\widetilde{v}_n; \ n \in \mathbb{N}\}$ is relatively compact in $\widetilde{C}_b([a, +\infty); Y)$, by virtue of (q_3), it follows that $\{q(u_n, \widetilde{v}_n); n \in \mathbb{N}\}$ is relatively compact in $\mathcal{Y}$. So, $\{q(u_n, \widetilde{v}_n)(0); n \in \mathbb{N}\}$ is relatively compact in Y. Using

242   *Delay Differential Evolutions Subjected to Nonlocal Initial Conditions*

once again Theorem 1.8.5, we deduce that $\{\widetilde{v}_n;\ n \in \mathbb{N}\}$ is relatively compact in $\widetilde{C}_b(\mathbb{R}_+; Y)$. Furthermore, again by (q_3), it readily follows that $\{\widetilde{v}_n;\ n \in \mathbb{N}\}$ is relatively compact in $\widetilde{C}_b([-\tau, +\infty); Y)$. So, there exists $\widetilde{v} \in C_b([-\tau, +\infty); Y)$ and a subsequence of $(\widetilde{v}_n)_n$ – denoted for simplicity again by $(\widetilde{v}_n)_n$ – such that

$$\lim_n \widetilde{v}_n = \widetilde{v} \ \text{ in } \ \widetilde{C}_b([-\tau, +\infty); Y).$$

At this point, let us observe that, in fact, we have the stronger conclusion, i.e.,

$$\lim_n \widetilde{v}_n = \widetilde{v} \ \text{ in } \ C_b([-\tau, +\infty); Y). \tag{6.9.10}$$

Indeed, let us fix $k_\varepsilon \in \mathbb{N}$ such that $k_\varepsilon \geq 1 + 1/\varepsilon$. Next, let $\alpha > 0$ be arbitrary and let $n(\alpha, \varepsilon) = n(\alpha) \in \mathbb{N}$ (we recall that here $\varepsilon > 0$ is fixed) be such that

$$\|\widetilde{v}_n(t) - \widetilde{v}(t)\| \leq \alpha$$

for each $n \in \mathbb{N}$, $n \geq n(\alpha)$ and $t \in [0, k_\varepsilon]$. Clearly, if $t \geq k_\varepsilon$, we have

$$h_n(s) = h(s) = 0$$

for $s \in [k_\varepsilon, t]$ and thus

$$\|\widetilde{v}_n(t) - \widetilde{v}(t)\| \leq e^{-\gamma k_\varepsilon} \|\widetilde{v}_n(k_\varepsilon) - \widetilde{v}(k_\varepsilon)\| + \int_{k_\varepsilon}^t e^{-\gamma(t-s)} \|h_n(s) - h(s)\|\, ds$$

$$\leq \|\widetilde{v}_n(k_\varepsilon) - \widetilde{v}(k_\varepsilon)\| \leq \alpha.$$

So, for each $\alpha > 0$ there exists $n(\alpha) \in \mathbb{N}$ such that, for each $n \in \mathbb{N}$ with $n \geq n(\alpha)$, we have

$$\|\widetilde{v}_n - \widetilde{v}\|_{C_b(\mathbb{R}_+; Y)} \leq \alpha.$$

So, we get (6.9.10).

Let us consider now $\widetilde{u}$ the unique C^0-solution of the problem

$$\begin{cases} \widetilde{u}'(t) \in A\widetilde{u}(t) + F(t, \widetilde{u}_t, \widetilde{v}_t), & t \in \mathbb{R}_+, \\ \widetilde{u}(t) = p(\widetilde{u}, \widetilde{v})(t), & t \in [-\tau, 0], \end{cases}$$

where $\widetilde{v}$ is as above.

By virtue of (6.9.10), (F_1) in (H_F) and (p_1) in (H_p), we can apply the C_b-continuity part in Theorem 4.4.1, by choosing $f_n : \mathbb{R}_+ \times \mathcal{X} \to X$, and $g_n : C_b(\mathbb{R}_+; \overline{D(A)}) \to \mathcal{D}$ in (4.4.2) as

$$f_n(t, \cdot) = F(t, \cdot, \widetilde{v}_{n_t}), \quad \text{and} \quad g_n(\cdot) = p(\cdot, \widetilde{v}_n)$$

for $n \in \mathbb{N}$ and $t \in \mathbb{R}_+$. We conclude that

$$\lim_n \widetilde{u}_n = \widetilde{u} \ \text{ in } \ C_b([-\tau, +\infty); X), \tag{6.9.11}$$

where $\widetilde{u}$ is the unique C^0-solution of the preceding problem.

Evolution Systems with Nonlocal Initial Conditions 243

To complete the proof, let us define $\widetilde{h}(t) = G_\varepsilon(t, \widetilde{u}_t, \widetilde{v}_t)$ for $t \in \mathbb{R}_+$. It remains to show that

$$\lim_n \widetilde{h}_n = \widetilde{h} \text{ in } C_b(\mathbb{R}_+; Y). \tag{6.9.12}$$

Since, by (6.9.10) and (6.9.11), there exists both limits $\lim_n \widetilde{v}_n = \widetilde{v}$ in $C_b([-\tau, +\infty); Y)$ and $\lim_n \widetilde{u}_n = \widetilde{u}$ in $C_b([-\tau, +\infty); X)$, p satisfies (p_1) and q satisfies (q_1), it follows that

$$\lim_n p(\widetilde{u}_n, \widetilde{v}_n) = p(\widetilde{u}, \widetilde{v})$$
$$\lim_n q(u_n, \widetilde{v}_n) = q(u, \widetilde{v}),$$

in X and Y, respectively. Recalling that

$$\widetilde{h}_n(t) = \eta_\varepsilon(t) G(t, \widetilde{u}_{n_t}, \widetilde{v}_{n_t})$$

for each $n \in \mathbb{N}$ and $t \in \mathbb{R}_+$ and G is continuous, from (6.9.10) and (6.9.11), we deduce (6.9.12). Analogously

$$\lim_n F(t, \widetilde{u}_{n_t}, \widetilde{v}_{n_t}) = F(t, \widetilde{u}_t, \widetilde{v}_t)$$

uniformly for $t \in \mathbb{R}_+$. Passing to the limit in (6.9.8) and (6.9.9) and using (6.9.12), we conclude that $(\widetilde{u}, \widetilde{v})$ is a C^0-solution of

$$\begin{cases} \widetilde{u}'(t) \in A\widetilde{u}(t) + F(t, \widetilde{u}_t, \widetilde{v}_t), & t \in \mathbb{R}_+, \\ \widetilde{v}'(t) \in B\widetilde{v}(t) + \widetilde{h}(t), & t \in \mathbb{R}_+, \\ \widetilde{u}(t) = p(\widetilde{u}, \widetilde{v})(t), & t \in [-\tau, 0], \\ \widetilde{v}(t) = q(u, \widetilde{v})(t), & t \in [-\tau, 0] \end{cases}$$

and

$$\widetilde{h}(t) = \eta_\varepsilon(t) G(t, \widetilde{u}_t, \widetilde{v}_t)$$

for each $t \in \mathbb{R}_+$. But this means that

$$\lim_n \Gamma_\varepsilon(u_n, h_n) = (\widetilde{u}, \widetilde{h}) = \Gamma_\varepsilon(u, h)$$

and this completes the proof. $\qquad\qquad\square$

LEMMA 6.9.3 *If* (H_A), (H_B), (H_F), (H_G), (H_p), (H_q) *and* $(\widetilde{H}_c)$ *are satisfied, then, for each* $\varepsilon \in (0, 1)$, *the set* $\Gamma_\varepsilon(\mathcal{K}_\varepsilon)$ *is relatively compact in the space* $C_b([-\tau, +\infty); X) \times C_b(\mathbb{R}_+; Y)$.

Proof. Let $((u_n, h_n))_n$ be an arbitrary sequence in $\mathcal{K}_\varepsilon$ and let

$$(\widetilde{u}_n, \widetilde{h}_n) = \Gamma_\varepsilon(u_n, h_n)$$

244   *Delay Differential Evolutions Subjected to Nonlocal Initial Conditions*

for $n \in \mathbb{N}$. This means that, for the unique C^0-solution, $\widetilde{v}_n$, of the problem (6.9.8), $\widetilde{u}_n$ is the unique C^0-solution of the problem (6.9.9), while

$$\widetilde{h}_n(t) = \eta_\varepsilon(t) G(t, \widetilde{u}_{n_t}, \widetilde{v}_{n_t})$$

for $n \in \mathbb{N}$ and for each $t \in \mathbb{R}_+$. Since, by (B_2) in (H_B), B generates a compact semigroup and $\{h_n; \ n \in \mathbb{N}\}$ is bounded in $C_b([-\tau, +\infty); Y)$, reasoning as in the proof of Lemma 6.5.1, we obtain that, on a subsequence at least, – denoted for simplicity again by $(\widetilde{v}_n)_n$ – $\lim_n \widetilde{v}_n = \widetilde{v}$ in $C_b([-\tau, +\infty); Y)$.

Taking $f_n : \mathbb{R}_+ \times X \to X$, $f_n(t, x) = F(t, x, v_{n_t})$ for each $(t, x) \in \mathbb{R}_+ \times X$ and $g_n : C_b(\mathbb{R}_+; \overline{D(A)}) \to \mathcal{D}$, $g_n(z) = p(z, \widetilde{v}_n)$ for each $z \in C_b(\mathbb{R}_+; \overline{D(A)})$ in (4.4.2), from the C_b–continuity part of Theorem 4.4.1, we deduce that

$$\lim_n \widetilde{u}_n = \widetilde{u}$$

in $C_b([-\tau, +\infty); X)$, where $\widetilde{u}$ is the unique C^0-solution of the problem

$$\begin{cases} \widetilde{u}'(t) \in A\widetilde{u}(t) + F(t, \widetilde{u}_t, \widetilde{v}_t), & t \in \mathbb{R}_+, \\ \widetilde{u}(t) = p(\widetilde{u}, \widetilde{v})(t), & t \in [-\tau, 0]. \end{cases}$$

Since we have both $\lim_n \widetilde{v}_n = \widetilde{v}$ in $C_b([-\tau, +\infty); Y)$ and $\lim_n \widetilde{u}_n = \widetilde{u}$ in $C_b([-\tau, +\infty); X)$, it follows that

$$\begin{cases} \lim_n \widetilde{u}_{n_t} = \widetilde{u}_t & \text{in } X \\ \lim_n \widetilde{v}_{n_t} = \widetilde{v}_t & \text{in } Y \end{cases}$$

uniformly for $t \in \mathbb{R}_+$. But, by (H_G), the function G is continuous from its domain, $\mathbb{R}_+ \times X \times Y$, to Y and thus we deduce that

$$\lim_n \widetilde{h}_n = \widetilde{h}$$

in $C_b(\mathbb{R}_+; Y)$, where $\widetilde{h}(t) = G_\varepsilon(t, \widetilde{u}_t, \widetilde{v}_t)$ for each $t \in \mathbb{R}_+$.

Hence, $\Gamma_\varepsilon(\mathcal{K}_\varepsilon)$ is relatively compact in $C_b([-\tau, +\infty); X) \times C_b(\mathbb{R}_+; Y)$. This completes the proof. $\qquad\square$

6.10   Proof of Theorem 6.7.1

Proof. Let $\varepsilon \in (0, 1)$ be arbitrary but fixed. By Lemmas 6.9.1 and 6.9.2, we know that $\Gamma_\varepsilon : \mathcal{K}_\varepsilon \to \mathcal{K}_\varepsilon$ satisfies all the hypotheses of the Schauder Fixed-Point Theorem 1.4.4. So, Γ_ε has at least one fixed point $(u_\varepsilon, h_\varepsilon)$. Clearly this means that the approximate problem (6.8.1) has at least one C^0-solution

Evolution Systems with Nonlocal Initial Conditions 245

$(u_\varepsilon, v_\varepsilon)$. For each $\varepsilon \in (0,1)$, let us fix such a solution and let us consider the set $\{(u_\varepsilon, v_\varepsilon); \ \varepsilon \in (0,1)\}$.

To complete the proof, it suffices to show that the set above is relatively compact in $\widetilde{C}_b([-\tau, +\infty); X) \times \widetilde{C}_b([-\tau, +\infty); Y)$. Then, if $\varepsilon_n \downarrow 0$, we can find a convergent sequence $((u_{\varepsilon_n}, v_{\varepsilon_n}))_n$, denoted for simplicity by $((u_n, v_n))_n$, to some (u, v), which turns out to be a C^0-solution of (6.7.1).

So, let $\varepsilon_n \downarrow 0$ and let $((u_n, v_n))_n$ be as above. Reasoning as in Lemma 6.5.1, we deduce that the set $\{v_n; \ n \in \mathbb{N}\}$ is relatively compact in $\widetilde{C}_b([-\tau, +\infty); Y)$. So, there exists $v \in C_b([-\tau, +\infty); Y)$ such that, on a subsequence at least, $\lim_n v_n = v$ in $\widetilde{C}_b([-\tau, +\infty); Y)$.

Now, let us observe that u_n is the unique C^0-solution of the problem (4.4.2) in Theorem 4.4.1, where $f_n : \mathbb{R}_+ \times X \to X$, $f_n(t, x) = F(t, x, v_{n_t})$ for each $(t, x) \in \mathbb{R}_+ \times X$ and $g_n : C_b(\mathbb{R}_+; \overline{D(A)}) \to \mathcal{D}$, $g_n(z) = p(z, \widetilde{v}_n)$ for each $z \in C_b(\mathbb{R}_+; \overline{D(A)})$. From (H_A), (H_F) and (H_p), it follows that we are in the hypotheses of the $\widetilde{C}_b$–continuity part in Theorem 4.4.1, from which we conclude that, on that subsequence at least, we have $\lim_n u_n = u$ in $\widetilde{C}_b([-\tau, +\infty); X)$ where u is the C^0-solution of the problem

$$\begin{cases} u'(t) \in Au(t) + F(t, u_t, v_t), & t \in \mathbb{R}_+, \\ u(t) = p(u, v)(t), & t \in [-\tau, 0]. \end{cases}$$

Since $h_n(t) = \eta_{\varepsilon_n}(t) G(t, u_{n_t}, v_{n_t})$ satisfies $\lim_n h_n(t) = G(t, u_t, v_t)$ uniformly for t in bounded intervals in $\mathbb{R}_+$, using (1.8.2) and the continuity property of the function q, we get that v is a C^0-solution of the problem

$$\begin{cases} v'(t) \in Bv(t) + G(t, u_t, v_t), & t \in \mathbb{R}_+, \\ v(t) = q(u, v)(t), & t \in [-\tau, 0]. \end{cases}$$

The proof of (6.7.2) follows the very same lines as those in the proof of Theorem 4.1.1, and so we do not provide details.

If the additional hypothesis (6.7.3) is satisfied, we write the system as an evolution equation in the product space $Z = X \times Y$, endowed with the maximum norm and we apply the last part of Theorem 4.1.1. This completes the proof of Theorem 6.7.1. $\qquad \square$

6.11 A nonlinear reaction–diffusion system in $L^1(\Omega)$

Let Ω be a nonempty, bounded domain in $\mathbb{R}^d$, $d \geq 1$, with C^1 boundary Σ, let $\tau \geq 0$, $\omega > 0$, $\gamma > 0$, let $Q_+ = \mathbb{R}_+ \times \Omega$, $\Sigma_+ = \mathbb{R}_+ \times \Sigma$, $Q_\tau = [-\tau, 0] \times \Omega$ and let $\varphi : D(\varphi) \subseteq \mathbb{R} \rightsquigarrow \mathbb{R}$ and $\psi : D(\psi) \subseteq \mathbb{R} \rightsquigarrow \mathbb{R}$ be maximal-monotone operators with $0 \in D(\varphi)$, $0 \in D(\psi)$, $0 \in \varphi(0)$, $0 \in \psi(0)$. Let $\mathcal{X}_1 = C([-\tau, 0]; L^1(\Omega))$ and let $F, G : \mathbb{R}_+ \times \mathcal{X}_1 \times \mathcal{X}_1 \to L^1(\Omega)$ be continuous.

246 *Delay Differential Evolutions Subjected to Nonlocal Initial Conditions*

Let μ_i, $i = 1, 2$, be two positive σ-finite and complete measures on the class of Borel measurable sets in $\mathbb{R}_+$, $k_i \in L^1(\mathbb{R}_+; \mu_i)$ with $\|k_i\|_{L^1(\mathbb{R}_+; \mu_i)} \leq 1$, $i = 1, 2$, let $W_i : \mathbb{R} \to \mathbb{R}$ be nonexpansive with $W_i(0) = 0$, $i = 1, 2$, and let $C : L^1(\Omega) \to \mathbb{R}$ be a linear continuous functional. We consider the following system subjected to mixed nonlocal initial conditions:

$$\begin{cases} \dfrac{\partial u}{\partial t}(t, x) = \Delta\varphi(u(t, x)) - \omega u(t, x) + F(t, u_t, v_t)(x), & \text{in } Q_+, \\[2mm] \dfrac{\partial v}{\partial t}(t, x) = \Delta\psi(u(t, x)) - \gamma v(t, x) + G(t, u_t, v_t)(x), & \text{in } Q_+, \\[2mm] \varphi(u(t, x)) = 0, \quad \psi(v(t, x)) = 0, & \text{on } \Sigma_+, \quad (6.11.1) \\[2mm] u(t, x) = \displaystyle\int_b^\infty k_1(s) W_1(v(t + s, x)) u(t + s, x) \, d\mu_1(s), & \text{in } Q_\tau, \\[2mm] v(t, x) = \displaystyle\int_b^\infty k_2(s) W_2(v(t + s, x)) C u(t + s, \cdot) \, d\mu_2(s), & \text{in } Q_\tau. \end{cases}$$

Our main result concerning the system (6.11.1) is:

THEOREM 6.11.1 *Let Ω be a nonempty, bounded, open subset in $\mathbb{R}^d$, $d \geq 1$, with C^1 boundary Σ and let $\varphi : D(\varphi) \subseteq \mathbb{R} \rightsquigarrow \mathbb{R}$, $\psi : D(\psi) \subseteq \mathbb{R} \rightsquigarrow \mathbb{R}$ be maximal-monotone operators with $0 \in D(\varphi)$, $0 \in D(\psi)$, $0 \in \varphi(0)$ and $0 \in \psi(0)$. Let $\tau \geq 0$, $\omega > 0$, $\gamma > 0$ and let $F, G : \mathbb{R}_+ \times \mathfrak{X}_1 \times \mathfrak{X}_1 \to L^1(\Omega)$. Furthermore, let $b > \tau$, let μ_i be positive σ-finite and complete measures defined on the class of Borel measurable sets in $\mathbb{R}_+$ with $\operatorname{supp} \mu_i \subseteq [b, +\infty)$, $i = 1, 2$, let $k_i \in L^\infty(\mathbb{R}_+; \mu_i)$ be nonnegative functions and let $W_i : \mathbb{R} \to \mathbb{R}_+$, $i = 1, 2$. Let us assume that*

(h_1) *$\psi : \mathbb{R} \to \mathbb{R}$ is continuous on $\mathbb{R}$ and C^1 on $\mathbb{R} \setminus \{0\}$ and there exist two constants $C > 0$ and $\alpha > 0$ if $d \leq 2$ and $\alpha > (d - 2)/d$ if $d \geq 3$ such that*

$$\psi'(r) \geq C|r|^{\alpha - 1}$$

for each $r \in \mathbb{R} \setminus \{0\}$

(h_2) *$F : \mathbb{R}_+ \times \mathfrak{X}_1 \times \mathfrak{X}_1 \to L^1(\Omega)$ is continuous and there exist $\ell > 0$ and $m > 0$ such that*

$$\|F(t, u, v) - F(t, \widetilde{u}, \widetilde{v})\|_{L^1(\Omega)} \leq \ell \max\{\|u - \widetilde{u}\|_{\mathfrak{X}_1}, \|v - \widetilde{v}\|_{\mathfrak{X}_1}\},$$

and

$$\|F(t, u, v)\|_{L^1(\Omega)} \leq \ell \|u\|_{\mathfrak{X}_1} + m,$$

for each $(t, u, v), (t, \widetilde{u}, \widetilde{v}) \in \mathbb{R}_+ \times \mathfrak{X}_1 \times \mathfrak{X}_1$

(h_3) *$G : \mathbb{R}_+ \times \mathfrak{X}_1 \times \mathfrak{X}_1 \to L^1(\Omega)$ is continuous and satisfies*

(h_4) *with ℓ and m given by (h_2), we have*

$$\|G(t, u, v)\|_{L^1(\Omega)} \leq \ell \max\{\|u\|_{\mathfrak{X}_1}, \|v\|_{\mathfrak{X}_1}\} + m,$$

for each $(t, u, v), (t, \widetilde{u}, \widetilde{v}) \in \mathbb{R}_+ \times \mathfrak{X}_1 \times \mathfrak{X}_1$

Evolution Systems with Nonlocal Initial Conditions

(h_5) $\|k_i\|_{L^\infty(\mathbb{R}_+;\mu_i)} \le 1$, $i = 1, 2$

(h_6) $|W_i(v) - W_i(\widetilde{v})| \le |v - \widetilde{v}|$, for each $v, \widetilde{v} \in \mathbb{R}$, $i = 1, 2$

(h_7) $W_i(0) = 0$ for $i = 1, 2$

(h_8) $C : L^1(\Omega) \to \mathbb{R}$ is a linear continuous functional whose norm satisfies $\|C\| \le 1$.

Let us assume also that $(\widetilde{H}_c)$ is satisfied. Then, (6.11.1) has at least one C^0-solution. If, in addition, G satisfies

$$\|G(t, u, v) - G(t, \widetilde{u}, \widetilde{v})\|_{L^1(\Omega)} \le \ell \max\{\|u - \widetilde{u}\|_{\mathfrak{X}_1}, \|v - \widetilde{v}\|_{\mathfrak{X}_1}\}$$

for each $(t, u, v), (t, \widetilde{u}, \widetilde{v}) \in \mathbb{R}_+ \times \mathfrak{X}_1 \times \mathfrak{X}_1$, then the C^0-solution of (6.11.1) is unique. Moreover, if both F and G are Lipschitz with respect to all three variables with Lipschitz constant ℓ, the unique C^0-solution of (6.11.1) is globally asymptotically stable.

Proof. In order to apply Theorem 6.7.1, we rewrite problem (6.11.1) as an abstract system of the form (6.7.1) in the function spaces $X = Y = L^1(\Omega)$ as follows. First, let $A : D(A) \subseteq L^1(\Omega) \to L^1(\Omega)$ be defined by

$$Au = \Delta\varphi(u) - \omega u$$

for each $u \in D(A) = D(\Delta\varphi)$, where

$$\begin{cases} D(\Delta\varphi) = \{u \in L^1(\Omega); \ \exists w \in \mathcal{S}_\varphi(u) \cap W_0^{1,1}(\Omega), \ \Delta w \in L^1(\Omega)\}, \\ \Delta\varphi(u) = \{\Delta w; \ w \in \mathcal{S}_\varphi(u) \cap W_0^{1,1}(\Omega)\} \cap L^1(\Omega) \text{ for } u \in D(\Delta\varphi), \end{cases}$$

$\mathcal{S}_\varphi(u)$ being defined by $\mathcal{S}_\varphi(u) = \{w \in L^1(\Omega); \ w(x) \in \varphi(u(x)), \text{ a.e. for } x \in \Omega\}$. Next, we define $B : D(B) \subseteq L^1(\Omega) \to L^1(\Omega)$ in the very same manner as we did in the case of the operator A, by simply replacing φ by ψ.

From Theorem 1.9.6, we know that A is ω-m-dissipative, $0 \in D(A), 0 \in A0$, $\overline{D(A)} = L^1(\Omega)$ and so, $\overline{D(A)}$ is convex. Also from Theorem 1.9.6, we conclude that B is γ-m-dissipative, $0 \in D(B)$ and $0 \in B0$ and B generates a compact semigroup in $\overline{D(B)} = L^1(\Omega)$. So, (H_A) and (H_B) are satisfied. From (h_2), it follows that F satisfies (H_F). In addition, by (h_3), G is continuous, while from (h_4), both have affine growth with constants ℓ and m. So, G satisfies (H_G).

Finally, from $(h_5) \sim (h_7)$, we deduce that the functions

$$p, q : C_b(\mathbb{R}_+; L^1(\Omega)) \times C_b(\mathbb{R}_+; L^1(\Omega)) \to \mathfrak{X}_1,$$

defined by

$$\begin{cases} p(u, v)(t)(x) = \displaystyle\int_b^\infty k_1(s) W_1(v(t + s)(x)) u(t + s)(x) \, d\mu_1(s), \\ q(u, v)(t)(x) = \displaystyle\int_b^\infty k_2(s) W_2(v(t + s)(x)) C u(t + s)(\cdot) \, d\mu_2(s), \end{cases}$$

248   *Delay Differential Evolutions Subjected to Nonlocal Initial Conditions*

for each $(u, v) \in C_b(\mathbb{R}_+; L^1(\Omega)) \times C_b(\mathbb{R}_+; L^1(\Omega))$, each $t \in [-\tau, 0]$ and a.e. for $x \in \Omega$, satisfy (H_p) and (H_q) – we notice that in our case $a = b - \tau > 0$. Consequently, Theorem 6.7.1 applies, from which the conclusion follows. $\quad\square$

6.12   Bibliographical notes and comments

Section 6.1 A general existence result for evolution systems without delay and subjected to coupled nonlocal initial conditions expressed in terms of Stieltjes integrals was obtained by Bolojan-Nica, Infante and Precup [40].

A single-valued perturbed system with delay similar to (6.1.1) subjected to some nonlocal initial conditions having linear growth but more general than the discrete mean conditions here considered, and by assuming that F and G satisfy an affine growth condition with respect to both variables, was studied by Burlică, Roşu and Vrabie [62].

Section 6.2 We emphasize that Theorem 6.2.1, which is a specific case of a general result due to Burlică and Roşu [60], is inspired by both Vrabie [256] and Burlică and Roşu [58]. For previous related results, see Burlică, Roşu and Vrabie [62].

The importance of such kind of systems rests in the simple observation that they include reaction–diffusion systems for which one unknown function is subjected to a time-periodic-like condition, while the other one to a mean condition over $\mathbb{R}_+$, a situation that is of great practical interest.

Theorem 6.2.2 is new and appears for the first time here.

Section 6.3 The idea of the proof is different from the one used in Burlică, Roşu and Vrabie [62] to solve the case in which $g(0) = 0$.

Section 6.4 Lemma 6.4.1 and Corollary 6.4.1 are also new, although not surprising. Remark 6.4.1 is perhaps known, but we have decided to include it here with some detailed arguments because it is one of the main ingredients in the study of abstract reaction–diffusion systems.

Section 6.5 Lemma 6.5.1 appears for the first time here.

Section 6.6 Theorem 6.6.1 is a specific form of a result in Burlică, Roşu and Vrabie [62].

Section 6.7 The system (6.7.1) was considered in Burlică, Roşu and Vrabie [62] under similar hypotheses. Theorem 6.7.1 is from Burlică, Roşu and Vrabie [62], while Theorem 6.7.2 is a simple consequence of the latter.

Section 6.8 The idea of the proof is also from Burlică, Roşu and Vrabie [62] and it was used for the first time there for the case of single-valued perturbed reaction–diffusion systems.

Evolution Systems with Nonlocal Initial Conditions 249

Section 6.9 Lemmas 6.9.1, 6.9.2 as well as Lemma 6.9.3 slightly differ from the corresponding ones in Burlică, Roşu and Vrabie [62].

Section 6.10 The proof of Theorem 6.7.1 is a simplified version of its corresponding counterpart in Burlică, Roşu and Vrabie [62].

Section 6.11 The example in this section, as well as Theorem 6.11.1, are new. We notice that, working in $L^\infty(\Omega)$, which is invariant with respect to the semigroup generated by $\Delta\psi$ and proceeding as in Díaz and Vrabie [96], [97], [98], one may relax (h_1) to the weaker assumption

(h_1') $\psi : \mathbb{R} \to \mathbb{R}$ *is continuous and strictly increasing on* $\mathbb{R}$.

Chapter 7

Delay Evolution Inclusions

Overview

In this chapter, we study a class of nonlinear functional differential evolution inclusions of the form

$$\begin{cases} u'(t) \in Au(t) + f(t), & t \in \mathbb{R}_+, \\ f(t) \in F(t, u_t), & t \in \mathbb{R}_+, \\ u(t) = g(u)(t), & t \in [-\tau, 0]. \end{cases}$$

Here, X is a Banach space, $A : D(A) \subseteq X \rightsquigarrow X$ is an ω-m-dissipative operator for a certain $\omega > 0$, $\tau \geq 0$, $\mathcal{X} = C([-\tau, 0]; X)$ and $\mathcal{D} = \{\varphi \in \mathcal{X}; \; \varphi(0) \in \overline{D(A)}\}$, $F : \mathbb{R}_+ \times \mathcal{X} \rightsquigarrow X$ is a nonempty, convex, and weakly compact-valued multifunction, and the function $g : C_b(\mathbb{R}_+; \overline{D(A)}) \to \mathcal{D}$ is nonexpansive.

7.1 The problem to be studied

In this chapter, we prove an existence result for bounded C^0-solutions to a class of nonlinear functional differential evolution inclusions subjected to nonlocal initial conditions of the form

$$\begin{cases} u'(t) \in Au(t) + f(t), & t \in \mathbb{R}_+, \\ f(t) \in F(t, u_t), & t \in \mathbb{R}_+, \\ u(t) = g(u)(t), & t \in [-\tau, 0], \end{cases} \tag{7.1.1}$$

where X is a Banach space, $\tau \geq 0$, the operator $A : D(A) \subseteq X \rightsquigarrow X$ is ω-m-dissipative for a certain $\omega > 0$,

$$\mathcal{X} = C([-\tau, 0]; X) \quad \text{and} \quad \mathcal{D} = \{\varphi \in \mathcal{X}; \; \varphi(0) \in \overline{D(A)}\},$$

$F : \mathbb{R}_+ \times \mathcal{X} \rightsquigarrow X$ is nonempty, convex, weakly compact-valued and almost strongly-weakly u.s.c., and $g : C_b(\mathbb{R}_+; \overline{D(A)}) \to \mathcal{D}$ is nonexpansive.

251

252 *Delay Differential Evolutions Subjected to Nonlocal Initial Conditions*

DEFINITION 7.1.1 By a C^0-*solution* of problem (7.1.1), we mean a function $u \in C([-\tau, +\infty); X)$ for which there exists $f \in L^1_{loc}(\mathbb{R}_+; X)$ satisfying $f(t) \in F(t, u_t)$ a.e. for $t \in \mathbb{R}_+$, $u(t) = g(u)(t)$ for each $t \in [-\tau, 0]$ and such that u is a C^0-solution of the problem

$$u'(t) \in Au(t) + f(t)$$

on $[0, T]$ in the sense of Definition 1.8.2 for each $T > 0$.

We begin by considering the auxiliary problem

$$\begin{cases} z'(t) \in Az(t), & t \in \mathbb{R}_+, \\ z(t) = g(z)(t), & t \in [-\tau, 0]. \end{cases} \tag{7.1.2}$$

The next lemma, which will prove useful later, is a specific case of Lemma 4.2.1.

LEMMA 7.1.1 *Let us assume that* $\tau \geq 0$ *and* $A : D(A) \subseteq X \rightsquigarrow X$ *is* ω-*m*-*dissipative for some* $\omega > 0$, $0 \in D(A)$ *and* $0 \in A0$. *Let us assume, in addition, that there exists* $a > 0$ *such that* $g : C_b(\mathbb{R}_+; \overline{D(A)}) \to \mathcal{D}$ *is nonexpansive from* $C_b([a, +\infty); \overline{D(A)})$ *to* X. *Then the problem* (7.1.2) *has a unique* C^0-*solution* $z \in C_b([-\tau, +\infty); X)$.

The assumptions we need in what follows are listed below.

(H_A) $A : D(A) \subseteq X \rightsquigarrow X$ is an operator with the following properties:

(A_1) $0 \in D(A)$, $0 \in A0$ and A is ω-m-dissipative for some $\omega > 0$

(A_2) the semigroup generated by A on $\overline{D(A)}$ is compact

(A_6) A is of complete continuous type.

(H_F) $F : \mathbb{R}_+ \times \mathcal{X} \rightsquigarrow X$ is nonempty, convex, weakly compact-valued, almost strongly-weakly u.s.c., and satisfies

(F_3) there exists $r > 0$ such that for each $t \in \mathbb{R}_+$, for each $v \in \mathcal{X}$, with $\|v - z_t\|_{\mathcal{X}} = r$ and for all $f \in F(t, v)$, we have $[v(0) - z(t), f]_+ \leq 0$, where z is the unique C^0-solution of the auxiliary problem (7.1.2)

(F'_3) there exists $r > 0$ such that for each $t \in \mathbb{R}_+$, for each $v \in \mathcal{X}$ with $\|v(0) - z(t)\| > r$ and for all $f \in F(t, v)$, we have $[v(0) - z(t), f]_+ \leq 0$, where z is the unique C^0-solution of the auxiliary problem (7.1.2)

(F_4) there exists a nonnegative function $\ell \in L^\infty(\mathbb{R}_+) \cap L^1(\mathbb{R}_+)$ such that a.e. for $t \in \mathbb{R}_+$ and for each $v \in \mathcal{X}$ satisfying

$$\|v(0) - z(t)\| \leq r,$$

where r is given by (F_3), and each $f \in F(t, v)$, we have

$$\|f\| \leq \ell(t)$$

Delay Evolution Inclusions 253

(F_4') there exists a nonnegative function $\ell \in L^\infty(\mathbb{R}_+) \cap L^1(\mathbb{R}_+)$ such that

$$\|f\| \le \ell(t)$$

a.e. for $t \in \mathbb{R}_+$, for each $v \in \mathcal{X}$ and each $f \in F(t, v)$.

(H_g) $g : C_b(\mathbb{R}_+; \overline{D(A)}) \to \mathcal{D}$ satisfies

(g_1) there exists $a > 0$ such that for each $u, v \in C_b(\mathbb{R}_+; \overline{D(A)})$, we have

$$\|g(u) - g(v)\|_\mathcal{X} \le \|u - v\|_{C_b([a, +\infty); X)}$$

(g_4) g is continuous from $C_b([a, +\infty); \overline{D(A)})$ endowed with the $\widetilde{C}_b([a, +\infty); X)$ topology to $\mathcal{X}$, where a is given by (g_1).

REMARK 7.1.1 The hypothesis (F_3) is simply an invariance condition for the C^0-solutions of the problem

$$\begin{cases} u'(t) \in Au(t) + f(t), & t \in \mathbb{R}_+, \\ f(t) \in F(t, u_t), & t \in \mathbb{R}_+. \end{cases}$$

Namely, it implies that each C^0-solution u of the problem above issuing from an initial history $\psi \in \mathcal{D}$, satisfying the constraints $\psi(t) - z(t) \in D(0, r)$ for each $t \in [-\tau, 0]$, where z is the unique C^0-solution of (7.1.2), satisfy the very same constraints as long as it exists, i.e., $u(t) - z(t) \in D(0, r)$ for each $t \in [-\tau, +\infty)$.

Conditions (g_1) and (g_4) are satisfied by all functions g of the general form specified in Remark 3.2.4.

7.2 The main results and the idea of the proof

We may now proceed to the statement of the main result in this chapter.

THEOREM 7.2.1 *If (H_A), (F_3), (F_4) in (H_F) and (H_g) are satisfied, then the problem (7.1.1) has at least one C^0-solution $u \in C_b([-\tau, +\infty); X)$. In addition*

$$u(t) - z(t) \in D(0, r)$$

for each $t \in \mathbb{R}_+$, where z is the unique C^0-solution of the auxiliary problem (7.1.2) and r is given by (F_3).

In order to say a few words about the case without delay, let us consider the problem

$$\begin{cases} u'(t) \in Au(t) + f(t), & t \in \mathbb{R}_+, \\ f(t) \in F(t, u(t)), & t \in \mathbb{R}_+, \\ u(0) = g(u), \end{cases} \qquad (7.2.1)$$

254 *Delay Differential Evolutions Subjected to Nonlocal Initial Conditions*

where $F : \mathbb{R}_+ \times X \rightsquigarrow X$ is nonempty, convex, weakly compact-valued and almost strongly-weakly u.s.c., and $g : C_b(\mathbb{R}_+; \overline{D(A)}) \to \overline{D(A)}$ is nonexpansive.

In order to reformulate the hypotheses (H_F), (F_3), (F_3'), (F_4), (F_4') and (H_g) in the case $\tau = 0$, we need to consider the nondelayed auxiliary problem

$$\begin{cases} w'(t) \in Aw(t), & t \in \mathbb{R}_+, \\ w(0) = g(w), \end{cases} \tag{7.2.2}$$

which has a unique C^0-solution $w \in C_b(\mathbb{R}_+; \overline{D(A)})$.

$(H_F^{[\tau=0]})$ $F : \mathbb{R}_+ \times X \rightsquigarrow X$ is nonempty, convex, weakly compact-valued, almost strongly-weakly u.s.c., and satisfies

$(F_3^{[\tau=0]})$ there exists $r > 0$ such that for each $(t, v) \in \mathbb{R}_+ \times X$, satisfying $\|v - w(t)\| = r$ and each $f \in F(t, v)$, we have $[\, v - w(t), f \,]_+ \le 0$, where w is the unique C^0-solution of the problem (7.2.2)

$(F_3'^{[\tau=0]})$ there exists $r > 0$ such that for each $(t, v) \in \mathbb{R}_+ \times X$, satisfying $\|v - w(t)\| > r$ and each $f \in F(t, v)$, we have $[\, v - w(t), f \,]_+ \le 0$, where w is the unique C^0-solution of the problem (7.2.2)

$(F_4^{[\tau=0]})$ there exists a nonnegative function $\ell \in L^\infty(\mathbb{R}_+) \cap L^1(\mathbb{R}_+)$ such that

$$\|f\| \le \ell(t)$$

a.e. for $t \in \mathbb{R}_+$, for each $v \in X$ with $\|v - w(t)\| \le r$ and each $f \in F(t, v)$, where r is given by $(F_3'^{[\tau=0]})$

$(F_4'^{[\tau=0]})$ there exists a nonnegative function $\ell \in L^\infty(\mathbb{R}_+) \cap L^1(\mathbb{R}_+)$ such that

$$\|f\| \le \ell(t)$$

a.e. for $t \in \mathbb{R}_+$, for each $v \in X$ and $f \in F(t, v)$.

$(H_g^{[\tau=0]})$ $g : C_b(\mathbb{R}_+; \overline{D(A)}) \to \overline{D(A)}$ satisfies

$(g_1^{[\tau=0]})$ there exists $a > 0$ such that for each $u, v \in C_b(\mathbb{R}_+; \overline{D(A)})$, we have

$$\|g(u) - g(v)\|_X \le \|u - v\|_{C_b([\, a, +\infty); X)}$$

$(g_4^{[\tau=0]})$ g is continuous from $C_b([\, a, +\infty); \overline{D(A)})$, endowed with the $\widetilde{C}_b([\, a, +\infty); X)$ topology, to X, where a is given by $(g_1^{[\tau=0]})$.

From Theorem 7.2.1, we deduce:

Delay Evolution Inclusions 255

THEOREM 7.2.2 *If* (H_A), $(F_3^{[\tau=0]})$, $(F_4^{[\tau=0]})$ *in* $(H_F^{[\tau=0]})$ *and* $(H_g^{[\tau=0]})$ *are satisfied, then the problem* (7.2.1) *has at least one* C^0-*solution* $u \in C_b(\mathbb{R}_+; \overline{D(A)})$. *In addition,*

$$u(t) - w(t) \in D(0, r)$$

for each $t \in \mathbb{R}_+$, *where* w *is the unique* C^0-*solution of the auxiliary problem* (7.2.2) *and* r *is given by* $(F_3^{[\tau=0]})$.

We will prove Theorem 7.2.1 with the help of the following:

THEOREM 7.2.3 *If* (H_A), (F_3'), (F_4') *in* (H_F) *and* (H_g) *are satisfied, then the problem* (7.1.1) *has at least one* C^0-*solution* $u \in C_b([-\tau, +\infty); X)$. *In addition,*

$$u(t) - z(t) \in D(0, r)$$

for each $t \in \mathbb{R}_+$, *where* z *is the unique* C^0-*solution of the auxiliary problem* (7.1.2) *and* r *is given by* (F_3').

7.2.1 The idea of the proof of Theorem 7.2.3

Since the proof is rather technical, for the sake of simplicity, we divide it into five steps, which we will label later on as five lemmas.

The first step. We show that, for each $\varepsilon \in (0, 1)$ and $f \in L^1(\mathbb{R}_+; X)$, the problem

$$\begin{cases} u'(t) \in Au(t) - \varepsilon[\,u(t) - z(t)\,] + f(t), & t \in \mathbb{R}_+, \\ u(t) = g(u)(t), & t \in [\,-\tau, 0\,], \end{cases} \tag{7.2.3}$$

has a unique C^0-solution $u_\varepsilon^f \in C_b([-\tau, +\infty); X)$.

The second step. We prove that, for each $t \in \mathbb{R}_+$, $u_\varepsilon^f(t)$ remains in a ball of radius $r > 0$ and centered at $z(t)$.

The third step. We show that for each arbitrary but fixed $\varepsilon \in (0, 1)$, the operator $f \mapsto u_\varepsilon^f$, which associates to f the unique C^0-solution u_ε^f of the problem (7.2.3), is compact from the set $\mathcal{F}$, defined by

$$\mathcal{F} = \{f \in L^\infty(\mathbb{R}_+; X) \cap L^1(\mathbb{R}_+; X); \ \|f(t)\| \le \ell(t) \text{ a.e. for } t \in \mathbb{R}_+\},$$

to $\widetilde{C}_b([-\tau, +\infty); X)$.

The fourth step. We exploit the fact that F is almost strongly-weakly u.s.c. More precisely, for the same arbitrary but fixed $\varepsilon > 0$ as in the latter step, by Definition 1.5.3, there exists $E_\varepsilon \subseteq \mathbb{R}_+$ whose Lebesgue measure $\lambda(E_\varepsilon) \le \varepsilon$ and such that the restriction of F to $(\mathbb{R}_+ \setminus E_\varepsilon) \times \mathcal{X}$, denoted as usual

256   *Delay Differential Evolutions Subjected to Nonlocal Initial Conditions*

by $F_{|(\mathbb{R}_+ \setminus E_\varepsilon) \times \mathfrak{X}}$, is strongly-weakly u.s.c. Next, we construct an approximation for F as follows. Let

$$D(F) = \mathbb{R}_+ \times \mathfrak{X},$$
$$D_\varepsilon(F) = (\mathbb{R}_+ \setminus E_\varepsilon) \times \mathfrak{X}$$

and let us define the multifunction $F_\varepsilon : \mathbb{R}_+ \times \mathfrak{X} \rightsquigarrow X$, by

$$F_\varepsilon(t, v) = \begin{cases} F(t, v), & (t, v) \in D_\varepsilon(F), \\ \{0\}, & (t, v) \in D(F) \setminus D_\varepsilon(F). \end{cases} \tag{7.2.4}$$

Further, we prove that the multifunction $f \mapsto \operatorname{Sel} F_\varepsilon(\cdot, u^f_{\varepsilon(\cdot)})$, where

$$\operatorname{Sel} F_\varepsilon(\cdot, u^f_{\varepsilon(\cdot)}) = \{h \in L^1(\mathbb{R}_+; X); \ h(t) \in F_\varepsilon(t, u^f_{\varepsilon t}) \text{ a.e. } t \in \mathbb{R}_+\},$$

maps some nonempty, convex and weakly compact set $\mathcal{K}_\varepsilon \subseteq L^1(\mathbb{R}_+; X)$ into itself and its graph is weakly×weakly sequentially closed. At this point, thanks to Theorem 1.5.4, we conclude that this multifunction has at least one fixed point which, by means of $f \mapsto u^f_\varepsilon$, produces a C^0-solution for the approximate problem

$$\begin{cases} u'(t) \in Au(t) - \varepsilon[\, u(t) - z(t)\,] + f(t), & t \in \mathbb{R}_+, \\ f(t) \in F_\varepsilon(t, u_t), & t \in \mathbb{R}_+, \\ u(t) = g(u)(t), & t \in [-\tau, 0\,], \end{cases} \tag{7.2.5}$$

where F_ε is defined by (7.2.4).

The fifth step. For each $\varepsilon \in (0, 1)$, we fix a C^0-solution u_ε of the problem (7.2.5) and we show that there exists a sequence $\varepsilon_n \downarrow 0$ such that $(u_{\varepsilon_n})_n$ converges in $\widetilde{C}_b([-\tau, +\infty); X)$ to a C^0-solution of the problem (7.1.1).

7.3 Proof of Theorem 7.2.1

As we already noted, for the sake of convenience and clarity, we have divided the proof of the Theorem 7.2.3 into five steps, which are labeled as five lemmas.

LEMMA 7.3.1 *Let us assume that (A_1) in (H_A) and (g_1) in (H_g) are satisfied. Then, for each $\varepsilon > 0$ and each $f \in L^1(\mathbb{R}_+; X)$, the problem (7.2.3) has a unique C^0-solution $u \in C_b([-\tau, +\infty); X)$.*

Proof. The proof is somehow similar to that of the first part of Lemma 4.2.1. Namely, let us first observe that from (g_1) in (H_g), we are in the hypotheses of Remark 3.2.3, from which we deduce that, for each $u \in C_b([-\tau, +\infty); X)$,

Delay Evolution Inclusions 257

$g(u)$ depends merely on the restriction $u_{|[a,+\infty)}$. Thus, we may assume with no loss of generality that

$$g : C_b([a, +\infty); \overline{D(A)}) \to \mathcal{D}.$$

Next, we will use a fixed-point argument. Let z be the unique C^0-solution of the auxiliary problem (7.1.2). First, we will prove that the problem

$$\begin{cases} u'(t) \in A_\varepsilon u(t) + f_\varepsilon(t), & t \in \mathbb{R}_+, \\ u(0) = g(u)(0), \end{cases} \tag{7.3.1}$$

where $A_\varepsilon = A - \varepsilon I$ and $f_\varepsilon(t) = f(t) + \varepsilon z(t)$ for $t \in \mathbb{R}_+$, has a unique C^0-solution $u \in C_b(\mathbb{R}_+; \overline{D(A)})$. Then, we have only to observe that $\overline{u} : [-\tau, +\infty) \to X$, defined by

$$\overline{u}(t) = \begin{cases} u(t), & t \in \mathbb{R}_+, \\ g(u)(t), & t \in [-\tau, 0), \end{cases}$$

is the unique C^0-solution of (7.2.3).

To prove that the problem (7.3.1) has a unique C^0-solution, we will use a fixed-point argument. Namely, let $v \in C_b([a, +\infty); \overline{D(A)})$ and let us consider the problem

$$\begin{cases} u'(t) \in A_\varepsilon u(t) + f_\varepsilon(t), & t \in \mathbb{R}_+, \\ u(0) = g(v)(0). \end{cases} \tag{7.3.2}$$

Since $g(v)(t) \in \mathcal{D}$ for each $t \in \mathbb{R}_+$, it follows that $g(v)(0) \in \overline{D(A)}$ and so, in view of Theorem 1.8.1, (7.3.2) has a unique C^0-solution $u \in C_b(\mathbb{R}_+; \overline{D(A)})$.

Let us consider the operator

$$P_\varepsilon : C_b([a, +\infty); \overline{D(A)}) \to C_b([a, +\infty); \overline{D(A)}),$$

defined by

$$P_\varepsilon(v) = u_{|[a,+\infty)},$$

where u is the unique C^0-solution of the problem (7.3.2).

Clearly, u is a C^0-solution of (7.3.1) if and only if $u_{|[a,+\infty)}$ is a fixed point of P_ε. In order to prove that P_ε has a unique fixed point, we will show that it is a strict contraction. To this end, let $v, w \in C_b([a, +\infty); \overline{D(A)})$ be arbitrary and let $t \in [a, +\infty)$. Since $A_\varepsilon + \varepsilon I$ is dissipative (in fact A_ε is ε-m-dissipative), from Theorem 1.8.1 and (ii) in Proposition 1.8.1, we have

$$\|P_\varepsilon(v)(t) - P_\varepsilon(w)(t)\| \le e^{-\varepsilon t} \|P_\varepsilon(v)(a) - P_\varepsilon(w)(a)\|$$

$$+ \int_a^t e^{-\varepsilon(t-s)} [P_\varepsilon(v)(s) - P_\varepsilon(w)(s), f_\varepsilon(s) - f_\varepsilon(s)]_+ \, ds$$

$$\le e^{-\varepsilon a} \|P_\varepsilon(v)(a) - P_\varepsilon(w)(a)\|$$

258   *Delay Differential Evolutions Subjected to Nonlocal Initial Conditions*

for each $t \in [a, +\infty)$. So,

$$\|P_\varepsilon(v) - P_\varepsilon(w)\|_{C_b([a,+\infty);X)} \leq e^{-\varepsilon a} \|P_\varepsilon(v)(a) - P_\varepsilon(w)(a)\|. \qquad (7.3.3)$$

Next, recalling that $P_\varepsilon(v) = u_{|[a,+\infty)}$ and $P_\varepsilon(w) = \widetilde{u}_{|[a,+\infty)}$, where u and $\widetilde{u}$ are the unique C^0-solutions of (7.3.2) corresponding to v and respectively to w, using again Theorem 1.8.1 and (g_1), we get

$$\|P_\varepsilon(v)(a) - P_\varepsilon(w)(a)\| = \|u(a) - \widetilde{u}(a)\| \leq e^{-\varepsilon a} \|u(0) - \widetilde{u}(0)\|$$

$$\leq \|g(v)(0) - g(w)(0)\| \leq \|v - w\|_{C_b([a,+\infty);X)}.$$

From this inequality and (7.3.3), we deduce

$$\|P_\varepsilon(v) - P_\varepsilon(w)\|_{C_b([a,+\infty);X)} \leq e^{-\varepsilon a} \|v - w\|_{C_b([a,+\infty);X)}.$$

Since v and w are arbitrary, the last inequality shows that P_ε is a contraction of constant $e^{-\varepsilon a}$ and this completes the proof. $\qquad\square$

LEMMA 7.3.2 *Let us assume that (A_1) in (H_A) and (g_1) in (H_g) are satisfied. Then, for each $\varepsilon > 0$ and each $f \in L^\infty(\mathbb{R}_+; X)$, the unique C^0-solution u_ε^f of the problem (7.2.3) belongs to $C_b([-\tau, +\infty); X)$ and satisfies*

$$\|u_\varepsilon^f - z\|_{C_b([-\tau,+\infty);X)} \leq \frac{1}{\varepsilon} \|f\|_{L^\infty(\mathbb{R}_+;X)}, \qquad (7.3.4)$$

where z is the unique C^0-solution of the auxiliary problem (7.1.2).

Proof. First, let us observe that (7.1.2) can be equivalently rewritten as

$$\begin{cases} z'(t) \in A_\varepsilon z(t) + h_\varepsilon(t), & t \in \mathbb{R}_+, \\ z(t) = g(z)(t), & t \in [-\tau, 0], \end{cases} \qquad (7.3.5)$$

where $A_\varepsilon = A - \varepsilon I$ and $h_\varepsilon(t) = \varepsilon z(t)$, for $t \in \mathbb{R}_+$. Then, for each $t \in (0, +\infty)$, the unique C^0-solution u_ε^f of (7.2.3) and the unique solution z of (7.3.5) satisfy

$$\|u_\varepsilon^f(t) - z(t)\| \leq e^{-\varepsilon t} \|u_\varepsilon^f(0) - z(0)\| + \int_0^t e^{-\varepsilon(t-s)} [\, u_\varepsilon^f(s) - z(s), f_\varepsilon(s) - h_\varepsilon(s) \,]_+ \, ds$$

$$\leq e^{-\varepsilon t} \|u_\varepsilon^f(0) - z(0)\| + \int_0^t e^{-\varepsilon(t-s)} \|f(s)\| \, ds$$

$$\leq e^{-\varepsilon t} \|u_\varepsilon^f(0) - z(0)\| + \frac{1 - e^{-\varepsilon t}}{\varepsilon} \|f\|_{L^\infty(\mathbb{R}_+;X)}$$

for each $t \in \mathbb{R}_+$. This clearly shows that

$$\sup_{t \in \mathbb{R}_+} \|u_\varepsilon^f(t) - z(t)\| < +\infty.$$

Delay Evolution Inclusions 259

Since, by Lemma 7.1.1, it follows that $z \in C_b([-\tau, +\infty); X)$, the last inequality shows that $u_\varepsilon^f \in C_b(\mathbb{R}_+; X)$. Hence

$$\|u_\varepsilon^f(t) - z(t)\| \le e^{-\varepsilon t}\|u_\varepsilon^f - z\|_{C_b(\mathbb{R}_+;X)} + \frac{1 - e^{-\varepsilon t}}{\varepsilon}\|f\|_{L^\infty(\mathbb{R}_+;X)} \qquad (7.3.6)$$

for each $t \in (0, +\infty)$.

Now, if $\|u_\varepsilon^f - z\|_{C_b(\mathbb{R}_+;X)}$ attains its maximum at $t = 0$, i.e.,

$$\|u_\varepsilon^f - z\|_{C_b(\mathbb{R}_+;X)} = \|u_\varepsilon^f(0) - z(0)\| = \|g(u_\varepsilon^f)(0) - g(z)(0)\|,$$

from (g_1) in (H_g), we get

$$\|u_\varepsilon^f - z\|_{C_b(\mathbb{R}_+;X)} \le \|u_\varepsilon^f - z\|_{C_b([a,+\infty);X)}.$$

So, either there exists $t > 0$ such that

$$\|u_\varepsilon^f - z\|_{C_b(\mathbb{R}_+;X)} = \|u_\varepsilon^f(t) - z(t)\|,$$

or there exists $(t_n)_n$ with $\lim_n t_n = +\infty$ and

$$\|u_\varepsilon^f - z\|_{C_b(\mathbb{R}_+;X)} = \lim_n \|u_\varepsilon^f(t_n) - z(t_n)\|.$$

In any case, from (7.3.6), it readily follows that

$$\|u_\varepsilon^f - z\|_{C_b(\mathbb{R}_+;X)} \le \frac{1}{\varepsilon}\|f\|_{L^\infty(\mathbb{R}_+;X)}.$$

Finally, if $t \in [-\tau, 0]$, from (g_1) in (H_g), we deduce

$$\|u_\varepsilon^f(t) - z(t)\| = \|g(u_\varepsilon^f)(t) - g(z)(t)\|$$

$$\le \|u_\varepsilon^f - z\|_{C_b([a,+\infty);X)} \le \|u_\varepsilon^f - z\|_{C_b(\mathbb{R}_+;X)}.$$

This shows that (7.3.4) holds true and $u_\varepsilon^f \in C_b([-\tau, +\infty); X)$, as claimed. $\square$

LEMMA 7.3.3 *Let us assume that* (A_1), (A_2) *in* (H_A) *and* (H_g) *are satisfied, let* $\ell \in L^\infty(\mathbb{R}_+) \cap L^1(\mathbb{R}_+)$ *be a nonnegative function, and let* $\varepsilon > 0$ *be fixed. Then the operator* $f \mapsto u_\varepsilon^f$, *where* u_ε^f *is the unique* C^0-*solution of the problem* (7.2.3) *corresponding to* f, *maps the set*

$$\mathcal{F} = \{f \in L^\infty(\mathbb{R}_+; X) \cap L^1(\mathbb{R}_+; X); \ \|f(t)\| \le \ell(t) \text{ a.e. for } t \in \mathbb{R}_+\},$$

into a relatively compact set in $\widetilde{C}_b([-\tau, +\infty); X)$.

Proof. Let us observe that (7.3.4) implies the boundedness of the set $\{u_\varepsilon^f; \ f \in \mathcal{F}\}$ in $C_b(\mathbb{R}_+; X)$, which includes $C_b(\mathbb{R}_+; \overline{D(A)})$. Thus $\{u_\varepsilon^f(0); \ f \in \mathcal{F}\}$ is bounded in $\overline{D(A)}$. Since $\mathcal{F}$ is uniformly integrable in $L^1(\mathbb{R}_+; X)$, from (A_2) and Theorem 1.8.6, we conclude that, for all $k = 1, 2, \dots$ and every

260   *Delay Differential Evolutions Subjected to Nonlocal Initial Conditions*

$\delta \in (0, k)$, $\{u_\varepsilon^f; f \in \mathcal{F}\}$ is relatively compact in $C([\delta, k]; X)$. Thanks to (g_1) and (g_4) in (H_g), we deduce that the set $\{g(u_\varepsilon^f); f \in \mathcal{F}\}$ is relatively compact in $\mathcal{D}$. Thus

$$\{g(u_\varepsilon^f)(0); f \in \mathcal{F}\} = \{u_\varepsilon^f(0); f \in \mathcal{F}\}$$

is relatively compact in $\overline{D(A)}$. Again, from (g_1) in (H_g) and the second part of Theorem 1.8.6, it follows that the set $\{u_\varepsilon^f; f \in \mathcal{F}\}$ is relatively compact in $\widetilde{C}_b([-\tau, +\infty); X)$ and this completes the proof. $\hfill\square$

LEMMA 7.3.4 *Let us assume that (H_A), (F_4') in (H_F) and (H_g) are satisfied. Then, for each $\varepsilon > 0$, the problem (7.2.5) has at least one C^0-solution u_ε.*

Proof. Let $\ell \in L^\infty(\mathbb{R}_+) \cap L^1(\mathbb{R}_+)$ be the nonnegative function given by (F_4') and let $\mathcal{F}$ be defined as in Lemma 7.3.3. First, let us remark that $\mathcal{F}$ is nonempty, bounded and convex. Moreover, it is closed in both the $L^\infty(\mathbb{R}_+; X)$-topology and in the $L^1(\mathbb{R}_+; X)$-topology as well. So, for $k = 1, 2, \ldots$, $\mathcal{F}$ is uniformly integrable in $L^1(0, k; X)$ being bounded in $L^\infty(0, k; X)$. See either (ii) or (iii) in Remark 1.4.3. Accordingly, from (A_2) in (H_A) combined with Theorem 1.8.6, we conclude that the set

$$C_\varepsilon^k = \overline{\left\{ u_{\varepsilon t}^f; \ f \in \mathcal{F}, \ t \in [0, k] \right\}}$$

is compact in X. Further, since the restriction of the multifunction F_ε to $([0, k] \setminus E_\varepsilon) \times X$ is strongly-weakly u.s.c., and has weakly compact values, from Lemma 1.5.1 and Krein–Šmulian Theorem 1.4.3, we deduce that the set

$$G_\varepsilon^k = \overline{\mathrm{conv}}\, F_\varepsilon(([0, k] \setminus E_\varepsilon) \times C_\varepsilon^k)$$

is weakly compact X. Hence

$$H_\varepsilon^k = \overline{\mathrm{conv}}\, F_\varepsilon([0, k] \times C_\varepsilon^k) = \overline{\mathrm{conv}} \left[F_\varepsilon(([0, k] \setminus E_\varepsilon) \times C_\varepsilon^k) \cup \{0\} \right]$$

is nonempty, convex and weakly compact in X. Let

$$\mathcal{K}_\varepsilon = \bigcap_{k=1}^{\infty} \{f \in \mathcal{F}; \ f(t) \in H_\varepsilon^k, \ \text{a.e. for } t \in (0, k)\},$$

which is nonempty (because H_ε^k is convex for each $k = 1, 2 \ldots$) and weakly compact in $L^1(\mathbb{R}_+; X)$. Indeed, to verify the weak compactness of the set $\mathcal{K}_\varepsilon$, just apply Theorem 1.4.13 with $\Omega = \mathbb{R}_+$, $\Omega_k = \Omega_{\gamma,k} = [0, k]$, $C_{\gamma,k} = H_\varepsilon^k$, $k = 1, 2, \ldots$, for $\gamma > 0$ and μ the Lebesgue measure on $\mathbb{R}_+$.

Now, let us define the operator $Q_\varepsilon : \mathcal{K}_\varepsilon \rightsquigarrow L^1(\mathbb{R}_+; X)$ by

$$Q_\varepsilon f = \mathrm{Sel}\, F_\varepsilon(\cdot, u_{\varepsilon(\cdot)}^f),$$

where u_ε^f is the unique C^0-solution of the problem (7.2.5) corresponding to $f \in \mathcal{K}_\varepsilon$, while $\mathrm{Sel}\, F_\varepsilon(\cdot, u_{\varepsilon(\cdot)}^f)$ denotes the set of all strongly measurable selections

Delay Evolution Inclusions 261

of the multifunction $t \mapsto F_\varepsilon(t, u_{\varepsilon t}^f)$. From Lemma 1.5.2, we conclude that Q_ε is nonempty-valued and everywhere-defined. In view of (F_4'), it maps the set $\mathcal{K}_\varepsilon$ into itself. Thanks to (F_4') in (H_F), it follows that Q_ε has nonempty, convex and weakly compact values in $\mathcal{K}_\varepsilon$. More than this, its graph is weakly$\times$weakly sequentially closed. Indeed, let $((f_n, g_n))_n$ be a sequence in the graph of Q_ε, which is weakly$\times$weakly convergent to some $(f, g) \in L^1(\mathbb{R}_+; X) \times L^1(\mathbb{R}_+; X)$. Then, taking into account Lemma 7.3.3 and recalling that, in view of (A_6) in (H_A), A is of complete continuous type, we get

$$\lim_n u_\varepsilon^{f_n} = u_\varepsilon^f$$

in $\widetilde{C}_b(\mathbb{R}_+; \overline{D(A)})$ and

$$\lim_n u_{\varepsilon t}^{f_n} = u_{\varepsilon t}^f$$

in $\mathcal{X}$, for each $t \in \mathbb{R}_+$. Since $g_n(t) \in F_\varepsilon(t, u_{\varepsilon t}^{f_n})$ for each $n \in \mathbb{N}$ and a.e. for $t \in \mathbb{R}_+$, by Theorem 1.5.1, it follows that

$$g(t) \in F_\varepsilon(t, u_{\varepsilon t}^f) \tag{7.3.7}$$

a.e. for $t \in \mathbb{R}_+ \setminus E_\varepsilon$. On the other hand, $g_n(t) = g(t) = 0$ a.e. for $t \in E_\varepsilon$ and consequently (7.3.7) holds true a.e. for $t \in \mathbb{R}_+$. So, the graph of Q_ε is weakly$\times$weakly sequentially closed. By Theorem 1.5.4, Q_ε has at least a fixed point $f \in \mathcal{K}_\varepsilon$. Since by means of $f \mapsto u_\varepsilon^f$, this fixed point f produces a C^0-solution of the problem (7.2.5), this completes the proof. $\qquad\square$

LEMMA 7.3.5 *If (H_A), (F_3'), (F_4') in (H_F) and (H_g) are satisfied, then, for each $\varepsilon \in (0, 1)$, each C^0-solution u_ε of the problem (7.2.5) satisfies*

$$\|u_\varepsilon - z\|_{C_b(\mathbb{R}_+; X)} \le r, \tag{7.3.8}$$

where z is the unique C^0-solution of the auxiliary problem (7.1.2) and r is given by (F_3').

Proof. Let us observe that, if $0 \le t < \widetilde{t}$, we have

$$\|u_\varepsilon(\widetilde{t}) - z(\widetilde{t})\| \le \|u_\varepsilon(t) - z(t)\|$$

$$+ \int_t^{\widetilde{t}} [\, u_\varepsilon(s) - z(s), f(s) \,]_+ \, ds - \varepsilon \int_t^{\widetilde{t}} \|u_\varepsilon(s) - z(s)\| \, ds. \tag{7.3.9}$$

Let us assume by contradiction that there exists $t \in \mathbb{R}_+$ such that

$$\|u_\varepsilon(t) - z(t)\| > r.$$

We distinguish between two complementary cases.

Case 1. There exists $\bar{t} \in \mathbb{R}_+$ such that

$$r < \|u_\varepsilon - z\|_{C_b(\mathbb{R}_+; X)} = \|u_\varepsilon(\bar{t}) - z(\bar{t})\|. \tag{7.3.10}$$

262   *Delay Differential Evolutions Subjected to Nonlocal Initial Conditions*

We recall that, by Lemma 7.3.2, $u_\varepsilon - z \in C_b([-\tau, +\infty); X)$ and therefore

$$\sup_{t \in \mathbb{R}_+} \|u_\varepsilon(t) - z(t)\| = \|u_\varepsilon - z\|_{C_b(\mathbb{R}_+; X)} < +\infty.$$

If $\bar{t} = 0$, then

$$r < \|u_\varepsilon - z\|_{C_b(\mathbb{R}_+; X)} = \|u_\varepsilon(0) - z(0)\| = \|g(u_\varepsilon)(0) - g(z)(0)\|$$

$$\leq \|u_\varepsilon - z\|_{C_b([a, +\infty); X)} \leq \|u_\varepsilon - z\|_{C_b(\mathbb{R}_+; X)}$$

and so

$$\|u_\varepsilon - z\|_{C_b(\mathbb{R}_+; X)} = \|u_\varepsilon - z\|_{C_b([a, +\infty); X)}.$$

Therefore, we can always confine ourselves to analyze the case when, in (7.3.10), either $\bar{t} \in (0, +\infty)$ or there is no $\bar{t} \in (0, +\infty)$ satisfying the equality in (7.3.10).

So, if there exists $\bar{t} \in (0, +\infty)$ such that (7.3.10) holds true, then the mapping $t \mapsto \|u_\varepsilon(t) - z(t)\|$ cannot be constant on $(0, \bar{t})$. Indeed, if we assume that

$$\|u_\varepsilon(s) - z(s)\| = \|u_\varepsilon(\bar{t}) - z(\bar{t})\|$$

for each $s \in (0, \bar{t})$, then, taking $t \in (0, \bar{t})$ and $\tilde{t} = \bar{t}$ in (7.3.9) and using (F_3') with $v(0) = u_{\varepsilon s}(0) = u_\varepsilon(s)$, we get

$$r < r - \varepsilon(\bar{t} - t)r < r, \tag{7.3.11}$$

which is impossible. Consequently, there exists $t_0 \in (0, \bar{t})$ such that

$$r < \|u_\varepsilon(t_0) - z(t_0)\| < \|u_\varepsilon(s) - z(s)\| \leq \|u_\varepsilon(\bar{t}) - z(\bar{t})\| = \|u_\varepsilon - z\|_{C_b(\mathbb{R}_+; X)}$$

for each $s \in (t_0, \bar{t})$. Since

$$\|u_\varepsilon(s) - z(s)\| \leq \|u_{\varepsilon s} - z_s\|_{\mathcal{X}},$$

for each $s \in \mathbb{R}_+$, we have

$$r < \|u_{\varepsilon s} - z_s\|_{\mathcal{X}}$$

for each $s \in (t_0, \bar{t})$ and then, using again (7.3.9) and (F_3'), we get

$$r < \|u_\varepsilon(\bar{t}) - z(\bar{t})\| \leq \|u_\varepsilon(t_0) - z(t_0)\| - \varepsilon(\bar{t} - t_0)r,$$

which implies the very same contradiction as before, i.e., (7.3.11).

Case 2. There is no $\bar{t} \in \mathbb{R}_+$ such that (7.3.10) holds true. Then, there exists at least one sequence $(t_k)_k$ such that

$$\begin{cases} \lim_k t_k = +\infty, \\ \lim_k \|u_\varepsilon(t_k) - z(t_k)\| = \|u_\varepsilon - z\|_{C_b(\mathbb{R}_+; X)}. \end{cases}$$

Delay Evolution Inclusions 263

To go ahead with the proof in this case, let us observe that, if there exists $\widetilde{t} \in \mathbb{R}_+$ such that $\|u_\varepsilon(\widetilde{t}) - z(\widetilde{t})\| = r$, then $\|u_\varepsilon(t) - z(t)\| \le r$ for each $t \in [\widetilde{t}, +\infty)$. Indeed, if we assume the contrary, there would exist $[t, \widetilde{t}] \subseteq \mathbb{R}_+$ such that

$$\|u_\varepsilon(t) - z(t)\| = r$$

and

$$r < \|u_\varepsilon(s) - z(s)\|$$

for each $s \in (t, \widetilde{t}]$. Then, using once again (7.3.9) and (F_3'), we get

$$r < \|u_\varepsilon(\widetilde{t}) - z(\widetilde{t})\| \le \|u_\varepsilon(t) - z(t)\| - \varepsilon(\widetilde{t} - t)r$$

$$\le r - \varepsilon(\widetilde{t} - t)r$$

leading to (7.3.11), which is impossible.

So, when both $r < \|u_\varepsilon - z\|_{C_b(\mathbb{R}_+; X)}$ and $\|u_\varepsilon(t) - z(t)\| < \|u_\varepsilon - z\|_{C_b(\mathbb{R}_+; X)}$ hold true for each $t \in \mathbb{R}_+$, we necessarily have

$$\|u_\varepsilon(t) - z(t)\| > r$$

for each $t \in \mathbb{R}_+$.

If this is the case, let us remark that we may assume with no loss of generality, by extracting a subsequence if necessary, that

$$t_{k+1} - t_k \ge 1$$

for $k = 0, 1, 2, \ldots$. Then we have

$$r < \|u_\varepsilon(t_{k+1}) - z(t_{k+1})\|$$

$$\le \|u_\varepsilon(t_k) - z(t_k)\| + \int_{t_k}^{t_{k+1}} [\, u_\varepsilon(s) - z(s), f(s) - \varepsilon(u_\varepsilon(s) - z(s))\,]_+ \, ds$$

$$\le \|u_\varepsilon(t_k) - z(t_k)\| - \varepsilon \int_{t_k}^{t_{k+1}} \|u_\varepsilon(s) - z(s)\| \, ds$$

$$\le \|u_\varepsilon(t_k) - z(t_k)\| - \varepsilon(t_{k+1} - t_k)r \le \|u_\varepsilon(t_k) - z(t_k)\| - \varepsilon r$$

for each $k \in \mathbb{N}$. Passing to the limit for $k \to +\infty$ in the inequalities

$$\|u_\varepsilon(t_{k+1}) - z(t_{k+1})\| \le \|u_\varepsilon(t_k) - z(t_k)\| - \varepsilon r, \ k = 1, 2, \ldots$$

we get

$$\lim_k \|u_\varepsilon(t_{k+1}) - z(t_{k+1})\| \le \lim_k \|u_\varepsilon(t_k) - z(t_k)\| - \varepsilon r.$$

But

$$\lim_k \|u_\varepsilon(t_{k+1}) - z(t_{k+1})\| = \lim_k \|u_\varepsilon(t_k) - z(t_k)\| = \|u_\varepsilon - z\|_{C_b(\mathbb{R}_+; X)},$$

264 *Delay Differential Evolutions Subjected to Nonlocal Initial Conditions*

which is finite and we get

$$\|u_\varepsilon - z\|_{C_b(\mathbb{R}_+;X)} \le \|u_\varepsilon - z\|_{C_b(\mathbb{R}_+;X)} - \varepsilon r,$$

a contradiction that can be eliminated only if **Case 2** cannot hold. Thus, both **Case 1** and **Case 2** are impossible. In turn, this is a contradiction too, because at least one of these two cases should hold true. Therefore, the initial supposition, i.e., that $\|u_\varepsilon - z\|_{C_b(\mathbb{R}_+;X)} > r$, is necessarily false. It then follows that (7.3.8) holds true and this completes the proof. $\qquad\Box$

Now, we are ready to proceed to the proof of Theorem 7.2.3.

Proof. Let $(\varepsilon_n)_n$ be a sequence with $\varepsilon_n \downarrow 0$, let $(u_n)_n$ be a sequence of C^0-solutions of the problem (7.2.5) corresponding to $\varepsilon = \varepsilon_n$ for $n \in \mathbb{N}$ and let $(f_n)_n$ be such that

$$\begin{cases} u_n'(t) \in Au_n(t) - \varepsilon_n[\, u_n(t) - z(t)\,] + f_n(t), & t \in \mathbb{R}_+, \\ f_n(t) \in F_{\varepsilon_n}(t, u_{nt}), & t \in \mathbb{R}_+, \\ u_n(t) = g(u_n)(t), & t \in [-\tau, 0]. \end{cases}$$

In view of Remark 1.5.1, we may assume without loss of generality that $E_{\varepsilon_{n+1}} \subset E_{\varepsilon_n}$ for $n = 0, 1, \dots$. This means that

$$F_{\varepsilon_n}(t, v) = F_{\varepsilon_{n+1}}(t, v) \tag{7.3.12}$$

for each $t \in \mathbb{R}_+ \setminus E_{\varepsilon_n}$ and $v \in \mathfrak{X}$.

From (F_4'), we deduce that, for $k = 1, 2, \dots$, the set $\{f_n;\ n \in \mathbb{N}\}$ is uniformly integrable in $L^1(0, k; X)$. Then, from Lemma 7.3.5, (A_2) in (H_A) and Theorem 1.8.6, it follows that, for $k = 1, 2, \dots$, and each $\delta \in (0, k)$, the set $\{u_n;\ n \in \mathbb{N}\}$ is relatively compact in $C([\,\delta, k\,]; X)$. In view of (g_4) in (H_g), we deduce that the set $\{u_n;\ n \in \mathbb{N}\}$ is relatively compact in $\mathfrak{X}$. In particular, the set

$$\{u_n(0) = g(u_n)(0);\ n \in \mathbb{N}\}$$

is relatively compact in X. From the last part of Theorem 1.8.6, we conclude that the set $\{u_n;\ n \in \mathbb{N}\}$ is relatively compact in $C([\,0, k\,]; X)$ for $k = 1, 2, \dots$ and thus in $C([-\tau, k\,]; X)$. So, $\{u_n;\ n \in \mathbb{N}\}$ is relatively compact in $\widetilde{C}_b([-\tau, +\infty); X)$. Accordingly, for each $k = 1, 2, \dots$,

$$C_k = \overline{\{u_n(t);\ n \in \mathbb{N},\ t \in [\,0, k\,]\}}$$

is compact in X. Let $\gamma \in (0, 1)$ be arbitrary, let E_γ be the Lebesgue measurable set in $\mathbb{R}_+$ given by Definition 1.5.3. Let $k = 1, 2, \dots$, and let us define the set

$$D_{\gamma,k} = \overline{\bigcup_{n \in \mathbb{N}} \{(t, u_{n_t});\ t \in [\,0, k\,] \setminus E_\gamma\}}.$$

Clearly, $D_{\gamma,k}$ is compact in $\mathbb{R}_+ \times \mathfrak{X}$. Furthermore, for each $\gamma \in (0, 1)$ and each $k = 1, 2, \dots$, let us define

$$C_{\gamma,k} = F_\gamma(D_{\gamma,k}) = F(D_{\gamma,k}) \cup \{0\},$$

Delay Evolution Inclusions 265

which is weakly compact since $D_{\gamma,k}$ is compact and $F_{|D_{\gamma,k}}$ is strongly-weakly u.s.c. See Lemma 1.5.1. Next, the family $\mathcal{F} = \{f_n; \ n = 0, 1, \dots\} \subseteq L^1(\mathbb{R}_+; X)$ satisfies the hypotheses of Theorem 1.4.13. To prove the last assertion, let $k = 1, 2, \dots$, let $\gamma \in (0, 1)$, let $\Omega = \mathbb{R}_+$, $\mu = \lambda$ the Lebesgue measure on $\mathbb{R}_+$, let $\Omega_k = [0, k]$, $\Omega_{\gamma,k} = \Omega_k \setminus E_\gamma$ and $C_{\gamma,k}$ as above. Clearly, we have $\lambda(\Omega_k \setminus \Omega_{\gamma,k}) \leq \gamma$ and

$$f_n(\Omega_{\gamma,k}) \subseteq \bigcup_{t \in [0,k] \setminus E_\gamma} F_{\varepsilon_n}(t, u_{n_t}) \subseteq F(D_{\gamma,k}) \cup \{0\} = C_{\gamma,k}.$$

From (F'_4), it follows that

$$\|f_n(t)\| \leq \ell(t)$$

for $n = 0, 1, \dots$ and a.e. for $t \in \mathbb{R}_+$. Since $\ell \in L^1(\mathbb{R}_+)$, we necessarily have

$$\lim_k \int_{\Omega \setminus \Omega_k} \|f_n(t)\| \, dt \leq \lim_k \int_k^{+\infty} \ell(t) \, dt = 0$$

and thus (1.4.1) holds true.

Hence, we are in the hypotheses of Theorem 1.4.13, from which we deduce that the family $\{f_n; \ n = 0, 1, \dots\}$ is weakly relatively compact in $L^1(\mathbb{R}_+; X)$. So, on a subsequence at least, we have

$$\begin{cases} \lim_n f_n = f \ \text{ weakly in } \ L^1(\mathbb{R}_+; X), \\ \lim_n u_n = u \ \text{ in } \ \widetilde{C}_b([-\tau, +\infty); X), \\ \lim_n u_{n_t} = u_t \ \text{in} \ X \text{ for each } t \in \mathbb{R}_+. \end{cases}$$

From Theorem 1.5.1 combined with (7.3.12), we get

$$f(t) \in F_{\varepsilon_n}(t, u_t)$$

for each $n \in \mathbb{R}$ and a.e. $t \in \mathbb{R}_+ \setminus E_{\varepsilon_n}$. Since $\lim_n \lambda(E_{\varepsilon_n}) = 0$, it follows that

$$f(t) \in F(t, u_t)$$

a.e. $t \in \mathbb{R}_+$. But A is of complete continuous type, from which it follows that u is a C^0-solution of the problem (7.1.1) corresponding to the selection f of the mapping $t \mapsto F(t, u_t)$.

Finally, it suffices to observe that, from (7.3.8) in Lemma 7.3.5, it follows that $u(t) - z(t) \in D(0, r)$ for each $t \in \mathbb{R}_+$. $\qquad\square$

We can now proceed to the proof of Theorem 7.2.1.

Proof. Let $r > 0$ be given by (F_3) and let us define the set

$$\mathcal{K}_r = \{(t, v) \in \mathbb{R}_+ \times \mathcal{X}; \ \|v(0) - z(t)\| \leq r\}.$$

266   *Delay Differential Evolutions Subjected to Nonlocal Initial Conditions*

Clearly, $\mathcal{K}_r$ is nonempty and closed in $\mathbb{R}_+ \times X$ and, for each $t \in \mathbb{R}_+$, the cross section of $\mathcal{K}_r$ at t, i.e.,

$$\mathcal{K}_r(t) = \{v \in X; \ (t, v) \in \mathcal{K}_r\},$$

is convex. Let $\pi : \mathbb{R}_+ \times X \to \mathbb{R}_+ \times X$ be defined by

$$\pi(t, v) = \begin{cases} (t, v) & \text{if } \|v(0) - z(t)\| \le r, \\ \left(t, \dfrac{r}{\|v - z_t\|_X}(v - z_t) + z_t\right) & \text{if } \|v(0) - z(t)\| > r. \end{cases}$$

Let us observe that π is continuous, π restricted to $\mathcal{K}_r$ is the identity operator, and π maps $\mathbb{R}_+ \times X$ into $\mathcal{K}_r$. The first two properties mentioned are obvious. To prove the fact that π maps $\mathbb{R}_+ \times X$ into $\mathcal{K}_r$, we have merely to observe that if $\|v(0) - z(t)\| > r$, then

$$\left\| \frac{r}{\|v - z_t\|_X}(v - z_t) + z_t - z_t \right\|_X = r$$

and so, in this case, $\pi(t, v) \in \mathcal{K}_r$. If $\|v(0) - z(t)\| \le r$, then $\pi(t, v) = (t, v)$ and thus, π maps $\mathbb{R}_+ \times X$ into $\mathcal{K}_r$.

Then, we can define the multifunction $F_\pi : \mathbb{R}_+ \times X \rightsquigarrow X$ by

$$F_\pi(t, v) = F(\pi(t, v)),$$

for each $(t, v) \in \mathbb{R}_+ \times X$. As π is continuous, it follows that F_π is nonempty, closed, convex, weakly compact valued, and strongly-weakly u.s.c. Moreover, one can easily verify that it satisfies (F_4'). In addition, since $\pi(\mathbb{R}_+ \times X) \subseteq \mathcal{K}_r$, we easily conclude that F_π satisfies (F_3') too. Indeed, let $(t, v) \in \mathbb{R}_+ \times X$ satisfy

$$\|v(0) - z(t)\| > r \tag{7.3.13}$$

and let $f \in F(\pi(t, v))$.

From the definition of π, it follows that the projection P_2 of $\pi(t, v)$ on the second component, i.e.,

$$P_2(\pi(t, v)) = \begin{cases} v & \text{if } \|v(0) - z(t)\| \le r, \\ \dfrac{r}{\|v - z_t\|_X}(v - z_t) + z_t & \text{if } \|v(0) - z(t)\| > r, \end{cases}$$

satisfies

$$\|P_2(\pi(t, v)) - z_t\|_X = \begin{cases} r & \text{if } \|v(0) - z(t)\| > r, \\ \|v - z_t\|_X & \text{if } \|v(0) - z(t)\| \le r. \end{cases}$$

Therefore, if (t, v) satisfies (7.3.13), it follows that

$$\|P_2(\pi(t, v)) - z_t\|_X = r.$$

Delay Evolution Inclusions 267

So, by (v) in Proposition 1.8.1 combined with (F_3), we have

$$[v(0) - z(t), f]_+ = [P_2(\pi(t, v))(0) - z(t), f]_+ \le 0,$$

which proves that F_π satisfies (F_3').

Hence, by virtue of Theorem 7.2.3, the problem

$$\begin{cases} u'(t) = Au(t) + f(t), & t \in \mathbb{R}_+, \\ f(t) \in F_\pi(t, u_t), & t \in \mathbb{R}_+, \\ u(t) = g(u)(t), & t \in [-\tau, 0] \end{cases}$$

has at least one C^0-solution, $u \in C_b([-\tau, +\infty); X)$.

By (7.3.8), we have $\|u_t(0) - z(t)\| \le r$ for each $t \in \mathbb{R}_+$. So, $(t, u_t) \in \mathcal{K}_r$, which shows that

$$F_\pi(t, u_t) = F(t, u_t)$$

for each $t \in \mathbb{R}_+$. Thus u is a C^0-solution of (7.1.1), and this completes the proof of Theorem 7.2.1. $\qquad\square$

7.4 A nonlinear parabolic differential inclusion

Let Ω be a nonempty, bounded and open subset in $\mathbb{R}^d$, $d \ge 1$, with C^2 boundary Σ, let $\tau \ge 0$, $Q_+ = \mathbb{R}_+ \times \Omega$, $\Sigma_+ = \mathbb{R}_+ \times \Sigma$ and $Q_\tau = [-\tau, 0] \times \Omega$, let $p \in [2, +\infty)$, let $\lambda > 0$, let $\omega > 0$, and let us consider the nonlinear problem

$$\begin{cases} \dfrac{\partial u}{\partial t}(t, x) = \Delta_p^\lambda u(t, x) - \omega u(t, x) + f(t, x), & \text{in } Q_+, \\[2mm] f(t, x) \in F_0\left(t, u(t, x), \displaystyle\int_{-\tau}^0 u(t + s, x)\, ds\right) + \alpha(t)h(x), & \text{in } Q_+, \\[2mm] -\dfrac{\partial u}{\partial \nu_p}(t, x) \in \ \beta(u(t, x)), & \text{on } \Sigma_+, \\[2mm] u(t, x) = \displaystyle\int_\tau^{+\infty} \mathcal{N}(u(t + \theta, x))\, d\mu(\theta) + \psi(t)(x), & \text{in } Q_\tau. \end{cases} \tag{7.4.1}$$

Here Δ_p^λ is the p-Laplace operator (see Example 1.9.6), while the multifunction $F_0 : \mathbb{R}_+ \times \mathbb{R} \times \mathbb{R} \rightsquigarrow \mathbb{R}$ is defined by

$$F_0(t, u, v) = [f_1(t, u, v), f_2(t, u, v)]$$

for each $(t, u, v) \in \mathbb{R}_+ \times \mathbb{R} \times \mathbb{R}$, with

$$f_i : \mathbb{R}_+ \times \mathbb{R} \times \mathbb{R} \to \mathbb{R}$$

268 Delay Differential Evolutions Subjected to Nonlocal Initial Conditions

for $i = 1, 2$, $\alpha \in L^1(\mathbb{R}_+) \cap L^\infty(\mathbb{R}_+)$, $h \in L^2(\Omega)$, $\mathcal{N} : \mathbb{R} \to \mathbb{R}$, μ is a measure on $[\tau, +\infty)$, $\mathcal{X}_2 = C([-\tau, 0]; L^2(\Omega))$ and $\psi \in \mathcal{X}_2$.

In order to formulate our main result, let us consider the auxiliary problem:

$$\begin{cases} \dfrac{\partial z}{\partial t}(t, x) = \Delta_p^\lambda z(t, x) - \omega z(t, x), & \text{in } Q_+, \\[2mm] -\dfrac{\partial z}{\partial \nu_p}(t, x) \in \beta(z(t, x)), & \text{on } \Sigma_+, \\[2mm] z(t, x) = \displaystyle\int_\tau^{+\infty} \mathcal{N}(z(t + \theta, x))\, d\mu(\theta) + \psi(t)(x), & \text{in } Q_\tau, \end{cases} \qquad (7.4.2)$$

whose unique C^0-solution is denoted by $z \in C_b([-\tau, +\infty); L^2(\Omega))$.

THEOREM 7.4.1 *Let Ω be a nonempty, bounded and open subset in $\mathbb{R}^d$, $d \geq 1$, with C^2 boundary Σ, let $\tau \geq 0$, $\omega > 0$, $p \in [2, +\infty)$, $\lambda > 0$ and let $\beta : D(\beta) \subseteq \mathbb{R} \rightsquigarrow \mathbb{R}$ be a maximal-monotone operator with $0 \in D(\beta)$ and $0 \in \beta(0)$. Let $f_i : \mathbb{R}_+ \times \mathbb{R} \times \mathbb{R} \to \mathbb{R}$, $i = 1, 2$, be two given functions, $h \in L^2(\Omega)$, $\|h\|_{L^2(\Omega)} > 0$, $\alpha \in L^1(\mathbb{R}_+) \cap L^\infty(\mathbb{R}_+)$, $\mathcal{N} : \mathbb{R} \to \mathbb{R}$, let μ be a σ-finite and complete measure on $[\tau, +\infty)$ and let $\psi \in \mathcal{X}_2$. Let us assume that*

(α_1) $|\alpha(t)| \leq 1$ *for each $t \in \mathbb{R}_+$*

(h_1) $f_1(t, u, v) \leq f_2(t, u, v)$ *for each $(t, u, v) \in \mathbb{R}_+ \times \mathbb{R} \times \mathbb{R}$*

(h_2) f_1 *is l.s.c. and f_2 is u.s.c., and, for each $(t, u, v), (t, u, w) \in \mathbb{R}_+ \times \mathbb{R} \times \mathbb{R}$ with $v \leq w$, we have*

$$\begin{cases} f_1(t, u, v) \leq f_1(t, u, w), \\ f_2(t, u, v) \geq f_2(t, u, w) \end{cases}$$

(h_3) *there exist $c > 0$ and $k > 0$ such that, for all $(t, u, v) \in \mathbb{R}_+ \times \mathbb{R} \times \mathbb{R}$, we have*

$$[u - z(t, x)]f_1(t, u, v) \leq -c|u - z(t, x)|^2 + k|u - z(t, x)|$$

a.e. for $x \in \{y \in \Omega; u - z(t, y) \leq 0\}$ and

$$[u - z(t, x)]f_2(t, u, v) \leq -c|u - z(t, x)|^2 + k|u - z(t, x)|$$

a.e. for $x \in \{y \in \Omega; u - z(t, y) > 0\}$, z being the unique C^0-solution of the problem (7.4.2)

(h_4) *there exists a nonnegative function $\widetilde{\ell} \in L^1(\mathbb{R}_+) \cap L^\infty(\mathbb{R}_+)$ such that*

$$|f_i(t, u, v)| \leq \widetilde{\ell}(t)$$

for $i = 1, 2$ and for each $(t, u, v) \in \mathbb{R}_+ \times \mathbb{R} \times \mathbb{R}$

Delay Evolution Inclusions

269

(μ_1) *there exists* $b > \tau$ *such that* $\operatorname{supp}\mu \subseteq [\,b, +\infty)$

(μ_2) $\mu([\,b, +\infty)) = 1$

(n_1) $|\mathcal{N}(u) - \mathcal{N}(v)| \leq |u - v|$ *for each* $u, v \in \mathbb{R}$

(n_2) $\mathcal{N}(0) = 0$.

Then, (7.4.1) *has at least one* C^0*-solution* $u \in C_b([\,-\tau, +\infty); L^2(\Omega))$ *which, for each* $(\delta, T) \subseteq (0, +\infty)$*, satisfies* $u \in AC([\,0, T\,]; W^{1,p}(\Omega)) \cap W^{1,2}([\,\delta, T\,]; L^2(\Omega))$ *and*

$$\|u - z\|_{C_b([\,-\tau, +\infty); L^2(\Omega))} \leq r,$$

where r *is given by*

$$r = \frac{k\sqrt{|\Omega|} + \|h\|_{L^2(\Omega)}}{c} > 0,$$

c *and* k *being as in* (h_3) *and* $|\Omega|$ *is the Lebesgue measure of* Ω.

Proof. Let $A : D(A) \subseteq L^2(\Omega) \to L^2(\Omega)$ be defined by

$$\begin{cases} D(A) = \left\{ u \in W^{1,p}(\Omega); \Delta_p^\lambda u \in L^2(\Omega), -\dfrac{\partial u}{\partial \nu_p}(x) \in \beta(u(x)), \ \text{a.e. for } x \in \Sigma \right\}, \\ Au = \Delta_p^\lambda u - \omega u, \quad \text{for } u \in D(A). \end{cases}$$

By Theorem 1.9.8, we know that A is ω-m-dissipative on $L^2(\Omega)$. In addition, as $C_0^\infty(\Omega)$ is dense in $D(A)$, it follows that $\overline{D(A)} = L^2(\Omega)$ and so $0 \in \overline{D(A)}$ and $0 \in A0$. Moreover, A generates a compact semigroup of nonexpansive mappings on $L^2(\Omega)$. Since $L^2(\Omega)$ has uniformly convex dual – being a Hilbert space – the operator A is of complete continuous type. See Remark 1.8.2. Hence A satisfies (H_A). Next, let $F : \mathbb{R}_+ \times \mathcal{X}_2 \rightsquigarrow L^2(\Omega)$ be given by

$$F(t, v) = \{ f \in L^2(\Omega); \ f(x) \in [\,\widetilde{f}_1(t, x, v), \widetilde{f}_2(t, x, v)\,], \ \text{a.e. for } x \in \Omega \} + \alpha(t)h,$$

for each $(t, v) \in \mathbb{R}_+ \times \mathcal{X}_2$, where $\widetilde{f}_i : \mathbb{R}_+ \times \Omega \times \mathcal{X}_2 \to \mathbb{R}$, $i = 1, 2$, are defined by

$$\begin{cases} \widetilde{f}_1(t, x, v) = f_1\left(t, v(0)(x), \displaystyle\int_{-\tau}^0 v(s)(x)\, ds\right), \\ \widetilde{f}_2(t, x, v) = f_2\left(t, v(0)(x), \displaystyle\int_{-\tau}^0 v(s)(x)\, ds\right) \end{cases} \tag{7.4.3}$$

for each $(t, v) \in \mathbb{R}_+ \times \mathcal{X}_2$ and a.e. for $x \in \Omega$.

Since $\overline{D(A)} = L^2(\Omega)$, we have $\mathcal{D}_2 = \{ \varphi \in \mathcal{X}_2; \ \varphi(0) \in \overline{D(A)} \} = \mathcal{X}_2$. So, let us define $g : C_b(\mathbb{R}_+; L^2(\Omega)) \to \mathcal{X}_2$ by

$$[g(u)(t)](x) = \int_\tau^{+\infty} \mathcal{N}(u(t + \theta)(x))\, d\mu(\theta) + \psi(t)(x)$$

270   *Delay Differential Evolutions Subjected to Nonlocal Initial Conditions*

for $u \in C_b(\mathbb{R}_+; L^2(\Omega))$, each $t \in [-\tau, 0]$ and a.e. for $x \in \Omega$.

With A, F and g as above, the problem (7.4.1) can be rewritten in the form (7.1.1).

Further, from (h_2), Fatou's Lemma 1.2.1 and (h_4), we conclude that the functions $\widetilde{f}_1, \widetilde{f}_2 : \mathbb{R}_+ \times \Omega \times \mathcal{X}_2 \to \mathbb{R}$, defined by (7.4.3), satisfy the hypotheses of Lemma 1.5.4. More precisely, $\widetilde{f}_1$ is l.s.c. and $\widetilde{f}_2$ is u.s.c. in the sense of Definition 1.5.6. By (h_4), both have affine growth. It follows that F is a nonempty, convex and weakly compact-valued, strongly-weakly u.s.c. multifunction.

We prove next that F satisfies (F_4'). To this end, let $(t, v) \in \mathbb{R}_+ \times \mathcal{X}_2$ and let $f \in F(t, v)$. Then $f = f_0 + \alpha(t)h$, where $\widetilde{f}_1(t, x, v) \leq f_0(x) \leq \widetilde{f}_2(t, x, v)$ a.e. for $x \in \Omega$.

Thanks to (h_4), we have

$$\|f\|_{L^2(\Omega)} \leq \max_{i \in \{1,2\}} \left\{ \left[\int_\Omega \left| f_i\left(t, v(0)(x), \int_{-\tau}^0 v(s)(x)\, ds\right) + \alpha(t)h(x) \right|^2 dx \right]^{1/2} \right\}$$

$$\leq \sqrt{|\Omega|}\, \widetilde{\ell}(t) + |\alpha(t)| \|h\|_{L^2(\Omega)},$$

which shows that F satisfies (F_4') with

$$\ell(t) = \sqrt{|\Omega|}\, \widetilde{\ell}(t) + |\alpha(t)| \|h\|_{L^2(\Omega)}$$

for each $t \in \mathbb{R}_+$. From (α_1) and (h_4) we conclude that $\ell \in L^1(\mathbb{R}_+) \cap L^\infty(\mathbb{R}_+)$. Let

$$r = \frac{k\sqrt{|\Omega|} + \|h\|_{L^2(\Omega)}}{c} > 0,$$

where c and k are as in (h_3) and $|\Omega|$ is the Lebesgue measure of Ω.

We prove next that F satisfies (F_3'). Indeed, from (h_2), (h_3) and (h_4), we conclude that for each $(t, v) \in \mathbb{R}_+ \times \mathcal{X}_2$, satisfying

$$\|v(0)(\cdot) - z(t, \cdot)\|_{L^2(\Omega)} > r,$$

and every $f \in F(t, v)$, we have

$$[\, v(0)(\cdot) - z(t, \cdot), f(\cdot) \,]_+ \leq 0.$$

To show this, let us observe that in our case, i.e., $X = L^2(\Omega)$ is a Hilbert space, we have

$$[\, v(0)(\cdot) - z(t, \cdot), f(\cdot) \,]_+ = \|v(0)(\cdot) - z(t, \cdot)\|_{L^2(\Omega)}^{-1} \langle v(0)(\cdot) - z(t, \cdot), f \rangle_{L^2(\Omega)}$$

whenever $\|v(0)(\cdot) - z(t, \cdot)\|_{L^2(\Omega)} \neq 0$. Clearly, each $f \in F(t, v)$ satisfies

$$f_1(t, v(0)(x), w(t)(x)) + \alpha(t)h(x) \leq f(x) \leq f_2(t, v(0)(x), w(t)(x)) + \alpha(t)h(x),$$

with

$$w(t)(x) = \int_{-\tau}^0 v(t + s)(x)\, ds.$$

Delay Evolution Inclusions

271

If $\|v(0)(\cdot) - z(t, \cdot)\|_{L^2(\Omega)} = r$, by (h_3), we get

$$[v(0)(\cdot) - z(t, \cdot), f(\cdot)]_+ = \frac{1}{r} \int_\Omega (v(0)(x) - z(t, x)) f(x)\, dx$$

$$= \frac{1}{r} \left[\int_{\{y \in \Omega; v(0)(y) - z(t,y) \leq 0\}} (v(0)(x) - z(t, x)) f(x)\, dx \right.$$

$$\left. + \int_{\{y \in \Omega; v(0)(y) - z(t,y) > 0\}} (v(0)(x) - z(t, x)) f(x)\, dx \right]$$

$$\leq \frac{1}{r} \left[\int_{\{y \in \Omega; v(0)(y) - z(t,y) \leq 0\}} (v(0)(x) - z(t, x)) d_1(t, x)\, dx \right]$$

$$+ \frac{1}{r} \left[\int_{\{y \in \Omega; v(0)(y) - z(t,y) > 0\}} (v(0)(x) - z(t, x)) d_2(t, x)\, dx \right],$$

where

$$d_i(t, x) = f_i(t, v(0)(x), w(t)(x)) + \alpha(t) h(x)$$

for $i = 1, 2$, $t \in \mathbb{R}_+$ and a.e. for $x \in \Omega$. From this inequality, (α_1) and (h_3), it follows that

$$r[v(0)(\cdot) - z(t, \cdot), f(\cdot)]_+ \leq \int_\Omega \left[-c|v(0)(x) - z(t, x)|^2 + k|v(0)(x) - z(t, x)| \right]\, dx$$

$$+ \int_\Omega |v(0)(x) - z(t, x)| \cdot |\alpha(t)| \cdot |h(x)|\, dx \leq -c\|v(0) - z(t, \cdot)\|^2_{L^2(\Omega)}$$

$$+ \left(k\sqrt{|\Omega|} + \|h\|_{L^2(\Omega)} \right) \|v(0) - z(t, \cdot)\|_{L^2(\Omega)} \leq 0,$$

for each $(t, v) \in \mathbb{R}_+ \times \mathfrak{X}_2$ with $\|v(0)(\cdot) - z(t, \cdot)\|_{L^2(\Omega)} > r$ and each $f \in F(t, v)$. Consequently, F satisfies (F_3').

Next, since $\mathcal{N}$ is nonexpansive, $\operatorname{supp} \mu \subseteq [b, +\infty)$ and $\mu([b, +\infty)) = 1$ – see (μ_1) and (μ_2) – we deduce

$$\|g(u) - g(v)\|_{\mathfrak{X}_2}$$

$$\leq \sup_{t \in [-\tau, 0]} \left[\int_\Omega \left| \int_\tau^\infty (\mathcal{N}(u(t + \theta)(x)) - \mathcal{N}(v(t + \theta)(x)))\, d\mu(\theta) \right|^2 dx \right]^{1/2}$$

$$\leq \sup_{t \in [-\tau, 0]} \left[\int_\Omega \left(\int_\tau^\infty |u(t + \theta)(x) - v(t + \theta)(x)|\, d\mu(\theta) \right)^2 dx \right]^{1/2}$$

$$\leq \sup_{t \in [-\tau, 0]} \left[\int_\Omega \left(\int_\tau^\infty d\mu(\theta) \right) \left(\int_\tau^\infty |u(t + \theta)(x) - v(t + \theta)(x)|^2\, d\mu(\theta) \right) dx \right]^{1/2}$$

$$= \mu([b, +\infty))^{1/2} \left(\sup_{t \in [-\tau, 0]} \int_\tau^\infty \int_\Omega |u(t + \theta)(x) - v(t + \theta)(x)|^2 \, dx \, d\mu(\theta) \right)^{1/2}$$

$$\leq \sup_{t \in [-\tau, 0]} \left[\int_\tau^\infty \|u(t + \theta) - v(t + \theta)\|_{L^2(\Omega)}^2 \, d\mu(\theta) \right]^{1/2} \leq \|u - v\|_{C([a, +\infty); L^2(\Omega))}$$

for each $u, v \in C(\mathbb{R}_+; L^2(\Omega))$, where $a = b - \tau > 0$. So g satisfies (g_1) in (H_g). Moreover, one may easily check that g satisfies (g_4) in (H_g). An appeal to Theorem 7.2.1 completes the proof. $\qquad\square$

REMARK 7.4.1 Particularizing $\mathcal{N}$ and μ as in Remark 3.2.4, from Theorem 7.4.1 we deduce several existence results concerning periodic C^0-solutions, anti-periodic C^0-solutions, and C^0-solutions subjected to multi-point mean initial conditions.

7.5 The nonlinear diffusion in $L^1(\Omega)$

Let Ω be a nonempty, bounded and open subset in $\mathbb{R}^d$, $d \geq 1$, with C^1 boundary Σ, let $\tau \geq 0$, $Q_+ = \mathbb{R}_+ \times \Omega$, $\Sigma_+ = \mathbb{R}_+ \times \Sigma$ and $Q_\tau = [-\tau, 0] \times \Omega$, let $\varphi : D(\varphi) \subseteq \mathbb{R} \rightsquigarrow \mathbb{R}$ be maximal-monotone with $0 \in D(\varphi)$, $0 \in \varphi(0)$, and let $\omega > 0$. Let us consider the porous medium equation subjected to nonlocal initial conditions

$$\begin{cases} \dfrac{\partial u}{\partial t}(t, x) \in \Delta\varphi(u(t, x)) - \omega u(t, x) + f(t, x), & \text{in } Q_+, \\[2mm] f(t, x) \in F_0\left(t, u(t, x), \displaystyle\int_{-\tau}^0 u(t + s, x) \, ds\right) + \sigma(t)h(x), & \text{in } Q_+, \\[2mm] \varphi(u(t, x)) = 0, & \text{on } \Sigma_+, \\[2mm] u(t, x) = \displaystyle\int_\tau^{+\infty} \mathcal{N}(u(\theta + t, \cdot))(x) \, d\mu(\theta) + \psi(t)(x), & \text{in } Q_\tau. \end{cases} \quad (7.5.1)$$

Let us consider the auxiliary problem

$$\begin{cases} \dfrac{\partial z}{\partial t}(t, x) \in \Delta\varphi(z(t, x)) - \omega z(t, x), & \text{in } Q_+, \\[2mm] \varphi(z(t, x)) = 0, & \text{on } \Sigma_+, \\[2mm] z(t, x) = \displaystyle\int_\tau^{+\infty} \mathcal{N}(z(\theta + t, \cdot))(x) \, d\mu(\theta) + \psi(t)(x), & \text{in } Q_\tau. \end{cases} \quad (7.5.2)$$

Hereinafter in this section, $z \in C_b([-\tau, +\infty); L^1(\Omega))$ denotes the unique C^0-solution of (7.5.2).

Delay Evolution Inclusions 273

Here Δ is the Laplace operator in the sense of distributions over the domain Ω, $\sigma \in L^1(\mathbb{R}_+)$, $h \in L^1(\Omega)$, $\mathcal{N} : L^1(\Omega) \to L^1(\Omega)$ and μ is a measure on $[\tau, +\infty)$.

Before passing to the statement of the main existence result concerning (7.5.1), we need to introduce some notations and to give the precise definition of the multifunction F_0.

Let $f_i : \mathbb{R}_+ \times \mathbb{R} \times \mathbb{R} \to \mathbb{R}$, $i = 1, 2$, with $f_1(t, u, v) \leq f_2(t, u, v)$ for each $(t, u, v) \in \mathbb{R}_+ \times \mathbb{R} \times \mathbb{R}$. Let $F_0 : \mathbb{R}_+ \times \mathbb{R} \times \mathbb{R} \rightsquigarrow \mathbb{R}$ be defined by

$$F_0(t, u, v) = [\, f_1(t, u, v), f_2(t, u, v)\,]$$

for each $(t, u, v) \in \mathbb{R}_+ \times \mathbb{R} \times \mathbb{R}$.

Let $\mathcal{X}_1 = C([-\tau, 0]; L^1(\Omega))$ and let $F : \mathbb{R}_+ \times \mathcal{X}_1 \rightsquigarrow L^1(\Omega)$ be defined by

$$F(t, v) = \left\{ f \in L^1(\Omega); f(x) \in [\, \widetilde{f}_1(t, x, v), \widetilde{f}_2(t, x, v)\,], \text{ a.e. for } x \in \Omega \right\} + \sigma(t)h,$$

where $\widetilde{f}_i : \mathbb{R}_+ \times \Omega \times \mathcal{X}_1 \to \mathbb{R}$, $i = 1, 2$, are defined as

$$\begin{cases} \widetilde{f}_1(t, x, v) = f_1 \left(t, v(0)(x), \displaystyle\int_{-\tau}^0 v(s)(x)\, ds \right), \\[2mm] \widetilde{f}_2(t, x, v) = f_2 \left(t, v(0)(x), \displaystyle\int_{-\tau}^0 v(s)(x)\, ds \right) \end{cases}$$

for each $(t, v) \in \mathbb{R}_+ \times \mathcal{X}_1$, a.e. in Ω, $\sigma \in L^1(\mathbb{R}_+)$ and $h \in L^1(\Omega)$ satisfies $\|h\|_{L^1(\Omega)} \neq 0$.

The next result is a direct consequence of Theorem 7.2.1.

THEOREM 7.5.1 *Let Ω be a nonempty, bounded and open subset in $\mathbb{R}^d$, $d \geq 1$, with C^1 boundary Σ, let $\tau \geq 0$, $\omega > 0$, and let $\varphi : \mathbb{R} \to \mathbb{R}$ be continuous on $\mathbb{R}$ and C^1 on $\mathbb{R} \setminus \{0\}$ with $\varphi(0) = 0$ and for which there exist two constants $C > 0$ and $\alpha > 0$ if $d \leq 2$, and $\alpha > (d-2)/d$ if $d \geq 3$ such that*

$$\varphi'(r) \geq C|r|^{\alpha - 1}$$

for each $r \in \mathbb{R} \setminus \{0\}$. Let $f_i : \mathbb{R}_+ \times \mathbb{R} \times \mathbb{R} \to \mathbb{R}$, $i = 1, 2$, be two given functions, $h \in L^1(\Omega)$, $\|h\|_{L^1(\Omega)} > 0$, $\sigma \in L^1(\mathbb{R}_+)$ and let F be defined as above.

Let $\mathcal{N} : L^1(\Omega) \to L^1(\Omega)$, $\psi \in \mathcal{X}_1$ and let μ be a σ-finite and complete measure on $[\tau, +\infty)$. Let us assume that

(σ_1) *$\sigma \in L^1(\mathbb{R}_+)$ and $|\sigma(t)| \leq 1$ a.e. for $t \in \mathbb{R}_+$*

(h_1) *$f_1(t, u, v) \leq f_2(t, u, v)$ for each $(t, u, v) \in \mathbb{R}_+ \times \mathbb{R} \times \mathbb{R}$*

(h_2) *f_1 is l.s.c. and f_2 is u.s.c., and, for each $(t, u, v), (t, u, w) \in \mathbb{R}_+ \times \mathbb{R} \times \mathbb{R}$ with $v \leq w$, we have*

$$\begin{cases} f_1(t, u, v) \leq f_1(t, u, w), \\ f_2(t, u, v) \geq f_2(t, u, w) \end{cases}$$

274   *Delay Differential Evolutions Subjected to Nonlocal Initial Conditions*

(h_3) *there exists $c > 0$ such that, for all $(t, x, v) \in \mathbb{R}_+ \times \mathbb{R} \times \mathfrak{X}_1$, we have*

$$\text{sign}\,[\,v(0)(x) - z(t, x)\,]f_0(x) \le -c|v(0)(x) - z(t, x)|$$

for each $f_0(x) \in [\,\widetilde{f}_1(t, x, v), \widetilde{f}_2(t, x, v)\,]$, z being the unique C^0-solution of the problem (7.5.2)

(h_4) *there exists a nonnegative function $\widetilde{\ell} \in L^1(\mathbb{R}_+) \cap L^\infty(\mathbb{R}_+)$ such that*

$$|f_i(t, u, v)| \le \widetilde{\ell}(t)$$

for $i = 1, 2$ and for each $(t, u, v) \in \mathbb{R}_+ \times \mathbb{R} \times \mathbb{R}$

(h_5) *for each $t \in \mathbb{R}_+$ and each $v \in \mathbb{R}$, we have*

$$f_i(t, z(t, x), v) = 0$$

for $i = 1, 2$ and a.e. for $x \in \Omega$

(μ_1) *there exists $b > \tau$ such that $\operatorname{supp} \mu \subseteq [\,b, +\infty)$*

(μ_2) $\mu([\,b, +\infty)) = 1$

(n_1) $\|\mathcal{N}(u) - \mathcal{N}(v)\|_{L^1(\Omega)} \le \|u - v\|_{L^1(\Omega)}$ *for each $u, v \in L^1(\Omega)$*

(n_2) $\mathcal{N}(0) = 0$.

Then, the problem (7.5.1) has at least one C^0-solution $u \in C_b([\,-\tau, +\infty); L^1(\Omega))$ satisfying

$$\|u - z\|_{C_b(\mathbb{R}_+; L^1(\Omega))} \le c^{-1} \|h\|_{L^1(\Omega)}.$$

REMARK 7.5.1 Condition (h_5) is satisfied, for instance, if

$$f_i(t, u, v) = \theta(t, u) \cdot \overline{f}_i(t, u, v),$$

where θ is a positive, continuous and bounded function with $\theta(t, z(t, x)) = 0$, while $\overline{f}_i$ satisfy $(h_1) \sim (h_4)$, $i = 1, 2$. In the particular case in which $\psi \equiv 0$, it follows that $z \equiv 0$ and so, (h_5) reduces to

$$f_i(t, 0, v) = 0$$

for each $(t, v) \in \mathbb{R}_+ \times \mathbb{R}$.

Proof. Let $X = L^1(\Omega)$ and let us define $A : D(A) \subseteq L^1(\Omega) \to L^1(\Omega)$, by

$$Au = \Delta\varphi(u) - \omega u$$

for each $u \in D(A)$, where

$$D(A) = \Big\{ u \in L^1(\Omega); \ \varphi(u) \in W_0^{1,1}(\Omega), \ \Delta\varphi(u) \in L^1(\Omega) \Big\}.$$

Delay Evolution Inclusions

Since $\varphi(0) = 0$, it follows that $C_0^\infty(\Omega)$ is dense in $D(A)$ and so $\overline{D(A)} = L^1(\Omega)$.

Theorem 1.9.6 implies that A is ω-m-dissipative in $L^1(\Omega)$, $\overline{D(A)} = L^1(\Omega)$, $A0 = 0$, and A generates a compact semigroup which is of complete continuous type. Hence, A satisfies (H_A). Moreover, since $\overline{D(A)} = L^1(\Omega)$, it follows that $\mathcal{D}_1 = \{\varphi \in \mathcal{X}_1; \ \varphi(0) \in \overline{D(A)}\} = \mathcal{X}_1$. Let F be defined as above and let

$$g : C_b(\mathbb{R}_+; L^1(\Omega)) \to \mathcal{X}_1$$

be defined by

$$[g(u)(t)](x) = \int_\tau^{+\infty} \mathcal{N}(u(t + \theta))(x) \, d\mu(\theta) + \psi(t)(x)$$

for each $u \in C_b([-\tau, +\infty); L^1(\Omega))$, each $t \in [-\tau, 0]$, and a.e. for $x \in \Omega$.

From (σ_1), (h_1), (h_2), (h_4) and Lemma 1.5.4, using similar arguments as in the proof of the corresponding part in the preceding section, we conclude that F is nonempty, closed, convex, weakly compact valued, and strongly-weakly u.s.c.

We prove next that F satisfies (F_3') with

$$r = c^{-1}\|h\|_{L^1(\Omega)}, \tag{7.5.3}$$

i.e., for each $(t, v) \in \mathbb{R}_+ \times \mathcal{X}_1$, with

$$\|v(0)(\cdot) - z(t, \cdot)\|_{L^1(\Omega)} > r,$$

and every $f \in F(t, v)$, we have

$$[v(0)(\cdot) - z(t, \cdot), f]_+ \leq 0. \tag{7.5.4}$$

So, let $(t, v) \in \mathbb{R}_+ \times \mathcal{X}_1$ and let $f \in F(t, v)$. Clearly f is of the form

$$f(t, x) = f_0(x) + \sigma(t)h(x)$$

a.e. for $x \in \Omega$, where $f_0 \in L^1(\Omega)$ satisfies $\widetilde{f}_1(t, x, v) \leq f_0(x) \leq \widetilde{f}_2(t, x, v)$ a.e. for $x \in \Omega$. From Example 1.8.1, (v) and (vi) in Proposition 1.8.1, we deduce

$$[v(0)(\cdot) - z(t, \cdot), f]_+ = \int_{\{y \in \Omega; v(0)(y) - z(t,y) > 0\}} f(x) \, dx$$

$$- \int_{\{y \in \Omega; v(0)(y) - z(t,y) < 0\}} f(x) \, dx + \int_{\{y \in \Omega; v(0)(y) - z(t,y) = 0\}} |f(x)| \, dx$$

$$\leq \int_{\{y \in \Omega; v(0)(y) - z(t,y) > 0\}} f_0(x) \, dx - \int_{\{y \in \Omega; v(0)(y) - z(t,y) < 0\}} f_0(x) \, dx$$

$$+ \int_{\{y \in \Omega; v(0)(y) - z(t,y) = 0\}} |f_0(x)| \, dx + \int_{\{y \in \Omega; v(0)(y) - z(t,y) > 0\}} \sigma(t)h(x) \, dx$$

276 Delay Differential Evolutions Subjected to Nonlocal Initial Conditions

$$- \int_{\{y\in\Omega; v(0)(y)-z(t,y)<0\}} \sigma(t)h(x)\,dx + \int_{\{y\in\Omega; v(0)(y)-z(t,y)=0\}} |\sigma(t)|\cdot|h(x)|\,dx.$$

Next, taking into account that, from (h_5), we have $f_0(x) = 0$ a.e. for those $x \in \Omega$ for which $v(0)(x) = z(t,x)$, the last inequality, (σ_1), (h_3), (7.5.3) and (7.5.4) yield

$$[\, v(0)(\cdot) - z(t,\cdot), f\,]_+ \leq \int_\Omega \text{sign}\,[\,v(0)(x) - z(t,x)\,]f_0(x)\,dx + \int_\Omega |\sigma(t)|\cdot|h(x)|\,dx$$

$$\leq -c\int_\Omega |v(0)(x) - z(t,x)|\,dx + \int_\Omega |h(x)|\,dx$$

$$= -c\|v(0) - z(t,\cdot)\|_{L^1(\Omega)} + \|h\|_{L^1(\Omega)} < -cr + \|h\|_{L^1(\Omega)} = 0.$$

So, F satisfies (F_3'). Clearly (F_4') follows from (h_3). Since the proof of (H_g) is very similar to the proof of Theorem 7.4.1, we do not provide details. So, we are in the hypotheses of Theorem 7.2.3, from which comes the conclusion. $\square$

7.6 The case when F has affine growth

In this section, we reconsider the nonlinear functional differential evolution inclusion subjected to nonlocal initial conditions

$$\begin{cases} u'(t) \in Au(t) + f(t), & t \in \mathbb{R}_+, \\ f(t) \in F(t, u_t), & t \in \mathbb{R}_+, \\ u(t) = g(u)(t), & t \in [-\tau, 0] \end{cases} \tag{7.6.1}$$

under different assumptions on the forcing term F.

Comparing the results in this section with those in Section 7.2, we will observe that here, we mainly assume a quite simple affine growth condition on F – see (F_2) in (H_F) below – instead of the purely invariance hypothesis (F_3). Although this affine growth condition along with the hypothesis (H_c) imply in fact the invariance of a suitably defined closed ball, it turns out to be much easier to verify in practical situations. Just have a glance at hypothesis (h_3) in Theorems 7.4.1 and 7.5.1, as well as the corresponding hypothesis (h_4) in Theorems 7.8.1 and 7.9.1 in this chapter. Of course, the price paid for this simpler condition is the stability hypothesis (H_c).

The assumptions we need in what follows are listed below.

(H_A) $A\colon D(A) \subseteq X \rightsquigarrow X$ is an operator with the following properties:

(A_1) $0 \in D(A)$, $0 \in A0$ and A is ω-m-dissipative for some $\omega > 0$

Delay Evolution Inclusions

(A_2) the semigroup generated by A on $\overline{D(A)}$ is compact

(A_6) A is of complete continuous type.

(H_F) $F : \mathbb{R}_+ \times \mathcal{X} \rightsquigarrow X$ is nonempty, convex and weakly compact-valued, almost strongly-weakly u.s.c., and satisfies

(F_2) there exist $\ell > 0$ and $m > 0$ such that

$$\|f\| \le \ell \|v\|_{\mathcal{X}} + m$$

a.e. for $t \in \mathbb{R}_+$, for each $v \in \mathcal{X}$ and $f \in F(t, v)$.

(H_c) The constants ℓ and ω satisfy the nonresonance condition

$$\ell < \omega.$$

(H_g) $g : C_b(\mathbb{R}_+; \overline{D(A)}) \to \mathcal{D}$ satisfies

(g_1) there exists $a > 0$ such that for each $u, v \in C_b(\mathbb{R}_+; \overline{D(A)})$, we have

$$\|g(u) - g(v)\|_{\mathcal{X}} \le \|u - v\|_{C_b([a, +\infty); X)}$$

(g_4) g is continuous from $C_b([a, +\infty); \overline{D(A)})$ endowed with the $\widetilde{C}_b([a, +\infty); X)$ topology to $\mathcal{X}$, where a is given by (g_1).

REMARK 7.6.1 As we will see later, the condition (F_2) along with (H_c) ensure the existence of a certain $r > 0$ with the property that each C^0-solution of the problem

$$\begin{cases} u'(t) \in Au(t) + f(t), & t \in \mathbb{R}_+, \\ f(t) \in F(t, u_t), & t \in \mathbb{R}_+, \end{cases}$$

whose initial history, $\varphi \in \mathcal{D}$, is $D(0, r)$-valued, i.e., $\varphi : [-\tau, 0] \to D(0, r)$, does not escape from $D(0, r)$ as long as it exists.

In order to formulate the main result in this section, we need to consider the auxiliary problem

$$\begin{cases} z'(t) \in Az(t), & t \in \mathbb{R}_+, \\ z(t) = g(z)(t), & t \in [-\tau, 0]. \end{cases} \tag{7.6.2}$$

Under the hypotheses (A_1) in (H_A) and (g_1) in (H_g), from Lemma 4.2.1, we deduce that (7.6.2) has a unique C^0-solution $z \in C_b([-\tau, +\infty); \overline{D(A)})$.

We are now ready to proceed to the statement of the main result in this chapter.

278   *Delay Differential Evolutions Subjected to Nonlocal Initial Conditions*

THEOREM 7.6.1 *If (H_A), (H_F), (H_g) and (H_c) are satisfied, then the problem (7.6.1) has at least one C^0-solution, $u \in C_b([-\tau, +\infty); X)$, satisfying*

$$\|u - z\|_{C_b([-\tau, +\infty); X)} \le \frac{m}{\omega - \ell} + \frac{\ell}{\omega - \ell} \cdot \frac{m_0 e^{\omega a}}{e^{\omega a} - 1},$$

where $m_0 = \|g(0)\|_X$.

As far as the nondelayed case is concerned, i.e., when $\tau = 0$, $\mathcal{X}$ and $\mathcal{D}$ reduce to X and to $\overline{D(A)}$, respectively, we consider the problem

$$\begin{cases} u'(t) \in Au(t) + f(t), & t \in \mathbb{R}_+, \\ f(t) \in F(t, u(t)), & t \in \mathbb{R}_+, \\ u(0) = g(u). \end{cases} \qquad (7.6.3)$$

Here $F : \mathbb{R}_+ \times X \rightsquigarrow X$ is nonempty, convex, weakly compact-valued and almost strongly-weakly u.s.c., while $g : C_b(\mathbb{R}_+; \overline{D(A)}) \to \overline{D(A)}$ is nonexpansive.

For this specific case, we need to reformulate the hypotheses $(H_F) \sim (H_g)$. First, let us consider the comparison problem

$$\begin{cases} w'(t) \in Aw(t), & t \in \mathbb{R}_+, \\ w(0) = g(w) \end{cases} \qquad (7.6.4)$$

and let us denote by $w \in C_b(\mathbb{R}_+; \overline{D(A)})$ its unique C^0-solution.

$(H_F^{[\tau=0]})$ $F : \mathbb{R}_+ \times X \rightsquigarrow X$ is a nonempty, convex and weakly compact-valued, almost strongly-weakly u.s.c. multifunction and satisfies

$(F_2^{[\tau=0]})$ there exist $\ell > 0$ and $m > 0$ such that

$$\|f\| \le \ell \|v\| + m$$

a.e. for $t \in \mathbb{R}_+$, for each $v \in \overline{D(A)}$ and $f \in F(t, v)$.

$(H_g^{[\tau=0]})$ $g : C_b(\mathbb{R}_+; \overline{D(A)}) \to \overline{D(A)}$ satisfies

$(g_1^{[\tau=0]})$ there exists $a > 0$ such that, for each $u, v \in C_b(\mathbb{R}_+; \overline{D(A)})$, we have

$$\|g(u) - g(v)\| \le \|u - v\|_{C_b([a, +\infty); X)}$$

$(g_4^{[\tau=0]})$ g is continuous from $C_b([a, +\infty); \overline{D(A)})$ endowed to the topology of $\widetilde{C}_b([a, +\infty); X)$ to X, where a is given by $(g_1^{[\tau=0]})$.

From Theorem 7.6.1, we deduce

Delay Evolution Inclusions 279

THEOREM 7.6.2 *If* (H_A), $(H_F^{[\tau=0]})$, $(H_g^{[\tau=0]})$ *and* (H_c) *are satisfied, then the problem* (7.6.3) *has at least one* C^0*-solution,* $u \in C_b([-\tau, +\infty); X)$, *satisfying*

$$\|u - w\|_{C_b(\mathbb{R}_+;X)} \leq \frac{m}{\omega - \ell} + \frac{\ell}{\omega - \ell} \cdot \frac{m_0 e^{\omega a}}{e^{\omega a} - 1},$$

where w *is the unique* C^0*-solution of the problem* (7.6.4) *and* $m_0 = \|g(0)\|$.

Before describing the main steps in the proof of Theorem 7.6.1, for easy reference, we begin by recalling the procedure in Section 7.2 showing how the almost strongly-weakly u.s.c. condition (H_F) should be exploited. More precisely, let $\varepsilon \in (0, 1)$ be arbitrary but fixed. By Definition 1.5.3, there exists a measurable set $E_\varepsilon \subseteq \mathbb{R}_+$ – which can always be taken to be open – whose Lebesgue measure $\lambda(E_\varepsilon) \leq \varepsilon$ and such that $F_{|(\mathbb{R}_+ \setminus E_\varepsilon) \times \mathcal{X}}$ is strongly-weakly u.s.c. Let us define

$$D(F) = \mathbb{R}_+ \times \mathcal{X},$$
$$D_\varepsilon(F) = (\mathbb{R}_+ \setminus E_\varepsilon) \times \mathcal{X}$$

and the multifunction $F_\varepsilon : \mathbb{R}_+ \times \mathcal{X} \rightsquigarrow X$, by

$$F_\varepsilon(t, v) = \begin{cases} F(t, v), & (t, v) \in D_\varepsilon(F), \\ \{0\}, & (t, v) \in D(F) \setminus D_\varepsilon(F). \end{cases} \quad (7.6.5)$$

Now, we can briefly explain the idea of the proof. First, from Lemma 4.2.1, we deduce that, for each $h \in L^\infty(\mathbb{R}_+; X)$, the problem

$$\begin{cases} u'(t) \in Au(t) + h(t), & t \in \mathbb{R}_+, \\ u(t) = g(u)(t), & t \in [-\tau, 0] \end{cases} \quad (7.6.6)$$

has a unique C^0-solution, $u^h \in C_b([-\tau, +\infty); X)$, satisfying (4.2.2).

Second, we prove that the ε-approximate problem

$$\begin{cases} u'(t) \in Au(t) + f(t), & t \in \mathbb{R}_+, \\ f(t) \in F_\varepsilon(t, u_t), & t \in \mathbb{R}_+, \\ u(t) = g(u)(t), & t \in [-\tau, 0], \end{cases} \quad (7.6.7)$$

where the approximate multifunction F_ε is given by (7.6.5), has at least one C^0-solution, $u_\varepsilon \in C_b([-\tau, +\infty); X)$.

In order to show this, we will use a fixed-point argument. More precisely, let us define $Q_\varepsilon : L^\infty(\mathbb{R}_+; X) \rightsquigarrow L^\infty(\mathbb{R}_+; X)$, by

$$Q_\varepsilon(h) = \mathcal{S}(F_\varepsilon \circ u^h)$$

for each $h \in L^\infty(\mathbb{R}_+; X)$, where $\mathcal{S}(F_\varepsilon \circ u^h)$ is the set of all strongly measurable and essentially bounded selections of the multifunction $t \mapsto F_\varepsilon(t, u_t^h)$, i.e.,

$$\mathcal{S}(F_\varepsilon \circ u^h) = \{f \in L^\infty(\mathbb{R}_+; X); \ f(t) \in F_\varepsilon(t, u_t^h), \text{ a.e. } t \in \mathbb{R}_+\}, \quad (7.6.8)$$

280   *Delay Differential Evolutions Subjected to Nonlocal Initial Conditions*

where, for each $h \in L^\infty(\mathbb{R}_+; X)$, u^h is the unique C^0-solution of (7.6.6). The fact that $\mathcal{S}(F_\varepsilon \circ u^h)$ has nonempty values follows from Lemma 1.5.2.

Clearly, h is a fixed point of Q_ε, i.e., $h \in Q_\varepsilon(h)$, if and only if u^h is a C^0-solution of the ε-approximate problem (7.6.7).

Moreover, from Theorem 1.5.1, we deduce that Q_ε is closed valued. Obviously, it is convex-valued too. We will next show that Q_ε maps a suitably defined nonempty, convex and weakly compact subset $\mathcal{K}$ in $L^1(\mathbb{R}_+; e^{-t}dt; X)$ into itself, and its graph is strongly-weakly sequentially closed. In view of Theorem 1.5.4, Q_ε has at least one fixed point h and so u^h is a C^0-solution of (7.6.7).

Third, for each $\varepsilon \in (0, 1)$, we pick up a C^0-solution, u_ε, of the problem (7.6.7) and we prove that the set $\{u_\varepsilon; \ \varepsilon \in (0, 1)\}$ is relatively compact in $\widetilde{C}_b([-\tau, +\infty); X)$. So, there exists a sequence $\varepsilon_n \downarrow 0$ such that $(u_{\varepsilon_n})_n$ converges in $\widetilde{C}_b([-\tau, +\infty); X)$ to a C^0-solution of the problem (7.6.1).

7.7   Proof of Theorem 7.6.1

For the sake of convenience and clarity, we begin by proving three auxiliary lemmas.

LEMMA 7.7.1 *If* (H_A) *and* (H_g) *are satisfied, then the operator* $h \mapsto u^h$, *where* u^h *is the unique solution of the problem* (7.6.6) *corresponding to* h, *is compact from each bounded subset in* $L^\infty(\mathbb{R}_+; X)$ *endowed with the* $L^1_{\mathrm{loc}}(\mathbb{R}_+; X)$ *topology to* $\widetilde{C}_b([-\tau, +\infty); X)$. *In addition,* $h \mapsto u^h$ *is continuous from each bounded subset in* $L^\infty(\mathbb{R}_+; X)$ *endowed with the weak* $L^1_{\mathrm{loc}}(\mathbb{R}_+; X)$ *topology to* $\widetilde{C}_b([-\tau, +\infty); X)$.

Proof. Let $\mathcal{F}$ be a bounded subset in $L^\infty(\mathbb{R}_+; X)$. From (4.2.2), it follows that $\{u^h; \ h \in \mathcal{F}\}$ is bounded in $C_b([-\tau, +\infty); X)$. Consequently, the set $\{u^h(0); \ h \in \mathcal{F}\}$ is bounded in $\overline{D(A)}$. Since $\mathcal{F}$ is uniformly integrable in $L^1(0, k; X)$ for $k = 1, 2 \ldots$, from (A_2) and Theorem 1.8.6, we conclude that $\{u^h; h \in \mathcal{F}\}$ is relatively compact in $C([\delta, k]; X)$, for each $k = 1, 2, \ldots$ and each $\delta \in (0, k)$. So, $\{u^h; h \in \mathcal{F}\}$ is relatively compact in $\widetilde{C}_b([a, +\infty); X)$. Thanks to (g_4) in (H_g), we deduce that the set $\{g(u^h); \ h \in \mathcal{F}\}$ is relatively compact in $\mathcal{X}$. Thus, $\{g(u^h)(0); \ h \in \mathcal{F}\} = \{u^h(0); \ h \in \mathcal{F}\}$ is relatively compact in $\overline{D(A)}$. Again, from (A_2) and the last part of Theorem 1.8.6, it follows that $\{u^h; \ h \in \mathcal{F}\}$ is relatively compact in $C([-\tau, k]; X)$ for $k = 1, 2, \ldots$ and thus it is relatively compact in $\widetilde{C}_b([-\tau, +\infty); X)$.

To prove the continuity of $h \mapsto u^h$ from each bounded subset $\mathcal{B}$ in $L^\infty(\mathbb{R}_+; X)$ endowed with the induced weak topology of $L^1_{\mathrm{loc}}(\mathbb{R}_+; X)$ to

Delay Evolution Inclusions 281

$\widetilde{C}_b([-\tau, +\infty); X)$, let $(h_n)_n$ be a sequence in $\mathcal{B}$ such that

$$\lim_n h_n = h$$

weakly in $L^1_{\mathrm{loc}}(\mathbb{R}_+; X)$. Set $u_n = u^{h_n}$ for $n \in \mathbb{N}$. As we have already seen, $\{u_n; \ n \in \mathbb{N}\}$ is relatively compact in $\widetilde{C}_b([-\tau, +\infty); X)$. So, to conclude the proof, it suffices to show that the only limit point of $(u_n)_n$ in the $\widetilde{C}_b([-\tau, +\infty); X)$ topology is simply the unique C^0-solution, u, of the problem (7.6.6), corresponding to h. To see that this is certainly the case, we have merely to recall that, from (A_6) in (H_A), the operator A is of complete continuous type. See Definition 1.8.5. Accordingly, if

$$\lim_n h_n = h$$

weakly in $L^1(0, T; X)$ and

$$\lim_n u_n = u$$

strongly in $C([0, T]; X)$, it follows that u is a C^0-solution of the problem

$$u'(t) \in Au(t) + h(t)$$

on $[0, T]$, for each $T > 0$. From (g_4) in (H_g) it then follows that

$$\lim_n g(u_n) = g(u)$$

in $\mathcal{X}$. As a consequence, u is the unique C^0-solution of the problem (7.6.6) corresponding to h, as claimed. This completes the proof. $\qquad\square$

Now, let us fix a sufficiently large $r > 0$ such that

$$\frac{\ell}{\omega} r + \ell \|z\|_{C_b([-\tau, +\infty); X)} + m \leq r, \tag{7.7.1}$$

where ω is given by (A_1) in (H_A), ℓ and m are given by (F_2) and z is the unique C^0-solution of the problem (7.6.2). We may always find such an r because, by (H_c), $\ell < \omega$.

LEMMA 7.7.2 *Let $\mathcal{F}$ be defined by*

$$\mathcal{F} = \{h \in L^\infty(\mathbb{R}_+; X); \ \|h\|_{L^\infty(\mathbb{R}_+; X)} \leq r\},$$

where r is defined in (7.7.1) and let $\varepsilon \in (0, 1)$. If (A_1) in (H_A), (H_F) and (H_g) are satisfied, then the operator $Q_\varepsilon : L^\infty(\mathbb{R}_+; X) \rightsquigarrow L^\infty(\mathbb{R}_+; X)$, given by

$$Q_\varepsilon(h) = \mathcal{S}(F_\varepsilon \circ u^h),$$

where $\mathcal{S}(F_\varepsilon \circ u^h)$ is defined in (7.6.8), maps $\mathcal{F}$ into itself.

282   *Delay Differential Evolutions Subjected to Nonlocal Initial Conditions*

Proof. Let $h \in \mathcal{F}$. By Lemma 4.2.1, the problem (7.6.6) has a unique C^0-solution $u^h \in C_b([-\tau, +\infty); \overline{D(A)})$ satisfying

$$\|u^h - z\|_{C_b([-\tau,+\infty);X)} \leq \frac{1}{\omega} \|h\|_{L^\infty(\mathbb{R}_+;X)}.$$

Hence

$$\|u^h - z\|_{C_b([-\tau,+\infty);X)} \leq \frac{r}{\omega}.$$

Now, let $f \in Q_\varepsilon(h)$. From (F_2) and the last inequality, we get

$$\|f\|_{L^\infty(\mathbb{R}_+;X)} \leq \ell \sup_{t \in \mathbb{R}_+} \|u_t^h\|_X + m$$

$$\leq \ell \left[\|u^h - z\|_{C_b([-\tau,+\infty);X)} + \|z\|_{C_b([-\tau,+\infty);X)} \right] + m$$

$$\leq \frac{\ell}{\omega} r + \ell \|z\|_{C_b([-\tau,+\infty);X)} + m \leq r.$$

From (7.7.1) it follows that

$$\|f\|_{L^\infty(\mathbb{R}_+;X)} \leq r,$$

which shows that $f \in \mathcal{F}$. The proof is complete. $\qquad\square$

LEMMA 7.7.3 *Let us assume that (H_A), (H_F), (H_g) and (H_c) are satisfied. Then, for each $\varepsilon > 0$ the problem (7.6.7) has at least one C^0-solution u_ε satisfying*

$$\|u_\varepsilon - z\|_{C_b([-\tau,+\infty);X)} \leq \frac{m}{\omega - \ell} + \frac{\ell}{\omega - \ell} \cdot \frac{m_0 e^{\omega a}}{e^{\omega a} - 1}, \qquad (7.7.2)$$

where z is the unique C^0-solution of the problem (7.6.2) and $m_0 = \|g(0)\|_X$. In addition, the corresponding selection f_ε in (7.6.7) satisfies

$$\|f_\varepsilon\|_{L^\infty(\mathbb{R}_+;X)} \leq r. \qquad (7.7.3)$$

Proof. Let $\varepsilon \in (0,1)$ be arbitrary but fixed and let $\mathcal{F}$ and Q_ε be defined as in Lemma 7.7.2. We know that Q_ε maps $\mathcal{F}$ into itself. In addition, from Lemma 7.7.1, it follows that $\{u^h; h \in \mathcal{F}\}$ is relatively compact in $\widetilde{C}_b([-\tau, +\infty); X)$. Therefore, for each $k = 1, 2, \ldots$, the set

$$\{(t, u_t^h); h \in \mathcal{F}, t \in [0, k] \setminus E_\varepsilon\}$$

is weakly relatively compact in $\mathbb{R} \times X$. We recall that here $E_\varepsilon \subseteq \mathbb{R}_+$ is the Lebesgue measurable set given by Definition 1.5.3. We emphasize that, without loss of generality, we can assume that E_ε is open. Since, by (A_1) in (H_A), $\overline{D(A)}$ is convex, from Mazur's Theorem 1.4.2, it follows that the set

$$D_\varepsilon^k = \overline{\text{conv}} \{(t, u_t^h); h \in \mathcal{F}, t \in [0, k] \setminus E_\varepsilon\},$$

Delay Evolution Inclusions

which is included in $[0,k] \times X$, is closed, convex and compact. Since $0 \in A0$, we conclude that the set

$$C_\varepsilon^k = \overline{\text{conv}\{(t, u_t^h); \ h \in \mathcal{F}, \ t \in [0,k] \setminus E_\varepsilon\} \cup \{0\}}$$

is closed, convex, compact in $[0,k] \times X$. Recalling that the multifunction $F_\varepsilon = F_{|(\mathbb{R}_+ \setminus E_\varepsilon) \times X}$ is strongly-weakly u.s.c., from Lemma 1.5.1, we deduce that

$$F(C_\varepsilon^k) = F_\varepsilon(C_\varepsilon^k) \cup \{0\}$$

is weakly compact too.

Let us define

$$\mathcal{K}_\varepsilon = \overline{\text{conv}} \, Q_\varepsilon(\mathcal{F}).$$

Obviously $Q_\varepsilon(\mathcal{K}_\varepsilon) \subseteq \mathcal{K}_\varepsilon \subseteq \mathcal{F}$. Clearly, $\mathcal{K}_\varepsilon$ is convex, bounded and closed in the norm topology of the space $L^1(\mathbb{R}_+; e^{-t}dt; X)$.

We will show next that $\mathcal{K}_\varepsilon$ is weakly compact in $L^1(\mathbb{R}_+; e^{-t}dt; X)$ and the graph of Q_ε is weakly$\times$weakly sequentially closed in the product space $L^1(\mathbb{R}_+; e^{-t}dt; X) \times L^1(\mathbb{R}_+; e^{-t}dt; X)$. To prove that $\mathcal{K}_\varepsilon$ is weakly compact in $L^1(\mathbb{R}_+; e^{-t}dt; X)$, we make use of Theorem 1.4.13. Namely, take $\Omega = \mathbb{R}_+$, $\mu = e^{-t}dt$, $\Omega_k = \Omega_{\gamma,k} = [0,k]$, $C_{\gamma,k} = F(C_\varepsilon^k)$ independent of $\gamma > 0$, and let us observe that we are in the hypotheses of Theorem 1.4.13, which shows that $\mathcal{K}_\varepsilon$ is weakly compact in $L^1(\mathbb{R}_+; e^{-t}dt; X)$.

Now, we prove that the graph of Q_ε is weakly$\times$weakly sequentially closed in the product space $L^1(\mathbb{R}_+; e^{-t}dt; X) \times L^1(\mathbb{R}_+; e^{-t}dt; X)$. To this end, let $((h_n, f_n))_n$ be an arbitrary sequence in the graph of Q_ε for which we suppose that there exists $(h, f) \in L^1(\mathbb{R}_+; e^{-t}dt; X) \times L^1(\mathbb{R}_+; e^{-t}dt; X)$ such that

$$\lim_n (h_n, f_n) = (h, f)$$

weakly in $L^1(\mathbb{R}_+; e^{-t}dt; X) \times L^1(\mathbb{R}_+; e^{-t}dt; X)$. This implies that

$$\lim_n (h_n, f_n) = (h, f)$$

weakly in $L^1(0, k; X) \times L^1(0, k; X)$ for each $k = 1, 2, \ldots$. Then, taking into account Lemma 7.7.1 and the fact that A is of complete continuous type – see (H_A) – we get

$$\lim_n u^{h_n} = u^h$$

in $\widetilde{C}_b(\mathbb{R}_+; X)$ and

$$\lim_n u_t^{h_n} = u_t^h$$

in X. Since $f_n(t) \in F_\varepsilon(t, u_t^{h_n})$ for each $n \in \mathbb{N}$ and a.e. $t \in \mathbb{R}_+$, by Theorem 1.5.1, it follows that

$$f(t) \in F_\varepsilon(t, u_t^h) \tag{7.7.4}$$

a.e. $t \in \mathbb{R}_+ \setminus E_\varepsilon$. On the other hand, $f_n(t) = f(t) = 0$ a.e. for $t \in E_\varepsilon$ and consequently (7.7.4) holds true a.e. for $t \in \mathbb{R}_+$. So, the graph of Q_ε is weakly$\times$weakly sequentially closed.

284   *Delay Differential Evolutions Subjected to Nonlocal Initial Conditions*

By Theorem 1.5.4, Q_ε has at least a fixed point $f_\varepsilon \in \mathcal{F}$ which produces a C^0-solution, u_ε, of the problem (7.6.7). Clearly, (7.7.3) follows from the definition of $\mathcal{F}$ – see Lemma 7.7.2. As far as (7.7.2) is concerned, let us observe that, by Lemma 7.7.2 and (F_2), we have

$$\|u_\varepsilon - z\|_{C_b([-\tau,+\infty);X)} \leq \frac{1}{\omega}\|f_\varepsilon\|_{L^\infty(\mathbb{R}_+;X)}$$

$$\leq \frac{\ell}{\omega}\left(\|u_\varepsilon - z\|_{C_b([-\tau,+\infty);X)} + \|z\|_{C_b([-\tau,+\infty);X)}\right) + \frac{m}{\omega}$$

and thus

$$\|u_\varepsilon - z\|_{C_b([-\tau,+\infty);X)} \leq \frac{m}{\omega - \ell} + \frac{\ell}{\omega - \ell}\|z\|_{C_b([-\tau,+\infty);X)}. \tag{7.7.5}$$

On the other hand, from Remark 3.2.2, we get

$$\|z\|_X \leq \|g(z)\|_X \leq \|z\|_{C_b([a,+\infty);X)} + m_0.$$

So, for $t \in (0,+\infty)$, we have

$$\|z(t)\| \leq e^{-\omega t}\|z\|_X \leq e^{-\omega t}\left(\|z\|_{C_b([a,+\infty);X)} + m_0\right).$$

Therefore

$$\|z(t)\| \leq e^{-\omega a}(\|z\|_{C_b([a,+\infty);X)} + m_0)$$

for each $t \in [a,+\infty)$ and so

$$\|z\|_{C_b([a,+\infty);X)} \leq e^{-\omega a}(\|z\|_{C_b([a,+\infty);X)} + m_0)$$

or

$$\|z\|_{C([a,+\infty);X)} \leq \frac{m_0 e^{\omega a}}{e^{\omega a} - 1}.$$

But

$$\|z\|_{C_b([-\tau,+\infty);X)} \leq \|z\|_X \leq \frac{m_0 e^{\omega a}}{e^{\omega a} - 1}.$$

From (7.7.5), it follows that

$$\|u_\varepsilon - z\|_{C_b([-\tau,+\infty);X)} \leq \frac{m}{\omega - \ell} + \frac{\ell}{\omega - \ell} \cdot \frac{m_0 e^{\omega a}}{e^{\omega a} - 1}.$$

This completes the proof. $\qquad\square$

We can now move on the proof of Theorem 7.6.1.

Proof. Let $(\varepsilon_n)_n$ be a sequence with $\varepsilon_n \downarrow 0$, let $(E_{\varepsilon_n})_n$ be a sequence of open subsets in $\mathbb{R}_+$ satisfying the condition in Definition 1.5.3 and, in addition,

$$E_{\varepsilon_{n+1}} \subseteq E_{\varepsilon_n} \tag{7.7.6}$$

Delay Evolution Inclusions 285

for each $n \in \mathbb{N}$. Let $(u_n)_n$ be a sequence of C^0-solutions of the problem (7.6.7) corresponding to $\varepsilon = \varepsilon_n$ and let $(F_n)_n$ be such that

$$
\begin{cases}
u'_n(t) \in Au_n(t) + f_n(t), & t \in \mathbb{R}_+, \\
f_n(t) \in F_{\varepsilon_n}(t, u_{n_t}), & t \in \mathbb{R}_+, \\
u_n(t) = g(u_n)(t), & t \in [-\tau, 0].
\end{cases}
$$

From (7.7.3), we deduce that the set $\{f_n; \ n \in \mathbb{N}\}$ is uniformly integrable in $L^1(0, k; X)$ for $k = 1, 2, \dots$. From this simple observation, (A_2) in (H_A), and Theorem 1.8.6, it follows that, for each $k = 1, 2, \dots$ and each $\delta \in (0, k)$, the set $\{u_n; \ n \in \mathbb{N}\}$ is relatively compact in $C([\delta, k]; X)$. In view of (g_4) in (H_g), we deduce that the set

$$
\{u_n(0); \ n \in \mathbb{N}\} = \{g(u_n)(0); \ n \in \mathbb{N}\}
$$

is relatively compact in $\overline{D(A)}$. From the second part of Theorem 1.8.6, we conclude that $\{u_n; \ n \in \mathbb{N}\}$ is relatively compact in $C([0, k]; X)$. So,

$$
C_k = \overline{\{u_{n_t}; \ n \in \mathbb{N}, \ t \in [0, k]\}}
$$

is compact in $\mathfrak{X}$. Since the restriction of the approximate multifunction F_{ε_n} to $([0, k] \setminus E_{\varepsilon_n}) \times \mathfrak{X}$ is strongly-weakly u.s.c., again by Lemma 1.5.1 combined with Krein–Smulian Theorem 1.4.3, we deduce that the set

$$
\overline{\mathrm{conv}}\, F_{\varepsilon_n}([0, k] \times C_k) = \overline{\mathrm{conv}}\, [F_{\varepsilon_n}(([0, k] \setminus E_{\varepsilon_n}) \times C_k) \cup \{0\}]
$$

is weakly compact in X. By Theorem 1.4.13, it follows that $\{f_n; \ n \in \mathbb{N}\}$ is sequentially relatively compact in $L^1(\mathbb{R}_+; e^{-t}dt; X)$. So, on a subsequence at least, we have

$$
\begin{cases}
\lim_n f_n = f & \text{weakly in } \ L^1(\mathbb{R}_+; e^{-t}dt; X), \\
\lim_n u_n(t) = u(t) & \text{uniformly for } \ t \in [0, k], \ k = 1, 2, \dots, \\
\lim_n g(u_n)(t) = g(u)(t) & \text{uniformly for } \ t \in [-\tau, 0].
\end{cases}
$$

In view of (7.7.4) and (7.7.6), we have $f_k(t) \in F_{\varepsilon_n}(t, u_{k_t})$ for each $n \in \mathbb{N}$, each $k \in \mathbb{N}$ with $k \geq n$, and a.e. for $t \in \mathbb{R}_+ \setminus E_{\varepsilon_n}$. Hence, from Theorem 1.5.1, it follows that $f(t) \in F_{\varepsilon_n}(t, u_t)$ for each $n \in \mathbb{N}$ and a.e. for $t \in \mathbb{R}_+ \setminus E_{\varepsilon_n}$. As $\lim_n \lambda(E_{\varepsilon_n}) = 0$, we finally obtain that $f(t) \in F(t, u_t)$ a.e. for $t \in \mathbb{R}_+$.

Since A is of complete continuous type, it follows that u is a C^0-solution of the problem (7.6.1) corresponding to the selection f of $t \mapsto F(t, u_t)$ and this concludes the proof of Theorem 7.6.1. $\qquad \square$

7.8 A differential inclusion governed by the p-Laplacian

We reconsider the case analyzed in Section 7.4 under different hypotheses. Namely, let Ω be a nonempty, bounded and open subset in $\mathbb{R}^d$, $d \geq 1$, with C^2 boundary Σ, let $p \in [2, +\infty)$, let $\tau \geq 0$, $\lambda > 0$ and $\omega > 0$. Set $Q_+ = \mathbb{R}_+ \times \Omega$, $\Sigma_+ = \mathbb{R}_+ \times \Sigma$ and $Q_\tau = [-\tau, 0] \times \Omega$. Let us consider the nonlinear problem

$$\begin{cases} \dfrac{\partial u}{\partial t}(t, x) = \Delta_p^\lambda u(t, x) - \omega u(t, x) + f(t, x), & \text{in } Q_+, \\[2mm] f(t, x) \in F(t, u_t)(x), & \text{in } Q_+, \\[2mm] -\dfrac{\partial u}{\partial \nu_p}(t, x) \in \beta(u(t, x)), & \text{on } \Sigma_+, \\[2mm] u(t, x) = \displaystyle\int_\tau^{+\infty} \mathcal{N}(u(t + \theta, \cdot))(x)\, d\mu(\theta) + \psi(t)(x), & \text{in } Q_\tau. \end{cases} \quad (7.8.1)$$

As in Section 7.4, we denote by $\mathcal{X}_2 = C([-\tau, 0]; L^2(\Omega))$. Here Δ_p^λ is defined as in Example 1.9.6, $F : \mathbb{R}_+ \times \mathcal{X}_2 \rightsquigarrow L^2(\Omega)$ is given by

$$F(t, v)(x) = [\, f_1(t, x, v) + h(x),\ f_2(t, x, v) + h(x)\,]$$

for each $(t, v) \in \mathbb{R}_+ \times \mathcal{X}_2$ and a.e. for $x \in \Omega$, while $f_1, f_2 : \mathbb{R}_+ \times \Omega \times \mathcal{X}_2 \to \mathbb{R}$, $h \in L^2(\Omega)$, $\mathcal{N} : L^2(\Omega) \to L^2(\Omega)$ is a nonexpansive function, μ is a σ-finite and complete measure on $[\tau, +\infty)$ and $\psi \in \mathcal{X}_2$.

Let us also consider

$$\begin{cases} \dfrac{\partial z}{\partial t}(t, x) = \Delta_p^\lambda z(t, x) - \omega z(t, x), & \text{in } Q_+, \\[2mm] -\dfrac{\partial z}{\partial \nu_p}(t, x) \in \beta(z(t, x)), & \text{on } \Sigma_+, \\[2mm] z(t, x) = \displaystyle\int_\tau^{+\infty} \mathcal{N}(z(t + \theta, \cdot))(x)\, d\mu(\theta) + \psi(t)(x), & \text{in } Q_\tau, \end{cases} \quad (7.8.2)$$

which, by Lemma 7.1.1, has a unique C^0-solution $z \in C_b([-\tau, +\infty); L^2(\Omega))$.

The main result in this section is:

THEOREM 7.8.1 *Let Ω be a nonempty, bounded and open subset in $\mathbb{R}^d$, $d \geq 1$, with C^1 boundary Σ, let $\tau \geq 0$, $\omega > 0$, $p \in [2, +\infty)$, $\lambda > 0$ and let $\beta : D(\beta) \subseteq \mathbb{R} \rightsquigarrow \mathbb{R}$ be a maximal-monotone operator with $0 \in D(\beta)$ and $0 \in \beta(0)$. Let $f_i : \mathbb{R}_+ \times \Omega \times \mathcal{X}_2 \to \mathbb{R}$, $i = 1, 2$, be two given functions, let $h \in L^2(\Omega)$, $\|h\|_{L^2(\Omega)} > 0$, let $\mathcal{N} : L^2(\Omega) \to L^2(\Omega)$ and let μ be a σ-finite and complete measure on $[\tau, +\infty)$ with $\operatorname{supp} \mu \subseteq [b, +\infty)$, where $b > \tau$. Let us assume that*

(h_1) *$f_1(t, x, v) \leq f_2(t, x, v)$ for each $(t, x, v) \in D(f_1, f_2)$, where $D(f_1, f_2) = \mathbb{R}_+ \times \Omega \times \mathcal{X}_2$*

Delay Evolution Inclusions 287

(h_2) *there exist* $\ell > 0$ *and* $m > 0$ *such that*

$$|f_i(t,x,v)| \leq \frac{\ell}{\sqrt{|\Omega|}}\|v\|_{\mathcal{X}_2} + \frac{m}{\sqrt{|\Omega|}}$$

for $i = 1, 2$ *and each* $(t, x, v) \in D(f_1, f_2)$

(h_3) f_1 *is l.s.c. and* f_2 *is u.s.c.*

(h_4) *the constants* ℓ *and* ω *satisfy* $\ell < \omega$

(μ_1) $\mu([b, +\infty)) = 1$

(μ_2) $\lim_{\delta \downarrow 0} \mu([\tau, \tau + \delta]) = 0$

(n_1) $\|\mathcal{N}(u) - \mathcal{N}(v)\|_{L^2(\Omega)} \leq \|u - v\|_{L^2(\Omega)}$ *for each* $u, v \in L^2(\Omega)$

(n_2) $\mathcal{N}(0) = 0.$

Then, (7.8.1) *has at least one* C^0-*solution,* $u \in C_b([-\tau, +\infty); L^2(\Omega))$, *which, for each* $T > 0$ *and* $\delta \in (0, T)$, *satisfies*

$$u \in AC([0, T]; W^{1,p}(\Omega)) \cap W^{1,2}([\delta, T]; L^2(\Omega)).$$

In addition,

$$\|u - z\|_{C_b([-\tau, +\infty); L^2(\Omega))} \leq \frac{m}{\omega - \ell} + \frac{\ell}{\omega - \ell} \cdot \frac{\|\psi\|_{\mathcal{X}_2} e^{\omega a}}{e^{\omega a} - 1},$$

where z *is the unique* C^0-*solution of the problem* (7.8.2) *and* $a = b - \tau > 0$.

We can now proceed to the proof of Theorem 7.8.1.

Proof. Let $A : D(A) \subseteq L^2(\Omega) \to L^2(\Omega)$ be defined by

$$\begin{cases} D(A) = \left\{ u \in W^{1,p}(\Omega); \Delta_p^\lambda u \in L^2(\Omega), -\dfrac{\partial u}{\partial \nu_p}(x) \in \beta(u(x)), \text{ a.e. for } x \in \Sigma \right\}, \\ Au = \Delta_p^\lambda u - \omega u, \text{ for } u \in D(A). \end{cases}$$

As in the proof of Theorem 7.4.1, it follows that A is m-dissipative on $L^2(\Omega)$, $\overline{D(A)} = L^2(\Omega)$ and $0 \in A0$. Moreover, it generates a compact semigroup of nonexpansive mappings on $L^2(\Omega)$ and is of complete continuous type. Hence A satisfies (H_A). Moreover, $\mathcal{D}_2 = \{\varphi \in \mathcal{X}_2; \varphi(0) \in \overline{D(A)}\} = \mathcal{X}_2$. Further, the multifunction $F : \mathbb{R}_+ \times \mathcal{X}_2 \rightsquigarrow L^2(\Omega)$ is given by

$$F(t, v) = \{f \in L^2(\Omega); f_1(t, x, v) + h(x) \leq f(x) \leq f_2(t, x, v) + h(x) \text{ a.e. for } x \in \Omega\}$$

for each $(t, v) \in \mathbb{R}_+ \times \mathcal{X}_2$. Let us define

$$g : C_b(\mathbb{R}_+; L^2(\Omega)) \to \mathcal{X}_2$$

288　*Delay Differential Evolutions Subjected to Nonlocal Initial Conditions*

by

$$[g(u)(t)](x) = \int_{\tau}^{+\infty} \mathcal{N}(u(t+\theta, \cdot))(x)\, d\mu(\theta) + \psi(t)(x)$$

for each $u \in C_b(\mathbb{R}_+; L^2(\Omega))$, each $t \in [-\tau, 0]$ and a.e. for $x \in \Omega$.

Clearly, with A, F and g as above, the problem (7.8.1) can be rewritten in the form (7.6.1). In order to appeal to Theorem 7.6.1, it remains to observe that, from $(h_1) \sim (h_3)$ and Lemma 1.5.3, it follows that F is a nonempty, convex and weakly compact-valued, strongly-weakly u.s.c. multifunction. So, F satisfies (H_F). From (h_2) and (h_4), we conclude that (H_c) also holds true. To show that g satisfies (H_g), except some obvious modifications, we have to proceed as in the proof of Theorem 7.4.1.

So, all the hypotheses of Theorem 7.6.1 are satisfied and this completes the proof. $\qquad\square$

7.9　A nonlinear diffusion inclusion in $L^1(\Omega)$

Let Ω be a nonempty, bounded and open subset in $\mathbb{R}^d$, $d \geq 1$, with C^1 boundary Σ, let $\tau \geq 0$, $\omega > 0$, let $\varphi : D(\varphi) \subseteq \mathbb{R} \rightsquigarrow \mathbb{R}$ be maximal-monotone with $0 \in D(\varphi)$, $0 \in \varphi(0)$ and let $\psi \in C([-\tau, 0]; L^1(\Omega))$. Let $Q_+ = \mathbb{R}_+ \times \Omega$, $\Sigma_+ = \mathbb{R}_+ \times \Sigma$ and $Q_\tau = [-\tau, 0] \times \Omega$, let us consider the porous medium equation subjected to nonlocal initial conditions

$$\begin{cases} \dfrac{\partial u}{\partial t}(t, x) \in \Delta\varphi(u(t, x)) - \omega u(t, x) + f(t, x), & \text{in } Q_+, \\[2mm] f(t, x) \in F(t, u_t)(x), & \text{in } Q_+, \\[2mm] \varphi(u(t, x)) = 0, & \text{on } \Sigma_+, \\[2mm] u(t, x) = \displaystyle\int_{\tau}^{+\infty} \mathcal{N}(u(\theta + t, \cdot))(x)\, d\mu(\theta) + \psi(t)(x), & \text{in } Q_\tau \end{cases} \tag{7.9.1}$$

and let us denote by $\mathcal{X}_1 = C([-\tau, 0]; L^1(\Omega))$. Here, for each $(t, v) \in \mathbb{R}_+ \times \mathcal{X}_1$

$$F(t, v) = \{ f \in L^1(\Omega);\ f_1(t, x, v) + h(x) \leq f(x) \leq f_2(t, x, v) + h(x) \text{ a.e. for } x \in \Omega \}, \tag{7.9.2}$$

$f_1, f_2 : \mathbb{R}_+ \times \Omega \times \mathcal{X}_1 \to \mathbb{R}$, $h \in L^1(\Omega)$, $\mathcal{N} : L^1(\Omega) \to L^1(\Omega)$ and μ is a σ-finite and complete measure on $[\tau, +\infty)$.

Let us also consider the unperturbed problem

$$\begin{cases} \dfrac{\partial z}{\partial t}(t, x) \in \Delta\varphi(z(t, x)) - \omega z(t, x), & \text{in } Q_+, \\[2mm] \varphi(z(t, x)) = 0, & \text{on } \Sigma_+, \\[2mm] z(t, x) = \displaystyle\int_{\tau}^{+\infty} \mathcal{N}(z(\theta + t, \cdot))(x)\, d\mu(\theta) + \psi(t)(x), & \text{in } Q_\tau. \end{cases} \tag{7.9.3}$$

Delay Evolution Inclusions 289

THEOREM 7.9.1 *Let Ω be a nonempty, bounded and open subset in $\mathbb{R}^d$, $d \geq 1$, with C^1 boundary Σ, let $\tau \geq 0$, $\omega > 0$, and let $\varphi : \mathbb{R} \to \mathbb{R}$ be continuous on $\mathbb{R}$ and C^1 on $\mathbb{R} \setminus \{0\}$ with $\varphi(0) = 0$ and for which there exist two constants $C > 0$ and $\alpha > 0$ if $d \leq 2$ and $\alpha > (d-2)/d$ if $d \geq 3$ such that*

$$\varphi'(r) \geq C|r|^{\alpha-1}$$

for each $r \in \mathbb{R} \setminus \{0\}$. Let $f_i : \mathbb{R}_+ \times \Omega \times \mathfrak{X}_1 \to \mathbb{R}$, $i = 1, 2$, two given functions, $h \in L^1(\Omega)$, $\|h\|_{L^1(\Omega)} > 0$, $\mathcal{N} : L^1(\Omega) \to L^1(\Omega)$, and let μ be a σ-finite and complete measure on $[\tau, +\infty)$ with $\operatorname{supp}\mu \subseteq [b, +\infty)$, where $b > \tau$. Let us assume that

(h_1) $f_1(t, x, v) \leq f_2(t, x, v)$ *for each $(t, x, v) \in D(f_1, f_2)$, where*
$\quad\quad D(f_1, f_2) = \mathbb{R}_+ \times \Omega \times \mathfrak{X}_1$

(h_2) *there exist $\ell > 0$ and $m > 0$ such that*

$$|f_i(t, x, v)| \leq \ell\|v\|_{\mathfrak{X}_1} + m$$

for $i = 1, 2$ and each $(t, x, v) \in D(f_1, f_2)$

(h_3) f_1 *is l.s.c. and f_2 is u.s.c.*

(h_4) *the constants ℓ and ω satisfy the nonresonance condition $\ell < \omega$*

(μ_1) $\mu([b, +\infty)) = 1$

(μ_2) $\lim_{\delta\downarrow 0} \mu([\tau, \tau + \delta]) = 0$

(n_1) $\|\mathcal{N}(u) - \mathcal{N}(v)\|_{L^1(\Omega)} \leq \|u - v\|_{L^1(\Omega)}$ *for each $u, v \in L^1(\Omega)$*

(n_2) $\mathcal{N}(0) = 0$.

Then (7.9.1) has at least one C^0-solution, $u \in C_b([-\tau, +\infty); L^1(\Omega))$, satisfying

$$\|u - z\|_{C_b([-\tau, +\infty); L^1(\Omega))} \leq \frac{m}{\omega - \ell} + \frac{\ell}{\omega - \ell} \cdot \frac{\|\psi\|_{\mathfrak{X}_1} e^{\omega a}}{e^{\omega a} - 1},$$

where z is the unique C^0-solution of the problem (7.9.3) and $a = b - \tau > 0$.

Proof. Let $X = L^1(\Omega)$ and let us define $A : D(A) \subseteq L^1(\Omega) \to L^1(\Omega)$, by

$$\begin{cases} D(A) = \left\{ u \in L^1(\Omega); \ \varphi(u) \in W_0^{1,1}(\Omega), \ \Delta\varphi(u) \in L^1(\Omega) \right\}, \\ Au = \Delta\varphi(u) - \omega u \text{ for each } u \in D(A). \end{cases}$$

Theorem 1.9.6 implies that A is m-dissipative in $L^1(\Omega)$, $A0 = 0$, A generates a compact semigroup and is of complete continuous type on $\overline{D(A)} = L^1(\Omega)$. Hence A satisfies (H_A). Clearly, $\mathcal{D}_1 = \{\varphi \in \mathfrak{X}_1; \ \varphi(0) \in \overline{D(A)}\} = \mathfrak{X}_1$. Let $F : \mathbb{R}_+ \times \mathfrak{X}_1 \rightsquigarrow L^1(\Omega)$ be defined as in (7.9.2) and let $g : C_b(\mathbb{R}_+; L^1(\Omega)) \to \mathfrak{X}_1$ be given by

290 *Delay Differential Evolutions Subjected to Nonlocal Initial Conditions*

$$[g(u)(t)](x) = \int_{\tau}^{+\infty} \mathcal{N}(u(t+\theta))(x) \, d\mu(\theta) + \psi(t)(x)$$

for each $u \in C_b(\mathbb{R}_+; L^1(\Omega))$, each $t \in [-\tau, 0]$ and a.e. for $x \in \Omega$. But $(h_1) \sim (h_3)$ show that we are in the hypotheses of Lemma 1.5.4, from which we conclude that F satisfies the general assumption in (H_F). From (h_2) and (h_4), we deduce that F satisfies (F_2) and also that (H_c) holds true. Since the proof of (H_g) follows the very same lines as in the case of Theorem 7.4.1, we do not provide details. So, with A, F and g defined as above, (7.9.1) can be written in the abstract form (7.6.1). Then Theorem 7.6.1 applies and this completes the proof. □

7.10   Bibliographical notes and comments

Section 7.1 The existence problem for differential inclusions without delay and subjected to initial-value conditions was studied by many authors starting with the pioneering works of Zaremba [270] and Ważewski [264]. For other important contributions in this topic, see Bothe [43], Kryszewski [159] and the references therein as well as the monograph of Cârjă, Necula and Vrabie [74]. A special case of the problem here considered, i.e., on the standard compact interval $[0, 2\pi]$, in the simplest case when $\tau = 0$, i.e., when the delay is absent, was studied by Paicu and Vrabie [204]. In this case $\mathcal{D}$ identifies with $\overline{D(A)}$, $F(t, u)$ identifies with a multifunction F from $[0, 2\pi] \times X$ to X and so, Paicu and Vrabie [204] have considered the problem

$$\begin{cases} u'(t) \in Au(t) + f(t), & t \in [0, 2\pi], \\ f(t) \in F(t, u(t)), & t \in [0, 2\pi], \\ u(0) = g(u). \end{cases} \qquad (7.B.1)$$

By using an interplay between compactness arguments and invariance techniques, they have proved an existence result handling periodic, anti-periodic, mean-value evolution inclusions subjected to initial conditions expressed by an integral with respect to a Radon measure μ. Classical nonlinear delay evolution initial-value problems of the type

$$\begin{cases} u'(t) \in Au(t) + f(t), & t \in [0, 2\pi], \\ f(t) \in F(t, u_t), & t \in [0, 2\pi], \\ u(t) = \varphi(t), & t \in [-\tau, 0], \end{cases} \qquad (7.B.2)$$

were studied by Mitidieri and Vrabie [185] and [186], also by using compactness arguments. It should be emphasized that in Mitidieri and Vrabie [185] and

Delay Evolution Inclusions 291

[186], the general assumptions on the forcing term F are very general allowing – in a certain specific case when A is a second-order elliptic operator – the dependence on Au as well. As we can easily see, the general problem (7.1.1) includes, as particular cases, both (7.B.1) and (7.B.2). There is a very long list of papers referring either to (7.B.1) or (7.B.2). For an existence and stability result for (7.B.2) based on topological methods involving the Hausdorff measure of noncompactness, see Tran Dinh Ke [153]. A very important specific case of (7.B.1) concerns T-periodic problems, which corresponds to the choice of g as $g(u) = u(T)$, was studied by Castaing and Monteiro-Marques [71], Kryszewski and Plaskacz [160], Lakshmikantham and Papageorgiou [165], Paicu [201], and Papageorgiou [206]. A thorough analysis of the nondelayed case (7.B.2) can be found in Aubin and Cellina [13], and Vrabie [252]. The case F single-valued was studied by Aizicovici, Papageorgiou and Staicu [4], Cașcaval and Vrabie [73], Hirano [146], Hirano and Shioji [147], Paicu [202], and Vrabie [251]. A very good survey concerning periodic, anti-periodic, quasi-periodic and almost periodic solutions to differential inclusions is that of Andres [10]. For a topological perspective on such problems, see also Andres [9]. As far as differential inclusions subjected to general nonlocal initial conditions without delay are concerned, i.e., problems of the type (7.B.1), we mention the papers of Aizicovici and Staicu [6] and Paicu and Vrabie [204].

The case of retarded equations and inclusions subjected to nonlocal initial conditions

$$\begin{cases} u'(t) \in Au(t) + f(t), & t \in \mathbb{R}_+, \\ f(t) \in F(t, u(t), u(t - \tau_1), u(t - \tau_2), \ldots, u(t - \tau_n)), & t \in \mathbb{R}_+, \\ u(t) = g(u)(t), & t \in [-\tau, 0], \end{cases}$$

with $\tau = \max\{\tau_1, \tau_2, \ldots, \tau_n\}$, were studied very recently by Vrabie [255]. See also Chen, Wang and Zhou [81].

Section 7.2 The main results in this section, i.e., Theorem 7.2.1 and 7.2.2, are from Necula and Vrabie [197] and extend previous results in Vrabie [256] and Paicu and Vrabie [204]. The novelty in both results rests in that here we allow $g(0)$ to be different than 0, a case which encompasses a broader class of nontrivial applications. Condition (F_3) is simply a sufficient condition in order that the moving set $t \mapsto \mathcal{K}(t) = \{v \in \mathfrak{X}; \|v - z_t\|_{\mathfrak{X}} = \|v(0) - z(t)\| \le r\}$ be invariant with respect to the differential inclusion

$$\begin{cases} u'(t) \in Au(t) + f(t), & t \in \mathbb{R}_+, \\ f(t) \in F(t, u_t), & t \in \mathbb{R}_+, \end{cases}$$

and it reduces to that in Vrabie [256] whenever g has linear growth. For more information on invariance and viability techniques for delay differential inclusions, see Necula and Popescu [191] for the case in which A is linear and Necula, Popescu and Vrabie [194] for the general case. The viability problem for delay differential inclusions subjected to a nonlocal implicit initial condition was considered for the first time in Necula, Popescu and Vrabie [195].

292 *Delay Differential Evolutions Subjected to Nonlocal Initial Conditions*

Section 7.3 The proofs of Theorems 7.2.1 and 7.2.3, although not identical, are both inspired by the proofs of Theorems 3.1 and respectively 3.2 in Vrabie [256]. Lemmas 7.3.1 and 7.3.3 are from Necula and Vrabie [197]. The proof of Lemma 7.3.5 is inspired by Lemma 4.4 in Paicu and Vrabie [204].

Section 7.4 Theorem 7.4.1 is a variant of a result proved in Vrabie [256], in the case $\psi \neq 0$.

Section 7.5 Theorem 7.5.1 is also a variant of similar result established in Vrabie [256], in the case $\psi \neq 0$.

Section 7.6 A specific but very important case of (7.6.1) concerns T-periodic problems, which corresponds to the choice of g as $g(u)(t) = u(t+T)$. From the very long list of contributors to the study of such problems we mention Castaing and Monteiro-Marques [71], Lakshmikantham and Papageorgiou [165], Paicu [201], and Papageorgiou [206]. For anti-periodic problems, i.e., when $g(u) = -u(T)$, see Aizicovici, Pavel and Vrabie [5] and the references therein. A good reference concerning these kinds of special solutions including periodic, anti-periodic, and almost periodic is Andres [10]. Theorems 7.6.1 and 7.6.2 extend to the case in which g has affine rather than linear growth – the latter, of course being not excluded – see the main results in Vrabie [258]. The idea of the proof is similar to that in Vrabie [258] but there are several new difficulties to overcome.

Section 7.7 Although Lemmas 7.7.1~7.7.3 are inspired by Vrabie [258], the approximating equations used, i.e., (7.6.7), are simpler than the corresponding ones considered in Vrabie [258, Equation 3.6, p. 485]. The proof of Lemma 7.7.2 is inspired by Vrabie [257].

REMARK. If (A_1) in (H_A) and (F_2) are satisfied, $r = \frac{m}{\omega-\ell}$ and A is strongly dissipative, i.e., $[\,x_1 - x_2, y_1 - y_2\,]_+ \leq 0$ for each $x_i \in D(A)$ and $y_i \in Ax_i$ then $A + F$ satisfies the tangency condition with respect to $D(0,r) \cap \overline{D(A)}$, i.e., $[\,u, y + f\,]_+ \leq 0$ for each $u \in D(A)$ with $\|u\| = r$, each $y \in Au$, $t \in \mathbb{R}_+$, $v \in \mathcal{D}$ with $\|v\|_X \leq r$ and each $f \in F(t,v)$. This simply follows from (vi) in Proposition 1.8.1. We notice that, in a Hilbert space frame, the condition above was used in Caşcaval and Vrabie [73]. Also, under the same circumstances we have

$$\lim_{\rho \to +\infty} \sup_{\|v\|_X \leq \rho} \frac{\|F(t,v)\|}{\|u\|} < \omega.$$

Section 7.8 Theorem 7.8.1 is an extension of Theorem 5.1 in Vrabie [258], to the case in which g has affine rather than linear growth.

Section 7.9 Theorem 7.9.1 is an extension of Theorem 5.3 in Vrabie [258] – referring to nonlocal functions g having linear growth – to the case in which g has affine growth.

Chapter 8

Multivalued Reaction–Diffusion Systems

Overview

In this chapter, we reconsider the problem studied in Chapter 6 in the more general case of a nonlinear multi-valued reaction–diffusion system with delay. Using compactness arguments coupled with metric fixed-point techniques, we prove some sufficient conditions for the global existence of a bounded C^0-solution.

8.1 The problem to be studied

Let X and Y be two real Banach spaces, let $A : D(A) \subseteq X \rightsquigarrow X$ be an ω-m-dissipative operator for some $\omega > 0$ and let $B : D(B) \subseteq Y \rightsquigarrow Y$ be an γ-m-dissipative operator for some $\gamma > 0$. Let $\tau \geq 0$ and let us denote

$$\mathfrak{X} = C([-\tau, 0]; X) \quad \text{and} \quad \mathfrak{Y} = C([-\tau, 0]; Y),$$

$$\mathcal{D} = \{\varphi \in \mathfrak{X}; \ \varphi(0) \in \overline{D(A)}\} \quad \text{and} \quad \mathcal{E} = \{\varphi \in \mathfrak{Y}; \ \varphi(0) \in \overline{D(B)}\}.$$

Let $F : \mathbb{R}_+ \times \mathfrak{X} \times \mathfrak{Y} \to X$ be a continuous function, and let $G : \mathbb{R}_+ \times \mathfrak{X} \times \mathfrak{Y} \rightsquigarrow Y$ be a nonempty, convex and weakly compact-valued multifunction which is weakly u.s.c. Finally, let

$$p : C_b(\mathbb{R}_+; \overline{D(A)}) \times C_b(\mathbb{R}_+; \overline{D(B)}) \to \mathcal{D}$$

and

$$q : C_b(\mathbb{R}_+; \overline{D(A)}) \times C_b(\mathbb{R}_+; \overline{D(B)}) \to \mathcal{E}$$

be two nonexpansive mappings.

294 *Delay Differential Evolutions Subjected to Nonlocal Initial Conditions*

In this chapter, we consider a class of nonlinear delay differential reaction–diffusion systems with nonlocal initial conditions

$$
\begin{cases}
u'(t) \in Au(t) + F(t, u_t, v_t), & t \in \mathbb{R}_+, \\
v'(t) \in Bv(t) + g(t), & t \in \mathbb{R}_+, \\
g(t) \in G(t, u_t, v_t), & t \in \mathbb{R}_+, \\
u(t) = p(u, v)(t), & t \in [-\tau, 0], \\
v(t) = q(u, v)(t), & t \in [-\tau, 0]
\end{cases}
\tag{8.1.1}
$$

and we prove some sufficient conditions for the global existence of bounded C^0-solutions.

Customarily, if $u \in C([-\tau, +\infty); X)$, $v \in C([-\tau, +\infty); Y)$ and $t \in \mathbb{R}_+$, $u_t \in \mathcal{X}$ and $v_t \in \mathcal{Y}$ are defined by

$$
\begin{cases}
u_t(s) = u(t+s), & s \in [-\tau, 0], \\
v_t(s) = v(t+s), & s \in [-\tau, 0].
\end{cases}
$$

If $u \in C_b([-\tau, +\infty); X)$ and $v \in C_b([-\tau, +\infty); Y)$ with $(u(t), v(t)) \in \overline{D(A)} \times \overline{D(B)}$ for each $t \in \mathbb{R}_+$, we denote by

$$
\begin{cases}
p(u, v) = p(u_{|\mathbb{R}_+}, v_{|\mathbb{R}_+}), \\
q(u, v) = q(u_{|\mathbb{R}_+}, v_{|\mathbb{R}_+}).
\end{cases}
$$

DEFINITION 8.1.1 By a C^0-*solution* of (8.1.1) we mean a continuous function $(u, v) : [-\tau, +\infty) \to X \times Y$ satisfying $(u(t), v(t)) = (p(u, v)(t), q(u, v)(t))$ for each $t \in [-\tau, 0]$ and there exists $g \in L^1_{\mathrm{loc}}(\mathbb{R}_+; Y)$ with $g(t) \in G(t, u_t, v_t)$ a.e. for $t \in \mathbb{R}_+$ such that, for each $T > 0$, (u, v) is a C^0-solution on $[0, T]$ in the sense of Definition 1.8.2 in the space $X \times Y$, for

$$
\begin{cases}
u'(t) \in Au(t) + f(t), & t \in \mathbb{R}_+, \\
v'(t) \in Bv(t) + g(t), & t \in \mathbb{R}_+,
\end{cases}
$$

where $f(t) = F(t, u_t, v_t)$ for $t \in [0, T]$.

8.2 The main result

We denote by $\widetilde{L}^1(\mathbb{R}_+; Y)$ the space $L^1_{\mathrm{loc}}(\mathbb{R}_+; Y)$ endowed with the family of semi-norms $\{\| \cdot \|_{k, x^*}; \ x^* \in L^1(0, k; Y)^*, \ k = 1, 2, \dots\}$, defined by

$$
\|g\|_{k, x^*} = |x^*(g)|
$$

for $g \in L^1_{\mathrm{loc}}(\mathbb{R}_+; Y)$. Endowed with this family of semi-norms, $\widetilde{L}^1(\mathbb{R}_+; Y)$ is a locally convex, separated vector space. Moreover, the convergence in $\widetilde{L}^1(\mathbb{R}_+; Y)$ is simply the weak convergence in $L^1(0, k; Y)$ for $k = 1, 2, \dots$.

Multivalued Reaction–Diffusion Systems

We begin by formulating the main hypotheses that will be used in what follows.

(H_A) The operator $A : D(A) \subseteq X \rightsquigarrow X$ satisfies

    (A_1) $0 \in D(A)$, $0 \in A0$ and A is ω-m-dissipative for some $\omega > 0$

    (A_3) $\overline{D(A)}$ is convex.

(H_B) The operator $B : D(B) \subseteq Y \rightsquigarrow Y$ satisfies

    (B_1) $0 \in D(B)$, $0 \in B0$ and B is γ-m-dissipative for some $\gamma > 0$

    (B_2) B generates a compact semigroup on $\overline{D(B)}$

    (B_4) B is of complete continuous type.

(H_F) The function $F : \mathbb{R}_+ \times X \times \mathcal{Y} \to X$ is continuous and satisfies

    (F_1) there exists $\ell > 0$ such that

$$\|F(t, u, v) - F(t, \widetilde{u}, \widetilde{v})\| \leq \ell \max\{\|u - \widetilde{u}\|_X, \|v - \widetilde{v}\|_{\mathcal{Y}}\}$$

    for each $(t, u, v), (t, \widetilde{u}, \widetilde{v}) \in \mathbb{R}_+ \times X \times \mathcal{Y}$

    (F_2) there exists $m > 0$ such that

$$\|F(t, u, v)\| \leq \ell \|u\|_X + m$$

    for each $(t, u, v) \in \mathbb{R}_+ \times X \times \mathcal{Y}$, where ℓ is given by (F_1).

(H_G) The multifunction $G : \mathbb{R}_+ \times X \times \mathcal{Y} \rightsquigarrow Y$ has nonempty, convex and weakly compact values and is strongly-weakly u.s.c., and

    (G_1) with ℓ and m given by (F_1) and (F_2), we have

$$\|y\| \leq \ell \max\{\|u\|_X, \|v\|_{\mathcal{Y}}\} + m$$

    for each $(t, u, v) \in \mathbb{R}_+ \times X \times \mathcal{Y}$ and each $y \in G(t, u, v)$.

(H_c) The constants ℓ and $\delta = \min\{\omega, \gamma\}$ satisfy the nonresonance condition

$$\ell < \delta.$$

(H_p) There exists $a > 0$ such that, with m given by (F_2), the function

$$p : C_b(\mathbb{R}_+; \overline{D(A)}) \times C_b(\mathbb{R}_+; \overline{D(B)}) \to \mathcal{D}$$

    satisfies

    (p_1) for each $(u, v), (\widetilde{u}, \widetilde{v}) \in C_b(\mathbb{R}_+; \overline{D(A)}) \times C_b(\mathbb{R}_+; \overline{D(B)})$, we have

$$\|p(u, v) - p(\widetilde{u}, \widetilde{v})\|_X$$

$$\leq \max\{\|u - \widetilde{u}\|_{C_b([a, +\infty); X)}, \|v - \widetilde{v}\|_{C_b([a, +\infty); Y)}\}$$

296   *Delay Differential Evolutions Subjected to Nonlocal Initial Conditions*

(p_2) for each $(u, v) \in C_b(\mathbb{R}_+; \overline{D(A)}) \times C_b(\mathbb{R}_+; \overline{D(B)})$, we have

$$\|p(u, v)\|_X \leq \|u\|_{C_b([a, +\infty); X)} + m.$$

(H_q) With a given by (H_p) and m given by (F_2), the function

$$q : C_b(\mathbb{R}_+; \overline{D(A)}) \times C_b(\mathbb{R}_+; \overline{D(B)}) \to \mathcal{E}$$

satisfies

(q_1) for each $(u, v), (\widetilde{u}, \widetilde{v}) \in C_b(\mathbb{R}_+; \overline{D(A)}) \times C_b(\mathbb{R}_+; \overline{D(B)})$, we have

$$\|q(u, v) - q(\widetilde{u}, \widetilde{v})\|_Y$$

$$\leq \max\{\|u - \widetilde{u}\|_{C_b([a, +\infty); X)}, \|v - \widetilde{v}\|_{C_b([a, +\infty); Y)}\}$$

(q_2) for each $(u, v) \in C_b(\mathbb{R}_+; \overline{D(A)}) \times C_b(\mathbb{R}_+; \overline{D(B)})$, we have

$$\|q(u, v)\|_Y \leq \|v\|_{C_b([a, +\infty); Y)} + m$$

(q_3) for each bounded set $\mathcal{U}$ in $C_b(\mathbb{R}_+; \overline{D(A)})$ and each $\mathcal{V}$ in $C_b([a, +\infty); \overline{D(B)})$ that is relatively compact in $\widetilde{C}_b([a, +\infty); Y)$, the set $q(\mathcal{U}, \mathcal{V})$ is relatively compact in $\mathcal{Y}$.

Let us also consider the unperturbed system

$$\begin{cases} z'(t) \in Az(t), & t \in \mathbb{R}_+, \\ w'(t) \in Bw(t), & t \in \mathbb{R}_+, \\ z(t) = p(z, w)(t), & t \in [-\tau, 0], \\ w(t) = q(z, w)(t), & t \in [-\tau, 0], \end{cases} \tag{8.2.1}$$

which, in view of Lemma 6.4.1, has a unique C^0-solution (z, w) satisfying

$$(z, w) \in C_b([-\tau, +\infty); X) \times C_b([-\tau, +\infty); Y).$$

REMARK 8.2.1 In fact, Lemma 6.4.1 follows from Lemma 4.2.1. Hence the latter can be applied directly for the system (8.2.1).

The main result in this chapter is:

THEOREM 8.2.1 *Let us assume that* (H_A), (H_B), (H_F), (H_G), (H_c), (H_p) *and* (H_q) *are satisfied. Then the problem* (8.1.1) *has at least one* C^0-*solution,* $(u, v) \in C_b([-\tau, +\infty); X) \times C_b([-\tau, +\infty); Y)$ *with* $(u(t), v(t)) \in \overline{D(A)} \times \overline{D(B)}$ *for each* $t \in \mathbb{R}_+$. *In addition, each* C^0-*solution* (u, v) *of* (8.1.1) *satisfies*

$$\begin{cases} \|u - z\|_{C_b([-\tau, +\infty); X)} \leq \dfrac{m}{\delta - \ell} + \dfrac{\ell}{\delta - \ell} \cdot \dfrac{me^{\delta a}}{e^{\delta a} - 1}, \\ \\ \|v - w\|_{C_b([-\tau, +\infty); Y)} \leq \dfrac{m}{\delta - \ell} + \dfrac{\ell}{\delta - \ell} \cdot \dfrac{me^{\delta a}}{e^{\delta a} - 1}, \end{cases} \tag{8.2.2}$$

where (z, w) *is the unique* C^0-*solution of* (8.2.1).

Multivalued Reaction–Diffusion Systems

The specific form of (8.1.1) in the nondelayed case, i.e., when $\tau = 0$, is

$$\begin{cases} u'(t) \in Au(t) + F(t, u(t), v(t)), & t \in \mathbb{R}_+, \\ v'(t) \in Bv(t) + g(t), & t \in \mathbb{R}_+, \\ g(t) \in G(t, u(t), v(t)), & t \in \mathbb{R}_+, \\ u(0) = p(u, v), \\ v(0) = q(u, v). \end{cases} \tag{8.2.3}$$

As far as (8.2.3) is concerned, we need the hypotheses below.

$(H_F^{[\tau=0]})$ The function $F : \mathbb{R}_+ \times X \times Y \to X$ is continuous and satisfies

$(F_1^{[\tau=0]})$ there exists $\ell > 0$ such that

$$\|F(t, u, v) - F(t, \widetilde{u}, \widetilde{v})\| \leq \ell \max\{\|u - \widetilde{u}\|, \|v - \widetilde{v}\|\}$$

for each $t \in \mathbb{R}_+$, each $u, \widetilde{u} \in X$ and each $v, \widetilde{v} \in Y$

$(F_2^{[\tau=0]})$ there exists $m > 0$ such that

$$\|F(t, u, v)\| \leq \ell \|u\| + m$$

for each $t \in \mathbb{R}_+$ and each $(u, v) \in X \times Y$, where ℓ is given by $(F_1^{[\tau=0]})$.

$(H_G^{[\tau=0]})$ The multifunction $G : \mathbb{R}_+ \times X \times Y \leadsto Y$ has nonempty, convex and weakly compact values, is strongly-weakly u.s.c., and satisfies

$(G_1^{[\tau=0]})$ with ℓ and m given by $(F_1^{[\tau=0]})$ and $(F_2^{[\tau=0]})$, we have

$$\|y\| \leq \ell \max\{\|u\|, \|v\|\} + m$$

for each $t \in \mathbb{R}_+$, $u \in X$, $v \in Y$ and each $y \in G(t, u, v)$.

$(H_p^{[\tau=0]})$ $p : C_b(\mathbb{R}_+; \overline{D(A)}) \times C_b(\mathbb{R}_+; \overline{D(B)}) \to \overline{D(A)}$ satisfies

$(p_1^{[\tau=0]})$ there exists $a > 0$ such that for each $u, \widetilde{u} \in C_b(\mathbb{R}_+; \overline{D(A)})$ and each $v, \widetilde{v} \in C_b(\mathbb{R}_+; \overline{D(B)})$, we have

$$\|p(u, v) - p(\widetilde{u}, \widetilde{v})\| \leq \max\{\|u - \widetilde{u}\|_{C_b([a, +\infty); X)}, \|v - \widetilde{v}\|_{C_b([a, +\infty); Y)}\}$$

$(p_2^{[\tau=0]})$ with m given by $(F_2^{[\tau=0]})$, for each $u \in C_b(\mathbb{R}_+; \overline{D(A)})$ and each $v \in C_b(\mathbb{R}_+; \overline{D(B)})$, we have

$$\|p(u, v)\| \leq \|u\|_{C_b(\mathbb{R}_+; X)} + m.$$

$(H_q^{[\tau=0]})$ $q : C_b(\mathbb{R}_+; \overline{D(A)}) \times C_b(\mathbb{R}_+; \overline{D(B)}) \to \overline{D(B)}$ satisfies

$(q_1^{[\tau=0]})$ with a given by $(p_2^{[\tau=0]})$, for each $u, \widetilde{u} \in C_b(\mathbb{R}_+; \overline{D(A)})$ and each $v, \widetilde{v} \in C_b(\mathbb{R}_+; \overline{D(B)})$, we have

$$\|q(u,v) - q(\widetilde{u}, \widetilde{v})\| \leq \max\{\|u - \widetilde{u}\|_{C_b([a,+\infty);X)}, \|v - \widetilde{v}\|_{C_b([a,+\infty);Y)}\}$$

$(q_2^{[\tau=0]})$ with m given by $(F_2^{[\tau=0]})$, for each $u \in C_b(\mathbb{R}_+; \overline{D(A)})$ and each $v \in C_b(\mathbb{R}_+; \overline{D(B)})$, we have

$$\|q(u,v)\| \leq \|v\| + m$$

$(q_3^{[\tau=0]})$ for each bounded set $\mathcal{U}$ in $C_b(\mathbb{R}_+; \overline{D(A)})$ and each $\mathcal{V}$ in $C_b([a,+\infty); \overline{D(B)})$ that is relatively compact in $\widetilde{C}_b([a,+\infty); Y)$, the set $q(\mathcal{U}, \mathcal{V})$ is relatively compact in $\overline{D(B)}$.

Let us consider

$$\begin{cases} \widetilde{z}'(t) \in A\widetilde{z}(t), & t \in \mathbb{R}_+, \\ \widetilde{w}'(t) \in B\widetilde{w}(t), & t \in \mathbb{R}_+, \\ \widetilde{z}(0) = p(\widetilde{z}, \widetilde{w}), \\ \widetilde{w}(0) = q(\widetilde{z}, \widetilde{w}). \end{cases} \tag{8.2.4}$$

REMARK 8.2.2 Let us observe that if (H_A), (B_1) in (H_B), $(p_1^{[\tau=0]})$, $(p_2^{[\tau=0]})$ in $(H_p^{[\tau=0]})$ and $(q_1^{[\tau=0]})$, $(q_2^{[\tau=0]})$ in $(H_q^{[\tau=0]})$ are satisfied, then (8.2.4) has a unique C^0-solution

$$(\widetilde{z}, \widetilde{w}) \in C_b(\mathbb{R}_+; \overline{D(A)}) \times C_b(\mathbb{R}_+; \overline{D(B)}).$$

This follows either from Lemma 6.4.1 or by Lemma 4.2.1 by taking $\tau = 0$. See Remark 8.2.1.

THEOREM 8.2.2 *Let us assume that* (H_A), (H_B), $(H_F^{[\tau=0]})$, $(H_G^{[\tau=0]})$, $(H_p^{[\tau=0]})$, $(H_q^{[\tau=0]})$ *and* (H_c) *are satisfied. Then the problem* (8.2.3) *has at least one C^0-solution,* $(u,v) \in C_b(\mathbb{R}_+; \overline{D(A)}) \times C_b(\mathbb{R}_+; \overline{D(B)})$. *In addition, each C^0-solution* (u,v) *of* (8.2.3) *satisfies*

$$\begin{cases} \|u - \widetilde{z}\|_{C_b(\mathbb{R}_+;X)} \leq \dfrac{m}{\delta - \ell} + \dfrac{\ell}{\delta - \ell} \cdot \dfrac{me^{\delta a}}{e^{\delta a} - 1}, \\[3mm] \|v - \widetilde{w}\|_{C_b(\mathbb{R}_+;Y)} \leq \dfrac{m}{\delta - \ell} + \dfrac{\ell}{\delta - \ell} \cdot \dfrac{me^{\delta a}}{e^{\delta a} - 1}, \end{cases}$$

where $\delta = \min\{\omega, \gamma\} > 0$ *and* $(\widetilde{z}, \widetilde{w})$ *is the unique C^0-solution of* (8.2.4).

By an appropriate choice of p and q, i.e., $p(u,v) = u(T)$ and $q(u,v) = v(T)$, from Theorem 8.2.2, we deduce an existence result for the T-periodic problem

$$\begin{cases} u'(t) \in Au(t) + F(t, u(t), v(t)), & t \in \mathbb{R}_+, \\ v'(t) \in Bv(t) + g(t), & t \in \mathbb{R}_+, \\ g(t) \in G(t, u(t), v(t)), & t \in \mathbb{R}_+, \\ u(t) = u(t+T), & t \in \mathbb{R}_+, \\ v(t) = v(t+T), & t \in \mathbb{R}_+. \end{cases} \tag{8.2.5}$$

Multivalued Reaction–Diffusion Systems

Clearly, if $p(u,v) = u(T)$ and $q(u,v) = v(T)$, the unique C^0-solution of the problem (8.2.4) is $(0,0)$. So, from Theorem 8.2.2, we get:

THEOREM 8.2.3 *Let us assume that* (H_A), (H_B), $(H_F^{[\tau=0]})$, $(H_G^{[\tau=0]})$ *and* (H_c) *are satisfied,* $F : \mathbb{R}_+ \times \overline{D(A)} \times \overline{D(B)} \to X$, $G : \mathbb{R}_+ \times \overline{D(A)} \times \overline{D(B)} \rightsquigarrow Y$ *and* F, G *are* T-periodic with respect to their first argument, i.e.,*

$$\begin{cases} F(t, u, v) = F(t + T, u, v) \\ G(t, u, v) = G(t + T, u, v) \end{cases}$$

for each $t \in \mathbb{R}_+$, *each* $(u,v) \in \overline{D(A)} \times \overline{D(B)}$. *Then* (8.2.5) *has at least one* C^0-solution, $(u,v) \in C_b(\mathbb{R}_+; \overline{D(A)}) \times C_b(\mathbb{R}_+; \overline{D(B)})$. *In addition, each* C^0-*solution* (u,v) *of* (8.2.5) *satisfies*

$$\begin{cases} \|u\|_{C_b(\mathbb{R}_+;X)} \le \dfrac{m}{\delta - \ell} + \dfrac{\ell}{\delta - \ell} \cdot \dfrac{m e^{\delta a}}{e^{\delta a} - 1}, \\[3mm] \|v\|_{C_b(\mathbb{R}_+;Y)} \le \dfrac{m}{\delta - \ell} + \dfrac{\ell}{\delta - \ell} \cdot \dfrac{m e^{\delta a}}{e^{\delta a} - 1}. \end{cases}$$

8.3 Idea of the proof of Theorem 8.2.1

Let $(f, g) \in C_b(\mathbb{R}_+; X) \times L^\infty(\mathbb{R}_+; Y)$ and let us consider the system

$$\begin{cases} u'(t) \in Au(t) + f(t), & t \in \mathbb{R}_+, \\ v'(t) \in Bv(t) + g(t), & t \in \mathbb{R}_+, \\ u(t) = p(u, v)(t), & t \in [-\tau, 0], \\ v(t) = q(u, v)(t), & t \in [-\tau, 0], \end{cases} \qquad (8.3.1)$$

which is simply a "perturbed" variant of the system (8.2.1). In view of Lemma 4.2.1, if (A_1), (B_1), (p_1), (p_2), (q_1) and (q_2) are satisfied, then, for each $(f, g) \in L^\infty(\mathbb{R}_+; X) \times L^\infty(\mathbb{R}_+; Y)$, the system (8.3.1) has a unique C^0-solution $(u^f, v^g) \in C_b([-\tau, \infty); X) \times C_b([-\tau, \infty); Y)$ whose restriction to $\mathbb{R}_+$ satisfies $(u^f_{|\mathbb{R}_+}, v^g_{|\mathbb{R}_+}) \in C_b(\mathbb{R}_+; \overline{D(A)}) \times C_b(\mathbb{R}_+; \overline{D(B)})$. Moreover, the mapping $(f, g) \mapsto (u^f, v^g)$ is Lipschitz continuous from $L^\infty(\mathbb{R}_+; X) \times L^\infty(\mathbb{R}_+; Y)$ to $C_b([-\tau, \infty); X) \times C_b([-\tau, \infty); Y)$, with Lipschitz constant $L = \delta^{-1}$, where $\delta = \min\{\omega, \gamma\}$, both domain and range being endowed with the max-norm of the corresponding factors.

Let $\varepsilon > 0$ be arbitrary but fixed, let $\chi_{[0,1/\varepsilon]} : \mathbb{R}_+ \to \{0, 1\}$ be the characteristic function of $[0, 1/\varepsilon]$, i.e.,

$$\chi_{[0,1/\varepsilon]} = \begin{cases} 1, & \text{if } t \in [0, 1/\varepsilon], \\ 0, & \text{if } t > 1/\varepsilon \end{cases}$$

300   *Delay Differential Evolutions Subjected to Nonlocal Initial Conditions*

and let us consider the approximate problem

$$\begin{cases} u'(t) \in Au(t) + F(t, u_t, v_t), & t \in \mathbb{R}_+, \\ v'(t) \in Bv(t) + g(t), & t \in \mathbb{R}_+, \\ g(t) \in \chi_{[0,1/\varepsilon]}(t)G(t, u_t, v_t), & t \in \mathbb{R}_+, \\ u(t) = p(u, v)(t), & t \in [-\tau, 0], \\ v(t) = q(u, v)(t), & t \in [-\tau, 0]. \end{cases} \qquad (8.3.2)$$

Let us further consider the two auxiliary systems

$$\begin{cases} \widetilde{v}'(t) \in B\widetilde{v}(t) + g(t), & t \in \mathbb{R}_+, \\ \widetilde{v}(t) = q(u, \widetilde{v})(t), & t \in [-\tau, 0] \end{cases} \qquad (8.3.3)$$

and

$$\begin{cases} \widetilde{u}'(t) \in A\widetilde{u}(t) + F(t, \widetilde{u}_t, \widetilde{v}_t), & t \in \mathbb{R}_+, \\ \widetilde{u}(t) = p(\widetilde{u}, \widetilde{v})(t), & t \in [-\tau, 0]. \end{cases} \qquad (8.3.4)$$

We will use a fixed-point argument which we describe below.

Let $(u, g) \in C_b(\mathbb{R}_+; \overline{D(A)}) \times L^\infty(\mathbb{R}_+; Y)$ be arbitrary. By Remark 2.3.1 combined with Lemma 4.2.1, we conclude that (8.3.3) has a unique C^0-solution $\widetilde{v} \in C_b([-\tau, +\infty); Y)$ with $\widetilde{v}_{|\mathbb{R}_+} \in C_b(\mathbb{R}_+; \overline{D(B)})$. Next, with this $\widetilde{v}$ fixed in (8.3.4), from Theorem 4.1.1, it follows that the problem (8.3.4) has a unique C^0-solution, $\widetilde{u} \in C_b([-\tau, +\infty); X)$, with $\widetilde{u}_{|\mathbb{R}_+} \in C_b(\mathbb{R}_+; \overline{D(A)})$.

Now we define the multifunction

$$\Gamma_\varepsilon : C_b(\mathbb{R}_+; \overline{D(A)}) \times L^\infty(\mathbb{R}_+; Y) \rightsquigarrow C_b(\mathbb{R}_+; \overline{D(A)}) \times L^\infty(\mathbb{R}_+; Y)$$

by

$$\Gamma_\varepsilon(u, g) = \{ (\widetilde{u}_{|\mathbb{R}_+}, \widetilde{g}); \ \widetilde{g} \in L^\infty(\mathbb{R}_+; Y), \ \widetilde{g}(t) \in G_\varepsilon(t, \widetilde{u}_t, \widetilde{v}_t) \text{ a.e. for } t \in \mathbb{R}_+ \}, \qquad (8.3.5)$$

for each $(u, g) \in C_b(\mathbb{R}_+; \overline{D(A)}) \times L^\infty(\mathbb{R}_+; Y)$, where

$$G_\varepsilon(t, \varphi, \psi) = \chi_{[0,1/\varepsilon]}(t)G(t, \varphi, \psi),$$

for each $(t, \varphi, \psi) \in \mathbb{R}_+ \times \mathcal{X} \times \mathcal{Y}$. Obviously (8.3.2) has a C^0-solution if and only if the multifunction Γ_ε has a fixed-point. Indeed, (u, g) is a fixed-point of Γ_ε if and only if $u = \widetilde{u}_{|\mathbb{R}_+}$ and $g = \widetilde{g}$, where $(\widetilde{u}, \widetilde{v})$ is a C^0-solution of the systems (8.3.3) and (8.3.4) and $\widetilde{g}(t) \in G_\varepsilon(t, \widetilde{u}_t, \widetilde{v}_t)$ a.e. for $t \in \mathbb{R}_+$. But this shows that $(\widetilde{u}, \widetilde{v})$ is a C^0-solution of the system (8.3.2).

Finally, we consider a family $\{(u_\varepsilon, v_\varepsilon); \ \varepsilon \in (0, 1)\}$ of C^0-solutions for the problem (8.3.2) and we show that we can pick up a sequence $((u_{\varepsilon_n}, v_{\varepsilon_n}))_n$, such that $\lim_n \varepsilon_n = 0$, and $((u_{\varepsilon_n}, v_{\varepsilon_n}))_n$ converges to (u, v) – a C^0-solution of (8.1.1) – in the topology of $C_b([-\tau, +\infty); X) \times \widetilde{C}_b([-\tau, +\infty); Y)$.

8.4 A first auxiliary lemma

We begin with the following variant of Lemma 6.5.1.

LEMMA 8.4.1 *Let us assume that* (H_A), (B_1) *in* (H_B), (H_F), (H_G), (H_c), (H_p) *and* (H_q) *are satisfied. Then there exist two constants* $\rho > 0$ *and* $r > 0$ *such that for each* $(u, g) \in C_b(\mathbb{R}_+; \overline{D(A)}) \times L^\infty(\mathbb{R}_+; Y)$ *satisfying*

$$\begin{cases} \|u\|_{C_b(\mathbb{R}_+;X)} \leq r, \\ \|g\|_{L^\infty(\mathbb{R}_+;Y)} \leq \rho, \end{cases} \tag{8.4.1}$$

the pair $(\widetilde{u}, \widetilde{v})$, *where* $\widetilde{v}$ *is the unique* C^0-*solution of* (8.3.3) *and* $\widetilde{u}$ *the unique* C^0-*solution of* (8.3.4), *satisfies*

$$\begin{cases} \|\widetilde{u}\|_{C_b(\mathbb{R}_+;X)} \leq r, \\ \|\widetilde{g}(t)\| \leq \rho \end{cases} \tag{8.4.2}$$

a.e. for $t \in \mathbb{R}_+$ *and for each* $\widetilde{g} : \mathbb{R}_+ \to Y$, $\widetilde{g}(t) \in G(t, \widetilde{u}_t, \widetilde{v}_t)$ *a.e. for* $t \in \mathbb{R}_+$.

Proof. Since $\ell < \delta$, there exists $\rho > 0$, large enough, such that

$$\begin{cases} \ell \left(\dfrac{\rho}{\delta} + m \right) + m \leq \rho, \\ \dfrac{m}{\delta - \ell} + \left[\dfrac{\delta}{\delta - \ell} \left(\dfrac{1}{e^{\delta a} - 1} + \dfrac{\ell}{\delta} \right) + 1 \right] m \leq \dfrac{\rho}{\delta} + m. \end{cases} \tag{8.4.3}$$

Let us define

$$\begin{cases} r = \dfrac{m}{\delta - \ell} + \left[\dfrac{\delta}{\delta - \ell} \left(\dfrac{1}{e^{\delta a} - 1} + \dfrac{\ell}{\delta} \right) + 1 \right] m, \\ \widetilde{r} = \dfrac{\rho}{\delta} + m. \end{cases} \tag{8.4.4}$$

Clearly

$$r \leq \widetilde{r}. \tag{8.4.5}$$

Next, from (8.4.1), we conclude that for $t \in (0, +\infty)$, we have

$$\|\widetilde{v}(t)\| \leq e^{-\delta t} \|\widetilde{v}(0)\| + \frac{1 - e^{-\delta t}}{\delta} \rho \leq e^{-\delta t} \|\widetilde{v}\|_{C_b(\mathbb{R}_+;Y)} + \frac{1 - e^{-\delta t}}{\delta} \rho.$$

Therefore, reasoning as in **Case 2** and **Case 3** in Lemma 4.2.1, we get

$$\|\widetilde{v}\|_{C_b(\mathbb{R}_+;Y)} \leq \frac{\rho}{\delta} \leq \widetilde{r}.$$

302 *Delay Differential Evolutions Subjected to Nonlocal Initial Conditions*

On the other hand, if $t \in [-\tau, 0]$, from the second inequality in (8.4.4), we get

$$\|\widetilde{v}(t)\| = \|q(u, \widetilde{v})(t)\| \le \|\widetilde{v}\|_{C_b([a, +\infty); Y)} + m \le \frac{\rho}{\delta} + m = \widetilde{r}.$$

From these inequalities, we conclude that

$$\|\widetilde{v}\|_{C_b([-\tau, +\infty); Y)} \le \widetilde{r}. \tag{8.4.6}$$

Let us observe that, for a fixed $\widetilde{v} \in C([-\tau, +\infty); Y)$, the mappings

$$\begin{cases} f(s, z) = F(s, z, \widetilde{v}_s), & \text{for each } (s, z) \in \mathbb{R}_+ \times \mathfrak{X}, \\ g(w) = p(w, \widetilde{v}_{|\mathbb{R}_+}), & \text{for each } w \in C_b(\mathbb{R}_+; \overline{D(A)}) \end{cases}$$

satisfy the hypotheses (H_f) and (H_g) in Theorem 4.1.1. It then follows that the problem (8.3.4) has a unique C^0-solution $\widetilde{u}$ satisfying

$$\|\widetilde{u}\|_{C_b([-\tau, +\infty); X)} \le \frac{m}{\delta - \ell} + \left[\frac{\delta}{\delta - \ell} \cdot \left(\frac{1}{e^{\delta a} - 1} + \frac{\ell}{\delta} \right) + 1 \right] m = r,$$

where r is defined by (8.4.4). Clearly, this implies the first inequality in (8.4.2). By (G_2) in (H_G), the definition of $\widetilde{r}$ in (8.4.4), (8.4.3) and (8.4.5), we get

$$\|\widetilde{g}(t)\| \le \ell \max\{r, \widetilde{r}\} + m \le \ell \left(\frac{\rho}{\delta} + m \right) + m \le \rho,$$

for each $\widetilde{g} : \mathbb{R}_+ \to Y$ satisfying $\widetilde{g}(t) \in G(t, \widetilde{u}_t, \widetilde{v}_t)$ a.e. for $t \in \mathbb{R}_+$, thereby proving the second inequality in (8.4.2). The proof is complete. $\qquad\square$

8.5 The operator Γ_ε

LEMMA 8.5.1 *Let us suppose that the hypotheses* (H_A), (H_B), (H_F), (H_G), (H_p), (H_q) *and* (H_c) *are satisfied and let* r *and* ρ *be as in Lemma 8.4.1. Let*

$$K_\varepsilon = K_r \times K_\rho,$$

where K_r *is the closed ball with center* 0 *and radius* r *in* $C_b(\mathbb{R}_+; \overline{D(A)})$ *and* K_ρ *is the closed ball with center* 0 *and radius* ρ *in* $L^\infty(\mathbb{R}_+; Y)$ *multiplied by* $\chi_{[0, 1/\varepsilon]}$.
Then K_ε *is nonempty, closed and convex in* $C_b(\mathbb{R}_+; X) \times \widetilde{L}^1(\mathbb{R}_+; Y)$ *and the operator* Γ_ε *defined by (8.3.5) maps* K_ε *into itself and its graph is sequentially closed with respect to the norm topology on* $C_b(\mathbb{R}_+; X)$ *and the locally convex topology on* $\widetilde{L}^1(\mathbb{R}_+; Y)$.

Multivalued Reaction–Diffusion Systems 303

Proof. Since by (H_A), $\overline{D(A)}$ is closed and convex and $0 \in \overline{D(A)}$, it follows that K_r is nonempty, closed and convex. Since K_ρ is clearly nonempty, closed and convex in $L^1(0, 1/\varepsilon; Y)$, which is a closed subspace in $\widetilde{L}^1(\mathbb{R}_+; Y)$, we conclude that K_ε is nonempty, closed and convex.

If $(u, g) \in K_\varepsilon$, from Lemma 8.4.1, we deduce that the pair $(\widetilde{u}, \widetilde{v})$, where $\widetilde{v}$ is the unique C^0-solution of (8.3.3) and $\widetilde{u}$ the unique C^0-solution of (8.3.4), satisfies (8.4.2). Let us observe first that Γ_ε maps K_ε into itself. Indeed, let $(u, g) \in K_\varepsilon$ and let $(\widetilde{u}, \widetilde{g}) \in \Gamma_\varepsilon(u, g)$. Then, from (8.4.2), we deduce that $(\widetilde{u}, \widetilde{g}) \in K_\varepsilon$ and thus Γ_ε maps K_ε into itself.

To prove that the graph of the multifunction Γ_ε is sequentially closed with respect to the norm topology on $C_b(\mathbb{R}_+; X)$ and the locally convex topology on $\widetilde{L}^1(\mathbb{R}_+; Y)$, let us consider two arbitrary sequences: $((u_n, g_n))_n$ in K_ε and $((\widetilde{u}_n, \widetilde{g}_n))_n$ with $(\widetilde{u}_n, \widetilde{g}_n) \in \Gamma_\varepsilon(u_n, g_n)$ for each $n \in \mathbb{N}$ and satisfying

$$\begin{cases} \lim_n (u_n, g_n) = (u, g), \\ \lim_n (\widetilde{u}_n, \widetilde{g}_n) = (\widetilde{u}, \widetilde{g}), \end{cases} \tag{8.5.1}$$

in $C_b(\mathbb{R}_+; X) \times \widetilde{L}^1(\mathbb{R}_+; Y)$. That means that, for each for $n \in \mathbb{N}$, the unique C^0-solution $\widetilde{v}_n$ of the problem

$$\begin{cases} \widetilde{v}_n'(t) \in B\widetilde{v}_n(t) + g_n(t), & t \in \mathbb{R}_+, \\ \widetilde{v}_n(t) = q(u_n, \widetilde{v}_n)(t), & t \in [-\tau, 0], \end{cases} \tag{8.5.2}$$

and the unique C^0-solution $\widetilde{u}_n$ of the problem

$$\begin{cases} \widetilde{u}_n(t) \in A\widetilde{u}_n(t) + F(t, \widetilde{u}_{n_t}, \widetilde{v}_{n_t}), & t \in \mathbb{R}_+, \\ \widetilde{u}_n(t) = p(\widetilde{u}_n, \widetilde{v}_n)(t), & t \in [-\tau, 0], \end{cases} \tag{8.5.3}$$

satisfy

$$\widetilde{g}_n(t) \in G_\varepsilon(t, \widetilde{u}_{n_t}, \widetilde{v}_{n_t})$$

a.e. for $t \in \mathbb{R}_+$. See the definition of Γ_ε.

The set $\{\widetilde{v}_n; \ n \in \mathbb{N}\}$ is relatively compact in $C_b([-\tau, +\infty); Y)$. Indeed, since $\widetilde{v}_n$ is the C^0-solution of (8.5.2), from (8.4.6), we get

$$\|\widetilde{v}_n(0)\| \le \|\widetilde{v}_n\|_{C_b([-\tau, +\infty); Y)} \le \widetilde{r}$$

for each $n \in \mathbb{N}$. Since the set $\{g_n; n \in \mathbb{N}\}$ is uniformly integrable, we are in the hypotheses of Theorem 1.8.5, from which it follows that $\{\widetilde{v}_n; \ n \in \mathbb{N}\}$ is relatively compact in $C([\sigma, k]; Y)$ for $k = 1, 2, \ldots$ and $\sigma \in (0, k)$ and thus in $\widetilde{C}_b([a, +\infty); Y)$. Since $(u_n)_n$ is bounded in $C_b(\mathbb{R}_+; X)$, by (q_3), it follows that $\{\widetilde{v}_n; \ n \in \mathbb{N}\}$ is relatively compact in $\mathcal{Y}$. Hence $\{\widetilde{v}_n(0); \ n \in \mathbb{N}\}$ is relatively compact in Y and, from the second part of Theorem 1.8.5, we conclude that $\{\widetilde{v}_n; \ n \in \mathbb{N}\}$ is relatively compact in $C([0, k]; Y)$ for $k = 1, 2, \ldots$ So, $\{\widetilde{v}_n; \ n \in \mathbb{N}\}$ is relatively compact in $\widetilde{C}_b([-\tau, +\infty); Y)$. Then,

304   Delay Differential Evolutions Subjected to Nonlocal Initial Conditions

there exist $\tilde{v} \in C_b([-\tau, +\infty); Y)$ and a subsequence of $(\tilde{v}_n)_n$ – denoted for simplicity again by $(\tilde{v}_n)_n$ – such that

$$\lim_n \tilde{v}_n = \tilde{v} \text{ in } \widetilde{C}_b([-\tau, +\infty); Y). \tag{8.5.4}$$

We will prove next that $(\tilde{v}_n)_n$ converges to $\tilde{v}$ even in the norm topology of $C_b([-\tau, +\infty); Y)$. Indeed, since $g_n \in K_\rho$, which is weakly closed in $\widetilde{L}^1(\mathbb{R}_+; Y)$ and

$$\lim_n g_n = g \text{ weakly in } \widetilde{L}^1(\mathbb{R}_+; Y), \tag{8.5.5}$$

we deduce that $g \in K_\rho$. From (8.5.4) combined with (8.5.5) and with the fact that, by (B_4) in (H_B), B is of complete continuous type, we obtain that $\tilde{v}$ is a C^0-solution of the problem

$$\tilde{v}'(t) \in B\tilde{v}(t) + g(t), \quad t \in \mathbb{R}_+.$$

Now, let us fix $k_\varepsilon \in \mathbb{N}$, with $k_\varepsilon \geq 1 + 1/\varepsilon$, and let $\alpha > 0$ be arbitrary. Taking into account that the sequence $(\tilde{v}_n)_n$ is uniformly convergent to $\tilde{v}$ on $[0, k_\varepsilon]$, we deduce that, for each $\alpha > 0$, there exists $n_\varepsilon(\alpha) \in \mathbb{N}$ such that

$$\|\tilde{v}_n(t) - \tilde{v}(t)\| \leq \alpha$$

for each $n \in \mathbb{N}$, $n \geq n_\varepsilon(\alpha)$ and each $t \in [0, k_\varepsilon]$.

From the definition of G_ε, we deduce that $\|g_n(s) - g(s)\| = 0$ for each $t \geq k_\varepsilon$ and a.e. for $s \in [k_\varepsilon, t]$ and then, using (1.8.2) in Theorem 1.8.1, we get

$$\|\tilde{v}_n(t) - \tilde{v}(t)\| \leq e^{-\gamma(t - k_\varepsilon)} \|\tilde{v}_n(k_\varepsilon) - \tilde{v}(k_\varepsilon)\| + \int_{k_\varepsilon}^t e^{-\gamma(t-s)}) \|g_n(s) - g(s)\| \, ds$$

$$\leq \|\tilde{v}_n(k_\varepsilon) - \tilde{v}(k_\varepsilon)\| \leq \alpha$$

for each $n \in \mathbb{N}$, $n \geq n_\varepsilon(\alpha)$ and each $t \geq k_\varepsilon$. From this inequality, taking into account (8.5.4), we deduce that

$$\lim_n \tilde{v}_n = \tilde{v} \text{ in } C_b([-\tau, +\infty); Y),$$

as claimed.

Now, let us consider the problem

$$\begin{cases} \tilde{u}'(t) \in A\tilde{u}(t) + F(t, \tilde{u}_t, \tilde{v}_t), & t \in \mathbb{R}_+, \\ \tilde{u}(t) = p(\tilde{u}, \tilde{v})(t), & t \in [-\tau, 0], \end{cases} \tag{8.5.6}$$

where $\tilde{v}$ is as above. By Theorem 4.1.1, it follows that the problem (8.5.6) has a unique C^0-solution $\tilde{u}$. Since, for each $n \in \mathbb{N}$, $\tilde{u}_n$ is the C^0-solution of (8.5.3) and, by (F_1), (p_1) and (p_2), both functions $f_n : \mathbb{R}_+ \times X \to X$, defined by $f_n(t, x) = F(t, x, \tilde{v}_{n_t})$ for all $(t, x) \in \mathbb{R}_+ \times X$ and $g_n : C_b(\mathbb{R}_+; \overline{D(A)}) \to \mathcal{D}$,

Multivalued Reaction–Diffusion Systems 305

defined by $g_n(z) = p(z, \tilde{v}_n)$ for each $z \in C_b(\mathbb{R}_+; \overline{D(A)})$, satisfy the hypotheses of Theorem 4.4.1, we deduce that

$$\lim_n \tilde{u}_n = \tilde{u} \quad \text{in} \quad C_b([-\tau, +\infty); X).$$

To complete the proof, it suffices to show that

$$\tilde{g}(t) \in G_\varepsilon(t, \tilde{u}_t, \tilde{v}_t)$$

a.e. for $t \in \mathbb{R}_+$. For $k = 1, 2, \ldots$, we have $\tilde{g}_n(t) \in G_\varepsilon(t, \tilde{u}_{n_t}, \tilde{v}_{n_t})$ for each $n \in \mathbb{N}$ and a.e. for $t \in [0, k]$. In addition, from (8.5.1), we conclude that

$$\lim_n \tilde{g}_n = \tilde{g}$$

weakly in $L^1([0, k]; Y)$ for $k = 1, 2, \ldots$ and

$$\lim_n (\tilde{u}_{n_t}, \tilde{v}_{n_t}) = (\tilde{u}_t, \tilde{v}_t)$$

in $\mathcal{D} \times \mathcal{E}$ uniformly for $t \in [0, k]$. Since, by (H_G), G_ε is strongly-weakly u.s.c., from Theorem 1.5.1, we deduce that $\tilde{g}(t) \in G_\varepsilon(t, \tilde{u}_t, \tilde{v}_t)$ a.e. for $t \in \mathbb{R}_+$. $\qquad \square$

We will need also the following compactness result.

LEMMA 8.5.2 *Let us suppose that the hypotheses (H_A), (H_B), (H_F), (H_G), (H_p), (H_q) and (H_c) are satisfied. Then, for each $\varepsilon \in (0, 1)$, the set $\Gamma_\varepsilon(K_\varepsilon)$, defined in Lemma 8.5.1, is relatively compact in the product space*

$$C_b(\mathbb{R}_+; X) \times \tilde{L}^1(\mathbb{R}_+; Y).$$

Proof. Let $((\tilde{u}_n, \tilde{g}_n))_n$ be an arbitrary sequence in $\Gamma_\varepsilon(K_\varepsilon)$ and $((u_n, g_n))_n$ be a sequence in K_ε such that

$$(\tilde{u}_n, \tilde{g}_n) \in \Gamma_\varepsilon(u_n, g_n)$$

for each $n \in \mathbb{N}$. So, $\tilde{v}_n$ is the unique C^0-solution of the problem

$$\begin{cases} \tilde{v}'_n(t) \in B\tilde{v}_n(t) + g_n(t), & t \in \mathbb{R}_+, \\ \tilde{v}_n(t) = q(u_n, \tilde{v}_n)(t), & t \in [-\tau, 0] \end{cases}$$

and $\tilde{u}_n$ is the unique C^0-solution of the problem

$$\begin{cases} \tilde{u}'_n(t) \in A\tilde{u}_n(t) + F(t, \tilde{u}_{n_t}, \tilde{v}_{n_t}), & t \in \mathbb{R}_+, \\ \tilde{u}_n(t) = p(\tilde{u}_n, \tilde{v}_n)(t), & t \in [-\tau, 0], \end{cases}$$

for each $n \in \mathbb{N}$.

Reasoning as in Lemma 8.5.1, there exists $\tilde{v} \in C_b([-\tau, +\infty); Y)$ such that, at least on a sub-subsequence, $\lim_n \tilde{v}_n = \tilde{v}$ in $C_b([-\tau, +\infty); Y)$. Furthermore,

306 *Delay Differential Evolutions Subjected to Nonlocal Initial Conditions*

there exists $\widetilde{u} \in C_b([-\tau, +\infty); X)$ such that $\lim_n \widetilde{u}_n = \widetilde{u}$ in $C_b([-\tau, +\infty); X)$. So, for each $k = 1, 2, \ldots$, the set

$$C_k = \{(t, \widetilde{u}_{n_t}, \widetilde{v}_{n_t}); \ n \in \mathbb{N}, \ t \in [0, k]\}$$

is relatively compact in $\mathbb{R}_+ \times X \times Y$. Since G_ε is strongly-weakly u.s.c., from Lemma 1.5.1, it follows that, for each $k = 1, 2, \ldots$, $G_\varepsilon(C_k)$ is weakly relatively compact in Y. Since for each $k = 1, 2, \ldots$, $\widetilde{g}_n(t) \in G_\varepsilon(C_k)$ a.e. for $t \in [0, k]$ and for each $n \in \mathbb{N}$, from Corollary 1.4.1, it follows that $\{\widetilde{g}_n; \ n \in \mathbb{N}\}$ is weakly relatively compact in $L^1(0, k; Y)$. So, on a subsequence, denoted for simplicity again by $(\widetilde{g}_n)_n$,

$$\lim_n \widetilde{g}_n = \widetilde{g} \text{ weakly in } \widetilde{L}^1(\mathbb{R}_+; Y).$$

As K_ρ is weakly closed in $\widetilde{L}^1(\mathbb{R}_+; Y)$, it follows that $\widetilde{g} \in K_\rho$.

To complete the proof, it suffices to show that

$$\widetilde{g}(t) \in G_\varepsilon(t, \widetilde{u}_t, \widetilde{v}_t) \tag{8.5.7}$$

a.e. for $t \in \mathbb{R}_+$. We have

$$\widetilde{g}_n(t) \in G_\varepsilon(t, \widetilde{u}_{n_t}, \widetilde{v}_{n_t})$$

for each $n \in \mathbb{N}$ and a.e. for $t \in [0, k]$ and $k = 1, 2, \ldots,$

$$\lim_n \widetilde{g}_n = \widetilde{g}$$

weakly in $L^1([0, k]; Y)$,

$$\lim_n (\widetilde{u}_{n_t}, \widetilde{v}_{n_t}) = (\widetilde{u}_t, \widetilde{v}_t)$$

in $X \times Y$ a.e. for $t \in [0, k]$, and $k = 1, 2, \ldots$ Since G_ε is strongly-weakly u.s.c., by Theorem 1.5.1, we deduce (8.5.7). So, $\Gamma_\varepsilon(K_\varepsilon)$ is relatively compact in $C_b([-\tau, +\infty); X) \times \widetilde{L}^1(\mathbb{R}_+; Y)$, as claimed. $\qquad\square$

8.6 Proof of Theorem 8.2.1

We can now proceed to prove Theorem 8.2.1.

Proof. Let $\varepsilon \in (0, 1)$ be arbitrary but fixed. By Lemma 8.5.1, Γ_ε maps K_ε into itself and, by Lemma 8.5.2, it follows that $\Gamma_\varepsilon(K_\varepsilon)$ is relatively compact. Moreover, again by Lemma 8.5.1, the graph of Γ_ε is sequentially closed. Since, in a Banach space, the weak closure of a weakly relatively compact set coincides with its weak sequential closure – see Theorem 1.4.7 – using a standard diagonal process, we deduce that the graph of Γ_ε is even closed with

Multivalued Reaction–Diffusion Systems

respect to the norm topology on $C_b(\mathbb{R}_+; X)$ and the locally convex topology on $\widetilde{L}^1(\mathbb{R}_+; Y)$. From Theorem 1.5.3, we deduce that Γ_ε has at least one fixed point $(u_\varepsilon, g_\varepsilon)$. Clearly, this means that the approximate problem (8.3.2) has at least one C^0-solution $(u_\varepsilon, v_\varepsilon)$. Let $\varepsilon \in (0, 1)$ be arbitrary and let us fix such a solution. Thus, we obtain the set $\{(u_\varepsilon, v_\varepsilon); \ \varepsilon \in (0, 1)\}$.

Reasoning as in the proof of Lemma 8.5.1, we deduce first that the set $\{v_\varepsilon; \ \varepsilon \in (0, 1)\}$ is relatively compact in $\widetilde{C}_b([-\tau, +\infty); Y)$, then that the set $\{u_\varepsilon; \ \varepsilon \in (0, 1)\}$ is relatively compact in $C_b([-\tau, +\infty); X)$. Consequently, it follows that $\{(u_\varepsilon, v_\varepsilon); \ \varepsilon \in (0, 1)\}$ is relatively compact in $C_b([-\tau, +\infty); X) \times \widetilde{C}_b([-\tau, +\infty); Y)$. Then, if $\varepsilon_n \downarrow 0$, there exists at least one subsequence of $((u_{\varepsilon_n}, v_{\varepsilon_n}))_n$, denoted for simplicity by $((u_n, v_n))_n$, convergent in $C_b([-\tau, +\infty); X) \times \widetilde{C}_b([-\tau, +\infty); Y)$ to some (u, v) that turns out to be a C^0-solution of (8.1.1).

So, let $\varepsilon_n \downarrow 0$ and let $((u_n, v_n))_n$ be as above. Again by the very same arguments as those in Lemma 8.5.2, we deduce that the set $\{v_n; \ n \in \mathbb{N}\}$ is relatively compact in $\widetilde{C}_b([-\tau, +\infty); Y)$ and the corresponding set $\{g_n; \ n \in \mathbb{N}\}$ with $g_n(t) \in \chi_{[0,1/\varepsilon_n]}(t)G(t, u_{n_t}, v_{n_t})$ a.e. for $t \in \mathbb{R}_+$, is weakly relatively compact in $L^1(0, k; Y)$, for $k = 1, 2, \ldots$. So, on a subsequence, we have both $\lim_n v_n = v$ in $\widetilde{C}_b([-\tau, +\infty); Y)$ and $\lim_n g_n = g$ weakly in $\widetilde{L}^1(\mathbb{R}_+; Y)$. Now, let us observe that u_n is the unique C^0-solution of the problem (4.4.2) in Theorem 4.4.1, where $f_n : \mathbb{R}_+ \times \mathcal{X} \to X$, $f_n(t, x) = F(t, x, v_{n_t})$ for each $(t, x) \in \mathbb{R}_+ \times \mathcal{X}$ and $g_n : C_b(\mathbb{R}_+; \overline{D(A)}) \to \mathcal{D}$, $g_n(z) = p(z, v_n)$ for each $z \in C_b(\mathbb{R}_+; \overline{D(A)})$. From (H_A), (H_F) and (H_p), it follows that we are in the hypotheses of the $\widetilde{C}_b$–continuity part in Theorem 4.4.1, from which we conclude that, on that subsequence at least, we have $\lim_n u_n = u$ in $\widetilde{C}_b([-\tau, +\infty); X)$ where u is the C^0-solution of the problem

$$\begin{cases} u'(t) \in Au(t) + F(t, u_t, v_t), & t \in \mathbb{R}_+, \\ u(t) = p(u, v)(t), & t \in [-\tau, 0]. \end{cases}$$

Recalling that by (B_4), B is of complete continuous type, it follows that (u, v) is a C^0-solution of the problem

$$\begin{cases} u'(t) \in Au(t) + F(t, u_t, v_t), & t \in \mathbb{R}_+, \\ v'(t) \in Bv(t) + g(t), & t \in \mathbb{R}_+, \\ u(t) = p(u, v)(t), & t \in [-\tau, 0], \\ v(t) = q(u, v)(t), & t \in [-\tau, 0]. \end{cases}$$

So, to conclude that (u, v) is a C^0-solution of the problem (8.1.1), we have merely to observe that, thanks to Theorem 1.5.1, we have $g(t) \in G(t, u_t, v_t)$ a.e. for $t \in \mathbb{R}_+$. Finally, as (8.2.2) follows by applying Theorem 7.6.1 to the system written in the product space, the proof is complete. $\qquad\square$

8.7 A reaction–diffusion system in $L^1(\Omega)$

Let Ω be a nonempty, bounded domain in $\mathbb{R}^d$, $d \geq 1$, with C^1 boundary Σ, let $\tau \geq 0$, $\omega > 0$, $\gamma > 0$, let $Q_+ = \mathbb{R}_+ \times \Omega$, $\Sigma_+ = \mathbb{R}_+ \times \Sigma$ and $Q_\tau = [-\tau, 0] \times \Omega$. Let $\varphi : D(\varphi) \subseteq \mathbb{R} \rightsquigarrow \mathbb{R}$ and $\psi : D(\psi) \subseteq \mathbb{R} \rightsquigarrow \mathbb{R}$ be maximal-monotone operators with $0 \in D(\varphi)$, $0 \in D(\psi)$, $0 \in \varphi(0)$, $0 \in \psi(0)$ and let $\mathfrak{X}_1 = C([-\tau, 0]; L^1(\Omega))$. Let $F : \mathbb{R}_+ \times \mathfrak{X}_1 \times \mathfrak{X}_1 \to L^1(\Omega)$ be continuous and $g_i : \mathbb{R}_+ \times \mathfrak{X}_1 \times \mathfrak{X}_1 \to L^1(\Omega)$, $i = 1, 2$, be two given functions such that $g_1(t, u, v)(x) \leq g_2(t, u, v)(x)$ for each (t, u, v) in $\mathbb{R}_+ \times \mathfrak{X}_1 \times \mathfrak{X}_1$ and a.e. for $x \in \Omega$. Let $b > \tau$, let μ be a positive σ-finite and complete measure on the class of Borel measurable sets in $\mathbb{R}_+$ with $\operatorname{supp} \mu \subseteq [a, +\infty)$, where $a = b - \tau$, let $k \in L^1(\mathbb{R}_+; \mu)$ be a nonnegative function with $\|k\|_{L^1(\mathbb{R}_+; \mu)} \leq 1$, let $W : \mathbb{R} \to \mathbb{R}$ be nonexpansive with $W(0) = 0$ and let $\eta \in \mathfrak{X}_1$. We consider the following reaction–diffusion system:

$$
\begin{cases}
\dfrac{\partial u}{\partial t}(t, x) = \Delta\varphi(u(t, x)) - \omega u(t, x) + F(t, u_t, v_t)(x), & \text{in } Q_+, \\[2mm]
\dfrac{\partial v}{\partial t}(t, x) = \Delta\psi(v(t, x)) - \gamma v(t, x) + g(t)(x), & \text{in } Q_+, \\[2mm]
g(t)(x) \in G(t, u_t, v_t)(x), & \text{in } Q_+, \\[2mm]
\varphi(u(t, x)) = 0, \quad \psi(v(t, x)) = 0, & \text{on } \Sigma_+, \\[2mm]
u(t, x) = \displaystyle\int_b^\infty k(s) W(v(t+s, x)) u(t+s, x)\, d\mu(s) + \eta(t)(x), & \text{in } Q_\tau, \\[2mm]
v(t, x) = v(t + T, x), & \text{in } Q_\tau,
\end{cases}
\tag{8.7.1}
$$

where $\tau \geq 0$ and $G : \mathbb{R}_+ \times \mathfrak{X}_1 \times \mathfrak{X}_1 \rightsquigarrow L^1(\Omega)$ is defined by

$$
G(t, u, v) = \{ h \in L^1(\Omega);\ g_1(t, u, v)(x) \leq h(x) \leq g_2(t, u, v)(x),\ \text{a.e. for } x \in \Omega \}
\tag{8.7.2}
$$

for each $(t, u, v) \in \mathbb{R}_+ \times \mathfrak{X}_1 \times \mathfrak{X}_1$.

THEOREM 8.7.1 *Let $\Omega \subseteq \mathbb{R}^d$ $(d \geq 1)$ be a nonempty, bounded and open set with C^1 boundary Σ, let $\varphi : D(\varphi) \subseteq \mathbb{R} \rightsquigarrow \mathbb{R}$ and $\psi : D(\psi) \subseteq \mathbb{R} \rightsquigarrow \mathbb{R}$ be maximal-monotone operators satisfying $0 \in D(\varphi)$, $0 \in D(\psi)$, $0 \in \varphi(0)$ and $0 \in \psi(0)$, let $F : \mathbb{R}_+ \times \mathfrak{X}_1 \times \mathfrak{X}_1 \to L^1(\Omega)$ be continuous and let $g_i : \mathbb{R}_+ \times \mathfrak{X}_1 \times \mathfrak{X}_1 \to L^1(\Omega)$, $i = 1, 2$, be two given functions satisfying*

$$
g_1(t, u, v)(x) \leq g_2(t, u, v)(x)
$$

for all $(t, u, v) \in \mathbb{R}_+ \times \mathfrak{X}_1 \times \mathfrak{X}_1$ and a.e. for $x \in \Omega$. Let $b > \tau$ and let μ be a positive σ-finite and complete measure defined on the class of Borel subsets in $\mathbb{R}_+$ with $\operatorname{supp} \mu \subseteq [a, +\infty)$, where $a = b - \tau$. Let $k \in L^1(\mathbb{R}_+; \mu)$ be a nonnegative function, let $W : \mathbb{R} \to \mathbb{R}$ and let $\eta \in \mathfrak{X}_1$. We assume that

Multivalued Reaction–Diffusion Systems

(h_1) $\psi : \mathbb{R} \to \mathbb{R}$ *is continuous on* $\mathbb{R}$ *and* C^1 *on* $\mathbb{R} \setminus \{0\}$ *and there exist two constants* $C > 0$ *and* $\alpha > 0$ *if* $d \leq 2$ *and* $\alpha > (d-2)/d$ *if* $d \geq 3$ *such that*

$$\psi'(r) \geq C|r|^{\alpha-1}$$

for each $r \in \mathbb{R} \setminus \{0\}$

(h_2) *there exist* $\ell > 0$ *and* $m > 0$ *such that*

$$\|F(t,u,v) - F(t,\widetilde{u},\widetilde{v})\|_{L^1(\Omega)} \leq \ell \max\{\|u - \widetilde{u}\|_{\mathfrak{X}_1}, \|v - \widetilde{v}\|_{\mathfrak{X}_1}\}$$

and

$$\|F(t,u,v)\|_{L^1(\Omega)} \leq \ell\|u\|_{\mathfrak{X}_1} + m,$$

for each $(t,u,v), (t,\widetilde{u},\widetilde{v}) \in \mathbb{R}_+ \times \mathfrak{X}_1 \times \mathfrak{X}_1$

(h_3) g_1 *is l.s.c and* g_2 *is u.s.c. with respect to the usual order on their domain in the sense of Definition 1.5.7*

(h_4) *there exist two nonnegative functions* $\alpha, \beta \in L^1(\mathbb{R}_+) \cap L^\infty(\mathbb{R}_+)$ *such that*

$$\|g_i(t,u,v)\|_{L^1(\Omega)} \leq \alpha(t) \max\{\|u\|_{\mathfrak{X}_1}, \|v\|_{\mathfrak{X}_1}\} + \beta(t)$$

for $i = 1, 2$ *and each* $(t,u,v) \in \mathbb{R}_+ \times \mathfrak{X}_1 \times \mathfrak{X}_1$

(h_5) *for each* $r > 0$, *there exists two nonnegative functions* $\sigma_r \in L^1(\mathbb{R}_+)$ *and* $\eta_r \in L^1(\Omega)$ *such that, a.e. for* $x \in \Omega$, *we have*

$$|g_i(t,u,v)(x)| \leq \sigma_r(t)\eta_r(x)$$

for $i = 1, 2$ *and each* $(t,u,v) \in \mathbb{R}_+ \times \mathfrak{X}_1 \times \mathfrak{X}_1$ *with*

$$\max\{\|u\|_{\mathfrak{X}_1}, \|v\|_{\mathfrak{X}_1}\} \leq r$$

(h_6) $\|k\|_{L^1(\mathbb{R}_+;\mu)} \leq 1$

(h_7) $|W(v) - W(\widetilde{v})| \leq |v - \widetilde{v}|$ *for each* $v, \widetilde{v} \in \mathbb{R}$

(h_8) $W(0) = 0$

(h_9) $\|\eta\|_{\mathfrak{X}_1} \leq m$, *where* m *is given by* (h_2).

Let us also assume that (H_c) *is satisfied. Then, the system* (8.7.1) *has at least one* C^0-*solution.*

Proof. The problem (8.7.1) can be rewritten as an abstract one of the form (8.1.1). Indeed, let $X = Y = L^1(\Omega)$ and let $A : D(A) \subseteq L^1(\Omega) \rightsquigarrow L^1(\Omega)$ be defined by

$$Au = \Delta\varphi(u) - \omega u$$

310   *Delay Differential Evolutions Subjected to Nonlocal Initial Conditions*

for each $u \in D(\Delta\varphi)$, where

$$\begin{cases} D(\Delta\varphi) = \{u \in L^1(\Omega); \ \exists w \in \mathcal{S}_\varphi(u) \cap W_0^{1,1}(\Omega), \ \Delta w \in L^1(\Omega)\}, \\ \Delta\varphi(u) = \{\Delta w; \ w \in \mathcal{S}_\varphi(u) \cap W_0^{1,1}(\Omega)\} \cap L^1(\Omega) \text{ for } u \in D(\Delta\varphi), \end{cases}$$

$\mathcal{S}_\varphi(u)$ being defined by

$$\mathcal{S}_\varphi(u) = \{w \in L^1(\Omega); \ w(x) \in \varphi(u(x)), \text{ a.e. for } x \in \Omega\}$$

and let $B : D(B) \subseteq L^1(\Omega) \rightsquigarrow L^1(\Omega)$ be defined similarly, i.e.,

$$Bv = \Delta\psi(v) - \gamma v$$

for each $v \in D(\Delta\psi)$, where

$$\begin{cases} D(\Delta\psi) = \{v \in L^1(\Omega); \ \exists w \in \mathcal{S}_\psi(v) \cap W_0^{1,1}(\Omega), \ \Delta w \in L^1(\Omega)\}, \\ \Delta\psi(v) = \{\Delta w; \ w \in \mathcal{S}_\psi(v) \cap W_0^{1,1}(\Omega)\} \cap L^1(\Omega) \text{ for } v \in D(\Delta\psi), \end{cases}$$

$\mathcal{S}_\psi(v)$ being defined by

$$\mathcal{S}_\psi(v) = \{w \in L^1(\Omega); \ w(x) \in \psi(v(x)), \text{ a.e. for } x \in \Omega\}.$$

From Theorem 1.9.6, it follows that both A and $A + \omega I$ are m-dissipative, that $0 \in D(A)$, $0 \in A0$ and $\overline{D(A)} = L^1(\Omega)$ (we recall that $0 \in \varphi(0)$ and thus $C_0^\infty(\Omega)$ is dense in both $D(A)$ and $L^1(\Omega)$), which is obviously convex. Also from Theorem 1.9.6, it follows that both B and $B + \gamma I$ are m-dissipative, $0 \in D(B)$, $0 \in B0$ and B generates a compact semigroup on $\overline{D(B)} = L^1(\Omega)$ and is of complete continuous type. So, (H_A) and (H_B) are satisfied. Clearly, from (h_2), it follows that F satisfies (H_F). Since, by (h_3), g_1 is l.s.c., g_2 is u.s.c., while from (h_4), both have affine growth, we conclude that G, defined as in (8.7.2), has affine growth with constants ℓ and m – see (h_9). Moreover, by (h_4) and Dunford Theorem 1.4.11, it follows that G, which has nonempty and convex values, maps bounded subsets in $\mathbb{R} \times \mathcal{X}_1 \times \mathcal{X}_1$ into weakly compact sets in $L^1(\Omega)$. Furthermore, by Lemma 1.5.5, we conclude that its graph is strongly$\times$weakly sequentially closed.

Finally, from $(h_6) \sim (h_9)$ and the fact that $\mathrm{supp}\,\mu \subseteq [a, +\infty)$, we deduce that the functions

$$p, q : C_b(\mathbb{R}_+; L^1(\Omega)) \times C_b(\mathbb{R}_+; L^1(\Omega)) \to \mathcal{X}_1,$$

defined by

$$\begin{cases} p(u,v)(t)(x) = \displaystyle\int_b^\infty k(s)W(v(t+s)(x))u(t+s)(x)\,d\mu(s) + \eta(t)(x), \\ q(u,v)(t)(x) = v(t+T)(x), \end{cases}$$

for each $(u, v) \in C_b(\mathbb{R}_+; L^1(\Omega)) \times C_b(\mathbb{R}_+; L^1(\Omega))$, each $t \in [-\tau, 0]$ and a.e. for $x \in \Omega$, satisfy (H_p) and (H_q) with $a = b - \tau > 0$ and so, all the hypotheses of the Theorem 8.2.1 are fulfilled, from which we get the conclusion. $\qquad\square$

Multivalued Reaction–Diffusion Systems 311

REMARK 8.7.1 Since u is not T-periodic, in spite of the fact that v satisfies the nonlocal condition $v(t,x) = v(t+T,x)$ for each $t \in [-\tau, 0]$ and a.e. for $x \in \Omega$, v may fail to be T-periodic.

8.8 A reaction–diffusion system in $L^2(\Omega)$

In this section we consider a semi-multivalued general variant of the system in Section 6.6, i.e., a system in which the forcing term of one equation is single-valued the one of the other equation is multivalued.

Let Ω be a nonempty, bounded domain in $\mathbb{R}^d$, $d \geq 1$, with C^1 boundary Σ, let $\tau \geq 0$, $\omega > 0$, $\gamma > 0$, let $Q_+ = \mathbb{R}_+ \times \Omega$, $\Sigma_+ = \mathbb{R}_+ \times \Sigma$, and let $Q_\tau = [-\tau, 0] \times \Omega$. Let $\alpha : D(\alpha) \subseteq \mathbb{R} \rightsquigarrow \mathbb{R}$ and $\beta : D(\beta) \subseteq \mathbb{R} \rightsquigarrow \mathbb{R}$ be maximal-monotone operators, let $\mathcal{X}_2 = C([-\tau, 0]; L^2(\Omega))$, let $F : \mathbb{R}_+ \times \mathcal{X}_2 \times \mathcal{X}_2 \to L^2(\Omega)$ be continuous and $G : \mathbb{R}_+ \times \mathcal{X}_2 \times \mathcal{X}_2 \rightsquigarrow L^2(\Omega)$ a multifunction to be defined later. Let $b > \tau$, let μ_i be positive σ-finite and complete measures on $\mathbb{R}_+$, with $\mathrm{supp}\, \mu_i \subseteq [b, +\infty)$, $i = 1, 2$. Let $k_i \in L^2(\mathbb{R}_+; \mu_i)$ be nonnegative functions, and let $W_i : \mathbb{R} \to \mathbb{R}$ be nonexpansive with $W_i(0) = 0$, $i = 1, 2$. Finally, let $C : L^2(\Omega) \to \mathbb{R}$ be a linear continuous functional and let $\xi_i \in \mathcal{X}_2$, $i = 1, 2$. We consider the following system subjected to nonlocal initial conditions:

$$
\begin{cases}
\dfrac{\partial u}{\partial t}(t,x) = \Delta u(t,x) - \omega u(t,x) + F(t, u_t, v_t)(x), & \text{in } Q_+, \\[2mm]
\dfrac{\partial v}{\partial t}(t,x) = \Delta v(t,x) - \gamma v(t,x) + g(t)(x), & \text{in } Q_+, \\[2mm]
g(t)(x) \in G(t, u_t, v_t)(x), & \text{in } Q_+, \\[2mm]
-\dfrac{\partial u}{\partial \nu}(t,x) \in \alpha(u(t,x)), \quad -\dfrac{\partial v}{\partial \nu}(t,x) \in \beta(u(t,x)), & \text{on } \Sigma_+, \\[2mm]
u(t,x) = \displaystyle\int_b^\infty k_1(s) W_1(v(t+s,x)) u(t+s,x)\, d\mu_1(s) + \xi_1(t)(x), & \text{in } Q_\tau, \\[2mm]
v(t,x) = \displaystyle\int_b^\infty k_2(s) W_2(v(t+s,x)) C u(t+s, \cdot)\, d\mu_2(s) + \xi_2(t)(x), & \text{in } Q_\tau,
\end{cases}
\tag{8.8.1}
$$

where $g_i : \mathbb{R}_+ \times \mathcal{X}_2 \times \mathcal{X}_2 \to L^2(\Omega)$, $i = 1, 2$, are two given functions with

$$
g_1(t, u, v)(x) \leq g_2(t, u, v)(x)
$$

for all $(t, u, v) \in \mathbb{R}_+ \times \mathcal{X}_2 \times \mathcal{X}_2$ and a.e. for $x \in \Omega$ and $G : \mathbb{R}_+ \times \mathcal{X}_2 \times \mathcal{X}_2 \rightsquigarrow L^2(\Omega)$ is defined by

$$
G(t, u, v) = \{\, h \in L^2(\Omega);\ g_1(t, u, v)(x) \leq h(x) \leq g_2(t, u, v)(x),\ \text{a.e. for } x \in \Omega \,\}
\tag{8.8.2}
$$

312   *Delay Differential Evolutions Subjected to Nonlocal Initial Conditions*

for each $(t, u, v) \in \mathbb{R}_+ \times \mathfrak{X}_2 \times \mathfrak{X}_2$.

Our main result concerning the system (8.8.1) is:

THEOREM 8.8.1 *Let Ω be a nonempty, bounded and open subset in $\mathbb{R}^d$ with $d \geq 1$ whose boundary Σ is of class C^1, let $\alpha : D(\alpha) \subseteq \mathbb{R} \rightsquigarrow \mathbb{R}$ and $\beta : D(\beta) \subseteq \mathbb{R} \rightsquigarrow \mathbb{R}$ be maximal-monotone operators with $0 \in D(\alpha)$, $0 \in \alpha(0)$, $0 \in D(\beta)$ and $0 \in \beta(0)$, let $F : \mathbb{R}_+ \times \mathfrak{X}_2 \times \mathfrak{X}_2 \to L^2(\Omega)$ be continuous, and let $g_i : \mathbb{R}_+ \times \mathfrak{X}_2 \times \mathfrak{X}_2 \to L^2(\Omega)$, $i = 1, 2$, be two given functions satisfying*

$$g_1(t, u, v)(x) \leq g_2(t, u, v)(x)$$

for each $(t, u, v) \in \mathbb{R}_+ \times \mathfrak{X}_2 \times \mathfrak{X}_2$ a.e. for $x \in \Omega$. Let $b > \tau$ and let μ_i be positive σ-finite and complete measures defined on the class of Borel subsets in $\mathbb{R}_+$, with $\operatorname{supp} \mu_i \subseteq [b, +\infty)$, $i = 1, 2$, let $k_i \in L^2(\mathbb{R}_+; \mu)$, $i = 1, 2$, let $W_i : \mathbb{R} \to \mathbb{R}$, $i = 1, 2$, and let $\xi_i \in \mathfrak{X}_2$, $i = 1, 2$. Let us assume that

(h_1) *there exist two nonnegative functions $\widetilde{\ell}, \widetilde{m} \in L^1(\mathbb{R}_+) \cap L^\infty(\mathbb{R}_+)$ such that*

$$\|F(t, u, v) - F(t, \widetilde{u}, \widetilde{v})\|_{L^2(\Omega)}$$

$$\leq \widetilde{\ell}(t) \max \{\|u - \widetilde{u}\|_{\mathfrak{X}_2}, \|v - \widetilde{v}\|_{\mathfrak{X}_2}\}$$

and

$$\|F(t, u, v)\|_{L^2(\Omega)} \leq \widetilde{\ell}(t)\|u\|_{\mathfrak{X}_2} + \widetilde{m}(t),$$

for each $(t, u, v), (t, \widetilde{u}, \widetilde{v}) \in \mathbb{R}_+ \times \mathfrak{X}_2 \times \mathfrak{X}_2$

(h_2) *g_1 is l.s.c and g_2 is u.s.c. with respect to the usual order relation on their domains in the sense of Definition 1.5.7*

(h_3) *there exist two nonnegative functions $\alpha, \beta \in L^1(\mathbb{R}_+) \cap L^\infty(\mathbb{R}_+)$ such that*

$$\|g_i(t, u, v)\|_{L^2(\Omega)} \leq \alpha(t) \max \{\|u\|_{\mathfrak{X}_2}, \|v\|_{\mathfrak{X}_2}\} + \beta(t)$$

for $i = 1, 2$ and each $(t, u, v) \in \mathbb{R}_+ \times \mathfrak{X}_2 \times \mathfrak{X}_2$

(h_4) *$\|k_i\|_{L^1(\mathbb{R}_+; \mu)} \leq 1$, $i = 1, 2$*

(h_5) *$|W_i(v) - W_i(\widetilde{v})| \leq |v - \widetilde{v}|$ for each $v, \widetilde{v} \in \mathbb{R}$, $i = 1, 2$*

(h_6) *$W_i(0) = 0$, $i = 1, 2$*

(h_7) *$C : L^2(\Omega) \to \mathbb{R}$ is nonexpansive.*

Let us assume also that $\ell = \|\widetilde{\ell}\|_{L^2(\mathbb{R}_+)}$ and $\delta = \min\{\omega, \gamma\}$ satisfy (H_c). Then, (8.8.1) has at least one C^0-solution.

Multivalued Reaction–Diffusion Systems 313

Proof. We will write (8.8.1) in the abstract form (8.1.1) and we will apply Theorem 8.2.1. So, let $X = Y = L^2(\Omega)$ and let $A : D(A) \subseteq X \to X$ and $B : D(B) \subseteq X \to X$ be defined by

$$\begin{cases} D(A) = \left\{ u \in H^2(\Omega), \; -\dfrac{\partial u}{\partial \nu}(x) \in \alpha(u(x)) \text{ a.e. for } x \in \Sigma \right\}, \\[2mm] Au = \Delta u - \omega u, \text{ for } u \in D(A) \end{cases}$$

and respectively by

$$\begin{cases} D(B) = \left\{ v \in H^2(\Omega), \; -\dfrac{\partial v}{\partial \nu}(x) \in \beta(v(x)) \text{ a.e. for } x \in \Sigma \right\}, \\[2mm] Bv = \Delta v - \gamma v, \text{ for } v \in D(B). \end{cases}$$

Since $0 \in \alpha(0)$ and $0 \in \beta(0)$, it follows that $C_0^\infty(\Omega)$ is dense in $D(A) \cap D(B)$ and consequently $\overline{D(A)} = X$ and $\overline{D(B)} = Y$. Moreover, A is ω-m-dissipative and B is γ-m-dissipative. In addition, both generate compact semigroups on $L^2(\Omega)$. This follows from Theorem 1.9.8 by observing that $A = \Delta_2^\omega$ and $B = \Delta_2^\gamma$. Also, since X and Y are Hilbert spaces, by Remark 1.8.2, we conclude that both are of complete continuous type.

From the remarks above and (h_1), we deduce that A, B and F satisfy (H_A), (H_B) and (H_F) in Theorem 8.2.1.

Next, let us define $G : \mathbb{R}_+ \times \mathfrak{X}_2 \times \mathfrak{X}_2 \rightsquigarrow L^2(\Omega)$ by (8.8.2) and let us observe that, by (h_2), (h_3) and Lemma 1.5.5, G satisfies (H_G) in Theorem 8.2.1.

Recalling that $\operatorname{supp} \mu_i \subseteq [a, +\infty)$ with $a = b - \tau$, $i = 1, 2$, from $(h_4) \sim (h_7)$, we deduce that the functions $p, q : C_b(\mathbb{R}_+; L^2(\Omega)) \times C_b(\mathbb{R}_+; L^2(\Omega)) \to \mathfrak{X}_2$, defined by

$$\begin{cases} p(u,v)(t)(x) = \displaystyle\int_b^\infty k_1(s) W_1(v(t+s)(x)) u(t+s)(x) \, d\mu_1(s) + \xi_1(t)(x), \\[3mm] q(u,v)(t)(x) = \displaystyle\int_b^\infty k_2(s) W_2(v(t+s)(x)) C u(t+s)(\cdot) \, d\mu_1(s) + \xi_2(t)(x) \end{cases}$$

for each $(u, v) \in C_b(\mathbb{R}_+; L^2(\Omega)) \times C_b(\mathbb{R}_+; L^2(\Omega))$, each $t \in [-\tau, 0]$ and a.e. for $x \in \Omega$, satisfy (H_p) and (H_q) in Theorem 8.2.1 with $a = b - \tau > 0$. The conclusion follows from Theorem 8.2.1 and this completes the proof. $\qquad\square$

8.9 Bibliographical notes and comments

Section 8.1 Various classes of reaction–diffusion systems without delay were studied earlier by Burlică [55], Burlică and Roşu [56], [57], and Díaz and Vrabie [97]. See also Necula and Vrabie [196], and Roşu [224], [225]. The

314   *Delay Differential Evolutions Subjected to Nonlocal Initial Conditions*

results in this chapter are new and inspired by Burlică, Roşu and Vrabie [62]. More precisely, they are based on the main result of Burlică and Roşu [59] and borrow some ideas from Vrabie [261].

Section 8.2 The main result in this section, i.e., Theorem 8.2.1, which is a multivalued version of Theorem 6.11.1, is new and appears for the first time here. It should be noted that a similar result was obtained by Burlică and Roşu [59], under the stronger assumption that both p and q have linear growth. Theorems 8.2.2 and 8.2.3 are also new.

Section 8.3 Although inspired by Burlică, Roşu and Vrabie [62] and Burlică and Roşu [59], the proof given here for Theorem 8.2.1 is somehow simpler due to the fact that we have adopted a different point of view. More precisely, we consider the system (8.1.1) as an evolution equation in a product space rather than a couple of two evolution equations.

Section 8.4 In Lemma 8.4.1, we have obtained some estimates in the product space $X \times Y$ by using only one constant, i.e., $\delta = \min\{\omega, \gamma\}$ in both equations, rather than using ω in the first equation and γ in the second one as was done in Burlică and Roşu [59].

Section 8.5 Lemmas 8.5.1 and 8.5.2 are from Burlică and Roşu [59].

Section 8.6 As we already have mentioned, the proof of Theorem 8.2.1 is partly inspired by Burlică and Roşu [59].

Section 8.7 The example in this section, as well as Theorem 8.7.1, are new. Unlike in Burlică and Roşu [59], here we allow p to have affine growth instead of linear growth and the conditions on both F and g_i, $i = 1, 2$, are expressed in the terms of the max-norm on $X \times Y$ rather than on the equivalent norm

$$\|(x, y)\|_1 = \|x\| + \|y\|$$

for each $(x, y) \in X \times Y$.

Section 8.8 The example in this section and Theorem 8.8.1 are also new.

Chapter 9

Viability for Nonlocal Evolution Inclusions

Overview

In this chapter, we consider a very special and challenging existence problem for abstract nonlinear inclusions with delay subjected to nonlocal initial conditions, i.e., the existence of C^0-solutions whose graphs are included in the graph of an a priori given multifunction.

9.1 The problem to be studied

Let $\tau \geq 0$, let X be a real Banach space, let $A : D(A) \subseteq X \rightsquigarrow X$ be an m-dissipative operator generating a nonlinear semigroup of contractions denoted by $\{S(t) : \overline{D(A)} \to \overline{D(A)}; \ t \in \mathbb{R}_+\}$, and let us denote by

$$\mathfrak{X} = C([-\tau, 0]; X) \quad \text{and} \quad \mathcal{D} = \{\varphi \in \mathfrak{X}; \ \varphi(0) \in \overline{D(A)}\}.$$

Let I be a given interval with nonempty interior, let $K : I \rightsquigarrow \overline{D(A)}$ and $F : \mathcal{K} \rightsquigarrow X$ be nonempty-valued multifunctions, where

$$\mathcal{K} = \{(t, \varphi) \in I \times \mathfrak{X}; \ \varphi(0) \in K(t)\}.$$

Our aim in this chapter is to find necessary, sufficient as well as necessary and sufficient conditions in order that $\mathcal{K}$ be viable with respect to $A + F$ in the sense of Definition 9.1.2 below. Let $(\sigma, \varphi) \in \mathcal{K}$ and let us consider the problem

$$\begin{cases} u'(t) \in Au(t) + F(t, u_t), & t \in [\sigma, T], \\ u(t) = \varphi(t - \sigma), & t \in [\sigma - \tau, \sigma]. \end{cases} \tag{9.1.1}$$

DEFINITION 9.1.1 *A function* $u \in C([\sigma - \tau, T]; X)$ *is said to be a* C^0-*solution of (9.1.1) on* $[\sigma - \tau, T] \subseteq I$ *if* $(t, u_t) \in \mathcal{K}$ *for* $t \in [\sigma, T]$, $u(t) = \varphi(t - \sigma)$ *for*

315

316   *Delay Differential Evolutions Subjected to Nonlocal Initial Conditions*

$t \in [\sigma - \tau, \sigma]$ and there exists $f \in L^1(\sigma, T; X)$ with $f(t) \in F(t, u_t)$ a.e. for $t \in [\sigma, T]$ such that u is a C^0-solution of the Cauchy problem

$$\begin{cases} u'(t) \in Au(t) + f(t), & t \in [\sigma, T], \\ u(\sigma) = \varphi(0). \end{cases} \qquad (9.1.2)$$

We say that the function $u : [\sigma - \tau, T) \to X$ is a C^0-*solution of* (9.1.1) *on* $[\sigma - \tau, T)$, if u is a C^0-solution on $[\sigma - \tau, \widetilde{T}]$ for all $\widetilde{T} < T$.

DEFINITION 9.1.2 We say that $\mathcal{K}$ is C^0-*viable with respect to* $A + F$ if, for each $(\sigma, \varphi) \in \mathcal{K}$, there exists $T > \sigma$ such that $[\sigma - \tau, T] \subseteq I$ and (9.1.1) has at least one C^0-solution $u : [\sigma - \tau, T] \to X$. If $T = \sup I$ for any $(\sigma, \varphi) \in \mathcal{K}$, we say that $\mathcal{K}$ is *globally* C^0-*viable with respect to* $A + F$.

We define the metric d on $\mathcal{K}$ by

$$d((\sigma, \varphi), (\theta, \psi)) = \max\{|\sigma - \theta|, \|\varphi - \psi\|_X\},$$

for all $(\sigma, \varphi), (\theta, \psi) \in \mathcal{K}$.

Let $(\sigma, \varphi) \in \mathcal{K}$, let $\eta \in X$ and let $E \subset X$ be a nonempty, bounded subset, let $h > 0$ and let $\mathcal{F}_E = \{f \in L^1_{\text{loc}}(I; X); \ f(s) \in E \text{ a.e. for } s \in I\}$. We denote by $u(\sigma + h, \sigma, \varphi(0), \mathcal{F}_E) = \{u(\sigma + h, \sigma, \varphi(0), f); \ f \in \mathcal{F}_E\}$, where $u(\cdot) = u(\cdot, \sigma, \varphi(0), f)$ is the unique C^0-solution of the problem (9.1.2).

DEFINITION 9.1.3 We say that E is A-*right-quasi-tangent* to $\mathcal{K}$ at (σ, φ) if

$$\liminf_{h \downarrow 0} \frac{1}{h} \text{dist} \left(u(\sigma + h, \sigma, \varphi(0), \mathcal{F}_E), K(\sigma + h) \right) = 0, \qquad (9.1.3)$$

where $\text{dist} \left(u(\sigma + h, \sigma, \varphi(0), \mathcal{F}_E), K(\sigma + h) \right)$ is defined as in (5.5.2).

9.2   Necessary conditions for viability

LEMMA 9.2.1 *Let* $f : [\sigma, T] \to X$ *be a strongly measurable function and let* $B, C \subset X$ *be two nonempty sets such that* $f(t) \in B + C$ *a.e. for* $t \in [\sigma, T]$. *Then, for all* $\varepsilon > 0$ *there exist three functions* $b : [\sigma, T] \to B$, $c : [\sigma, T] \to C$ *and* $r : [\sigma, T] \to S(0, \varepsilon)$, *all strongly measurable, such that*

$$f(t) = b(t) + c(t) + r(t)$$

a.e. for $t \in [\sigma, T]$.

Viability for Nonlocal Evolution Inclusions 317

Proof. Let $\varepsilon > 0$ be fixed. Let $\overline{f} : [\,\sigma, T\,] \to X$ be countably valued and such that $\|\overline{f}(t) - f(t)\| < \frac{\varepsilon}{2}$ a.e. for $t \in [\,\sigma, T\,]$. Then, we have

$$\overline{f}(t) \in B + C + S\left(0, \frac{\varepsilon}{2}\right) \text{ a.e. for } t \in [\,\sigma, T\,].$$

So, there exist three countably valued functions $b : [\,\sigma, T\,] \to B$, $c : [\,\sigma, T\,] \to C$ and $\overline{r} : [\,\sigma, T\,] \to S(0, \frac{\varepsilon}{2})$ such that

$$\overline{f}(t) = b(t) + c(t) + \overline{r}(t) \text{ a.e. for } t \in [\,\sigma, T\,].$$

The proof is complete once we take $r(t) = \overline{r}(t) + f(t) - \overline{f}(t)$ a.e. for $t \in [\,\sigma, T\,]$. $\square$

THEOREM 9.2.1 *If* $F : \mathcal{K} \rightsquigarrow X$ *is u.s.c., and* $\mathcal{K}$ *is* C^0*-viable with respect to* $A + F$ *then, for all* $(\sigma, \varphi) \in \mathcal{K}$*, we have*

$$\lim_{h\downarrow 0} \frac{1}{h} \mathrm{dist}\left(u(\sigma + h, \sigma, \varphi(0), \mathcal{F}_{F(\sigma,\varphi)}), K(\sigma + h)\right) = 0. \tag{9.2.1}$$

Proof. If $(\sigma, \varphi) \in \mathcal{K}$ and $u : [\,\sigma - \tau, T\,] \to X$ is a C^0-solution of (9.1.1), then there exists $f \in L^1(\sigma, T; X)$ such that $f(s) \in F(s, u_s)$ a.e. for $s \in [\,\sigma, T\,]$ and $u(t) = u(t, \sigma, \varphi(0), f)$ for all $t \in [\,\sigma, T\,]$. Let $\varepsilon > 0$ be arbitrary but fixed.

Since F is u.s.c. at (σ, φ) and $\lim_{t \to \sigma} u_t = u_\sigma = \varphi$ in $\mathcal{X}$, there exists $\delta > 0$ such that $f(s) \in F(s, u_s) \subseteq F(\sigma, \varphi) + S(0, \varepsilon)$ a.e. for $s \in [\,\sigma, \sigma + \delta\,]$.

Set $B = F(\sigma, \varphi)$ and $C = S(0, \varepsilon)$ in Lemma 9.2.1. Then there exist two integrable functions $g : [\,\sigma, \sigma + \delta\,] \to F(\sigma, \varphi)$ and $r : [\,\sigma, \sigma + \delta\,] \to S(0, 2\varepsilon)$ such that $f(s) = g(s) + r(s)$ a.e. for $s \in [\,\sigma, \sigma + \delta\,]$. Since $u(\sigma + h) \in K(\sigma + h)$, it follows that, for each $0 < h < \delta$, we have

$$\mathrm{dist}\left(u(\sigma + h, \sigma, \varphi(0), \mathcal{F}_{F(\sigma,\varphi)}), K(\sigma + h)\right)$$

$$\leq \mathrm{dist}\left(u(\sigma + h, \sigma, \varphi(0), g), u(\sigma + h, \sigma, \varphi(0), f)\right) \leq \int_\sigma^{\sigma+h} \|g(s) - f(s)\| ds \leq 2\varepsilon h$$

which shows that

$$\limsup_{h\downarrow 0} \frac{1}{h} \mathrm{dist}\left(u(\sigma + h, \sigma, \varphi(0), \mathcal{F}_{F(\sigma,\varphi)}), K(\sigma + h)\right) \leq 2\varepsilon.$$

As $\varepsilon > 0$ is arbitrary, this completes the proof. $\square$

Since (9.2.1) implies (9.1.3), from Theorem 9.2.1, we get:

THEOREM 9.2.2 *If* $F : \mathcal{K} \rightsquigarrow X$ *is u.s.c., and* $\mathcal{K}$ *is* C^0*-viable with respect to* $A + F$ *then* $F(\sigma, \varphi)$ *is* A*-right-quasi-tangent to* $\mathcal{K}$ *at* (σ, φ)*, i.e., the the tangency condition*

$$\liminf_{h\downarrow 0} \frac{1}{h} \mathrm{dist}\left(u(\sigma + h, \sigma, \varphi(0), \mathcal{F}_{F(\sigma,\varphi)}), K(\sigma + h)\right) = 0 \tag{9.2.2}$$

is satisfied for all $(\sigma, \varphi) \in \mathcal{K}$*.*

9.3 Sufficient conditions for viability

DEFINITION 9.3.1 We say that the multifunction $K : I \rightsquigarrow \overline{D(A)}$ is

(i) *closed from the left on I if for any sequence $((t_n, x_n))_{n \geq 1}$ in $I \times \overline{D(A)}$, with $x_n \in K(t_n)$ and $(t_n)_n$ nondecreasing and satisfying $\lim_n t_n = t \in I$ and $\lim_n x_n = x$, we have $x \in K(t)$*

(ii) *locally closed from the left if for each $(\sigma, \xi) \in I \times \overline{D(A)}$ with $\xi \in K(\sigma)$ there exist $T > \sigma$ and $\rho > 0$ such that the multifunction $t \rightsquigarrow K(t) \cap D(\xi, \rho)$ is closed from the left on $[\sigma, T]$.*

In what follows, $D_{\mathcal{X}}(\varphi, \rho)$ denotes the closed ball with center φ and radius $\rho > 0$ in $\mathcal{X}$.

DEFINITION 9.3.2 The multifunction $F : \mathcal{K} \rightsquigarrow X$ is called *locally bounded* if, for each (σ, φ) in $\mathcal{K}$, there exist $\delta > 0$, $\rho > 0$, and $M > 0$ such that for all (t, ψ) in $([\sigma - \delta, \sigma + \delta] \times D_{\mathcal{X}}(\varphi, \rho)) \cap \mathcal{K}$, we have $\|F(t, \psi)\| \leq M$.

THEOREM 9.3.1 *Let $K : I \rightsquigarrow \overline{D(A)}$ be locally closed from the left and let $F : \mathcal{K} \rightsquigarrow X$ be nonempty, convex and weakly compact valued. If F is strongly-weakly u.s.c., locally bounded and $A : D(A) \subseteq X \rightsquigarrow X$ is of complete continuous type and generates a compact semigroup, then a sufficient condition for $\mathcal{K}$ to be C^0-viable with respect to $A + F$ is the tangency condition (9.2.2) for all $(\sigma, \varphi) \in \mathcal{K}$. If, in addition, F is u.s.c., then the tangency condition (9.2.2) is also necessary in order that $\mathcal{K}$ be C^0-viable with respect to $A + F$.*

LEMMA 9.3.1 *Let $K : I \rightsquigarrow \overline{D(A)}$ be locally closed from the left, $F : \mathcal{K} \rightsquigarrow X$ be locally bounded and let $(\sigma, \varphi) \in \mathcal{K}$. Let us assume that the tangency condition (9.2.2) is satisfied. Let $\rho > 0$, $T > \sigma$ and $M > 0$ be such that $[\sigma, T] \subseteq I$, and*

(1) *the multifunction $t \rightsquigarrow K(t) \cap D(\varphi(0), \rho)$ is closed from the left on $[\sigma, T)$*

(2) $\|F(t, \psi)\| \leq M$ *for all $t \in [\sigma, T]$ and all $\psi \in D_{\mathcal{X}}(\varphi, \rho)$ with $(t, \psi) \in \mathcal{K}$*

(3) $\displaystyle \sup_{t \in [\sigma, T]} \|S(t - \sigma)\varphi(0) - \varphi(0)\| + \sup_{|t - s| \leq T - \sigma} \|\varphi(t) - \varphi(s)\| + (T - \sigma)(M + 1) < \rho.$

Then, for each $\varepsilon \in (0, 1)$, there exist a family $\mathcal{P}_T = \{[t_m, s_m); m \in \Gamma\}$ of disjoint intervals, with Γ finite or at most countable, and two functions $f \in L^1(\sigma, T; X)$ and $u \in C([\sigma - \tau, T]; X)$ such that:

(i) $\displaystyle \bigcup_{m \in \Gamma} [t_m, s_m) = [\sigma, T)$ *and $s_m - t_m \leq \varepsilon$ for all $m \in \Gamma$*

(ii) $u(t_m) \in K(t_m)$ *for all $m \in \Gamma$ and $u(T) \in K(T)$*

(iii) *we have both $f(s) \in F(t_m, u_{t_m})$ a.e. for $s \in [t_m, s_m)$, for all $m \in \Gamma$ and $\|f(s)\| \leq M$ a.e. for $s \in [\sigma, T]$*

Viability for Nonlocal Evolution Inclusions

(iv) $u(t) = \varphi(t - \sigma)$ for $t \in [\sigma - \tau, \sigma]$ and
$\|u(t) - u(t, t_m, u(t_m), f)\| \le (t - t_m)\varepsilon$ for $t \in [t_m, T]$ and $m \in \Gamma$

(v) $\|u_t - \varphi\|_{\mathcal{X}} < \rho$ for all $t \in [\sigma, T]$

(vi) $\|u(t) - u(t_m)\| \le \varepsilon$ for all $t \in [t_m, s_m)$ and all $m \in \Gamma$.

Proof. It is easy to see that if $(i) \sim (iv)$ are satisfied, then (v) is satisfied too, i.e., $\|u(t + s) - \varphi(s)\| < \rho$ for all $t \in [\sigma, T]$ and $s \in [-\tau, 0]$. Indeed, if $t + s \le \sigma$ then

$$\|u(t + s) - \varphi(s)\| = \|\varphi(t + s - \sigma) - \varphi(s)\| \le \sup_{|t_1 - t_2| \le T - \sigma} \|\varphi(t_1) - \varphi(t_2)\| < \rho.$$

If $t + s > \sigma$, then $|s| < T - \sigma$ and from (3), (iii) and (iv), we get

$$\|u(t + s) - \varphi(s)\| \le \|u(t + s) - u(t + s, \sigma, \varphi(0), f)\|$$

$$+ \|u(t + s, \sigma, \varphi(0), f) - u(t + s, \sigma, \varphi(0), 0)\|$$

$$+ \|u(t + s, \sigma, \varphi(0), 0) - \varphi(0)\| + \|\varphi(0) - \varphi(s)\|$$

$$\le (t + s - \sigma)\varepsilon + \int_\sigma^{t+s} \|f(\theta)\| d\theta + \|S(t + s - \sigma)\varphi(0) - \varphi(0)\| + \|\varphi(0) - \varphi(s)\|$$

$$\le (T - \sigma)(1 + M) + \|S(t + s - \sigma)\varphi(0) - \varphi(0)\| + \|\varphi(0) - \varphi(s)\| < \rho.$$

Let $\varepsilon \in (0, 1)$ be arbitrary, but fixed. First, we will show that there exist $\delta = \delta(\varepsilon)$ in (σ, T) and $\mathcal{P}_\delta$, f, u such that the conditions $(i) \sim (vi)$ hold true with T replaced by δ.

From the tangency condition (9.2.2), it follows that there exist three sequences: $(h_n)_n$ with $h_n \downarrow 0$, $(g_n)_n \subset \mathcal{F}_{F(\sigma, \varphi)}$ and $(p_n)_n \subset X$ with $\|p_n\| \to 0$ satisfying

$$u(\sigma + h_n, \sigma, \varphi(0), g_n) + p_n h_n \in K(\sigma + h_n)$$

for all $n \in \mathbb{N}$, $n \ge 1$. Take $n_0 \in \mathbb{N}$ and choose $\delta = \sigma + h_{n_0}$ such that $\delta \in (\sigma, T)$, $h_{n_0} < \varepsilon$ and $\|p_{n_0}\| < \varepsilon$.

We define $\mathcal{P}_\delta = \{[\sigma, \delta)\}$, $f(t) = g_{n_0}(t)$ and

$$u(t) = u(t, \sigma, \varphi(0), g_{n_0}) + (t - \sigma)p_{n_0}$$

for $t \in [\sigma, \delta]$. Obviously, $(i) \sim (v)$ are satisfied. Moreover, we may diminish $\delta > \sigma$ (increase n_0), if necessary, in order for (vi) to be satisfied too.

Let

$$\mathcal{U} = \{(\mathcal{P}_\delta, f, u); \ \delta \in (\sigma, T] \text{ and } (i) \sim (vi) \text{ are satisfied with } \delta \text{ instead of } T\}.$$

As we already have shown, $\mathcal{U} \ne \emptyset$. On $\mathcal{U}$ we define the binary relation "$\preceq$" by

$$(\mathcal{P}_{\delta_1}, f_1, u_1) \preceq (\mathcal{P}_{\delta_2}, f_2, u_2),$$

320 *Delay Differential Evolutions Subjected to Nonlocal Initial Conditions*

if $\delta_1 \leq \delta_2$, $\mathcal{P}_{\delta_1} \subseteq \mathcal{P}_{\delta_2}$, $f_1(s) = f_2(s)$ a.e. for $s \in [\sigma, \delta_1]$ and $u_1(s) = u_2(s)$ for all $s \in [\sigma, \delta_1]$. Obviously "$\preceq$" is a partial order on $\mathcal{U}$.

Next, we will prove that each nondecreasing sequence in $\mathcal{U}$ is bounded from above. Let $((\mathcal{P}_{\delta_j}, f_j, u_j))_{j \geq 1}$ be a nondecreasing sequence in $\mathcal{U}$ and let $\delta = \sup_{j \geq 1} \delta_j$. We distinguish between two complementary cases.

Case 1. If there exists $j_0 \in \mathbb{N}$ such that $\delta_{j_0} = \delta$, then $(\mathcal{P}_{\delta_{j_0}}, f_{j_0}, u_{j_0})$ is an upper bound for the sequence.

Case 2. If $\delta_j < \delta$, for all $j \geq 1$, we define $\mathcal{P}_\delta = \cup_{j \geq 1} \mathcal{P}_{\delta_j}$, $f(t) = f_j(t)$ and $u(t) = u_j(t)$ for all $j \geq 1$ and $t \in [\sigma, \delta_j)$. Clearly, $f \in L^1(\sigma, \delta; X)$ and $u \in C([\sigma, \delta); X)$.

Let us observe that, in view of (iv), we have

$$\|u(t) - u(s)\| \leq \|u(t) - u(t, \delta_j, u(\delta_j), f)\|$$

$$+ \|u(t, \delta_j, u(\delta_j), f) - u(s, \delta_j, u(\delta_j), f)\| + \|u(s, \delta_j, u(\delta_j), f) - u(s)\|$$

$$\leq (t - \delta_j)\varepsilon + \|u(t, \delta_j, u(\delta_j), f) - u(s, \delta_j, u(\delta_j), f)\| + (s - \delta_j)\varepsilon$$

$$\leq 2(\delta - \delta_j)\varepsilon + \|u(t, \delta_j, u(\delta_j), f) - u(s, \delta_j, u(\delta_j), f)\|$$

for all $j \geq 1$ and all $t, s \in [\delta_j, \delta)$. As $\lim_j \delta_j = \delta$ and $u(\cdot, \delta_j, u(\delta_j), f)$ is continuous at $t = \delta$, we conclude that u satisfies the Cauchy condition for the existence of the limit at $t = \delta$. So, u can be extended by continuity to $[\sigma, \delta]$. On the other hand, since $u(\delta) = \lim_{t \uparrow \delta} u(t) = \lim_j u(\delta_j) = \lim_j u_j(\delta_j)$, $u_j(\delta_j) \in D(\varphi(0), \rho) \cap K(\delta_j)$ for $j \geq 1$ and the latter is closed from the left, we deduce that $u(\delta) \in D(\varphi(0), \rho) \cap K(\delta)$. The rest of the conditions in the lemma being obviously satisfied, it follows that $(\mathcal{P}_\delta, f, u)$ is an upper bound for the sequence.

As a consequence, $(\mathcal{U}, \preceq)$ and $\mathcal{N} : (\mathcal{U}, \preceq) \to R$, defined by $\mathcal{N}(\mathcal{P}_\delta, f, u) = \delta$, for each $(\mathcal{P}_\delta, f, u) \in \mathcal{U}$, satisfy the hypotheses of the Brezis–Browder Ordering Principle Theorem 1.14.1. Accordingly, there exists an $\mathcal{N}$-maximal element in $\mathcal{U}$, i.e., there exists $(\mathcal{P}_{\delta^*}, f^*, u^*) \in \mathcal{U}$ such that, whenever $(\mathcal{P}_{\delta^*}, f^*, u^*) \preceq (\mathcal{P}_{\overline{\delta}}, \overline{f}, \overline{u})$, we necessarily have $\mathcal{N}(\mathcal{P}_{\delta^*}, f^*, u^*) = \mathcal{N}(\mathcal{P}_{\overline{\delta}}, \overline{f}, \overline{u})$. In order to complete the proof, it is enough to show that $\delta^* = T$. To this end, let us assume by contradiction that $\delta^* < T$.

As $(\delta^*, u_{\delta^*}^*) \in \mathcal{K}$, from the tangency condition (9.2.2), we deduce the existence of three sequences: $(h_n)_n$ with $h_n \downarrow 0$, $(g_n)_n \subset \mathcal{F}_{F(\sigma, \varphi)}$ and $(p_n)_n \subset X$ with $\|p_n\| \to 0$ and satisfying

$$u(\delta^* + h_n, \delta^*, u^*(\delta^*), g_n) + p_n h_n \in K(\delta^* + h_n)$$

for all $n \in \mathbb{N}$, $n \geq 1$. Let $n_0 \in \mathbb{N}$ and $\overline{\delta} = \delta^* + h_{n_0}$ with $\overline{\delta} \in (\delta^*, T)$, $h_{n_0} < \varepsilon$ and $\|p_{n_0}\| < \varepsilon$. Let $\mathcal{P}_{\overline{\delta}} = \mathcal{P}_{\delta^*} \cup \{[\delta^*, \overline{\delta}]\}$,

$$\overline{f}(t) = \begin{cases} f^*(t), & t \in [\sigma, \delta^*], \\ f_{n_0}(t), & t \in (\delta^*, \overline{\delta}] \end{cases}$$

Viability for Nonlocal Evolution Inclusions

and

$$\overline{u}(t) = \begin{cases} u^*(t), & t \in [\sigma, \delta^*], \\ u(t, \delta^*, u^*(\delta^*), f_{n_0}) + (t - \delta^*)p_{n_0}, & t \in (\delta^*, \overline{\delta}]. \end{cases}$$

By (v), we have

$$\|u_{\delta^*}^* - \varphi\|_X < \rho.$$

So, (2) implies that $\|\overline{f}(s)\| \leq M$ a.e. for $s \in (\sigma, \overline{\delta})$ and thus $(i) \sim (iii)$ are satisfied. In order to prove (iv) we will consider only the case $t_m \leq \delta^* \leq t$, the other cases being obvious. Using the evolution property, i.e., $u(t, a, \xi, f) = u(t, b, u(b, a, \xi, f), f)$ for $\sigma \leq a \leq b \leq t \leq T$, we get

$$\|\overline{u}(t) - u(t, t_m, u^*(t_m), \overline{f})\|$$

$$\leq \|u(t, \delta^*, u^*(\delta^*), \overline{f}) - u(t, t_m, u^*(t_m), \overline{f})\| + (t - \delta^*)\varepsilon$$

$$= \|u(t, \delta^*, u^*(\delta^*), \overline{f}) - u(t, \delta^*, u(\delta^*, t_m, u^*(t_m), \overline{f}), \overline{f})\| + (t - \delta^*)\varepsilon$$

$$\leq \|u^*(\delta^*) - u(\delta^*, t_m, u^*(t_m), \overline{f})\| + (t - \delta^*)\varepsilon$$

$$\leq (\delta^* - t_m)\varepsilon + (t - \delta^*)\varepsilon = (t - t_m)\varepsilon,$$

which proves (iv). As we already have mentioned at the beginning of the proof, $(i) \sim (iv)$ imply (v).

Finally, let us observe that we can diminish $\overline{\delta}$ (increase n_0), if necessary, in order that (vi) be satisfied too.

So, $(\mathcal{P}_{\overline{\delta}}, \overline{f}, \overline{u}) \in \mathcal{U}$, $(\mathcal{P}_{\delta^*}, f^*, u^*) \preceq (\mathcal{P}_{\overline{\delta}}, \overline{f}, \overline{u})$, but $\delta^* < \overline{\delta}$ which contradicts the maximality of $(\mathcal{P}_{\delta^*}, f^*, u^*)$. Hence $\delta^* = T$, and $\mathcal{P}_{\delta^*}$, f^* and u^* satisfy all the conditions $(i) \sim (vi)$. The proof is complete. $\qquad \square$

DEFINITION 9.3.3 Let $\varepsilon > 0$. An element $(\mathcal{P}_T, f, u)$ satisfying $(i) \sim (vi)$ in Lemma 9.3.1, is called an *ε-approximate C^0-solution of* (9.1.1).

We can now proceed to the proof of Theorem 9.3.1.

Proof. The necessity follows from Theorem 9.2.2. As long as the proof of the sufficiency is concerned, let $\rho > 0$, $T > \sigma$ and $M > 0$ be as in Lemma 9.3.1. Let $\varepsilon_n \in (0, 1)$, with $\varepsilon_n \downarrow 0$. Let $((\mathcal{P}_T^n, f_n, u_n))_n$ be a sequence of ε_n-approximate C^0-solutions of (9.1.1) given by Lemma 9.3.1. If $\mathcal{P}_T^n = \{[t_m^n, s_m^n); \ m \in \Gamma_n\}$ with Γ_n finite or at most countable, we denote by $a_n : [\sigma, T) \to [\sigma, T)$ the step function, defined by $a_n(s) = t_m^n$ for all $n \in \mathbb{N}$, $n \geq 1$ and each $s \in [t_m^n, s_m^n)$. Clearly, $\lim_n a_n(s) = s$ uniformly for $s \in [\sigma, T)$, while from (vi), we deduce that $\lim_n \|u_n(t) - u_n(a_n(t))\| = 0$, uniformly for $t \in [\sigma, T)$. From (iv), we get

$$\lim_n (u_n(t) - u(t, \sigma, \varphi(0), f_n)) = 0 \qquad (9.3.1)$$

uniformly for $t \in [\sigma, T]$. Since $\|f_n(t)\| \leq M$ for all $n \in \mathbb{N}$, $n \geq 1$ and a.e. for $t \in [\sigma, T]$ and the semigroup generated by A is compact, by Theorem 1.8.5, we deduce that the set $\{u(\cdot, \sigma, \varphi(0), f_n); \ n \geq 1\}$ is relatively compact in $C([\sigma, T]; X)$. From this remark and (9.3.1), we conclude that $(u_n)_n$ has at

322 *Delay Differential Evolutions Subjected to Nonlocal Initial Conditions*

least one uniformly convergent subsequence on $[\sigma, T]$ to some function u. For the sake of simplicity, we denote that subsequence again by $(u_n)_n$.

Since $a_n(t) \uparrow t$, $\lim_n u_n(a_n(t)) = u(t)$, uniformly for $t \in [\sigma, T)$ and the mapping $t \to K(t) \cap D(\varphi(0), \rho)$ is closed from the left, we get that $u(t) \in K(t)$ for all $t \in [\sigma, T]$. But $\lim_n (u_n)_{a_n(t)} = u_t$ in $\mathfrak{X}$, uniformly for $t \in [\sigma, T)$. Hence, the set

$$C = \overline{\{(a_n(t), (u_n)_{a_n(t)}); n \geq 1, \ t \in [\sigma, T)\}}$$

is compact in $I \times \mathfrak{X}$ and $C \subseteq \mathcal{K}$.

At this point, recalling that F is strongly-weakly u.s.c., and has weakly compact values, by Lemma 1.5.1, it follows that the set

$$B = \overline{\text{conv}} \left(\bigcup_{n \geq 1} \bigcup_{t \in [\sigma, T)} F(a_n(t), (u_n)_{a_n(t)}) \right)$$

is weakly compact. We notice that $f_n(s) \in B$ for all $n \geq 1$ and a.e. for $s \in [\sigma, T]$. An appeal to Theorem 1.4.12 shows that, at least on a subsequence, $\lim_n f_n = f$ weakly in $L^1(\sigma, T; X)$. As F is strongly-weakly u.s.c. with closed and convex values while, by (iii) in Lemma 9.3.1, for each $n \geq 1$, we have $f_n(s) \in F(a_n(s), (u_n)_{a_n(s)})$ a.e. for $s \in [\sigma, T]$, from Theorem 1.5.1, we conclude that $f(s) \in F(s, u_s)$ a.e. for $s \in [\sigma, T]$.

Finally, by (9.3.1) and the fact that A is of complete continuous type, we get $u(t) = u(t, \sigma, \varphi(0), f)$ for each $t \in [\sigma, T]$ and so, u is a C^0-solution of (9.1.1). $\qquad \square$

From Lemma 1.13.3 and Theorem 9.3.1, we get:

THEOREM 9.3.2 *Let $K : I \rightsquigarrow \overline{D(A)}$ be closed from the left and let $F : \mathcal{K} \rightsquigarrow X$ be nonempty, convex and weakly compact valued. If there exist $a, b \in C(I)$ such that*

$$\|F(t, \varphi)\| \leq a(t) + b(t)\|\varphi(0)\|$$

for all $(t, \varphi) \in \mathcal{K}$, F is strongly-weakly u.s.c., and $A : D(A) \subseteq X \rightsquigarrow X$ is of complete continuous type and generates a compact semigroup, then a sufficient condition in order that $\mathcal{K}$ be globally C^0-viable with respect to $A + F$ is the tangency condition (9.2.2). If, in addition, F is u.s.c., then the tangency condition (9.2.2) is also necessary in order that $\mathcal{K}$ be globally C^0-viable with respect to $A + F$.

9.4 A sufficient condition for null controllability

Let X be a reflexive Banach space, let $A : D(A) \subseteq X \rightsquigarrow X$ be an m-dissipative operator, let $\tau \geq 0$, $f : \mathbb{R}_+ \times \mathfrak{X} \to X$ a given function, and let $(\sigma, \varphi) \in \mathbb{R}_+ \times \mathcal{D}$. The *null controllability* problem consists of finding a strongly measurable

Viability for Nonlocal Evolution Inclusions

control $c(\cdot)$ taking values in $D(0,1)$ in order to reach the origin in some time $T > 0$ by C^0-solutions of the state equation

$$\begin{cases} u'(t) \in Au(t) + f(t, u_t) + c(t), & t \in [\sigma, \sigma + T], \\ u(t) = \varphi(t - \sigma), & t \in [\sigma - \tau, \sigma]. \end{cases} \tag{9.4.1}$$

With $F : \mathbb{R}_+ \times \mathcal{X} \rightsquigarrow X$, defined by

$$F(t, v) = av(0) + f(t, v) + D(0, 1)$$

for each $(t, v) \in \mathbb{R}_+ \times \mathcal{X}$, where $a \in \mathbb{R}$, the above problem can be equivalently reformulated as follows: find $T > 0$ and a C^0-solution of the problem

$$\begin{cases} u'(t) \in (A - aI)u(t) + F(t, u_t), & t \in [\sigma, \sigma + T], \\ u(t) = \varphi(t - \sigma), & t \in [\sigma - \tau, \sigma], \\ u(\sigma + T) = 0. \end{cases} \tag{9.4.2}$$

THEOREM 9.4.1 *Let X be a reflexive Banach space and let $A : D(A) \subseteq X \rightsquigarrow X$ be such that, for some $a \in \mathbb{R}$, $A - aI$ is an m-dissipative operator of a complete continuous type that is also the infinitesimal generator of a compact semigroup of contractions, $\{S(t) : \overline{D(A)} \to \overline{D(A)}; \ t \in \mathbb{R}_+\}$. Let $f : \mathbb{R}_+ \times \mathcal{X} \to X$ be a continuous function such that for some $L > 0$ we have*

$$\|f(t, v)\| \le L\|v(0)\|, \quad \text{for all } (t, v) \in \mathbb{R}_+ \times \mathcal{X}. \tag{9.4.3}$$

In addition, let us assume that $0 \in D(A)$ and $0 \in A0$.

Then, for each $(\sigma, \varphi) \in \mathbb{R}_+ \times \mathcal{X}$ with $\varphi(0) \in \overline{D(A)} \setminus \{0\}$, there exists a C^0-solution $u : [\sigma, +\infty) \to X$ of (9.4.2) satisfying

$$\|u(t)\| \le \|\varphi(0)\| - (t - \sigma) + (L + a) \int_\sigma^t \|u(s)\| ds, \tag{9.4.4}$$

for all $t \ge \sigma$ for which $u(t) \ne 0$.

Proof. Let $(\sigma, \varphi) \in \mathbb{R}_+ \times \mathcal{X}$ be arbitrary with $\varphi(0) \in \overline{D(A)} \setminus \{0\}$. The idea is to show that there exist $T \in (0, +\infty)$ and a noncontinuable C^0-solution $(z, u) : [\sigma, \sigma + T) \to \mathbb{R} \times X$ of the problem

$$\begin{cases} z'(t) = (L + a)\|u(t)\| - 1, & t \in [\sigma, \sigma + T), \\ u'(t) \in (A - aI)u(t) + F(t, u_t), & t \in [\sigma, \sigma + T), \\ z_\sigma = \|\varphi\| \quad \text{and} \quad u(t) = \varphi(t - \sigma), & t \in [\sigma - \tau, \sigma], \\ \|u(t)\| \le z(t), & t \in [\sigma, \sigma + T) \end{cases} \tag{9.4.5}$$

and then, from the first equation and the last inequality in (9.4.5), to get (9.4.4).

324   *Delay Differential Evolutions Subjected to Nonlocal Initial Conditions*

We begin by observing that, on the Banach space $Z = \mathbb{R} \times X$, the operator $\mathcal{A} = (0, A - aI)$ generates a compact semigroup of contractions

$$\{(1, S(t)); \ (1, S(t)) : \mathbb{R} \times \overline{D(A)} \to \mathbb{R} \times \overline{D(A)}; \ t \in \mathbb{R}_+\}.$$

We denote by $\mathcal{R} = C([-\tau, 0]; \mathbb{R})$, $\mathcal{X} = C([-\tau, 0]; X)$ and $\mathcal{Z} = C([-\tau, 0]; Z)$. Let K be the locally closed set defined as

$$K = \{(x_1, x_2) \in \mathbb{R}_+ \times (\overline{D(A)} \setminus \{0\}); \ \|x_2\| \le x_1\},$$

and let $\mathcal{K}$ be the associate set given by

$$\mathcal{K} = \{(t, \psi_1, \psi_2) \in \mathbb{R} \times \mathcal{Z}; \ \|\psi_2(0)\| \le \psi_1(0)\}.$$

We define the multifunction $\mathcal{F} : \mathcal{K} \rightsquigarrow Z$ by

$$\mathcal{F}(t, \psi_1, \psi_2) = ((L + a)\|\psi_2(0)\| - 1, a\psi_2(0) + f(t, \psi_2) + D(0, 1)),$$

for each $(t, \psi_1, \psi_2) \in \mathcal{K}$.

To verify that $\mathcal{F}$ satisfies the tangency condition (9.2.2) for all $(\sigma, \psi_1, \psi_2) \in \mathcal{K}$, we shall prove a stronger tangency condition, i.e.,

(STC) there exists $(\eta_1, \eta_2) \in \mathcal{F}(\sigma, \psi_1, \psi_2)$ such that

$$\liminf_{h \downarrow 0} \frac{1}{h} \text{dist}\,(\mathcal{U}(\sigma + h, \sigma, (\xi_1, \xi_2), (\eta_1, \eta_2)), K) = 0, \tag{9.4.6}$$

where $(\xi_1, \xi_2) = (\psi_1(0), \psi_2(0))$ and $\mathcal{U}(\cdot, \sigma, (\xi_1, \xi_2), (\eta_1, \eta_2))$ is the C^0-solution of the corresponding Cauchy problem for the operator $\mathcal{A}$, i.e.,

$$\mathcal{U}(t, \sigma, (\xi_1, \xi_2), (\eta_1, \eta_2)) = (\xi_1 + (t - \sigma)\eta_1, u(t, \sigma, \xi_2, \eta_2)),$$

$u(\cdot, \sigma, \xi_2, \eta_2)$ being the corresponding C^0-solution of the problem

$$\begin{cases} u'(t) \in Au(t) - au(t) + \eta_2, & t \in [\sigma, +\infty), \\ u(\sigma) = \xi_2. \end{cases}$$

To this end, it suffices to prove that there exist $(h_n)_n$ in $\mathbb{R}_+$, with $h_n \downarrow 0$, and $((\theta_n, p_n))_n$ in Z, with $(\theta_n, p_n) \to (0, 0)$, such that, for all $n \in \mathbb{N}$, $n \ge 1$, we have

$$\|u(\sigma + h_n, \sigma, \xi_2, \eta_2) + h_n p_n\| \le \xi_1 + h_n \eta_1 + h_n \theta_n. \tag{9.4.7}$$

Clearly,

$$\|u(\sigma + h, \sigma, \xi_2, \eta_2)\| \le \|\xi_2\| + \int_{\sigma}^{\sigma + h} [u(s, \sigma, \xi_2, \eta_2), \eta_2]_+ ds$$

for all $h > 0$. By (xi) in Proposition 1.8.1, the normalized semi-inner product

$$(x, y) \mapsto [x, y]_+ = \lim_{h \downarrow 0} \frac{1}{h} (\|x + hy\| - \|x\|)$$

Viability for Nonlocal Evolution Inclusions 325

is u.s.c. Hence, setting $\ell(s) = u(s, \sigma, \xi_2, \eta_2)$, we get

$$\liminf_{h \downarrow 0} \frac{1}{h} \int_\sigma^{\sigma+h} [\ell(s), \eta_2]_+ ds \leq \limsup_{h \downarrow 0} \frac{1}{h} \int_\sigma^{\sigma+h} [\ell(s), \eta_2]_+ ds \leq [\xi_2, \eta_2]_+.$$

Let

$$\begin{cases} \eta_1 = (L+a)\|\psi_2(0)\| - 1 = (L+a)\|\xi_2\| - 1, \\ \eta_2 = a\xi_2 + f(\sigma, \psi_2) - \dfrac{\xi_2}{\|\xi_2\|}. \end{cases}$$

Clearly, $\eta_2 \in a\xi_2 + f(\sigma, \psi_2) + D(0, 1)$ and so, $(\eta_1, \eta_2) \in \mathcal{F}(\sigma, \psi_1, \psi_2)$. From (9.4.3), we get

$$[\xi_2, \eta_2]_+ = a\|\xi_2\| + [\xi_2, f(\sigma, \psi_2)]_+ - 1 \leq (L+a)\|\xi_2\| - 1 = \eta_1$$

and hence

$$\liminf_{h \downarrow 0} \frac{1}{h} \left(\|u(\sigma + h, \sigma, \xi_2, \eta_2)\| - \|\xi_2\| \right) \leq \eta_1.$$

Since for $(\sigma, \psi_1, \psi_2) \in \mathcal{K}$ we have $\|\xi_2\| = \|\psi_2(0)\| \leq \psi_1(0) = \xi_1$, the last inequality proves (9.4.7) with $p_n = 0$. Thus we get (9.4.6). From Theorem 9.3.1, $\mathcal{K}$ is C^0-viable with respect to $\mathcal{A} + \mathcal{F}$. As $(\sigma, \|\varphi\|, \varphi) \in \mathcal{K}$, thanks to Brezis–Browder Ordering Principle Theorem 1.14.1, we obtain that there exist $T \in (0, +\infty]$ and a noncontinuable C^0-solution $(z, u) : [\sigma, \sigma + T) \to Z$ of (9.4.5) that satisfies $(z(t), u(t)) \in K$ for all $t \in [\sigma, \sigma + T)$. This means that (9.4.4) is satisfied for all $t \in [\sigma, \sigma + T)$. Since F has sublinear growth, u, as a solution of (9.4.2), can be continued to $\mathbb{R}_+$. So, $u(\sigma + T)$ exists, even though the solution (z, u) of (9.4.5) is defined merely on $[\sigma, \sigma + T)$ if T is finite. In this case, $u(\sigma + T) = 0$ since otherwise (z, u) can be continued to the right of $\sigma + T$, which is a contradiction. $\qquad\square$

The corollary below exhibits very sharp estimates for the time T at which the C^0-solution reaches the origin.

COROLLARY 9.4.1 *Under the hypothesis of Theorem 9.4.1, we have the following*:

(i) *If $L + a \leq 0$, for any $(\sigma, \varphi) \in \mathbb{R}_+ \times X$ with $\xi = \varphi(0) \in \overline{D(A)} \setminus \{0\}$, there exist a control $c(\cdot)$ and a C^0-solution of (9.4.1) that reaches the origin of X in some time $T \leq \|\xi\|$ and satisfies*

$$\|u(t)\| \leq \|\xi\| - (t - \sigma)$$

for all $\sigma \leq t \leq \sigma + T$.

(ii) *If $L + a > 0$, for all $(\sigma, \varphi) \in \mathbb{R}_+ \times X$ with $\xi = \varphi(0) \in \overline{D(A)} \setminus \{0\}$ satisfying $0 < \|\xi\| < 1/(L + a)$, there exist a control $c(\cdot)$ and a C^0-solution of (9.4.1) that reaches the origin of X in some time*

$$T \leq (L+a)^{-1} \log \left\{ [1 - (L+a)\|\xi\|]^{-1} \right\}$$

326   Delay Differential Evolutions Subjected to Nonlocal Initial Conditions

and

$$\|u(t)\| \le e^{(L+a)(t-\sigma)} \left[\|\xi\| - (L+a)^{-1}\right] + (L+a)^{-1}$$

for $t \in [\sigma, \sigma + T]$.

9.5   The case of nonlocal initial conditions

As before, let $K : \mathbb{R}_+ \rightsquigarrow \overline{D(A)}$, let

$$\mathfrak{X} = C([-\tau, 0]; X) \quad \text{and} \quad \mathfrak{D} = \{\varphi \in \mathfrak{X}; \ \varphi(0) \in \overline{D(A)}\}$$

and let

$$\mathfrak{K} = \{(t, \varphi) \in \mathbb{R}_+ \times \mathfrak{X}; \ \varphi(0) \in K(t)\}.$$

Let $f \colon \mathfrak{K} \to X$ be a continuous function which is Lipschitz with respect to its second argument. Let $g \colon C_b(\mathbb{R}_+; \overline{D(A)}) \to \mathfrak{D}$ and let us consider the problem

$$\begin{cases} u'(t) \in Au(t) + f(t, u_t), & t \in \mathbb{R}_+, \\ u(t) = g(u)(t), & t \in [-\tau, 0]. \end{cases} \tag{9.5.1}$$

DEFINITION 9.5.1 We say that $\mathfrak{K}$ is *globally C^0-viable with respect to* (A, f, g), if there exists at least one C^0-solution $u \in C_b([-\tau, +\infty); X)$ of (9.5.1) such that, for each $t \in \mathbb{R}_+$, $(t, u_t) \in \mathfrak{K}$.

Our aim here is to prove a necessary and a sufficient condition in order that $\mathfrak{K}$ be globally viable with respect to (A, f, g).

The hypotheses we need in what follows are listed below.

(H_A) The operator $A \colon D(A) \subseteq X \rightsquigarrow X$ satisfies

(A_1) $0 \in D(A)$, $0 \in A0$ and A is ω-m-dissipative for some $\omega > 0$

(A_2) A generates a compact semigroup on $\overline{D(A)}$.

(H_f) The function $f \colon \mathfrak{K} \to X$ is continuous and satisfies

(f_1) there exists $\ell > 0$ such that, for each $(t, v), (t, \widetilde{v}) \in \mathfrak{K}$, we have

$$\|f(t, v) - f(t, \widetilde{v})\| \le \ell \|v - \widetilde{v}\|_{\mathfrak{X}}$$

(f_2) there exists $m_0 > 0$ such that, for each $(t, v) \in \mathfrak{K}$, we have

$$\|f(t, v)\| \le \ell \|v\|_{\mathfrak{X}} + m_0,$$

where ℓ is given by (f_1).

(H_g) The function $g : C_b(\mathbb{R}_+; \overline{D(A)}) \to \mathfrak{D}$ satisfies

Viability for Nonlocal Evolution Inclusions

(g_1) there exists $a > 0$ such that

$$\|g(u) - g(\widetilde{u})\|_X \leq \|u - \widetilde{u}\|_{C_b([\,a,+\infty);X)}$$

for each $u, \widetilde{u} \in C_b(\mathbb{R}_+; \overline{D(A)})$ with $(t, u_t), (t, \widetilde{u}_t) \in \mathcal{K}$ for each $t \in \mathbb{R}_+$

(g_2) there exists $m \geq 0$ such that

$$\|g(u)\|_X \leq \|u\|_{C_b([\,a,+\infty);X)} + m$$

for each $u \in C_b(\mathbb{R}_+; X)$ with $(t, u_t) \in \mathcal{K}$ for each $t \in \mathbb{R}_+$

(g_4) g is continuous from $C_b([\,a, +\infty); \overline{D(A)})$ endowed with the $\widetilde{C}_b([\,a, +\infty); X)$ topology to X

(g_6) for each $u \in C_b(\mathbb{R}_+; X)$ with $u(t) \in K(t)$ for each $t \in \mathbb{R}_+$, we have

$$g(u)(0) \in K(0).$$

(H_c) The constants ω and ℓ satisfy

$$\ell < \omega.$$

(H_K) The multifunction $K : \mathbb{R}_+ \rightsquigarrow \overline{D(A)}$ satisfies

(K_1) there exists at least one function $u \in C_b(\mathbb{R}_+; \overline{D(A)})$ with the property

$$u(t) \in K(t)$$

for each $t \in \mathbb{R}_+$

(K_2) K is closed from the left in the sense of Definition 9.3.1

(K_3) for each $\xi \in K(0)$ and each $\lambda \in (0, 1)$, we have $\lambda \xi \in K(0)$.

(H_T) A and f satisfy the tangency condition below with respect to $\mathcal{K}$:

for each $(\sigma, \varphi) \in \mathcal{K}$, we have

$$\liminf_{h \downarrow 0} \frac{1}{h} \operatorname{dist}\left(u(\sigma + h, \sigma, \varphi(0), f(\sigma, \varphi)); K(\sigma + h)\right) = 0,$$

$\operatorname{dist}\left(u(\sigma + h, \sigma, \varphi(0), f(\sigma, \varphi)), K(\sigma + h)\right)$ being defined as in (5.5.2).

THEOREM 9.5.1 *If (H_A), (H_f), (H_g), (H_c), (H_K) and (H_T) are satisfied, then $\mathcal{K}$ is globally C^0-viable with respect to (A, f, g) in the sense of Definition 9.5.1.*

Let $\sigma \geq 0$, $T > \sigma$ and let us consider the following classical initial-value delay equation

$$\begin{cases} u'(t) \in Au(t) + f(t, u_t), & t \in [\,\sigma, T\,], \\ u(t) = \varphi(t - \sigma), & t \in [\,\sigma - \tau, \sigma\,], \end{cases} \tag{9.5.2}$$

328   *Delay Differential Evolutions Subjected to Nonlocal Initial Conditions*

DEFINITION 9.5.2 The set $\mathcal{K}$ is called C^0-*viable with respect to* $A + f$, if for each $(\sigma, \varphi) \in \mathcal{K}$, there exist $T > \sigma$ and a C^0-solution $u : [\sigma - \tau, T] \to \overline{D(A)}$ of (9.5.2) such that, for each $t \in [\sigma, T]$, $(t, u_t) \in \mathcal{K}$.

We state for easy reference the viability result below, which is a particular case of Theorem 9.3.1 that will play a crucial role in the proof of Theorem 9.5.1.

THEOREM 9.5.2 *Let us assume that* (H_A), (f_1) *in* (H_f) *and* (H_K) *are satisfied. Then, a necessary and sufficient condition in order that* $\mathcal{K}$ *be* C^0-*viable with respect to* $A + f$ *in the sense of Definition 9.5.2 is the tangency condition* (H_T). *If, in addition,* (f_2) *in* (H_f) *is satisfied, then each noncontinuable* C^0-*solution of* (9.5.2) *is global.*

9.6   An approximate equation

The idea of the proof is to show that we can construct a family of approximate problems such that each problem of the family has a unique C^0-solution $u \in C_b([-\tau, +\infty); X)$ satisfying $(t, u_t) \in \mathcal{K}$ for each $t \in \mathbb{R}_+$. Then, by using a compactness argument, we get a sequence of approximate C^0-solutions converging to a C^0-solution of the initial problem.

So, let $\varepsilon \in (0, 1)$ and let us consider the problem

$$\begin{cases} u'(t) \in Au(t) + f(t, u_t), & t \in \mathbb{R}_+, \\ u(t) = (1 - \varepsilon)g(u)(t), & t \in [-\tau, 0]. \end{cases} \tag{9.6.1}$$

THEOREM 9.6.1 *Let us assume that* (H_A), (H_f), (H_g), (H_c) *and* (H_K) *are satisfied and let* $\varepsilon \in (0, 1)$. *Then the problem* (9.6.1) *has a unique* C^0-*solution,* $u_\varepsilon \in C_b([-\tau, +\infty); X)$, *such that* $(t, u_{\varepsilon_t}) \in \mathcal{K}$ *for each* $t \in \mathbb{R}_+$.

Proof. Let us define the set

$$\mathcal{C} = \{v; \ v \in C_b(\mathbb{R}_+; \overline{D(A)}), \ g(v)(0) \in K(0)\}.$$

Thanks to (K_1) in (H_K) and (g_6) in (H_g), $\mathcal{C}$ is nonempty and closed. We define the operator $Q : \mathcal{C} \to C_b(\mathbb{R}_+; \overline{D(A)})$ by

$$Q(v) = u_{|\mathbb{R}_+}$$

for each $v \in \mathcal{C}$, where u is the unique C^0-solution of the problem

$$\begin{cases} u'(t) \in Au(t) + f(t, u_t), & t \in \mathbb{R}_+, \\ u(t) = (1 - \varepsilon)g(v)(t), & t \in [-\tau, 0], \end{cases}$$

whose existence is ensured by Theorem 9.5.2. Indeed, $g(v)(0) \in K(0)$ for each

Viability for Nonlocal Evolution Inclusions

$v \in \mathcal{C}$, from (K_3), we have $u(0) = (1 - \varepsilon)g(v)(0) \in K(0)$. Thus, we are in the hypotheses of Theorem 9.5.2.

Next, let us observe that the operator Q maps $\mathcal{C}$ into itself. Clearly, again from Theorem 9.5.2, it follows that if $v \in \mathcal{C}$, then $Q(v) \in C_b(\mathbb{R}_+; \overline{D(A)})$ and $Q(v)(t) \in K(t)$ for each $t \in \mathbb{R}_+$. From (g_6) in (H_g), we easily deduce that $g(Q(v))(0) \in K(0)$ for each $v \in \mathcal{C}$ and so, $Q(\mathcal{C}) \subseteq \mathcal{C}$.

We will show next that Q is a strict contraction. Let $v, \widetilde{v} \in C_b(\mathbb{R}_+; \overline{D(A)})$, let $Q(v) = u_{|\mathbb{R}_+}$, $Q(\widetilde{v}) = \widetilde{u}_{|\mathbb{R}_+}$, where $u, \widetilde{u} \in C_b([-\tau, +\infty); X)$ are the C^0-solutions of the above problem, corresponding to v and $\widetilde{v}$ respectively, and let us observe that at least one of the following three cases must hold true.

Case 1. There exists $t \in [-\tau, 0]$ such that

$$\|u - \widetilde{u}\|_{C_b([-\tau, +\infty); X)} = \|u(t) - \widetilde{u}(t)\|.$$

But $\|Q(v) - Q(\widetilde{v})\|_{C_b(\mathbb{R}_+; X)} \le \|u - \widetilde{u}\|_{C_b([-\tau, +\infty); X)}$ which implies

$$\|Q(v) - Q(\widetilde{v})\|_{C_b(\mathbb{R}_+; X)} \le (1 - \varepsilon)\|g(v)(t) - g(\widetilde{v})(t)\| \le (1 - \varepsilon)\|v - \widetilde{v}\|_{C_b([a, +\infty); X)}.$$

Thus

$$\|Q(v) - Q(\widetilde{v})\|_{C_b(\mathbb{R}_+; X)} \le (1 - \varepsilon)\|v - \widetilde{v}\|_{C_b(\mathbb{R}_+; X)}. \tag{9.6.2}$$

Case 2. There exists $t \in (0, +\infty)$ such that

$$\|u - \widetilde{u}\|_{C_b([-\tau, +\infty); X)} = \|Q(v)(t) - Q(\widetilde{v})(t)\|.$$

Then,

$$\|Q(v)(t) - Q(\widetilde{v})(t)\| = \|Q(v) - Q(\widetilde{v})\|_{C_b(\mathbb{R}_+; X)} \le e^{-\omega t}\|Q(v)(0) - Q(\widetilde{v})(0)\|$$

$$+ \frac{\ell}{\omega}(1 - e^{-\omega t})\|u - \widetilde{u}\|_{C_b([-\tau, +\infty); X)}$$

$$\le (1 - \varepsilon)e^{-\omega t}\|v - \widetilde{v}\|_{C_b(\mathbb{R}_+; X)} + \frac{\ell}{\omega}(1 - e^{-\omega t})\|Q(v) - Q(\widetilde{v})\|_{C_b(\mathbb{R}_+; X)}.$$

Set

$$\begin{cases} x = \|Q(v) - Q(\widetilde{v})\|_{C_b(\mathbb{R}_+; X)}, \\ y = \|v - \widetilde{v}\|_{C_b(\mathbb{R}_+; X)}, \\ \alpha = (1 - \varepsilon)e^{-\omega t}, \\ \beta = \frac{\ell}{\omega}(1 - e^{-\omega t}) \end{cases}$$

and let us observe that the inequality above rewrites as

$$x \le \alpha y + \beta x.$$

Since $0 < \alpha < 1$ and $0 < \beta < 1$, it follows

$$x \le \frac{\alpha}{1 - \beta} y. \tag{9.6.3}$$

330 *Delay Differential Evolutions Subjected to Nonlocal Initial Conditions*

Now, let us observe that

$$\frac{\alpha}{1-\beta} < 1-\varepsilon. \tag{9.6.4}$$

Indeed, the last inequality is equivalent to

$$\alpha < (1-\varepsilon)(1-\beta),$$

which in turn is equivalent to

$$(1-\varepsilon)e^{-\omega t} < (1-\varepsilon)[1 - \frac{\ell}{\omega}(1-e^{-\omega t})],$$

or even to

$$\omega(1-\varepsilon)e^{-\omega t} < (1-\varepsilon)(\omega - \ell + \ell e^{-\omega t})$$
$$e^{-\omega t}(\omega - \ell) < \omega - \ell,$$

which is obviously satisfied. From (9.6.3), (9.6.4) and the definition of x, y, we conclude that (9.6.2) holds true.

Case 3. There exists $(t_n)_n$ in $\mathbb{R}_+$ with $\lim_n t_n = +\infty$ and such that

$$\lim_n \|u(t_n) - \widetilde{u}(t_n)\| = \|u - \widetilde{u}\|_{C_b([-\tau,+\infty);X)}$$

$$= \lim_n \|Q(v)(t_n) - Q(\widetilde{v})(t_n)\| = \|Q(v) - Q(\widetilde{v})\|_{C_b(\mathbb{R}_+;X)}.$$

We have

$$\|Q(v)(t_n) - Q(\widetilde{v})(t_n)\| \le e^{-\omega t_n}\|Q(v)(0) - Q(\widetilde{v})(0)\|$$

$$+ \frac{\ell}{\omega}\left(1 - e^{-\omega t_n}\right)\|u - \widetilde{u}\|_{C_b([-\tau,+\infty);X)}$$

$$\le e^{-\omega t_n}\|Q(v)(0) - Q(\widetilde{v})(0)\| + \frac{\ell}{\omega}\left(1 - e^{-\omega t_n}\right)\|Q(v) - Q(\widetilde{v})\|_{C_b(\mathbb{R}_+;X)}$$

for each $n \in \mathbb{N}$, $n \ge 1$, and thus, passing to the limit for $n \to +\infty$ in this inequality, we get

$$\|Q(v) - Q(\widetilde{v})\|_{C_b(\mathbb{R}_+;X)} \le \frac{\ell}{\omega}\|Q(v) - Q(\widetilde{v})\|_{C_b(\mathbb{R}_+;X)}.$$

Since $\ell < \omega$, this shows that

$$\|Q(v) - Q(\widetilde{v})\|_{C_b(\mathbb{R}_+;X)} = 0.$$

So, in all possible cases, we have

$$\|Q(v) - Q(\widetilde{v})\|_{C_b(\mathbb{R}_+;X)} \le (1-\varepsilon)\|v - \widetilde{v}\|_{C_b(\mathbb{R}_+;X)}$$

for each $v, \widetilde{v} \in \mathcal{C}$.

In view of the Banach Fixed-Point Theorem, we conclude that Q has a unique fixed point $u_\varepsilon \in \mathcal{C}$, which is a C^0-solution of the problem (9.6.1). $\square$

Viabiliy for Nonlocal Evolution Inclusions

9.7 Proof of Theorem 9.5.1

Proof. Let $\varepsilon \in (0,1)$ and let us denote by $u_\varepsilon \in C_b([-\tau, +\infty); X)$ the unique C^0-solution of (9.6.1) whose existence is ensured by Theorem 9.6.1.

We will show that $\{u_\varepsilon; \ \varepsilon \in (0,1)\}$ is relatively compact in the locally convex space $\widetilde{C}_b([-\tau, +\infty); X)$. First, let us observe that, by Lemma 3.3.1, this set is bounded in $C_b([-\tau, +\infty); X)$. In view of (H_f) and (H_g), we conclude that $\{f(\cdot, u_{\varepsilon(\cdot)}); \ \varepsilon \in (0,1)\}$ is bounded in $C_b(\mathbb{R}_+; X)$ too, and thus its restriction to each compact interval in $\mathbb{R}_+$ is uniformly integrable. Since, by (A_2) in (H_A) the semigroup generated by A is compact, in view of Theorem 1.8.6, it follows that the family $\{u_\varepsilon; \ \varepsilon \in (0,1)\}$ is relatively compact in $\widetilde{C}_b([a, +\infty); X)$. By (g_4) in (H_g), we deduce that $\{u_\varepsilon; \ \varepsilon \in (0,1)\} = \{(1-\varepsilon)g(u_\varepsilon); \ \varepsilon \in (0,1)\}$ is relatively compact in X. As a consequence, the set

$$\{u_\varepsilon(0); \ \varepsilon \in (0,1)\} = \{(1-\varepsilon)g(u_\varepsilon)(0); \ \varepsilon \in (0,1)\}$$

is relatively compact in X. Using once again Theorem 1.8.6, we conclude that the family $\{u_\varepsilon; \ \varepsilon \in (0,1)\}$ is relatively compact in $\widetilde{C}_b(\mathbb{R}_+; X)$ and consequently in $\widetilde{C}_b([-\tau, +\infty); X)$. This implies that there exists at least one sequence $\varepsilon_n \downarrow 0$ and a function $u \in C_b([-\tau, +\infty); X)$ such that $\lim_n u_{\varepsilon_n} = u$ uniformly on compacta in $[-\tau, +\infty)$. Taking $\varepsilon = \varepsilon_n$ in the problem below

$$\begin{cases} u_\varepsilon'(t) \in Au_\varepsilon(t) + f(t, u_{\varepsilon_t}), & t \in \mathbb{R}_+, \\ u_\varepsilon(t) = (1-\varepsilon)g(u_\varepsilon)(t), & t \in [-\tau, 0], \end{cases}$$

and passing to the limit for $n \to +\infty$, we deduce that u is a C^0-solution of the problem (9.5.1). Since, by (K_2) in (H_K), $K(t)$ is closed and $u_{\varepsilon_n}(t) \in K(t)$ for each $t \in \mathbb{R}_+$, we conclude that $u(t) \in K(t)$ for each $t \in \mathbb{R}_+$. Thus $\mathcal{K}$ is globally C^0-viable with respect to (A, f, g) and this completes the proof of Theorem 9.5.1. $\qquad \square$

9.8 A comparison result for the nonlinear diffusion

Let $\Omega \subseteq \mathbb{R}^d$, $d = 1, 2, \ldots$, be a bounded domain with C^2 boundary Σ, let $\tau \geq 0$, $\omega > 0$, let $Q_+ = \mathbb{R}_+ \times \Omega$, $Q_\tau = [-\tau, 0] \times \Omega$, $\Sigma_+ = \mathbb{R}_+ \times \Sigma$, let $\psi : \mathbb{R} \to \mathbb{R}$ be a continuous and strictly increasing function with $\psi(0) = 0$ and

332 *Delay Differential Evolutions Subjected to Nonlocal Initial Conditions*

let us consider the nonlinear diffusion equation with delay

$$
\begin{cases}
\dfrac{\partial u}{\partial t}(t,x) = \Delta\psi(u(t,x)) - \omega u(t,x) + f(t,x,u_t), & \text{in } Q_+, \\[2mm]
u(t,x) = 0, & \text{on } \Sigma_+, \\[2mm]
u(t,x) = g(u)(t)(x), & \text{in } Q_\tau,
\end{cases}
\tag{9.8.1}
$$

where $\Delta\psi$ is the nonlinear diffusion operator,

$$
\mathcal{X}_1 = C([-\tau,0]; L^1(\Omega)) = \mathcal{D}_1
$$

$f : \mathbb{R}_+ \times \overline{\Omega} \times \mathcal{X}_1 \to \mathbb{R}_+$ and $g : C_b(\mathbb{R}_+; L^1(\Omega)) \to \mathcal{D}_1$.

Let $\widetilde{f} : \mathbb{R}_+ \times \overline{\Omega} \to \mathbb{R}_+$ be continuous such that $0 \le f(t,x,v) \le \widetilde{f}(t,x)$ for each $(t,x,v) \in \mathbb{R}_+ \times \overline{\Omega} \times \mathcal{X}_1$. Finally, let $\widetilde{u} : [-\tau,+\infty) \to L^1(\Omega)$ be a global positive C^0-solution of the comparison equation

$$
\begin{cases}
\dfrac{\partial \widetilde{u}}{\partial t}(t,x) = \Delta\psi(\widetilde{u}(t,x)) - \omega\widetilde{u}(t,x) + \widetilde{f}(t,x), & \text{in } Q_+, \\[2mm]
\widetilde{u}(t,x) = 0, & \text{on } \Sigma_+, \\[2mm]
\widetilde{u}(t,x) = g(\widetilde{u})(t)(x), & \text{in } Q_\tau.
\end{cases}
\tag{9.8.2}
$$

We will prove that, under some very general hypotheses on f, $\widetilde{f}$ and g, the nonlinear diffusion equation (9.8.1) has at least one global C^0-solution u satisfying $0 \le u(t,x) \le \widetilde{u}(t,x)$ for each $t \in [-\tau,+\infty)$ and a.e. for $x \in \Omega$.

THEOREM 9.8.1 *Let $\Omega \subseteq \mathbb{R}^d$, $d = 1,2,\ldots$, be a bounded domain with C^2 boundary Σ, let $\tau \ge 0$, $\omega > 0$ and let $\psi : \mathbb{R} \to \mathbb{R}$ be continuous on $\mathbb{R}$ and C^1 on $\mathbb{R}\setminus\{0\}$, with $\psi(0) = 0$, and for which there exist $C > 0$ and $\alpha > 0$ if $d \le 2$ and $\alpha > (d-2)/d$ if $d \ge 3$ such that*

$$
\psi'(r) \ge C|r|^{\alpha-1}
$$

for each $r \in \mathbb{R}\setminus\{0\}$.

Let us assume that $f : \mathbb{R}_+ \times \overline{\Omega} \times \mathcal{X}_1 \to \mathbb{R}$ and $\widetilde{f} : \mathbb{R}_+ \times \overline{\Omega} \to \mathbb{R}$ are continuous on their domains and satisfy

(f_1) *there exists $\ell > 0$ such that*

$$
\|f(t,\cdot,v) - f(t,\cdot,\widetilde{v})\|_{L^1(\Omega)} \le \ell\|v - \widetilde{v}\|_{\mathcal{X}_1}
$$

for each $t \in \mathbb{R}_+$ and $v, \widetilde{v} \in \mathcal{X}_1$

(f_2) *there exists $m_0 > 0$ such that, for each $(t,v) \in \mathbb{R}_+ \times \mathcal{X}_1$, we have*

$$
\|f(t,\cdot,v)\|_{L^1(\Omega)} \le \ell\|v\|_{\mathcal{X}_1} + m_0,
$$

where ℓ is given by (f_1)

Viability for Nonlocal Evolution Inclusions

(f_3) f is nonnegative and dominated by $\widetilde{f}$, i.e.,

$$0 \leq f(t,x,v) \leq \widetilde{f}(t,x)$$

for each $t \in \mathbb{R}_+$, each $v \in \mathfrak{X}_1$ and a.e. for $x \in \Omega$.

Further, $g : C_b(\mathbb{R}_+; L^1(\Omega)) \to \mathcal{D}_1$ satisfies

(g_1) there exists $a > 0$ such that

$$\|g(u) - g(w)\|_{\mathfrak{X}_1} \leq \|u - w\|_{C_b([a,+\infty);L^1(\Omega))}$$

for each $u, w \in C_b(\mathbb{R}_+; L^1(\Omega))$

(g_4) g is continuous from $\widetilde{C}_b([a,+\infty); L^1(\Omega))$ to $\mathfrak{X}_1$

(g_6) for each $u \in C_b([-\tau,+\infty); L^1(\Omega))$ satisfying

$$0 \leq u(t)(x) \leq \widetilde{u}(t)(x)$$

for each $t \in \mathbb{R}_+$ and a.e. for $x \in \Omega$, where $\widetilde{u}$ is the C^0-solution of (9.8.2), we have

$$0 \leq g(u)(0)(x) \leq \widetilde{u}(0)(x)$$

a.e. for $x \in \Omega$.

Let us assume that, in addition,

(H_c) the constants ω and ℓ satisfy

$$\ell < \omega.$$

Then, there exists at least one global C^0-solution, $u : [-\tau,+\infty) \to L^1(\Omega)$ of (9.8.1), satisfying

$$0 \leq u(t,x) \leq \widetilde{u}(t,x) \tag{9.8.3}$$

for each $t \in [-\tau,+\infty)$ and a.e. $x \in \Omega$.

Proof. To get the conclusion, we will make use of Theorem 9.5.1. First, since in our case ψ is single-valued, let us define $A : D(A) \subseteq X \to X$ by

$$\begin{cases} D(A) = \{u \in L^1(\Omega); \ \psi(u) \in W^{1,1}(\Omega), \ \Delta\psi(u) \in L^1(\Omega)\}, \\ Au = \Delta\psi(u) - \omega u \ \text{ for } u \in D(A). \end{cases}$$

In view of Theorem 1.9.6, A satisfies the hypotheses (H_A) in Theorem 9.5.1. Now, let us define the multifunction $K : \mathbb{R}_+ \rightsquigarrow L^1(\Omega)$ by

$$K(t) = \{u \in L^1(\Omega); \ 0 \leq u(x) \leq \widetilde{u}(t)(x) \ \text{ a.e. for } \ x \in \Omega\}$$

334   *Delay Differential Evolutions Subjected to Nonlocal Initial Conditions*

for each $t \in \mathbb{R}_+$. Furthermore, we define $\mathcal{K}$ by

$$\mathcal{K} = \{(t, \varphi) \in \mathbb{R}_+ \times \mathcal{X}_1; \ \varphi(0) \in K(t)\}.$$

Taking into account that f and g satisfy (H_f) and (H_g) and ℓ, ω satisfy (H_c), in order to apply Theorem 9.5.1, it remains only to check that f satisfies the tangency condition (H_T) with respect to $\mathcal{K}$.

So, let $(\sigma, \varphi) \in \mathcal{K}$ be arbitrary and let us observe that, to prove (H_T), it suffices to show that, for each $h > 0$, there exists $u^h \in L^1(\Omega)$ such that $(\sigma + h, u^h) \in \mathcal{K}$ and

$$\liminf_{h \downarrow 0} \frac{1}{h} \|u(\sigma + h, \sigma, \varphi(0), f(\sigma, \cdot, \varphi)) - u^h\|_{L^1(\Omega)} = 0. \tag{9.8.4}$$

Let $h > 0$ and let us define

$$u^h = u(\sigma + h, \sigma, \varphi(0), f(\sigma, \cdot, \varphi))$$

$$+ u(\sigma + h, \sigma, \widetilde{u}(\sigma), \widetilde{f}(\cdot, \cdot)) - u(\sigma + h, \sigma, \widetilde{u}(\sigma), \widetilde{f}(\sigma, \cdot)).$$

Since $0 \le \varphi(0)(x) \le \widetilde{u}(\sigma)(x)$ and $0 \le f(\sigma, x, \varphi) \le \widetilde{f}(\sigma, x)$ a.e. for $x \in \Omega$, in view of Lemma 1.9.2, we deduce

$$0 \le u^h(x) \le u(\sigma + h, \sigma, \widetilde{u}(\sigma), \widetilde{f}(\sigma, \cdot))(x)$$

$$+ u(\sigma + h, \sigma, \widetilde{u}(\sigma), \widetilde{f}(\cdot, \cdot))(x) - u(\sigma + h, \sigma, \widetilde{u}(\sigma), \widetilde{f}(\sigma, \cdot))(x) = \widetilde{u}(\sigma + h)(x)$$

a.e. for $x \in \Omega$. Thus $(\sigma + h, u^h) \in \mathcal{K}$. On the other hand, since $\widetilde{f}$ is continuous, we deduce

$$\lim_{h \downarrow 0} \frac{1}{h} \|u(\sigma + h, \sigma, \widetilde{u}(\sigma), \widetilde{f}(\cdot, \cdot)) - u(\sigma + h, \sigma, \widetilde{u}(\sigma), \widetilde{f}(\sigma, \cdot))\|_{L^1(\Omega)}$$

$$\le \lim_{h \downarrow 0} \frac{1}{h} \int_\sigma^{\sigma + h} \|\widetilde{f}(s, \cdot) - \widetilde{f}(\sigma, \cdot)\|_{L^1(\Omega)} \, ds = 0$$

and hence (9.8.4) is satisfied. Consequently, we have

$$\lim_{h \downarrow 0} \frac{1}{h} \mathrm{dist}\, (u(\sigma + h, \sigma, \varphi, f(\sigma, \cdot, \varphi)); K(\sigma + h)) = 0$$

and so, (H_T) is satisfied. In view of Theorem 9.5.1, $\mathcal{K}$ is globally C^0-viable with respect to (A, f, g). In other words, the problem (9.8.1) has at least one C^0-solution, $u \in C_b([-\tau, +\infty); L^1(\Omega))$ satisfying (9.8.3) for each $t \in \mathbb{R}_+$. Finally, using (g_6) and the fact that, due to (g_1), g depends only on the restriction of its argument to $[a, +\infty)$ – see Remark 3.2.3 – we conclude that u satisfies (9.8.3) for $t \in [-\tau, 0]$, too. The proof is complete. $\qquad \square$

REMARK 9.8.1 A glance at Theorem 9.5.1 shows that, in Theorem 9.8.1, we can substitute the condition that f is defined on $\mathbb{R}_+ \times \overline{\Omega} \times \mathcal{X}_1$ to $f \colon \mathcal{K} \to L^1(\Omega)$. However, in practice, it is rather complicated to consider functions f defined merely on $\mathcal{K}$.

Viability for Nonlocal Evolution Inclusions 335

Theorem 9.8.1 is sufficiently general to handle various nonlocal problems of great practical relevance. To illustrate this remark, we present a comparison result referring to T-periodic C^0-solutions.

So, let Ω, Σ, Ω_+, Σ_+, ω be as above and let $\tau = 0$. Let $f : \mathbb{R}_+ \times \overline{\Omega} \times L^1(\Omega) \to \mathbb{R}$, $\widetilde{f} : \mathbb{R}_+ \times \overline{\Omega} \to \mathbb{R}$ and let us consider the T-periodic problems

$$\begin{cases} \dfrac{\partial u}{\partial t}(t, x) = \Delta \psi(u(t, x)) - \omega u(t, x) + f(t, x, u(t, \cdot)), & \text{in } Q_+, \\ u(t, x) = 0, & \text{on } \Sigma_+, \\ u(t, x) = u(t + T, x), & \text{in } Q_+, \end{cases} \qquad (9.8.5)$$

and

$$\begin{cases} \dfrac{\partial \widetilde{u}}{\partial t}(t, x) = \Delta \psi(\widetilde{u}(t, x)) - \omega \widetilde{u}(t, x) + \widetilde{f}(t, x), & \text{in } Q_+, \\ \widetilde{u}(t, x) = 0, & \text{on } \Sigma_+, \\ \widetilde{u}(t, x) = \widetilde{u}(t + T, x), & \text{in } Q_+. \end{cases} \qquad (9.8.6)$$

Let us assume that $\widetilde{u} : \mathbb{R}_+ \to L^1(\Omega)$ is a T-periodic C^0-solution of (9.8.6). Since in this simple case when the delay is absent, i.e., $\tau = 0$, $C([-\tau, 0]; L^1(\Omega))$ identifies with $L^1(\Omega)$, from Theorem 9.8.1, we deduce:

THEOREM 9.8.2 *Let $\Omega \subseteq \mathbb{R}^d$, $d = 1, 2, \ldots$, be a bounded domain with C^2 boundary Σ, let $\omega > 0$, and let $\psi : \mathbb{R} \to \mathbb{R}$ be continuous on $\mathbb{R}$ and C^1 on $\mathbb{R} \setminus \{0\}$, with $\psi(0) = 0$, and for which there exist $C > 0$ and $\alpha > 0$ if $d \leq 2$ and $\alpha > (d-2)/d$ if $d \geq 3$ such that*

$$\psi'(r) \geq C|r|^{\alpha - 1}$$

for each $r \in \mathbb{R} \setminus \{0\}$.

Let us assume that $f : \mathbb{R}_+ \times \overline{\Omega} \times L^1(\Omega) \to \mathbb{R}$ and $\widetilde{f} : \mathbb{R}_+ \times \overline{\Omega} \to \mathbb{R}$ are continuous on their domains and satisfy

(f_1) *there exists $\ell > 0$ such that*

$$\|f(t, \cdot, v) - f(t, \cdot, \widetilde{v})\|_{L^1(\Omega)} \leq \ell \|v - \widetilde{v}\|_{L^1(\Omega)}$$

for each $t \in \mathbb{R}_+$ and $v, \widetilde{v} \in L^1(\Omega)$

(f_2) *there exists $m_0 > 0$ such that, for each $(t, v) \in \mathbb{R}_+ \times L^1(\Omega)$, we have*

$$\|f(t, \cdot, v)\|_{L^1(\Omega)} \leq \ell \|v\|_{L^1(\Omega)} + m_0,$$

where ℓ is given by (f_1)

(f_3) *f is nonnegative and dominated by $\widetilde{f}$, i.e.,*

$$0 \leq f(t, x, v) \leq \widetilde{f}(t, x)$$

for each $t \in \mathbb{R}_+$, each $v \in L^1(\Omega)$ and a.e. for $x \in \Omega$

336 Delay Differential Evolutions Subjected to Nonlocal Initial Conditions

(f_4) *f is T-periodic with respect to its first argument, i.e.,*

$$f(t, \cdot, v) = f(t + T, \cdot, v)$$

for each $(t, v) \in \mathbb{R}_+ \times L^1(\Omega)$.

Let us assume that, in addition,

(H_c) *the constants ω and ℓ satisfy*

$$\ell < \omega.$$

Then, there exists at least one T-periodic C^0-solution, $u \colon \mathbb{R}_+ \to L^1(\Omega)$ of (9.8.5), satisfying

$$0 \le u(t, x) \le \widetilde{u}(t, x),$$

for each $t \in \mathbb{R}_+$ and a.e. for $x \in \Omega$.

Proof. Let us observe that it would be enough to verify that the history function $g \colon C_b(\mathbb{R}_+; L^1(\Omega)) \to L^1(\Omega)$, corresponding to the T-periodicity condition and defined by

$$g(u) = u(T)$$

for each $u \in C_b(\mathbb{R}_+; L^1(\Omega))$, satisfies (g_1) and (g_4) in Theorem 9.8.1. Since this is an easy exercise, the proof of Theorem 9.8.2 is complete. $\qquad\square$

9.9 Bibliographical notes and comments

The viability problem for fully nonlinear evolution equations of the form $u'(t) \in Au(t) + f(t, u(t))$, with A m-dissipative and $f \colon I \times K \to X$ merely continuous, was considered for the first time by Vrabie [248]. He introduced the suitable tangency condition to apply also for points of the locally closed set $K \subseteq \overline{D(A)}$, which do not belong to $D(A)$, i.e.,

$$\lim_{h \downarrow 0} \frac{1}{h} \operatorname{dist} \left(u(t + h, t, \xi, f(t, \xi)); K \right) = 0, \tag{9.9.1}$$

for each $(t, \xi) \in I \times K$. He proved that if A is the infinitesimal generator of a compact semigroup of contractions and (9.9.1) holds *uniformly* with respect to $(t, \xi) \in I \times K$, then $I \times K$ is C^0-viable with respect to $A + f$. We emphasize that, whenever A is linear, (9.9.1) is equivalent to

$$\lim_{h \downarrow 0} \frac{1}{h} \operatorname{dist} \left(S(h)\xi + \int_{\tau}^{\tau+h} S(\tau + h - s) f(\tau, \xi) \, ds; K \right) = 0$$

Viability for Nonlocal Evolution Inclusions 337

which, in turn, reduces to the tangency condition introduced by Pavel [207]. Subsequent contributions in this context are due to Bothe [42], who allowed K to depend on t as well. In particular, in the case when K is independent of t, Bothe [42] showed that (9.9.1) with "lim inf" instead of "lim" is necessary and sufficient for viability. An application to reaction–diffusion systems was obtained by Necula and Vrabie [196].

The case of evolution inclusions without delay, i.e., in the simplest case when $\tau = 0$, was studied by Bothe [41]. Similar problems were previously considered by Kryszewski [159], Necula, Popescu and Vrabie [192] and [193]. The main tools used by the last three authors were the concepts of both tangent set and quasi-tangent set – introduced and studied by Cârjă, Necula and Vrabie [74], [75], [76] and [77]. For viability results referring to delay evolution equations and inclusions, we mention the pioneering papers of Pavel and Iacob [209] and Haddad [133]. For related results, see Gavioli and Malaguti [124], Lakshmikantham, Leela and Moauro [164], Leela and Moauro [168], Lupulescu and Necula [178], Ruess [229] and Ghavidel [125]. The semilinear case was very recently considered by Necula and Popescu [191]. Extensions to the fully nonlinear case of the results there obtained are due to Necula, Popescu and Vrabie [194]. Viability results for abstract multi-valued reaction–diffusion systems were obtained by Burlică [55].

Section 9.1 If K is a constant multifunction, the set E is right-quasi-tangent to $\mathcal{K}$ at (σ, φ) if and only if it is A-quasi-tangent to K at $\xi = \varphi(0)$ in the sense of Cârjă, Necula and Vrabie [74, Definition 11.1.3, p. 224], i.e.,

$$\liminf_{h\downarrow 0} \frac{1}{h}\text{dist}\left(u(\sigma + h, \sigma, \varphi(0), \mathcal{F}_E), K\right) = 0.$$

The latter concept has been introduced as a natural substitute of the notion of tangent vector in the sense of Bouligand and Severi simply because, in infinite dimensional Banach spaces, there are sets E which are 0-quasi-tangent to K at $\xi \in K$ but nevertheless they do not contain tangent vectors, $\eta \in E$, in the sense of Bouligand and Severi, i.e., satisfying

$$\liminf_{h\downarrow 0} \frac{1}{h}\text{dist}\left(\xi + h\eta, K\right) = 0.$$

See Cârjă, Necula and Vrabie [74, Definition 2.4.1, p. 35 and Example 2.4.1, p. 36]. The interesting features of tangent sets and A-quasi-tangent sets are twofold. First, they can be used in a very similar manner as tangent vectors to get viability results. Second, the tangency conditions expressed by means of tangent sets and A-quasi-tangent sets are rather easy to verify in concrete applications. See Sections 9.4 and 9.8.

Section 9.2 Lemma 9.2.1 was proved in Necula and Popescu [191]. The main necessary conditions for C^0-viability, i.e., Theorems 9.2.1 and 9.2.2 are also from Necula, Popescu and Vrabie [191] being extensions of previous results in Necula, Popescu and Vrabie [192] and [193], to the delay evolution

338   *Delay Differential Evolutions Subjected to Nonlocal Initial Conditions*

inclusions. The importance of these results rests in the fact that they completely close the viability problem once the tangency condition is proved to be sufficient in certain important cases such as those analyzed in Section 9.3.

Section 9.3 Lemma 9.3.1 is inspired by Cârjă and Vrabie [78] and is a typical tool of obtaining approximate C^0-solutions in order to prove viability results. Theorem 9.3.1 is from Necula, Popescu and Vrabie [194]. Previous viability results of a closed subset $\mathcal{K}_0$ of

$$\mathcal{K} = \{\varphi \in \mathcal{X};\ \varphi(0) \in \overline{D(A)} \cap K\},$$

where $K \subseteq X$ is closed, with $K \cap D(A) \neq \emptyset$, with respect to $A + f$, with f not depending on t, have been obtained by Ruess [229], by using a tangency condition of the form

$$\liminf_{h \downarrow 0} \frac{1}{h} \text{dist}(\varphi(0) + hf(\varphi); (I + hA)(D(A) \cap K)) = 0$$

for each $\varphi \in \mathcal{K}_0$, suitable for the use of ε-discrete schemes in order to obtain approximate solutions. It should be noted that Ruess [229] also considers the case of infinite delay, i.e., when $\tau = +\infty$. For related results see the reference list in Ruess [229] and the thesis of Ghavidel [125].

Section 9.4 Theorem 9.4.1 and Corollary 9.4.1 are from Necula, Popescu and Vrabie [194] and are simply "delay" versions of previous results of Cârjă, Necula and Vrabie [75, Theorem 12.1 and Corollary 12.1] showing that viability is a very efficient tool in obtaining null controllability conditions.

Section 9.5 Theorem 9.5.1 is from Necula, Popescu and Vrabie [195] and, to our knowledge, this is the first viability result referring to nonlocal initial-value problems for delay evolution equations. Theorem 9.5.2 is an immediate consequence of Theorem 9.3.1 but is important by itself.

Section 9.6 Theorem 9.6.1 is also from Necula, Popescu and Vrabie [195].

Section 9.7 The proof of Theorem 9.5.1 is essentially the same as that given in Necula, Popescu and Vrabie [195].

Section 9.8 Theorem 9.8.1 is from Necula, Popescu and Vrabie [195] and it is inspired by Theorem 11.1 in Cârjă, Necula and Vrabie [75]. It turn, the latter is a comparison result referring to classical Cauchy problems governed by multivalued perturbations of m-dissipative operators. Theorem 9.8.2 appears for the first time here.

Bibliography

[1] R. A. Adams, *Sobolev Spaces*, Academic Press, Boston San Diego New York London Sidney Tokyo Toronto, 1978.

[2] S. Aizicovici and H. Lee, Nonlinear nonlocal Cauchy problems in Banach spaces, *Appl. Math. Lett.*, **18** (2005), 401–407.

[3] S. Aizicovici and M. McKibben, Existence results for a class of abstract nonlocal Cauchy problems, *Nonlinear Anal.*, **39** (2000), 649–668.

[4] S. Aizicovici, N. S. Papageorgiou and V. Staicu, Periodic solutions of nonlinear evolution inclusions in Banach spaces, *J. Nonlinear Convex Anal.*, **7** (2006), 163–177.

[5] S. Aizicovici, N. H. Pavel and I. I. Vrabie, Anti-periodic solutions to strongly nonlinear evolution equations in Hilbert spaces, *An. Ştiinţ. Univ. Al. I. Cuza Iaşi (N.S.)*, **XLIV** (1998), 227–234.

[6] S. Aizicovici and V. Staicu, Multivalued evolution equations with nonlocal initial conditions in Banach spaces, *NoDEA Nonlinear Differential Equations Appl.*, **14** (2007), 361–376.

[7] R. A. Al-Omair and A. G. Ibrahim, Existence of mild solutions of a semilinear evolution differential inclusions with nonlocal conditions, *Electronic Journal of Differential Equations*, 2009, No. 42, 11 pp.

[8] L. Amerio and G. Prouse, *Almost Periodic Functions and Functional Equations*, Van Nostrand Reinhold Company, 1971.

[9] J. Andres, Topological principles for ordinary differential equations, in *Handbook of Differential Equations: Ordinary Differential Equations*, Vol. III, 1–101, A. Cañada, P. Drábek and A. Fonda Editors, Elsevier/North-Holland, Amsterdam, 2006.

[10] J. Andres, *Periodic-Type Solutions of Differential Inclusions*, Advances in Mathematical Research, Vol. 8 (A. R. Baswell, Editor), Nova Sciences Publishers, New York, 2009, 295–353.

[11] J. Appell, E. De Pascale, Nguyêñ Hông Thái and P. P. Zabreĭko, *Multi-Valued Superpositions*, Dissertationes Mathematicae, Polska Akademia Nauk, Instytut Matematyczny, **CCCXLY**, Warszawa, 1995.

Bibliography

[12] O. Arino, S Gautier and J. P. Penot, A fixed point theorem for sequentially continuous mappings with applications to ordinary differential equations, *Funkcial. Ekvac.*, **27** (1984), 273–279.

[13] J. P. Aubin and A. Cellina, *Differential Inclusions*, Springer-Verlag, Berlin Heidelberg New York Tokyo, 1984.

[14] G. Avalishvili and M. Avalishvili, On time nonlocal problems for some nonlinear evolution equations, *Bull. Georgian Natl. Acad. Sci. (N.S.)*, **2** (2008), 14–18.

[15] G. Avalishvili and M. Avalishvili, Nonclassical problems with nonlocal initial conditions for abstract second-order evolution equations, *Bull. Georgian Natl. Acad. Sci. (N.S.)*, **5** (2011), 17–24.

[16] M. Badii, J. I. Díaz and A. Tesei, Existence and attractivity results for a class of degenerate functional parabolic problems, *Rend. Semin. Mat. Univ. Padova*, **78** (1987), 109–124.

[17] K. Balachandran and J. Y. Park, Existence of a mild solution of a functional integrodifferential equation with nonlocal condition, *Bull. Korean Math. Soc.*, **38** (2001), 175–182.

[18] K. Balachandran and F. P. Samuel, Existence of solutions for quasilinear delay integrodifferential equations with nonlocal conditions, *Electronic Journal of Differential Equations*, 2009, No. 6, 7 pp.

[19] P. Baras, Compacité de l'opérateur $f \mapsto u$ solution d'une équation d'évolution non linéaire $(du/dt) + Au \ni f$, *C. R. Math. Acad. Sci. Paris*, **286** (1978), 1113–1116.

[20] P. Baras, J. C. Hassan and L. Veron, Compacité de l'opérateur définissant la solution d'une équation d'évolution non homogène, *C. R. Math. Acad. Sci. Paris*, **284** (1977), 779–802.

[21] V. Barbu, *Nonlinear Semigroups and Differential Equations in Banach Spaces*, Noordhoff, Leyden, 1976.

[22] V. Barbu, *Nonlinear Differential Equations of Monotone Type in Banach Spaces*, Springer Monographs in Mathematics, Springer-Verlag, 2010.

[23] M. M. Basova and V. V. Obukhovskii, On some boundary-value problems for functional-differential inclusions in Banach spaces, *J. Math. Sci.*, **149** (2008), 1376–1384.

[24] A. Bátkai and S. Piazzera, *Semigroups for Delay Equations*, Research Notes in Mathematics, 10. A. K. Peters, Ltd., Wellesley, MA, 2005.

[25] R. I. Becker, Periodic solutions of semilinear equations of evolution of compact type, *J. Math. Anal. Appl.*, **82** (1981), 33–48.

Bibliography   341

[26] R. Bellman, The stability of solutions of linear differential equations, *Duke Math. J.*, **10** (1943), 643–647.

[27] H. Benabdellah and C. Castaing, Weak compactness criteria and convergences in $L_E^1(\mu)$, *Collect. Math.*, **48** (1997), 423–448.

[28] M. Benchohra and S. Abbas, *Advances Functional Evolution Equations and Inclusions*, Springer, 2015.

[29] I. Benedetti, L. Malaguti and V. Taddei, Nonlocal semilinear evolution equations without strong compactness: Theory and applications, *Bound. Value Probl.*, 2013, **2013**: 60.

[30] I. Benedetti, L. Malaguti, V. Taddei and I. I. Vrabie, Semilinear delay evolution equations with measures subjected to nonlocal initial conditions, *Ann. Mat. Pura Appl.*, (2015) DOI 10.1007/s10231-015-0535-6.

[31] I. Benedetti and P. Rubbioni, Existence of solutions on compact and non-compact intervals for semilinear impulsive differential inclusions with delay, *Topol. Methods Nonlinear Anal.*, **32** (2008), 227–245.

[32] P. Benilan, *Équations d'évolution dans un espace de Banach quelconque et applications*, Thèse, Orsay (1972).

[33] P. Benilan and H. Brezis, Solutions faibles d'équations d'évolution dans les espaces de Hilbert, *Ann. Inst. Fourier (Grenoble)*, **22** (1972), 311–329.

[34] C. Berge, *Espaces Topologiques. Fonctions Multivoques, Collection Universitaire de Mathématiques*, Vol. III Dunod, Paris 1959, 272 pp.

[35] A. Bielecki, Une remarque sur la méthode de Banach–Cacciopoli–Tikhonov dans la théorie des équations différentielles ordinaires, *Bull. Pol. Acad. Sci. Math.*, Cl. III, **4** (1956), 261–264.

[36] S. Bochner, Beiträge zur Theorie der fastperiodischen Funktionen I, Funktionen einer Variablen, *Math. Ann.*, **96** (1927), 119–147.

[37] S. Bochner, Integration von Funktionen, deren Werte die Elemente eines Vektoraumes sind, *Fund. Math.*, **20** (1933), 262–276.

[38] S. Bochner and A. E. Taylor, Linear functionals on certain spaces of abstractly valued functions, *Ann. of Math.* (2), **39** (1938), 913–944.

[39] H. Bohr, Zur Theorie der fastperiodischen Funktionen I, *Acta Math.*, **45**, (1925), 29–127.

[40] O. Bolojan-Nica, G. Infante and R. Precup, Existence results for systems with coupled nonlocal initial conditions, *Nonlinear Anal.*, **94** (2014), 231–242.

342 Bibliography

[41] D. Bothe, Multivalued differential equations on graphs, *Nonlinear Anal.*, **18** (1992), 245–252.

[42] D. Bothe, Flow invariance for perturbed nonlinear evolution equations, *Abstr. Appl. Anal.*, **1** (1996), 417–433.

[43] D. Bothe, Multivalued perturbations of m-accretive differential inclusions, *Israel J. Math.*, **108** (1998), 109–138.

[44] A. Boucherif and R. Precup, Semilinear evolution equations with nonlocal initial conditions, *Dynam. Systems Appl.*, **16** (2007), 507–516.

[45] G. Bouligand, Sur la semi-continuité d'inclusions et quelques sujets connexes, *Ens. Math.*, **31** (1932), 14–22.

[46] F. Brauer and C. Castillo-Chavez, *Mathematical Models in Population Biology and Epidemiology*, second edition, Texts in Applied Mathematics **40**, 2012.

[47] D. W. Brewer, A nonlinear semigroup for a functional differential equation, *Trans. Amer. Math. Soc.*, **236** (1978), 173–191.

[48] D. W. Brewer, The asymptotic stability of a nonlinear functional differential equation of infinite delay, *Houston J. Math.*, **6** (1980), 321–330.

[49] H. Brezis, *Opérateurs Maximaux Monotones et Semi-Groupes de Contractions dans un Espace de Hilbert*, North Holland, 1973.

[50] H. Brezis, New results concerning monotone operators and nonlinear semigroups, *Analysis of Nonlinear Problems*, RIMS Kôkyûroku Bessatsu, (1974), 2–27 .

[51] H. Brezis, Propriétés régularisantes de certains semi-groupes non linéaires, *Israel J. Math.*, **9** (1971), 513–534.

[52] H. Brezis and F. E. Browder, A general principle on ordered sets in nonlinear functional analysis, *Adv. in Mathematics*, **21** (1976), 355–364.

[53] H. Brezis and W. A. Strauss, Semi-linear second-order elliptic equations in L^1, *J. Math. Soc. Japan*, **25** (1973), 565–590.

[54] F. E. Browder, Nonlinear equations of evolution, *Ann. of Math.*, **80** (1964), 485–523.

[55] M. D. Burlică, Viability for multi-valued semilinear reaction–diffusion systems, *Ann. Acad. Rom. Sci. Ser. Math.*, **2** (2010), 3–24.

[56] M. D. Burlică and D. Roşu, A viability result for semilinear reaction–diffusion systems, *An. Ştiinţ. Univ. Al. I. Cuza Iaşi (N.S.)*, **LIV** (2008), 361–382.

Bibliography

[57] M. D. Burlică and D. Roşu, The initial value and the periodic problems for a class of reaction–diffusion systems, *Dyn. Contin. Discrete Impuls. Syst. Ser. A Math. Anal.*, **15** (2008), 427–444.

[58] M. D. Burlică and D. Roşu, A class of nonlinear delay evolution equations with nonlocal initial conditions, *Proc. Amer. Math. Soc.*, **142** (2014), 2445–2458.

[59] M. D. Burlică and D. Roşu, A class of reaction–diffusion systems with nonlocal initial conditions, *An. Ştiinţ. Univ. Al. I. Cuza Iaşi (N.S.)*, **LXI** (2015), 59–78.

[60] M. D. Burlică and D. Roşu, Nonlinear delay reaction–diffusion systems with nonlocal initial conditions having affine growth, *Topol. Methods Nonlinear Anal.*, DOI: 10.12775/TMNA.2016.027.

[61] M. D. Burlică, D. Roşu and I. I. Vrabie, Continuity with respect to the data for a delay evolution equation with nonlocal initial conditions, *Libertas Math. (New series)*, **32** (2012), 37–48.

[62] M. D. Burlică, D. Roşu and I. I. Vrabie, Abstract reaction–diffusion systems with nonlocal initial conditions, *Nonlinear Anal.*, **94** (2014), 107–119.

[63] L. Byszewski, Theorem about existence and uniqueness of continuous solution of nonlocal problem for nonlinear hyperbolic equation, *Appl. Anal.*, **40** (1991), 173–180.

[64] L. Byszewski, Theorems about the existence and uniqueness of solutions of a semilinear evolution nonlocal Cauchy problem, *J. Math. Anal. Appl.*, **162** (1991), 494–505.

[65] L. Byszewski, *Differential and Functional-Differential Problems Together with Nonlocal Conditions*, Cracow University of Technology, Cracow, Poland 1995.

[66] L. Byszewski, On some application of the Bochenek theorem, *Univ. Iagel. Acta Math.*, **45** (2007), 147–153.

[67] L. Byszewski and H. Akca, On a mild solution of a semilinear functional-differential evolution nonlocal problem, *J. Appl. Math. Stoch. Anal.*, **10** (1997), 265–271.

[68] L. Byszewski and V. Lakshmikantham, Theorem about the existence and uniqueness of a solution of a nonlocal abstract Cauchy problem in a Banach space, *Appl. Anal.*, **40** (1991), 11–19.

[69] T. Cardinali, R. Precup, and P. Rubbioni, A unified existence theory for evolution equations and systems under nonlocal conditions, *arXiv*:1406.6825v1 [math.CA].

344 *Bibliography*

[70] T. Cardinali and P. Rubbioni, Impulsive mild solutions for semilinear differential inclusions with nonlocal conditions in Banach spaces, *Nonlinear Anal. T.M.A.*, **75** (2012), 871–879.

[71] C. Castaing and D. P. Monteiro-Marques, Periodic solutions of evolution problems associated with a moving convex set, *C. R. Math. Acad. Sci. Paris*, **321** (1995), 531–536.

[72] C. Castaing and M. Valadier, *Convex Analysis and Measurable Multifunctions*, Lecture Notes in Mathematics, **580**, Springer-Verlag, 1977.

[73] R. Caşcaval and I. I. Vrabie, Existence of periodic solutions for a class of nonlinear evolution equations, *Rev. Mat. Complut.*, **7** (1994), 325–338.

[74] O. Cârjă, M. Necula and I. I. Vrabie, *Viability, Invariance and Applications*, Elsevier Horth-Holland Mathematics Studies **207**, 2007.

[75] O. Cârjă, M. Necula and I. I. Vrabie, Necessary and sufficient conditions for viability for nonlinear evolution inclusions, *Set-Valued Anal.*, **16** (2008), 701–731.

[76] O. Cârjă, M. Necula and I. I. Vrabie, Necessary and sufficient conditions for viability for semilinear differential inclusions, *Trans. Amer. Math. Soc.*, **361** (2009), 343–390.

[77] O. Cârjă, M. Necula and I. I. Vrabie, Tangent sets, viability for differential inclusions and applications, *Nonlinear Anal.*, **71** (2009), e979–e990.

[78] O. Cârjă and I. I. Vrabie, Some new viability results for semilinear differential inclusions, *NoDEA Nonlinear Differential Equations Appl.*, **4** (1997), 401–424.

[79] M. Chandrasekaran, Nonlocal Cauchy problem for quasilinear integrodifferential equations in Banach spaces, *Electronic Journal of Differential Equations*, 2007, No. 33, 6 pp.

[80] J.-C. Chang and H. Liu, Existence of solutions for a class of neutral partial differential equations with nonlocal conditions in the α-norm, *Nonlinear Analysis*, **71** (2009), 3759–3768.

[81] D. H. Chen, R. N. Wang and Y. Zhou, Nonlinear evolution inclusions: Topological characterizations of solution sets and applications, *J. Funct. Anal.*, **265** (2013), 2039–2073.

[82] I. Ciorănescu, *Aplicaţii de Dualitate în Analiza Funcţională Neliniară*, Editura Academiei, Bucureşti, 1974. (Romanian)

[83] J. A. Clarkson, Uniformly convex spaces, *Trans. Amer. Math. Soc.*, **40** (1936), 396–414.

Bibliography

[84] C. Corduneanu, *Almost-Periodic Functions*, Interscience Tracts in Pure and Applied Mathematics, No. 22. Interscience Publishers [John Wiley & Sons], New York-London-Sydney, 1968.

[85] C. Corduneanu, *Functional Equations with Causal Operators*, Taylor & Francis, 2002.

[86] C. Corduneanu, *Almost Periodic Oscillations and Waves*, Springer, 2009.

[87] C. Corduneanu and V. Lakshmikantham, Equations with unbounded delay: survey, *Nonlinear Anal.*, **4** (1980), 831-877.

[88] M. G. Crandall, Nonlinear semigroups and evolution governed by accretive operators, in *Nonlinear Functional Analysis and Its Applications, F. E. Browder editor, Proc. Sympos. Pure. Math.*, **45-1**, 1986, 305–337.

[89] M. G. Crandall and L. C. Evans, On the relation of the operator $\partial/\partial s + \partial/\partial\tau$ to evolution governed by accretive operators, *Israel J. Math.*, **21** (1975), 261–278.

[90] M. G. Crandall and T. M. Liggett, Generation of semi-groups of nonlinear transformations in general Banach spaces, *Amer. J. Math.*, **93** (1971), 265–298.

[91] M. G. Crandall and A. Pazy, Nonlinear evolution equations in Banach spaces, *Israel J. Math.*, **11** (1972), 57–94.

[92] R. F. Curtain and H. J. Zwart, *An Introduction to Infinite-Dimensional Linear Systems Theory.* Texts in Applied Mathematics, 21. Springer-Verlag, New York, 1995.

[93] A. Ćwiszewski and P. Kokocki, Periodic solutions for nonlinear hyperbolic evolution systems, *J. Evol. Equ.*, **10** (2010), 677–710.

[94] M. Damak, K. Ezzinbi and L. Souden, Weighted pseudo-almost periodic solutions for some neutral partial functional differential equations, *Electron. J. of Differential Equations*, 2012, No. 47, 13 pp.

[95] K. Deng, Exponential decay of solutions of semilinear parabolic equations with nonlocal initial conditions, *J. Math. Anal. Appl.*, **179** (1993), 630–637.

[96] J. I. Díaz and I. I. Vrabie, Propriétes de compacité de l'opérateur de Green généralisé pour l'équation des millieux poreux, *C. R. Math. Acad. Sci. Paris*, **309** (1989), 221–223.

[97] J. I. Díaz and I. I. Vrabie, Existence for reaction diffusion systems. A compactness method approach, *J. Math. Anal. Appl.*, **188** (1994), 521–540.

346 *Bibliography*

[98] J. I. Díaz and I. I. Vrabie, Compactness of the Green operator of nonlinear diffusion equations: Application to Boussinesq type systems in fluid mechanics, *Topol. Methods Nonlinear Anal.*, **4** (1994), 399–416.

[99] J. Diestel, Remarks on weak compactness in $L_1(\mu; X)$, *Glasg. Math. J.*, **18** (1977), 87–91.

[100] J. Diestel, W. M. Ruess and W. Schachermayer, On weak compactness in $L^1(\mu; X)$, *Proc. Amer. Math. Soc.*, **118** (1993), 447–453.

[101] J. Diestel and J. J. Uhl Jr., *Vector Measures*, Mathematical Surveys, Amer. Math. Soc., **15**, 1977.

[102] N. Dinculeanu, *Vector Measures*, International Series of Monographs in Pure and Applied Mathematics, Vol. 95 Pergamon Press, Oxford-New York-Toronto, Ont.; VEB Deutscher Verlag der Wissenschaften, Berlin 1967.

[103] R. D. Driver, *Ordinary and Delay Differential Equations*, Applied Mathematical Sciences, **20**, Springer-Verlag, New York-Heidelberg, 1977.

[104] N. Dunford, A mean ergodic theorem, *Duke Math. J.*, **5** (1939), 635–646.

[105] N. Dunford and J. T. Schwartz, *Linear Operators Part I: General Theory*, Interscience Publishers, Inc. New York, 1958.

[106] J. Dyson and R. Villella Bressan, Functional differential equations and non-linear evolution operators, *Proc. Royal Soc. Edinburgh*, **75A** (1975/76), 223–234.

[107] J. Dyson and R. Villella Bressan, Some remarks on the asymptotic behavior of a nonautonomous, nonlinear functional differential equation, *J. Differential Equations*, **25** (1977), 275–287.

[108] J. Dyson, R. Villella Bressan and G. F. Webb, A singular transport equation modelling a proliferating maturity structured cell population, *Canad. Appl. Math. Quart.*, **4** (1996), 65–95.

[109] W. F. Eberlein, Weak compactness in Banach spaces, I, *Proc. Natl, Acad. Sci. USA*, **33** (1947), 51–53.

[110] R. E. Edwards, *Functional Analysis. Theory and Applications*. Holt, Rinehart and Winston, New York-Toronto-London 1965.

[111] L. C. Evans, *Partial Differential Equations*, second edition, Graduate Studies in Mathematics, American Mathematical Society, **19**, 2010.

[112] Zhenbin Fan, Qixiang Dong and Gang Li, Semilinear differential equations with nonlocal conditions in Banach spaces, *Int. J. Nonlinear Sci.*, **2** (2006), 131–139.

Bibliography 347

[113] Ky Fan, Fixed-point and minimax theorems in locally convex topological linear spaces, *Proc. Natl. Acad. Sci. USA*, **38** (1952), 121–126.

[114] P. Fatou, Séries trigonométriques et séries de Taylor, *Acta Math.*, **30** (1906), 335–400.

[115] W. Feller, On the generation of unbounded semi-groups of bounded linear operators, *Ann. of Math.*, **58** (1953), 166–174.

[116] W. E. Fitzgibbon, Semilinear functional differential equations in Banach space, *J. Differential Equations*, **29** (1978), 1–14.

[117] H. Flaschka and M. Leitman, On semigroups of nonlinear operators and the solution of the functional differential equation $x'(t) = F(x_t)$, *J. Math. Anal. Appl.*, **49** (1975), 649–658.

[118] M. Fréchet, Les fonctions asymptotiquement presque-périodiques continues, *C.R. Acad. Sci. Paris*, **213** (1941), 520–522.

[119] M. Fréchet, Les fonctions asymptotiquement presque-périodiques, *Rev. Sci.*, **79** (1941), 341–354.

[120] G. Fubini, Sugli integrali multipli, *Rend. dellÁcc. naz. dei Lincei, classe di scienze fisiche, mat. e natur.*, **XVII** (1907), 608–614.

[121] E. Gagliardo, Proprietà di alcune classi di funzioni in più variabili, *Ric. Mat.*, **7** (1958), 102–137.

[122] J. García–Falset, Existence results and asymptotic behaviour for nonlocal abstract Cauchy problems, *J. Math. Anal. Appl.*, **338** (2008), 639–652.

[123] J. García–Falset and S. Reich, Integral solutions to a class of nonlocal evolution equations, *Comm. Contemp. Math.*, **12** (2010), 1032–1054.

[124] A. Gavioli and L. Malaguti, Viable solutions of differential inclusions with memory in Banach spaces, *Portugal. Math.*, **57** (2000), 203–217.

[125] S. M. Ghavidel, *Existence and flow-invariance of solutions to non-autonomous partial differential delay equations*, Dissertation, Universität Duisburg-Essen, Essen, 2007.

[126] S. M. Ghavidel, Flow invariance for solutions to nonlinear nonautonomous partial differential delay equations, *J. Math. Anal. Appl.*, **345** (2008), 854–870.

[127] I. L. Glicksberg, A further generalization of the Kakutani fixed point theorem, with application to Nash equilibrium points, *Proc. Amer. Math. Soc.*, **3** (1952), 170–174.

[128] K. Gopalsamy and B. G. Zhang, On a neutral delay logistic equation, *Dyn. Stab. Syst.*, **2** (1988), 183–195.

348   *Bibliography*

[129] D. G. Gordeziani, On some initial conditions for parabolic equations, *Rep. of Enlarged Sess. of the Sem. of I. Vekua Inst. Appl. Math.*, **4** (1989), 57–60.

[130] D. Gordeziani, M. Avalishvili and G. Avalishvili, On the investigation of one nonclassical problem for Navier-Stokes equations, *Appl. Math. Inform.*, **7** (2002), 66–77.

[131] L. Górniewicz, *Topological Fixed Point Theory of Multivalued Mappings*, second edition, Topological Fixed-Point Theory and Its Applications, Volume 4, Springer-Verlag, 2006.

[132] A. Granas and J. Dugundji, *Fixed Point Theory*, Springer Monographs in Mathematics, Springer-Verlag, New York, 2003.

[133] G. Haddad, Monotone trajectories of differential inclusions and functional-differential inclusions with memory, *Israel J. Math.*, **39** (1981), 83–100.

[134] A. Halanay, *Differential Equations: Stability, Oscillations, Time Lags*, Academic Press, New York-London 1966.

[135] J. Hale, *Functional Differential Equations*, Applied Mathematical Sciences 3, Springer-Verlag, New York-Heidelberg, 1971.

[136] J. Hale, *Theory of Functional Differential Equations*, Applied Mathematical Sciences 3, Springer-Verlag, New York-Heidelberg, 1977.

[137] P. Halmos, How to write Mathematics, *L'Enseignement Mathématique*, **16** (1970), 123–152.

[138] F. Hausdorff, *Grundzüge der Mengenlehre*, Verlag von Veit&Comp. in Leipzig, 1914.

[139] H. R. Henríquez and C. Lizama, Periodic solutions of abstract functional differential equations with infinite delay, *Nonlinear Anal.*, **75** (2012), 2016–2023.

[140] H. R. Henríquez and C. H. Vásquez, Almost periodic solutions of abstract retarded functional differential equations with unbounded delay, *Acta Appl. Math.*, **57** (1999), 105–132.

[141] E. Hernández and M. L. Pelicer, Asymptotically almost periodic and almost periodic solutions for partial neutral differential equations, *Appl. Math. Lett.*, **18** (2005), 1265–1272.

[142] E. Hille, *Functional Analysis and Semi-Groups*, American Mathematical Society, New York, 1948.

[143] E. Hille and R. S. Phillips, *Functional Analysis and Semi-Groups*, Amer. Math. Soc. Colloq. Publ., **31**, Fourth Printing of Revised Edition, 1981.

Bibliography

[144] Y. Hino, S. Murakami and T. Naito, *Functional-Differential Equations with Infinite Delay*, Springer-Verlag, Berlin, 1991.

[145] Y. Hino, S. Murakami and T. Yoshizawa, Existence of almost periodic solutions of some functional-differential equations with infinite delay in a Banach space, *Tôhoku Math. J.*, (2) **49** (1997), 133–147.

[146] N. Hirano, Existence of periodic solutions for nonlinear evolution equations in Hilbert spaces, *Proc. Amer. Math. Soc.*, **120** (1994), 185–192.

[147] N. Hirano and N. Shioji, Invariant sets for nonlinear evolution equations, Cauchy problems and periodic problems, *Abstr. Appl. Anal.*, **2004** (2004), 183–203.

[148] B. Jia, L. Erbe and R. Mert, A Halanay-type inequality on time scales in higher dimensional spaces, *Math. Inequal. Appl.*, **17** (2014), 813–821.

[149] Jin-Mun Jeong, Dong-Hwa Kim and Jong-Yeoul Park, Nonlinear variational evolution inequalities with nonlocal conditions, *J. Korean Math. Soc.*, **41** (2004), 647–665.

[150] S. Kakutani, Weak topology, bicompact sets, and the principle of duality, *Proc. Imp. Acad. Tokyo,* **14** (1940), 63–67.

[151] S. Kakutani, A generalization of Brower's fixed point theorem, *Duke Math. J.*, **8** (1941), 457–459.

[152] T. Kato, Nonlinear semigroups and evolution equations, *J. Math. Soc. Japan.*, **19** (1967), 508–520.

[153] Tran Dinh Ke, Cauchy problems for functional evolution inclusions involving accretive operators, *Electronic Journal of Qualitative Theory of Differential Equations*, 2013, No. 75, 13 pp.

[154] Tran Dinh Ke, V. Obukhovskii, Ngai-Ching Wong and Jen-Chih Yao, On semilinear integro-differential equations with nonlocal conditions in Banach spaces, *Abstract and Applied Analysis*, **2012** (2012), Article ID 137576, 26 pages, doi:10.1155/2012/137576.

[155] K. Kobayasi, Y. Kobayashi and S. Oharu, Nonlinear evolution operators in Banach spaces, *Osaka J. Math.*, **21** (1984), 281–310.

[156] Y. Kobayashi, Difference approximation of Cauchy problems for quasi-dissipative operators and generation of nonlinear semigroups, *J. Math. Soc. Japan*, **27** (1975), 640–665.

[157] V. I. Kondrašov, On some properties of functions in the space L_p^ν, *Dokl. Akad. Nauk SSSR*, **48** (1945), 563–566. (Russian)

[158] M. Krein and V. Šmulian, On regulary convex sets in the space conjugate to a Banach space, *Ann. Math.*, **41** (1940), 556–583.

350 Bibliography

[159] W. Kryszewski, Topological structure of solution sets of differential inclusions: The constrained case, *Abstract and Applied Analysis*, **6** (2003), 325–351.

[160] W. Kryszewski and S. Plaskacz, Periodic solutions to impulsive differential inclusions with constraints, *Nonlinear Anal.*, **65** (2006), 1794–1804.

[161] Y. Kuang, *Delay Differential Equations with Applications in Population Dynamics*, Academic Press, Boston, 1993.

[162] C. Kuratowski, Les fonctions semi-continues dans éspace des ensembles fermés, *Fund. Math.*, **18** (1932), 148–159.

[163] V. Lakshmikantham and S. Leela, *Nonlinear Differential Equations in Abstract Spaces*, International Series in Nonlinear Mathematics, **2**, Pergamon Press, 1981.

[164] V. Lakshmikantham, S. Leela and V. Moauro, Existence and uniqueness of solutions of delay differential equations on a closed subset of a Banach space, *Nonlinear Anal.*, **2** (1978), 311–327.

[165] V. Lakshmikantham and N. S. Papageorgiou, Periodic solutions of nonlinear evolution inclusions, *Comput. Appl. Math.*, **52** (1994), 277–286.

[166] I. Lasiecka and R. Triggiani, *Control Theory for Partial Differential Equa tions: Continuous and Approximation Theories*, vol. **1** and **2**, Cambridge University Press, 2000.

[167] H. Lebesgue, Sur l'intégration des fonctions discontinues, *Annales scientifiques de l'École normale supérieure*, **27** (1910), 361–450.

[168] S. Leela and V. Moauro, Existence of solutions in a closed set for delay differential equations in Banach spaces *Nonlinear Anal.*, **2** (1978), 47–58.

[169] B. M. Levitan and V. V. Zhikov, *Almost Periodic Functions and Differential Equations*, Cambridge University Press, 1982.

[170] Yongxiang Li, Existence and uniqueness of periodic solution for a class of semilinear evolution equations, *J. Math. Anal. Appl.*, **349** (2009), 226–234.

[171] Yongxiang Li, Existence and asymptotic stability of periodic solution for evolution equations with delays, *J. Func. Anal.*, **261** (2011), 1309–1324.

[172] J. L. Lions, *Quelques méthodes de résolution des problèmes aux limites non Linéaires*, Dunod, Gauthier-Villars, Paris, 1969.

[173] Hsiang Liu and Jung-Chan Chang, Existence for a class of partial differential equations with nonlocal conditions, *Nonlinear Anal.*, **70** (2009), 3076–3083.

Bibliography

[174] A. Lorenzi and I. I. Vrabie, An identification problem for a semilinear evolution delay equation, *J. Inverse Ill-Posed Probl.*, **22** (2014), 209–244.

[175] A. Lorenzi and I. I. Vrabie, Identification of a source term in a semilinear evolution delay equation, *An. Ştiinţ. Univ. Al. I. Cuza Iaşi (N.S.)*, **LXI** (2015), 1–39.

[176] G. Lumer, Semi-inner-product spaces, *Trans. Amer. Math. Soc.*, **100** (1961), 29–43.

[177] G. Lumer and R. S. Phillips, Dissipative operators in a Banach space, *Pacific J. Math.*, **11** (1961), 679–698.

[178] V. Lupulescu and M. Necula, A viability result for nonconvex semilinear functional differential inclusions, *Discuss. Math. Differ. Incl. Control Optim.*, **25** (2005), 109–128.

[179] Md. Maqbul, Almost periodic solutions for neutral functional differential equations with Stepanov-almost periodic terms, *Electron. J. Differential Equations*, 2011, No. 72, 9 pp.

[180] V. G. Maz'ja, *Sobolev Spaces*. Translated from Russian by T. O. Shaposhnikova, Springer-Verlag, 1985.

[181] S. Mazur, Über convexe Mengen in linearen normierten Räumen, *Studia Math.*, **4** (1933), 70–84.

[182] M. McKibben, *Discovering Evolution Equations with Applications*. Vol. I *Deterministic Models*, Chapman & Hall/CRC Appl. Math. Nonlinear Sci. Ser., 2011.

[183] D. Milman, On some criteria for regularity of spaces of type (B), *Dokl. Akad. Nauk*, **20** (1938), 243–246.

[184] G. Minty, Monotone (nonlinear) operators in Hilbert spaces, *Duke Math. J.*, **29** (1962), 341–346.

[185] E. Mitidieri and I. I. Vrabie, Existence for nonlinear functional differential equations, *Hiroshima Math. J.*, **17** (1987), 627–649.

[186] E. Mitidieri and I. I. Vrabie, A class of strongly nonlinear functional differential equations, *Ann. Mat. Pura Appl.*, (4), **CLI** (1988), 125–147.

[187] I. Miyadera, Generation of a strongly continuous semi-groups of operators, *Tôhoku Math. J.*, **4** (1952), 109–114.

[188] I. Miyadera, *Nonlinear semigroups*, Transl. Math. Monogr., **109**, 1992.

[189] R. L. Moore, Concerning upper semi-continuous collections of continua, *Trans. Amer. Math. Soc.*, **27** (1925), 416–428.

Bibliography

[190] J. Moreau, Proximité et dualité dans un espace hilbertien, *Bull. Soc. Math. France*, **93** (1965), 273–299.

[191] M. Necula and M. Popescu, Viability of a time dependent closed set with respect to a semilinear delay evolution inclusion, *An. Științ. Univ. Al. I. Cuza Iași. Mat. (N.S.)*, **61** (2015), no. 1, 41–58.

[192] M. Necula, M. Popescu and I. I. Vrabie, Viability for differential inclusions on graphs, *Set-Valued Analysis*, **16** (2008), 961–981.

[193] M. Necula, M. Popescu and I. I. Vrabie, Evolution equations on locally closed graphs and applications, *Nonlinear Anal.*, **71** (2009), e2205–e2216.

[194] M. Necula, M. Popescu and I. I. Vrabie, Nonlinear delay evolution inclusions on graphs, *Proceedings of the IFIP TC7/2013 on System Modeling and Optimization, Klagenfurt*, Lecture Notes in Computer Science, Barbara Kaltenbacher, Clemens Heuberger, Christian Pötze and Franz Rendl Editors, 2014, 207–216.

[195] M. Necula, M. Popescu and I. I. Vrabie, Viability for delay evolution equations with nonlocal initial conditions, *Nonlinear Anal.*, **121** (2015), 164–172.

[196] M. Necula and I. I. Vrabie, A viability result for a class of fully nonlinear reaction–diffusion systems, *Nonlinear Anal.*, **69** (2008), 1732–1743.

[197] M. Necula and I. I. Vrabie, Nonlinear delay evolution inclusions with general nonlocal initial conditions, *Ann. Acad. Rom. Sci. Ser. Math.*, **7** (2015), 67–97.

[198] S. K. Ntouyas, Nonlocal initial and boundary value problems: A survey, *Handbook of Differential Equations: Ordinary Differential Equations*, **2**, 461–557, Elsevier B. V., Amsterdam, 2005.

[199] S. K. Ntouyas and P. Ch. Tsamatos, Global existence for semilinear evolution equations with nonlocal conditions, *J. Math. Anal. Appl.*, **210** (1997), 679–687.

[200] W. E. Olmstead and C. A. Roberts, The one-dimensional heat equation with a nonlocal initial condition, *Appl. Math. Lett.*, **10** (1997), 89–94.

[201] A. Paicu, Periodic solutions for a class of differential inclusions in general Banach spaces, *J. Math. Anal. Appl.*, **337** (2008), 1238–1248.

[202] A. Paicu, Periodic solutions for a class of nonlinear evolution equations in Banach spaces, *An. Științ. Univ. Al. I. Cuza Iași, (N.S.)*, **LV** (2009), 107–118.

[203] A. Paicu, *Probleme cu valori inițiale și periodice pentru ecuații de evoluție*, Doctoral Thesis, Iași, 2010. (Romanian)

Bibliography 353

[204] A. Paicu and I. I. Vrabie, A class of nonlinear evolution equations subjected to nonlocal initial conditions, *Nonlinear Anal.*, **72** (2010), 4091–4100.

[205] C. V. Pao, Reaction diffusion equations with nonlocal boundary and nonlocal initial conditions, *J. Math. Anal. Appl.*, **195** (1995), 702–718.

[206] N. S. Papageorgiou, Periodic trajectories for evolution inclusions associated with time dependent subdifferentials, *Ann. Univ. Sci. Budapest. Eötvös Sect. Math.*, **37** (1994), 139–155.

[207] N. H. Pavel, Nonlinear evolution equations governed by f-quasi-dissipative operators, *Nonlinear. Anal.*, **5** (1981), 449–468.

[208] N. H. Pavel, *Nonlinear Evolution Operators and Semigroups. Applications to Partial Differential Equations*, Lecture Notes in Math., **1260**, Springer-Verlag, Berlin, 1987.

[209] N. H. Pavel and F. Iacob, Invariant sets for a class of perturbed differential equations of retarded type, *Israel J. Math.*, **28** (1977), 254–264.

[210] A. Pazy, On the differentiability and compactness of semigroups of linear operators, *J. Math. Mech.*, **17** (1968), 1131–1141.

[211] A. Pazy, A class of semi-linear equations of evolution, *Israel J. Math.*, **20** (1975), 23–36.

[212] A. Pazy, *Semigroups of Linear Operators and Applications to Partial Differential Equations*, Applied Mathematical Sciences, **44**, Springer-Verlag, New York, 1983.

[213] G. Peano, Integrazione per le serie delle equationi differentiali lineari, *Atti. Reale Sci. Torino*, **22** (1887), 293–302.

[214] B. J. Pettis, On integration in vector spaces, *Trans. Amer. Math. Soc.*, **44** (1938), 277–304.

[215] B. J. Pettis, A note on regular Banach spaces, *Bull. Amer. Math. Soc. (N.S.)*, **44** (1938), 420–428.

[216] B. J. Pettis, A proof that every uniformly convex space is reflexive, *Duke Math. J.*, **5** (1939), 249–253.

[217] R. S. Phillips, On the generation of semigroups of linear operators, *Pacific J. Math.*, **2** (1952), 343–369.

[218] A. T. Plant, Nonlinear semigroups of translations in Banach space generated by functional differential equations, *J. Math. Anal. Appl.*, **60** (1977), 67–74.

Bibliography

[219] A. T. Plant, Stability of nonlinear functional differential equations using weighted norms, *Houston J. Math.*, **3** (1977), 99–108.

[220] J. Prüss, Periodic solutions of semilinear evolution equations, *Nonlinear Anal.*, **3** (1979), 601–612.

[221] F. Rabier, P. Courtier and M. Ehrendorfer, Four-dimensional data assimilation: Comparison of variational and sequential algorithms, *Quart. J. Roy. Meteorol. Sci.*, **118** (1992), 673–713.

[222] A. O. Rey and M. C. Mackey, Multistability and boundary layer development in a transport equation with delayed arguments, *Canad. Appl. Math. Quart.*, **1** (1993), 61–81.

[223] F. Rellich, Ein Satz über mittlere Konvergenz, *Nachrichten von der Gesellschaft der Wissenschaften zu Göttingen, Mathematisch-Physikalische Klasse* (1930), 30–35.

[224] D. Roşu, Viability for nonlinear multi-valued reaction–diffusion systems, *NoDEA Nonlinear Differential Equations Appl.*, **17** (2010), 479–496.

[225] D. Roşu, Viability for a nonlinear multi-valued system on locally closed graph, *An. Ştiinţ. Univ. Al. I. Cuza, Iaşi, (N.S.)*, **LVI** (2010), 343–362.

[226] T. Roubíček, *Nonlinear Partial Differential Equations with Applications*, International Series of Numerical Mathematics, **153**, Birkhäuser Verlag Basel, 2005.

[227] W. M. Ruess, The evolution operator approach to functional-differential equations with delay, *Proc. Amer. math. Soc.*, **119** (1993), 783–791.

[228] W. M. Ruess, Existence and stability of solutions to partial functional-differential equations with delay, *Adv. Differential Equations*, **4** (1999), 843–876.

[229] W. M. Ruess, Flow invariance for nonlinear partial differential delay equations, *Trans. Amer. Math. Soc.*, **361** (2009), 4367–4403.

[230] W. M. Ruess and W. H. Summers, Operator semigroups for functional-differential equations with delay, *Trans. Amer. Math. Soc.*, **341** (1994), 695–719.

[231] K. Sato, On the generators of non-negative contraction semigroups in Banach lattices, *J. Math. Soc. Japan*, **20** (1968), 423–436.

[232] J. Schauder, Der Fixpunktsatz in Funktionalräumen, *Studia Math.*, **2** (1930), 171–180.

[233] H. Schaefer, Über die Methode der a priori-Schranken, *Math. Ann.*, **129** (1955), 415–416.

Bibliography 355

[234] G. Scorza Dragoni, Un teorema sulle funzioni continue rispetto ad una e misurabili rispetto ad un'altra variabile, *Rend. Semin. Mat. Univ. Padova*, **17** (1948), 102–106.

[235] I. Segal, Non-linear semi-groups, *Ann. of Math.* (2), **78** (1963), 339–364.

[236] V. V. Shelukhin, A non-local in time model for radionuclides propagation in Stokes fluid, *Dynamics of Fluids with Free Boundaries, Inst. Hydrodynam*, **107** (1993), 180–193 (in Russian).

[237] V. V. Shelukhin, A time-nonlocal problem for equations of the dynamics of a barotropic ocean. (Russian) *Sibirsk. Mat. Zh.* **36** (1995), 701–724; translation in *Siberian Math. J.*, **36** (1995), 608–630.

[238] R. E. Showalter, *Montone Operators in Banach Space and Nonlinear Partial Differential Equations*, Math. Surveys Monogr., American Mathematical Society, Providence, RI, **49**, 1997.

[239] S. L. Sobolev, On a theorem of functional analysis, *Mat. Sb.* **46**, (1938), 39–68. English translation: *Amer. Math. Soc. Transl.*, **34** (1963), 39–68.

[240] S. L. Sobolev, *Applications of Functional Analysis in Mathematical Physics*, Leningrad: Izd. LGU im. A. A. Ždanova, 1950 (Russian). English translation: *Amer. Math. Soc. Transl.*, **7** (1963).

[241] V. Šmulian, Sur les ensambles régulièrement férmes et faibelment compacts à l'éspace du type (B), *Dokl. Akad. Nauk. SSSR*, **18** (1938), 405–407.

[242] T. Takahashi, Convergence of difference approximation of nonlinear evolution equations and generation of semigroups, *J. Math. Soc. Japan.*, **28** (1976), 96–113.

[243] H. R. Thieme and I. I. Vrabie, Relatively compact orbits and compact attractors for a class of nonlinear evolution equations, *J. Dynam. Differential Equations*, **15** (2003), 731–750.

[244] C. C. Travis and G. F. Webb, Existence and stability for partial functional differential equations, *Trans. Amer. Math. Soc.*, **200** (1974), 395–418.

[245] A. Tychonoff, Ein Fixpunktsatz, *Math. Ann.*, **111** (1935), 767–776.

[246] F. Vasilesco, *Essai sur les fonctions multiformes de variables réelles* (Thèse), Gauthier- Villars, Paris 1925.

[247] I. I. Vrabie, The nonlinear version of Pazy's local existence theorem, *Israel J. Math.*, **32** (1979), 221–235.

Bibliography

[248] I. I. Vrabie, Compactness methods and flow-invariance for perturbed nonlinear semigroups, *An. Ştiinţ. Univ. "Al. I. Cuza" Iaşi Secţ. I a Mat.*, **27** (1981), 117–125.

[249] I. I. Vrabie, Compactness methods for an abstract nonlinear Volterra integrodifferential equation, *Nonlinear Anal.*, **5** (1981), 355–371.

[250] I. I. Vrabie, A compactness criterion in $C(0, T; X)$ for subsets of solutions of nonlinear evolution equations governed by accretive operators, *Rend. Sem. Mat. Univ. Politec. Torino*, **43** (1985), 149–157.

[251] I. I. Vrabie, Periodic solutions for nonlinear evolution equations in a Banach space, *Proc. Amer. Math. Soc.*, **109** (1990), 653–661.

[252] I. I. Vrabie, *Compactness Methods for Nonlinear Evolutions*, Second Edition, Pitman Monographs and Surveys in Pure and Applied Mathematics **75**, Longman 1995.

[253] I. I. Vrabie, C_0-*Semigroups and Applications*, North-Holland Mathematics Studies, **191**, 2003.

[254] I. I. Vrabie, *Differential Equations. An Introduction to Basic Concepts, Results and Applications*, Second Edition, World Scientific Publishing Co. Pte. Ltd., Hackensack, NJ, 2011.

[255] I. I. Vrabie, Existence for nonlinear evolution inclusions with nonlocal retarded initial conditions, *Nonlinear Anal.*, **74** (2011), 7047–7060.

[256] I. I. Vrabie, Existence in the large for nonlinear delay evolution inclusions with nonlocal initial conditions, *J. Funct. Anal.*, **262** (2012), 1363–1391.

[257] I. I. Vrabie, Nonlinear retarded evolution equations with nonlocal initial conditions, *Dynam. Systems Appl.*, **21** (2012), 417–439.

[258] I. I. Vrabie, Global solutions for nonlinear delay evolution inclusions with nonlocal initial conditions, *Set-Valued Var. Anal.*, **20** (2012), 477–497.

[259] I. I. Vrabie, Almost periodic solutions for nonlinear delay evolutions with nonlocal initial conditions, *J. Evol. Equ.*, **13** (2013), 693–714.

[260] I. I. Vrabie, Semilinear delay evolution equations with nonlocal initial conditions, in *New Prospects in Direct, Inverse and Control Problems for Evolution Equations*, A. Favini, G. Fragnelli and R. Mininni editors, Springer INdAM Series Volume **10** (2014), 419–435.

[261] I. I. Vrabie, Delay evolution equations with mixed nonlocal plus local initial conditions, *Commun. Contemp. Math.*, **17** (2015) 1350035 (22 pages) DOI: 10.1142/S0219199713500351.

Bibliography 357

[262] I. I. Vrabie, A local existence theorem for a class of delay differential equations, *Topol. Methods Nonlinear Anal.*, DOI: 10.12775/TMNA.2016.023.

[263] R. N. Wang and P. X. Zhu, Non-autonomous evolution inclusions with nonlocal history conditions: Global integral solutions, *Nonlinear Anal.*, **85** (2013), 180–191.

[264] T. Ważewski, Sur un condition équivalent à l'équation au contingent, *Bull. Acad. Polon. Sci. Sér. Sci. Math. Astronom. et Phys.*, **9** (1961), 865–867.

[265] G. F. Webb, Autonomous nonlinear functional differential equations and nonlinear semigroups, *J. Math. Anal. Appl.*, **46** (1974), 1–12.

[266] G. F. Webb, Asymptotic stability for abstract nonlinear functional differential equations, *Proc. Amer. Math. Soc.*, **54** (1976), 225–230.

[267] K. Yosida, On the differentiability and the representation of one-parameter semi-groups of linear operators, *J. Math. Soc. Japan.*, **1** (1948), 15–21.

[268] K. Yosida, *Functional Analysis*, Grundlehren der Mathematischen Wissenschaften, **123**, 3rd edition, Springer-Verlag, 1971.

[269] S. Zaidman, *Almost-Periodic Functions in Abstract Spaces*, Research Notes in Mathematics **126**, Pitman Advanced Publishing Program (1985).

[270] S. C. Zaremba, Sur les équations au paratingent, *Bull. Sci. Math.*, (2) **60** (1936), 139–160.

[271] W. Zygmunt, On superpositional measurability of semi-Carathéodory multifunctions, *Comment. Math. Univ. Carolin.*, **35** (1994), 741–744.

Index

A-right-quasi-tangent set, 316
Arzelà–Ascoli Theorem, 11
Axiom of Dependent Choice, 58

Banach space
    reflexive, 2
    topological dual, 1
    uniformly convex, 2
Bochner integral, 4
Brezis–Browder Ordering Principle,
    54

C_0-group, 23
C_0-semigroup, 23
    compact, 28
    of contractions, 23
    of nonexpansive mappings, 23
    of type (M, a), 23
C^0-solution, 31, 56, 67, 92, 123, 127,
    252
    continuable, 72
    ε-approximate, 321
    globally uniformly
        asymptotically stable, 47,
        51
    non-continuable, 72
    saturated, 72
    stable, 47, 51, 97
C^0-viable with respect to
    $A + F$, 316
    $A + f$, 328
controlled porous media diffusion
    equation, 83
    with delay, 137
convex hull, 2

derivative

left directional
    of $\frac{1}{2}|\cdot|^2$, 29
    of the norm, 29
right directional
    of $\frac{1}{2}|\cdot|^2$, 28
    of the norm, 28
DS-solution, 56
dual norm, 1
Duhamel Principle, 26

evolution property, 33
evolution system of type γ, 44

family of functions
    equi-absolutely-continuous, 11
    equicontinuous, 11
    uniformly almost periodic, 194,
        213
    uniformly integrable, 11
fast diffusion, 83
Fatou Lemma, 6
function
    having affine growth, 17
    having linear growth, 17
    almost periodic, 194
    asymptotic almost periodic, 194
    asymptotically autonomous, 203
    Bochner integrable, 4
    compact, 163
    countably-valued, 3
    representation of a, 4
    Bochner integrable, 4
    l.s.c. at $x \in X$, 17
    l.s.c. on X, 18
    l.s.c. with respect to the usual
        order at (t, u, v), 20

360   *Index*

l.s.c. with respect to the usual
  order on its domain, 21
Lipschitz on bounded sets with
  respect to its second
  argument, 67
lower semicontinuous at $x \in X$,
  17
lower semicontinuous on X, 18
lower semicontinuous with
  respect to the usual order
  at (t, u, v), 20
measurable, 3
strongly measurable, 3
the Bochner integral of a, 5
u.s.c. at $x \in X$, 18
u.s.c. on X, 18
u.s.c. with respect to the usual
  order at (t, u, v), 21
u.s.c. with respect to the usual
  order on its domain, 21
upper semicontinuous at $x \in X$,
  18
upper semicontinuous on X, 18
upper semicontinuous with
  respect to the usual order
  at (t, u, v), 21
with affine growth, 95

globally C^0-viable with respect to
  (A, f, g), 326
globally C^0-viable with respect to
  $A + F$, 316

inequality
  Bellman, 52
  Gronwall, 51, 58
  Poincaré, 37
infinitesimal generator, 23

Lebesgue Dominated Convergence
  Theorem, 6

mapping
  asymptotically compact, 207
  contraction, 3
  duality, 2

nonexpansive, 3
strict contraction, 3
measure
  complete, 3
  space
    σ-finite, 3
    of totally bounded type, 12
    complete, 3
    finite, 3
    product, 7
metric space
  satisfying the First Axiom of
    Countability, 9
multifunction, 30
  (strongly-weakly) u.s.c. at a
    point, 14
  almost strongly-weakly u.s.c., 15
  closed from the left, 318
  locally bounded, 318
  locally closed from the left, 318
  strongly-weakly u.s.c. on a set,
    14

$\mathcal{N}$-maximal element, 53
nonlinear diffusion equation, 332
normal functions, 213
normalized semi-inner products on
  X, 29
null controllability, 322

operator, 30
  ω-m-dissipative, 31
  ω-dissipative, 31
  m-dissipative, 30
  p-Laplace operator, 42
  of complete continuous type, 36
  compact, 10
  dissipative, 30
  Laplace
    in $L^1(\Omega)$, 38
    in $L^2(\Omega)$, 37
  linear
    dissipative, 25
  maximal dissipative, 39
  maximal monotone, 39

Index

361

mild solution, 27
 monotone, 39
 of type ω, 31
 subdifferential, 43
 superposition, 8
  well-defined, 8

preorder relation, 53
purely nonlocal condition, 93

regular values, 24

σ-finite representation, 4
selection, 14, 16
semi-inner products on X, 29
semigroup
 compact, 34
 of contractions, 32, 33
 of nonexpansive mappings, 32, 33
 of type ω generated by A, 33
 of type gamma, 32
sequence
 of functions
  with uniform affine growth, 140
set
 compact, 9
 invariant, 204
 omega-limit, 203
 precompact, 9
 relatively compact, 9
 resolvent, 24
 sequentially compact, 9
 sequentially relatively compact, 9
 totally bounded, 9
slow diffusion, 83
solution
 C^1, 26
 ε-difference scheme, 31
 ε-DS, 31
 absolutely continuous, 26
 classical, 26, 60
 exact, xxi

globally asymptotically stable, 97
 mild, 26
 observed, xxi
 strong, 26, 31
subdifferential, 43

variation of constants formula, 26

wave operator, 38
weak topology, 2